I0043929

James C. Nagle

A Fieldmanual for Railroad Engineers

James C. Nagle

A Fieldmanual for Railroad Engineers

ISBN/EAN: 9783744678834

Printed in Europe, USA, Canada, Australia, Japan

Cover: Foto ©berggeist007 / pixelio.de

More available books at **www.hansebooks.com**

A
FIELD-MANUAL
FOR
RAILROAD ENGINEERS.

BY

J. C. NAGLE, M.A., M.C.E.,

Professor of Civil Engineering in the Agricultural and Mechanical College of Texas.

FIRST EDITION.

FIRST THOUSAND

NEW YORK
JOHN WILEY & SONS.
LONDON: CHAPMAN & HALL, LIMITED.
1897.

PREFACE.

EASE of reference and uniformity of notation are essential in a book that is to be consulted in the field. With this in mind an effort has been made in the following pages to secure a systematic arrangement of the subject-matter and uniformity of terms and notation. Except for a few cases Greek letters have been avoided and a single letter is used to designate an angle. In so far as practicable each figure is intended to be self-explanatory, so that the explanations necessary in connection with the problems have been reduced to a minimum. Algebraic equations stand each in a distinct line, thus rendering them more easily read.

A knowledge of the elements of geometry and trigonometry has been assumed, and only in the derivation of a few formulas in connection with the theory of transition-curves will any higher mathematics be needed. But these formulas may be accepted by the reader who is unfamiliar with the calculus without in any way affecting his ability to understand their applications or to follow subsequent reasoning.

One can most readily turn to what he wants in a book after having become familiar with its contents in the classroom. Keeping this in mind this book has been written so that it may be used as a text as well as for reference in the field. Wherever practicable solutions to problems have been given in a rigid, general form, followed by illustrative examples, so that the student need not lose sight of the principle involved while following the solution for a particular case. Wherever approximate solutions seemed preferable they have also been given and their limitations pointed out.

Free use has been made of the Table of Functions of a One-degree Curve, thus reducing the labor of field computations. By defining the degree of curve with reference to short chords for sharp curves--and, with tables of Radii, Long Chords, Mid-ordinates, etc., based on appropriate equations—the errors resulting from assuming the radius to vary inversely with the degree of curve will generally be found to be quite small.

Chapter I gives briefly the general method of making Reconnoissance; Chapter II treats of Preliminary Surveys; while Chapter III relates to Location.

Chapter IV, on Transition-curves, follows the method adopted by Professor Crandall, and enables one to locate the transition-curve with rigid accuracy where such is necessary. Approximate methods are also given by means of which the curve may be as easily located as any of the more limited easement curves ordinarily met with.

Chapter V, on Frogs and Switches, contains all that is necessary for their location. The formulas have been arranged to give the desired quantities in terms of the frog number whenever the resulting equations would be easier of application than the trigonometric ones usually given. The turnout tables are unusually full and give not only the theoretical lead but the stub lead as well, from which the practical lead can be at once found when the length of switch-rail is known.

Chapter VI, on Construction, tells how to set slope-stakes, and gives simple methods for computing areas and volumes either directly or by the use of tables. A short table of prismoidal corrections is given for end sections level, and also a formula for three-level sections, by means of which a suitable table may be computed if desired.

The tables at the end of this book have been arranged with a view to ease of reference, for, whatever the character of the text, the chief value of a field-book must depend upon the ease with which the tables may be consulted and upon their extent and accuracy. Table IX—Functions of a One-degree Curve—separates the logarithmic functions on the one side from the natural functions on the other and will be of assistance in locating these tables. Table XVI—Transition-curve Table—reading lengthwise of the page, likewise serves to separate the trigonometric tables from the miscellaneous tables that follow.

Some engineers object to the use of logarithmic tables in the field, but for them the natural functions are at hand; while for those who prefer logarithms the five-place tables of logarithmic sines, cosines, etc., will be found easy to consult and interpolate between.

All trigonometric tables are five-place, and others were carried to as many decimal places as their character demanded.

Tables I, III, IV, and V have been computed to agree with the definition of the degree of curve requiring curves sharper than 7° to be run with chords less than 100 feet in length, as described in the text. Tables XVII and XVIII were also computed expressly for this book.

Tables VI and XXVII are from electrotypes from Carhart's *Field Book for Civil Engineers* and were furnished by Ginn & Co. Electrotypes of Tables II, X, XII, XIII, XIX, XX, XXIV, XXV, XXVI, and also XVI — this last being from Crandall's book, *The Transition Curve*—were furnished by John Wiley & Sons.

Of the others, some were arranged from standard tables and others adapted in part and extended to increase their usefulness.

It will be noticed that vertical lines have been omitted wherever practicable, thus rendering it easier to refer to the tables.

Acknowledgments are due my associate, Professor D. W. Spence, for aid in making the tabular computations and in reading proof.

J. C. NAGLE.

COLLEGE STATION, TEXAS, May, 1897.

CONTENTS.

CHAPTER I.

RECONNOISSANCE.

ARTICLE 1. OBJECTS OF RECONNOISSANCE—HOW MADE.

CHAPTER II.

PRELIMINARY SURVEYS.

ARTICLE 2. OBJECTS; THE FIELD CORPS; DUTIES OF THE CHIEF.

ARTICLE 3. THE TRANSIT PARTY.

A. DUTIES OF THE MEMBERS.

B. TRANSIT ADJUSTMENTS—THE VERNIER.

CHAPTER III.

LOCATION.

B. LOCATING SIMPLE CURVES.

C. OBSTACLES.

D. CHANGE OF LOCATION.

ARTICLE 9. COMPOUND CURVES.

A. LOCATION PROBLEMS.

B. OBSTACLES.

C. CHANGE OF LOCATION.

ARTICLE 10. TRACK PROBLEMS.

CHAPTER IV.

TRANSITION-CURVES.

ARTICLE 11. THEORY OF THE TRANSITION-CURVE.

ARTICLE 12. FIELD-WORK.

A. FIELD FORMULAS.

B. SETTING OUT TRANSITION-CURVES.

ARTICLE 13. TRANSITION CURVE PROBLEMS.

CHAPTER V.

FROGS AND SWITCHES.

ARTICLE 14. TURNOUTS.

A. TURNOUTS FROM STRAIGHT LINES.

B. TURNOUTS FROM CURVES.

C. THE STUB LEAD.

ARTICLE 15. CROSSOVERS.

ARTICLE 16. CROSSING-FROGS AND CROSSING-SLIPS.

A. CROSSING-FROGS.

B. CROSSING-SLIPS.

CHAPTER VI.

CONSTRUCTION.

ARTICLE 17. DEFINITIONS ; GENERAL CONSIDERATIONS ; VERTICAL CURVES ; ELEVATION OF OUTER RAIL.

ARTICLE 18. EARTHWORK.

A. SETTING SLOPE-STAKES.

TABLES.

A FIELD-MANUAL FOR RAILROAD ENGINEERS.

CHAPTER I.

RECONNOISSANCE.

ARTICLE 1. OBJECTS OF RECONNOISSANCE—HOW MADE.

1. THE question of the selection of the proper route for a line of railway is essentially an economic one, involving not only the cost of construction, but of maintenance and operation, and a consideration of the immediate and future traffic likely to pass over the completed road.

The engineer upon whom devolves the duty of making the surveys for a railroad is not often called upon to determine whether it should or should not be built, though his preliminary estimate may decide those whose duty it is to do so: the problem confronting him is *how to secure the best line, answering a given purpose, for the least cost.* Keeping in mind the proper working of the completed road, the problem may be divided into two general parts:

First. The selection of the general route between terminal points, and in some cases the selection of the terminals themselves.

Second. The fitting of the line to the ground in such a manner as will render the cost of constructing and operating the road a minimum.

The first is by far the more important and difficult operation, requiring the highest grade of engineering skill—a fact too seldom recognized by those selecting engineers for this work. The acquirement of the necessary skill can result only from long practice and close observation, coupled with the ability to fully grasp and weigh all the complex features of the question. A passing reference only can be made to it in this little volume, which is intended to furnish hints and aids to the better execution of the second part. For the benefit of the beginner who has to do with the location and construction a few definitions and hints relating to reconnoissance will be given before going on to the special problems arising in the work of the railroad engineer.

2. The Reconnoissance is a rapid, general survey of the *area* through which the proposed railroad must pass, made only with such instruments as can be easily carried, and which should enable the engineer to restrict the more accurate instrumental work that follows to one or two general lines. The time required for this part of the work will in general be only a small fraction of the time consumed in location, involving the service of very few men; yet there is no part of the work more rapidly and improperly done—not always because the engineer in charge underestimates its importance, but because he is not usually allowed sufficient time in which to study thoroughly the area under consideration.

Properly the reconnoissance includes the determination of the terminal points of the road, but the locating engineer is usually relieved from the necessity of selecting these points, and the question reduces to that of finding *the best available line which admits of being built, maintained, and operated at the least cost between two given points.*

The reconnoissance must be made over an *area*—not a line or lines. Even what seems the most unpromising portion should be carefully studied, for the engineer can never be satisfied he has selected the best route until he has convinced himself by careful study that all others are inferior. Too much haste on reconnoissance means either a poor line or a much greater expenditure of time and money on the preliminary. No amount of notes or topography can take the place of an intimate personal knowledge of the problems to be encountered, and hence the reconnoissance and preliminary survey should be made by the engineer who is to locate the road.

3. The Instruments needed will rarely be more than a pocket-compass, hand-level, aneroid barometer, field-glasses, and sometimes a pedometer or an odometer.

(*a*) **The Pocket-compass** is used to obtain the magnetic bearings of lines and the angles they make with each other.

(*b*) **The Hand-level** enables one to obtain differences of elevation between points not far apart.

(*c*) **The Aneroid Barometer** gives approximate heights of the mercury column, and serves to roughly determine the difference of elevation of given points. In addition to the scale giving readings in inches, it should have also a scale graduated to give readings in feet. If two aneroids, which have been previously compared, are read simultaneously, one at each of the points whose difference of elevation is desired, or if the same aneroid is read at each successively at a short interval of time, during which the atmospheric pressure has not sensibly altered, we may find the difference of elevation by the formula*

$$d = 60000 (\log H - \log h)\left(1 + \frac{T + t - 60}{900}\right), \quad . \quad . \quad (1)$$

in which d is the difference of altitude in feet, H and h the barometric readings in inches—the logarithms being of the common or Briggs kind, T and t the temperatures of the two stations in Fahrenheit degrees.

If the sum of the temperatures, $T + t$, is taken as 105°, formula (1) reduces to

$$d = 63000 (\log H - \log h). \quad . \quad . \quad . \quad . \quad . \quad (1')$$

EXAMPLE.—The reading of the barometer at the foot of a mountain is 28.8 inches, and at the top 26.7 inches. Required the height of the mountain.

By (1'). $d = 63000 (\log 28.7 - \log 26.7) = 2071$ feet.

The effect of temperature on the metal of the instrument should be considered in the barometric formula when very precise work is to be done; but this correction, being small, may be neglected in the rough work of reconnoissance, particularly since the makers of the instrument construct it in such a way as to compensate, as closely as possible, for such changes of temperature.

(*d*) **The Pedometer** is an instrument which automatically counts the number of steps made by a person when the instrument is attached to his belt; then, knowing the average length of step, the distance passed over can be readily computed.

The Odometer registers the number of revolutions of a wheel to which it is attached, and the number of revolutions multiplied by the circumference of the wheel gives the space passed over.

---

* See Plympton's Aneroid Barometer, p. 38, for formula (1).

4. The Map.—Before beginning the reconnoissance the engineer should provide himself with the best available map of the region to be traversed; if this is a topographic one, he can at once determine from it the lines that are likely to justify an examination; and even if it is only a sketch-map, he can get material assistance by observing the courses of the streams and remembering that their positions indicate the relative elevations of the portion of the region through which they flow. Thus the large streams follow the lines of least elevation, and the manner in which the lateral streams unite with the principal one indicates the general trend of the terrain. Two streams flowing nearly parallel approach or recede from each other according as the intervening land diminishes or increases in altitude. Two streams flowing away from each other on opposite sides of a divide, and having their source therein, approach each other closest at the point of least elevation, and indicate the position of a pass or the lowest point of the dividing ridge. The study of any good contour map covering sufficient area will illustrate the laws governing the courses followed by streams.

The elevations of a few correctly mapped points, when obtainable, from the map or otherwise, serve as a guide in tentatively fixing on the maximum gradient to be employed and the amount of development needed.

A skillful engineer will thus be enabled to project his lines with sufficient accuracy to enable him to select on the ground the most feasible route or routes for his preliminaries in the least possible time. He should guard against the conviction, however, that it is unnecessary for him to look elsewhere than along the projected routes; for the inaccuracies of the map, local peculiarities, the nature of the excavation and embankment, the number and cost of bridges and other mechanical structures,—all these may conspire to make the most promising map-line inferior to some other whose advantages have to be sought for on the ground.

5. Having tentatively decided on the limiting grades and curvature to be employed, the engineer goes carefully over the ground, examining the entire area that seems likely to afford passage, in order to determine whether a suitable line may be secured for the grades and curves previously assumed. With his pocket-compass he takes the bearings of lines, and by means of the hand-level and aneroid determines differences of elevation;

Distances are estimated by the eye, paced, and the count taken from the pedometer, or, if the country admits of the use of a vehicle, taken from the odometer readings. If a well-gaited saddle-horse is used, very good results may be gotten by timing him, or by the use of the pedometer if his stride is uniform.

But in all cases much dependence must be placed on the ability to estimate with the eye differences of elevation and distances. The ability to do this with even reasonable accuracy comes only from long practice and careful observation, even to the most gifted in this respect. New and unexpected conditions sometimes deceive even the most practiced eye, but under ordinary conditions almost any one can train his eye to estimate horizontal distances fairly well. Vertical heights are more deceptive, possibly because we have less practice in this line, and the mind seems natnrally to exaggerate the vertical as compared with the horizontal; practice, however, will enable us to make allowance for the natural tendency to overestimate heights and slopes.

The ground should be gone over in both directions, for the appearance may be quite different when approached from different quarters. Ruling points, such as a pass in the mountains, the crossing of a large stream, or a town or city through which the road must be built, serve to reduce the problem to a number of special ones, each having its own solution.

In a mountainous region offering a limited number of possible routes, but heavy construction work, it may often happen that the location of a line is a much less difficult operation than in an open, rolling country offering a score of possible lines, between which the engineer making the reconnoissance must decide, selecting only those that in his judgment seem to justify an accurate instrumental survey.

The engineer must keep constantly in mind all the factors of the general problem of economic location and maintenance, and successful operation of trains. One line may cost more for construction and maintenance than another, but less for operation, or may invite less traffic. In all cases, however, the question of grades, curvature, length of line, earthwork, and mechanical structures are the controlling elements to be considered.

Having decided upon the route or routes over which to run preliminaries, these are marked on the map, and the engineering party organized and put in the field, with all the necessary instruments.

CHAPTER II.

PRELIMINARY SURVEYS.

ARTICLE 2. OBJECTS; THE FIELD CORPS; DUTIES OF THE CHIEF.

6. **The Objects** of the preliminary surveys are to secure all the data necessary to determine which one of the routes selected on reconnoissance is the most feasible, all things considered, and the approximate cost of construction. In rough country it will be economical to make two, or even three, surveys over the route selected for location before beginning to place the line in the position it is finally to occupy. The first of these is often omitted, and is called an "exploration-line"; it will frequently save the making of the more expensive "preliminary" over one or more of the routes.

7. **The Exploration-line** may be made with either transit or compass, and consists of a rapidly run line, made for the purpose of determining the maximum curvature and gradients with which to project the preliminary. It will not be necessary to make a detailed study of the region at this time, the distances and elevations, with such sketch topography as may be easily taken, being all that is needed. The magnetic bearing of lines is taken by the compassman, and the chainmen align each other with the flag set by the flagman. As the progress of the level party will be slower than that of the compass party, it will be economical to add an extra rodman, and sometimes a recorder. The compassman may sketch in the features adjacent to the line while waiting for his chainmen, who may be either in front of or behind the compass.

The stadia method of surveying—to be spoken of later—would seem to offer exceptional advantages for this work—only three or four men being needed in addition to the chief. With it, by setting the transit over alternate stations, very rapid progress may be made, and obstacles avoided with as much or greater ease than with the compass.

The exploration-line will more than pay for itself in showing what routes it will be unnecessary to make preliminaries over, and in indicating the most feasible one. It should be run over all the routes selected on reconnoissance.

8. The Preliminary Survey follows the exploration, or, when this is omitted, comes next after the reconnoissance. It may, with advantage, be made in two parts—first and second preliminary. It is made with such instrumental accuracy as the nature of the case may demand, sufficient data being obtained to determine the best line on which to locate and the approximate cost of construction. The rapidity with which this work can be done will depend on the care with which the reconnoissance was made. The preliminary line should approximate, as closely as the eye can determine, to the position the located line should occupy, and forms the base on which the topographic work rests. In reasonably easy country, where exploration-lines have been run, one preliminary should suffice for each route, but in difficult regions it will be best to run a second preliminary. If portions of the route are easy, followed by difficult parts, it will often be sufficient to "back up" and re-run the difficult portion until a reasonably satisfactory line has been obtained.

9. The Field Corps consists of a chief of party, transitman, leveler, rodman, two chainmen, rear rodman or "back-flag," stakeman, and two or more axemen. If a topographic party is added, as it should be in any but the easiest country, there will be also a topographer with two or more assistants. A cook and teamster will be needed with the camp outfit.

The corps is usually divided into the following parties:

(*a*) THE TRANSIT PARTY.

(*b*) THE LEVEL PARTY.

(*c*) THE TOPOGRAPHIC PARTY.

10. The Chief of Party receives his orders from the chief engineer, or such other officer as may be in charge, directs the motions of the surveying corps, and is responsible for their conduct and progress. He provides accommodations and supplies, pays all expenses, taking receipts or vouchers for all outlays—in duplicate when required. In the less thickly populated sections he must provide tents, wagons, cook, and all necessary camping outfit and supplies. He must direct the field operations in person, keeping in advance of the transit, establish turning-points or hubs, and direct the transitman in the proper course. He should keep a record—or direct the transitman and topographer to do so—of the character of earthwork likely to be encountered, the places where drains, culverts, bridges, cattle-guards, etc., are needed; the nature of material for embankment, piling, etc., adjacent to the line; the probable amount of clearing and grubbing, and all other features likely to affect the cost of construction. He should see that the names of property owners and residents along the line and the positions and bearings of property lines, when possible, are noted.

He should have authority to discharge assistants—except transitman, leveler, and topographer—whose services are unsatisfactory, and in many cases it will be best for him to have entire control, engaging or discharging any member of the corps as circumstances may require.

ARTICLE 3. THE TRANSIT PARTY.

A. Duties of the Members.

11. The Transit Party should consist of a transitman, head chainman, rear chainman, rear flagman, stakeman, and as many axemen as may be required—rarely less than two even for open country.

12. The Transitman cares for his instrument, keeping it in adjustment; directs the chainmen into line; notes the angle between successive tangents as read on plates; notes also the bearings of tangents, of highways, streams, and property lines (on location), with the plus at which the line crosses them. If there is no topographic party he must make sketches, on the right-hand page of note-book, of the surface features adjacent to the line; the red line down the middle of page represents the transit line, whether straight, broken, or curved, to which the sketches are adjusted. He must see that the axemen keep in line, in order that no unnecessary chopping may be done. Large trees need rarely be felled on preliminary, even when a given general course has to be followed, for small angles may be turned to avoid them, the deflections to right being made to approximately balance those to left.

When the chief of party is absent the transitman is ranking man, and will take temporary charge.

13. The Head Chainman carries a range-pole or "flag," and drags the chain, which he must see is straight and horizontal when setting a point for a stake. He directs the stakeman where to drive his stake, calling out the number after the rear chainman has read and called out the number on his stake; he keeps the axemen in line by setting his flag and going ahead, directing them where to cut by keeping them in line with the flag and transit. The speed of the party is dependent on the rapidity and accuracy with which he can set his flag in position, by ranging with stakes already set between him and transit, and in seeing that the axemen make all their work count.

14. The Rear Chainman must be careful to hold his end of the chain in the proper place, and that it is kept straight and taut when the head chainman is setting a stake. He must give all pluses, note the number on each stake as he comes up to it, and see that the stakeman has marked it correctly; he must make a note of pluses for roads, fences, streams, etc., to be given to the transitman later on.

15. The Stakeman must keep himself supplied with stakes about $1\frac{1}{2}'' \times 2'' \times 24''$, marking the number on them plainly, and driving them as directed by the head chainman.

If sawed stakes are not provided, he must cut the stakes and face them for the numbers. He must keep on hand a number of plugs or "hubs," to be driven flush with the ground and having the point where flag rested marked with a tack. About ten or twelve inches to the left of and facing the hub a guard stake is driven, on which is marked the station number, and which enables one to find the hub at any time.

16. The Axemen do all necessary clearing and chopping in order that the transit and level parties may have a clear sightway, and yet restrict the work of clearing to a minimum. One of them may be detailed to keep the stakeman supplied with stakes.

17. The Rear Flagman holds his flag on the last turning-point for the transitman to use in back-sighting.

18. The Instruments used by the party are the transit (or compass), one-hundred-foot chain or tape, range-poles, and the necessary axes and hatchet for axemen and stakeman.

B. Transit Adjustments—The Vernier.

19. For railroad work the transit is usually plain, but it is often convenient to have a clamp and tangent movement to telescope, a vertical circle, a level on telescope, stadia wires, and a gradienter; the solar attachment will rarely be needed.

20. To Adjust the Plate Levels.—The axis of the instrument is set at right angles to the plates by the manufacturer, so that when the axis is made vertical the plates will be horizontal.

In making adjustments remember that *a complete reversal always doubles any existing error.*

Place the bubble-tube parallel to a diagonal pair of leveling-screws, and bring the bubble to the centre of its run. Revolve the instrument 180° on the vertical axis, and the level-tube will be parallel to the same pair of leveling-screws as before, but reversed. If the bubble has moved from its central position bring it *half*-way back by means of the capstan-headed screws at the ends of the tube. Relevel and repeat until the bubble remains at the centre after reversal. Do the same for the other bubble. Both bubbles should remain at the centres of their tubes during a complete reversal.

21. Parallax is an apparent movement of the cross-wires with respect to the object sighted when the eye is moved from side to side of the eyepiece, and shows that the image does not fall in the plane of the cross-wires. In precise measurements it should be removed before making an observation with the telescope. To do this, first bring the cross-wires clearly into view when the object-glass is turned towards the sky, then, when sighting an object, note if there is any relative movement of cross-wires and image when the eye is moved from side to side at the eyepiece; if there is, refocus the object-glass until this movement disappears.

22. To Adjust the Line of Collimation is to make the line joining the intersection of cross-wires and optical center of objective describe a plane perpendicular to the horizontal axis of instrument.

FIRST METHOD.—Level the instrument and clamp the movements on vertical axis. Sight some well-defined object distant about the length of an average sight, and in the same horizontal plane as telescope. Reverse the telescope on its horizontal axis, and fix a point about as far from instrument as first point, and in the same horizontal plane. Revolve the instrument on its vertical axis and sight the first point; then reverse the telescope and note if line of sight cuts the second point. If not, loosen the capstan-headed screws holding cross-wire ring and move the vertical wire over *one fourth* the apparent error—since there were two reversals—remembering that the image of the cross-wires is inverted, while that of the object appears in its true position. Test by repetition.

SECOND METHOD.—If the limb graduations can be relied on they may be used in adjusting the vertical wire. With the instrument level sight a well-defined point, then revolve 180° by vernier-plate, reading both verniers; reverse telescope, and note if line of sight cuts the point. If not, correct *one half* the apparent error by moving diaphragm; then test by repetition.

The manufacturers adjust the object-glass slide so that the objective travels in the telescope axis, and this adjustment is not liable to serious derangement. It is well, however, to sometimes test by adjusting the line of collimation for both near and distant objects. If not correct for both, move the ring which guides the rear end of object-glass slide until the adjustment is correct for both positions.

Next make the vertical wire vertical by noting if it coincides throughout its length with a plumb-line, or by observing if it deviates from a point, on which the intersection has been fixed, when the telescope is elevated or depressed. Any error is corrected by turning the ring after slightly loosening the screws holding it.

The horizontal wire should also be adjusted so that the intersection of the cross-wires will be in the axis of the telescope; if the transit is to be used as a leveling instrument this adjustment is essential.

Drive a stake close to the instrument, and with the telescope clamped as nearly horizontal as can be conveniently done read a rod held on top of the stake; about 300 feet distant, and in line with first stake and instrument, drive a second stake and read the rod on it. Revolve 180° on vertical axis, reverse the telescope and bring the horizontal wire to the former reading when the rod is held on first stake; if the reading on the second stake is not the same as before, correct *one half* the apparent error by moving the cross-wire ring. Repeat as a test. The vertical wire should again be tested lest the movement of the ring may have caused it to change.

23. To Adjust the Standards is to make the plane described by the line of collimation vertical. Set up the transit about as far in front of some high building, or other tall object, as the highest point that can be sighted is above the base. Level the instrument and fix the intersection of the cross-wires on the highest point that can be easily sighted. Depress the telescope and fix a point near the base of the building at about the height of the telescope. Unclamp and revolve on the vertical axis until the telescope reversed cuts the lower point. Clamp the plates and raise the telescope until the cross-wires are at the height of the upper point. If they cut it the standards are in adjustment. If they do not, bring them *half-way* back by means of the adjustable screws at the top of one of the standards. Repeat as a test.

24. To Adjust the Level on Telescope is to make the bubble stand at the center of its run when the line of sight is horizontal. Bring the telescope as nearly horizontal as may be convenient, and take readings on the tops of two pegs in the same vertical plane with, and equidistant from, the instrument—say 300 feet. The difference of readings will equal the difference of elevation of the pegs; this difference may be obtained with the wye-level if preferred.

Move the instrument to a point beyond one of the pegs and in line with both. Set up as close to nearer peg as convenient, but not so close that the rod cannot be easily read. Bring the telescope as nearly horizontal as possible, and read on both pegs. If the difference of readings equals their difference of elevation the line of sight is horizontal, and the bubble may be brought to the center by means of the adjustable screws attaching the level-tube to the telescope. If this is not the case, we must set the telescope so the reading on second peg equals the reading on first peg plus the difference of elevation; then read again on first peg and proceed as before until the condition is satisfied. Or we may proceed as follows:

In Fig. 1 let the transit be at *O*, and *A* and *B* be the pegs. *AC* is a horizontal through *A*, so that *CB* is the difference of elevation

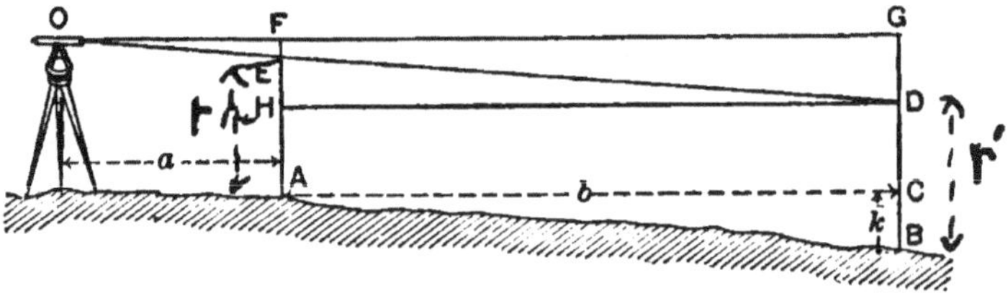

FIG. 1.

of *A* and *B*. Suppose line of sight to cut the rods at *E* and *D*, we must find *DG* so that the target may be set at the proper reading to make the line of sight horizontal. Let $OF = a$, $FG = b$, $EA = r$, $DB = r'$, $CB = k$. Draw DH parallel to CA and OG; then $EH = r + k - r'$.

From similar triangles

$$DG = EH\,\frac{a+b}{b} = (r + k - r')\left(\frac{a+b}{b}\right)$$

Set the target at a reading $GB = GD + r'$, sight to G, and the line of sight will be horizontal. Bring the bubble to the center of its run while the telescope is in this position, and the adjustment is complete.

If desired, a correction for the curvature of the earth and refraction may be introduced, but for short sights this is a useless refinement.

25. The Vernier is an auxiliary scale for measuring smaller divisions than those graduated on the limb. There are two classes, the direct-reading and the retrograde, according as the fractional parts of limb readings are taken on that side of the zero of vernier scale towards which the vernier has moved with respect to the limb, or the reverse. On the direct vernier a certain number of divisions on the vernier equals the same number of divisions on the limb, less one; on the retrograde there is one more division on limb than on vernier when the same space is covered by both.

26. The Least Count of a vernier is the smallest subdivision of limb graduation that can be read by it, and equals the difference of one space on limb and one on vernier.

Let l = value of one space on limb;
v = value of one space on vernier;
n = number of spaces on vernier.

Then for the direct vernier

$$nv = (n - 1)l;$$

from which we get the least count,

$$l - v = \frac{l}{n}.$$

For the retrograde vernier

$$nv = (n + 1)l,$$

from which the least count is

$$v - l = \frac{l}{n},$$

the same result as found for the direct vernier.

So, to find the least count: *Divide the value of one limb space by the number of spaces on the vernier.*

For example: If the limb of a transit is divided to half-degrees and the number of spaces on the vernier is 30, the least count will be $\frac{1}{2}$ divided by 30, or $\frac{1}{60}$ of a degree—that is, 1 minute.

27. To Read a Vernier, take the number of the last division on limb back of the vernier zero, then look along the vernier until a line is found to coincide with a line on the limb; add the number of this vernier line, multiplied by the least count, to the scale reading, and the result will be the required reading.

C. Accessories.

(1°) *The Gradienter.*

28. The Gradienter consists of a tangent-screw having a micrometer-head, attached to one of the standards of the transit and capable of being clamped to the horizontal axis of the telescope. It is used—as its name indicates—in running grades, and it accurately measures a small vertical angle in terms of its tangent. The screw is so cut that one revolution moves the telescope through an angle whose tangent at one hundred feet from the instrument has a certain value, usually one foot. The graduated head is divided into 100 parts, so that one division corresponds to 0.01 ft. at 100 feet from instrument.

To run a given gradient, bring the telescope level and read the micrometer-head of screw; then turn the screw as many divisions as there are hundredths of a foot rise or fall in 100 feet, and with a target set at the height of the horizontal axis, points on the surface corresponding to the given grade can be found.

For example: To run a 0.75 per cent grade, move the micrometer milled head 75 graduations from the horizontal.

When used as a **Telemeter,** we may either measure the space on the rod moved over by the line of sight for a given number of revolutions of the screw, or we may note the number of revolutions required to move the line of sight over a certain space on rod. The second method is the more accurate, particularly for long sights.

(2°) *The Stadia, or Telemeter.*

29. The Stadia is an instrument for determining the distance of a point from the observer by noting the space intercepted on a rod by a given visual angle, as determined by two auxiliary wires parallel to, and equidistant from, the horizontal wire of the transit telescope. When used with an ordinary leveling-rod the wires should be adjustable; if they are fixed (which for some reasons is preferable), the rod must be graduated to correspond. In addition to the distance of a point from the instrument, the difference of elevation is determined by observing the angle made by line of sight with the horizontal when the middle horizontal wire cuts a point on the rod as high above the ground as is the centre of the telescope.

The horizontal position of the point is determined from its magnetic bearing, or the azimuth of line of sight with reference to some fixed line, usually the north-south line.

30. Line of Sight Horizontal.—In Fig. 2 let a and b be the stadia wires, AB the intercept on the rod. The secondary axes

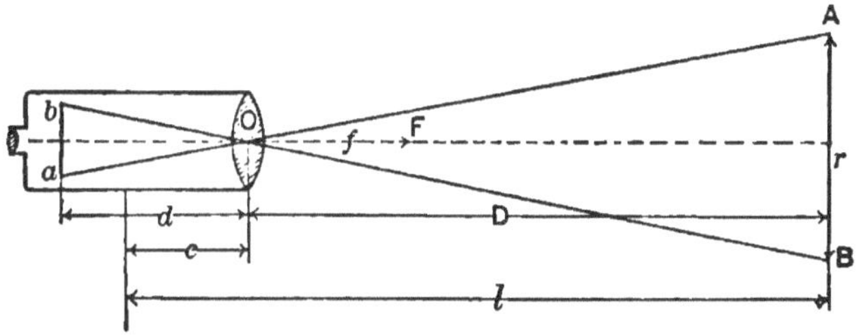

FIG. 2.

aA and bB pass through the optical center O. Let $h = ab$, $r = AB$, d = distance of cross-wires from objective, D = distance of rod from objective.

From similar triangles,

$$\frac{h}{d} = \frac{r}{D}.$$

From optics,

$$\frac{1}{d} + \frac{1}{D} = \frac{1}{f},$$

in which f is the focal length of objective.

Eliminating d from these two equations,

$$D = f + \frac{f}{h}r.$$

Let c be the *mean* distance of objective from center of instrument. Adding this to D gives, for the distance of the rod from the center of the instrument,

$$l = c + f + \frac{f}{h} r. \qquad (2)$$

$\frac{f}{h}$ may be made constant, when (2) becomes

$$l = a + kr. \qquad (2')$$

31. Line of Sight Inclined.—When the line of sight is not level it is difficult to hold the rod perpendicular thereto; hence the rod is held vertical, the angle of inclination measured, and a correction applied. In Fig. 3

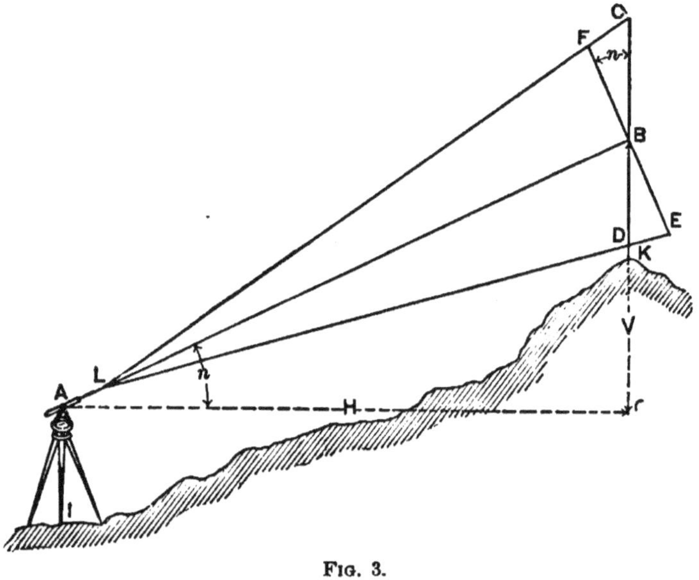

FIG. 3.

let $r = CD$ be the reading on rod held vertical;
$r' = FE$, the reading perpendicular to line of sight;
$H = AG$, the horizontal distance from A to B;
$V = BG$, the difference of elevation between A and B;
$n = BAG$, the angle of inclination of line of sight.

Assume angles AFB and $AEB = 90°$, from which they rarely differ more than 15′ to 17′. Then, since $FBC = n$,

$$r' = r \cos n.$$

By (2′). $AB = a + kr'.$

Hence $AB = a + kr \cos n.$

From triangle ABG

$$H = AB \cos n$$

$$\therefore H = a \cos n + kr \cos^2 n. \quad . \quad . \quad . \quad . \quad . \quad . \quad . \quad (3)$$

$$V = AB \sin n;$$

$$\therefore V = a \sin n + kr \sin n \cos n.$$

But $2 \sin n \cos n = \sin 2n.$

Hence $V = a \sin n + \frac{1}{2}kr \sin 2n. \quad . \quad . \quad . \quad . \quad . \quad . \quad (4)$

32. The Instrumental Constant a [$= c + f$ of (2)] may be found by measuring the distance from center of instrument to mean position of objective, which equals c; then focusing on a very distant object, preferably a star, and measuring from center of objective to plane of cross-wires, which equals f. The sum of these distances is a in formulas (3) and (4).

If the stadia wires are fixed, k may be found by measuring forward on level ground the distance a from plumb-line, and from this point a further distance b; then note carefully the stadia reading r when the telescope is level. Then, remembering (2)′,

$$a + b = a + kr.$$

$$\therefore k = \frac{b}{r}, \text{ a constant ratio.}$$

If the stadia wires are adjustable, we may so adjust k that any desired reading may be had for a given length of base. A convenient value of k is 100, which corresponds to an intercept of 1 foot on the rod at 100 feet from a point a feet in front of the instrument, 2 feet at 200 feet in front, etc.

33. A Stadia Table based on formulas (3) and (4) is published by the D. Van Nostrand Company in Winslow's *Stadia Surveying*, and can be used more rapidly than the formulas. Johnson's *Reduction Diagram*, by John Wiley & Sons, gives values of H and V graphically. Colby's *Slide-rule*, manufactured by Mahn & Co., St. Louis, gives values of V for distances in feet, yards, or meters to tenths of a foot, and can be used with great rapidity.

D. *Field-work.*

34. Station Numbers should begin with zero for the initial stake, and are marked on rear side of stake, from the top downward, the number of the preliminary, *A*, *B*, *C*, etc., being marked on the forward side. The marking should be with kiel, or crayon that will withstand the action of sun and rain. Stakes may be set every hundred feet or only at even stations, as preferred.

35. Hubs, or Plugs, are transit turning-points, and are short, flat-topped stakes driven into the ground flush with the surface. The flag is held on the top and carefully aligned, the position of the point being marked by a tack. A special tack with concave head offers a foothold for point of flag when used in backsighting.* About 10 inches to the left of and with numbered side facing the hub is driven a guard-stake to mark its position.

36. Reference-points are two or more hubs, with guard-stakes, in each of two lines making a good intersection angle at the point whose position they serve to locate. They should be driven beyond reach of disturbance, and are used in replacing a dislocated hub.

These need rarely be used on preliminary.

37. Alignment.—It is not intended that the preliminary and location lines occupy exactly the same position; hence considerable latitude is allowable in the size and number of angles turned, care being taken, however, that the maximum curvature need not be exceeded on location. Large trees and other obstructions may be avoided by turning a small angle until the obstacle has been passed, then making a deflection in the opposite sense. Bearings of tangents are taken with the needle, to serve as a check on the angle read on the plates.

In easy country not requiring a topographic party large angles should not be turned, a succession of small ones with short intervening tangents being substituted in order to make the preliminary profile approximate more closely to the location profile. These short tangents may conveniently be the long chords of the curve that is to follow.

---

* Such a tack is manufactured by the A. S. Aloe Co., St. Louis.

38. The Transit Notes may be kept in the form below, which shows both pages of the note-book. The notes run from the bottom up, the right-hand page being reserved for sketches; the red line up the middle of the page represents the transit line, whether straight or broken, to which the sketches must be adjusted.

Sta.	Angle.	Calculated Course.	Magnetic Course.	Remarks and Sketches.
68				
67 ⊙	20° 0′ L.	N. 1° 48′ W.	N. 1° 45′ W.	⊙
66				
65				
64				
63 ⊙	6° 2′ R.	N. 18° 12′ E.	N. 18° 15′ E.	⊙
62				
61				

39. Stadia Methods for Preliminary Surveys.—Preliminary lines are usually run with the transit, but the compass will answer nearly as well in most cases, besides admitting of more rapid work. The transit and stadia method might well be employed, and would effect considerable saving in the cost of preliminary surveys. For some reason railroad engineers have not regarded it with favor, though it is extensively employed in topographic surveying where the map is to be used for work that is often more precise than needed for railroad preliminaries.

Particularly is this method applicable to exploration lines. With the transit and stadia the entire surveying corps need not exceed five or six men, the instrument man acting as transitman, leveler, and topographer all in one. The only objection would seem to be in the amount of reduction the notes would need; however, with tables and slide-rule (see **33**) this work may be very rapidly done. For vertical angles of less than one degree the horizontal reduction can be neglected, and with side readings for topography the angle may be 5 or 10 degrees without necessitating the correction. Vertical heights are found by the slide-rule or by charts.

This method would really necessitate the making of a topographic map along a narrow strip of country, from which the profile could readily be taken. With a skilled observer and two to four rodmen the progress may be more rapid, and fully as good for the purpose intended as the more expensive method usually employed.

The transit need only be set at alternate stations (which may be any length within the reading limits of the wires), the bearings to other stations and points off the line being taken with the needle. The horizontal angle should also be read on the plates for points on stadia line, as a check on the bearings.

E. Obstacles in Tangent.

40. Obstructions to vision and measurement in tangent may be avoided in a number of ways, a few of which are given in the following problems. Other methods of avoiding them will suggest themselves in special cases.

The same devices may be used on location, but it is more important to maintain a clear sightway then; so, when possible, we should remove the obstruction.

41. To Pass an Obstacle by Means of Parallel Lines.—In Fig. 4, O is the obstruction, AB the obstructed line. At B set

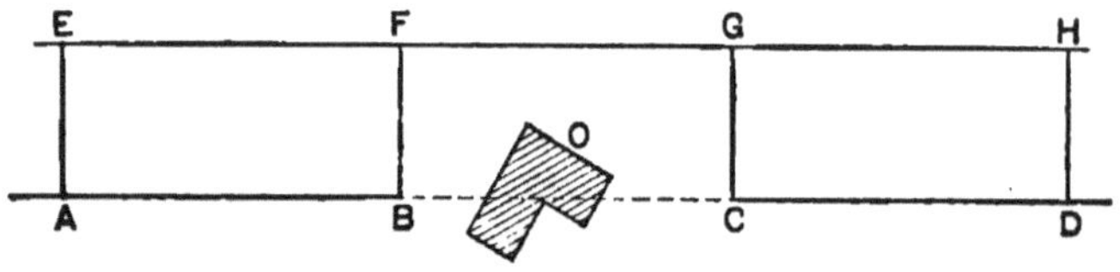

FIG. 4.

transit; turn 90° and measure BF long enough to clear obstruction. Set transit at F, make $BFG = 90°$, and measure FG. Move to G and backsight to F, making $FGC = 90°$. Measure $GC = FB$, and move to C, where the angle GCD is made equal to 90°. CD is the desired line, and $BC = FG$.

Otherwise, at A and B erect perpendiculars; take $BF = AE$; produce EF, and at G and H, beyond O, erect perpendiculars making $GC = HD = FB$. CD will be the desired line, and $BC = FG$.

42. To Pass an Obstacle by Angular Deflections.

GENERAL CASE. *Angle anything less than* 90°.

At B (Fig. 5) on the obstructed line deflect an angle a to one side and measure BC, taking C so that after deflecting $2a$ to the other side CD will clear the obstruction. Make $CD = BC$ and deflect an angle a to the same side as at B; DE will lie in AB produced. Draw CH perpendicular to BD; then

$$BD = BH + HD = 2BC \cos a. \quad . \quad . \quad . \quad . \quad (5)$$

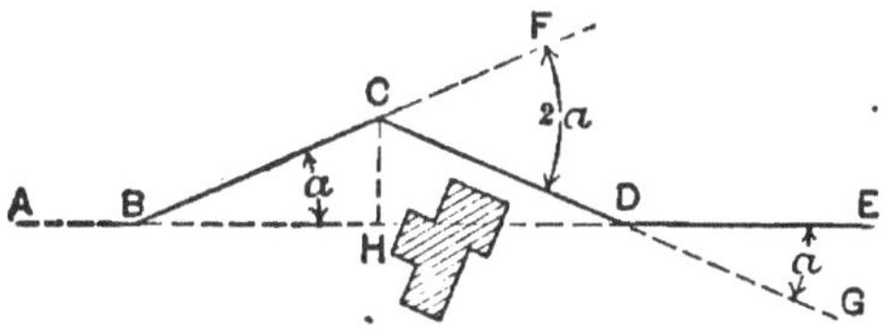

FIG. 5.

EXAMPLE.—Suppose $a = 14° 10'$, $BC = CD = 520$ ft.

$$BD = 2 \times 520 \times 0.96959 = 1008.37 \text{ feet.}$$

SPECIAL CASE. *Angle 60 degrees.*

In this case the triangle BDF (Fig. 6) is equilateral and $BF = BD = DF$.

Should it be inconvenient to run to D we may stop at C, having measured BC. At C deflect 60° and measure CE; at E again de-

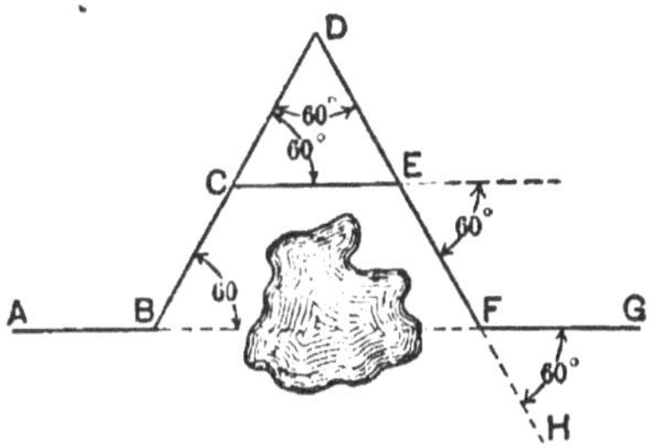

FIG. 6.

flect 60° and make $EF = BC$. At F a final deflection of 60° in the opposite sense will put the telescope in the desired line, FG, and

$$BF = BC + CE. \quad \quad (5a)$$

43. To Pass an Obstruction, such as a River, when the Preceding Methods are Inapplicable.

FIRST CASE. *Point beyond obstruction visible.*

In Fig. 7 let BC be required.

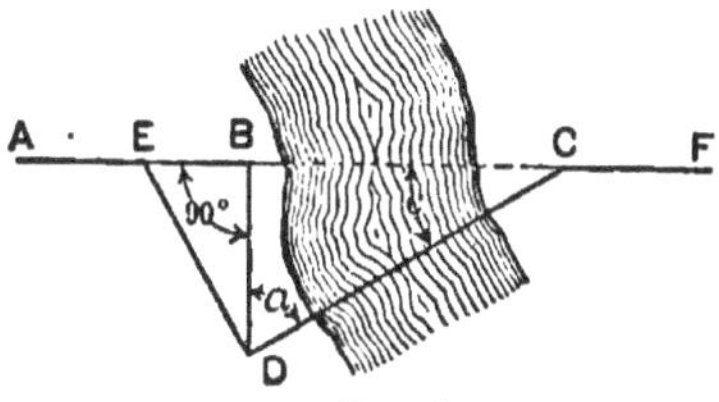

FIG. 7.

At B erect and measure the perpendicular BD; set instrument at D and measure angle $BDC = a$; then

$$BC = BD \tan a. \quad \quad (6)$$

Or, if a trigonometric table is not at hand, make $CDE = 90°$ and fix the point E where DE intersects AB; measuring EB there results, from similar triangles,

$$\frac{CB}{BD} = \frac{BD}{EB},$$

whence
$$CB = \frac{\overline{BD}^2}{EB}. \quad . \quad . \quad . \quad . \quad . \quad . \quad . \quad . \quad (6a)$$

Otherwise, if a right angle at B is not convenient, measure angles $CBD = b$, $BDC = a$, and side BD. Then $c = 180° - (a + b)$.

From triangle BDC,

$$BC = BD \frac{\sin a}{\sin c}. \quad . \quad . \quad . \quad . \quad . \quad . \quad . \quad (6b)$$

EXAMPLE.—$a = 56°$, $b = 70°$, $BD = 400$ feet.

By (6*b*), $$BC = 400 \frac{\sin 56°}{\sin 54°} = 409.8 \text{ feet.}$$

SECOND CASE. *Point beyond obstruction invisible.*

At B (Fig. 8) measure angle b and line BE; move to E and measure angle y, and set hubs on line EC so the line BC will pass

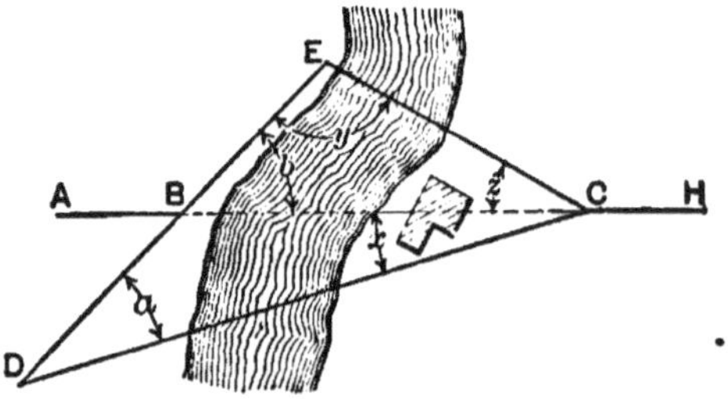

FIG. 8.

between them. Angle $z = ECB = 180 - (b + y)$. Then from triangle BEC

$$BC = BE \frac{\sin y}{\sin z}. \quad . \quad . \quad . \quad . \quad . \quad . \quad . \quad (7)$$

Produce EB to D, where DC will be sure to clear obstruction; measure BD.

From triangle BDC,

$$\frac{\tan \frac{1}{2}(a - x)}{\tan \frac{1}{2}(a + x)} = \frac{BC - BD}{BC + BD}.$$

But $a + x = b$, hence

$$\tan \tfrac{1}{2}(a - x) = \frac{BC - BD}{BC + BD} \cdot \tan \tfrac{1}{2}b. \quad . \quad . \quad . \quad (8)$$

The sum and difference of a and x are now known, so both may be readily found.

At D set off the angle a with the transit, and have the chainmen stretch a cord between the hubs set on line EC at C. Now signal the flagman to move his rod along this cord until the vertical wire cuts it at C. Set a hub here and place the transit over it. Sight to D or E, reverse telescope and deflect into CH.

ARTICLE 4.—THE LEVEL PARTY.

44. The Level Party consists generally of two members, the leveler and a rodman; sometimes an axeman is added to keep the rodman supplied with pegs for turning-points and in clearing the line of sight for the level. As the party follows the transit little or no clearing will be needed. The instruments used are a level, a rod, and a hand-axe or hatchet.

45. The Leveler makes all necessary observations with his instrument, keeping a neat, accurate record of readings and elevations; also the positions and elevations of benches and turning-points. He should work out elevations of stations while the rodman is going from one station to the next; he must see that the rodman gives him readings at points where the longitudinal slope changes suddenly, recording the plus. He must plot his profile at night, or at such times as the chief of party is likely to need it. The rodman's readings at turning-points should be checked.

46. The Rodman holds his rod at each station, calling out the number. If stakes are set only at even stations, he must hold his rod midway between stakes, the point being found by pacing the distance. Target-readings need only be taken at turning-points and benches, and the rodman should keep a record of these in his "peg-book," checking the calculations of leveler for heights of instrument and elevations of turning-points. At any marked surface change he will hold his rod, calling out the plus to leveler. He must assist the leveler in plotting up the notes.

A. Adjustments of the Level.

47. To Adjust the Line of Collimation is to bring the intersection of the cross-wires into the optical axis of the telescope.

Set up and level the instrument, then bring the vertical wire into coincidence with a plumb line or vertical edge of a building, at the mean length of sight, and note if the vertical wire is truly parallel thereto. If it is not, loosen the capstan-headed screws holding cross-wire ring and turn slightly so that the wire is parallel to the vertical line.

Loosen the wye-clips and bring the vertical wire into coincidence with the line and clamp the instrument. Rotate the telescope in the wyes 180° and note if the wire coincides with the line. If not, correct *one half* the error by loosening one and tightening the opposite, of the capstan-headed screws that hold the cross-wire ring in place, remembering that the image of the cross wires is inverted by the eyepiece.

Turn the telescope until the horizontal wire is parallel to the plumb-line or edge of building, and make the same test and correction. Repeat for both wires. The horizontal wire is the one on which the accuracy of leveling depends, but it is wise to have both adjusted. Their intersection should remain on a point during a complete rotation of the telescope in the wyes.

48. To Adjust the Level-bubble is to bring the axis of the level-tube into the same vertical plane with the line of collimation, and to make the bubble stand at the center when the line of sight is horizontal.

Since the axis of the telescope coincides with the line joining the center of the wye-rings (which requires these to be of the same size), it is sufficient to make the axis of the bubble parallel to this line.

(*a*) With the telescope over one diagonal pair of leveling-screws and the clips loosened, bring the bubble to the center of its run; then turn the telescope, in the wyes, a little to either side of the vertical plane through the telescope and note if the bubble remains at the center. If not, correct the error by means of the screw at end of the level-tube case arranged for lateral movement. Repeat until the tube may be rotated half an inch or more to either side of vertical without movement of the bubble. This adjustment is made merely to prevent error from failure to set level-tube vertically beneath telescope.

(*b*) With the wye-clips opened well out, again bring the bubble to the center of its run; remove the telescope from wyes and turn it end for end, then carefully replace it in the wyes. Should the bubble fail to remain at the center, bring it *half-way* back by raising the lower or depressing the higher end of tube at the points of attachment to telescope. Relevel and repeat as a test.

49. To Adjust the Wyes is to make the axis of the telescope perpendicular to the vertical axis. With the wye-clips closed place the telescope over one pair of leveling-screws and bring the bubble to the center of its run; then turn the telescope half-way round on its vertical axis, so that its ends have changed places. If there is any error, correct by bringing the bubble *half-way* back to center by means of the screws connecting wyes with level-bar. Repeat until the bubble remains in the center during a complete revolution.

B. Theory of Leveling.

50. When the level has been adjusted the line of collimation will describe a plane parallel to the horizontal plane tangent to the earth's surface at the point where the instrument is placed. A level surface, such as the surface of still water, will coincide with this plane only at the point of tangency, and will depart farther and farther therefrom as the point considered recedes from the instrument. For short sights this difference may be neglected in railroad work, as will presently be shown, but for long sights a correction must be applied.

The effect of curvature is to make objects appear lower than they really are, while the refraction of a beam of light, due to the greater density of the layers of air nearest the earth's surface, has a contrary effect. Experience shows the average error due to refraction to be about one seventh of that due to curvature.

51. The Error due to Curvature at any point is the deviation of a tangent line from true level, as the point recedes from the point of tangency.

Let O be the center of the earth, T the point of tangency, and N the point where the error due to curvature is desired. Let the notation be as shown in Fig. 9. From the right triangle OTP, we have

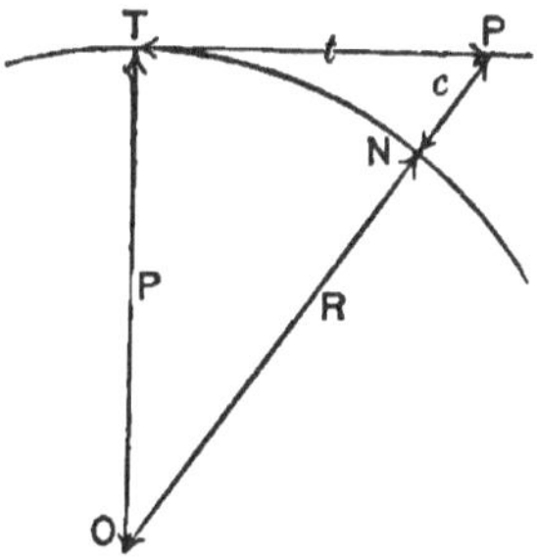

FIG. 9.

$$(R + c)^2 = R^2 + t^2.$$

From which

$$c = \frac{t^2}{2R + c}.$$

Now, since c is always very small compared with $2R$, the quotient resulting from the division of t^2 by $2R$ will not differ sensibly from that obtained by dividing by $2R + c$. Therefore we write

$$c = \frac{t^2}{2R}. \qquad (9)$$

For $t = 1$ mile, $R = 3963$ miles, $c =$ about 8 inches. Hence for any other distance in miles we have, for c,

$$c = 8 \times t^2 \text{ inches.} \qquad (9a)$$

The correction for refraction is about $\frac{1}{7}c$, hence we have, from (9),

$$C = c - \frac{1}{7}c = \frac{6}{7}c = \frac{3}{7}\frac{t^2}{R},$$

or, closely enough,

$$C = .85c. \qquad (10)$$

EXAMPLE.—What is the correction for a half-mile sight? For one eighth of a mile?

By (9a), $c = 8 \times (\frac{1}{2})^2 = 2''$ for first case,

and $c = 8 \times (\frac{1}{8})^2 = 0''.125$ for second case.

By (10) the final correction is

$$c = 0.85 \times 2 = 1''.7 \text{ for first case,}$$

$$c = 0.85 \times 0.125 = 0.106'' \text{ for second case.}$$

52. The Difference of Elevation between two points not so far apart but that a rod may be read on each from some intermediate point may be readily found from these rod-readings.

In Fig. 10 let the instrument be at I, A and B the points whose difference of elevation is desired. Let $r = AD$, $r' = BC$. Since the line of sight, DC, is horizontal, the difference of

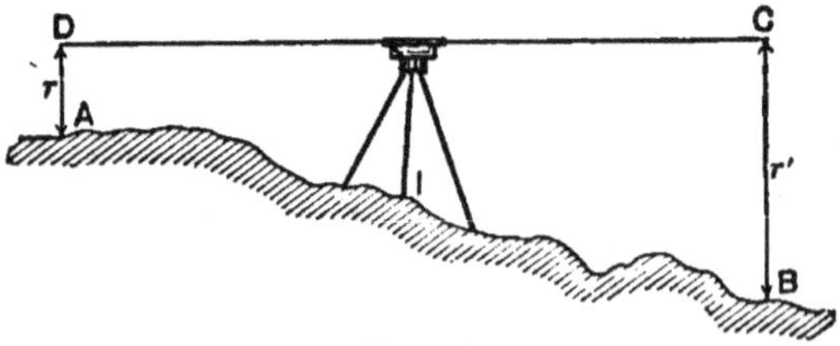

FIG. 10.

elevation will evidently be $r' - r$. When the distance from I to A equals that from I to B the errors due to curvature evidently balance.

When the points are so situated that the rod cannot be read on both from one intermediate position of the instrument, an

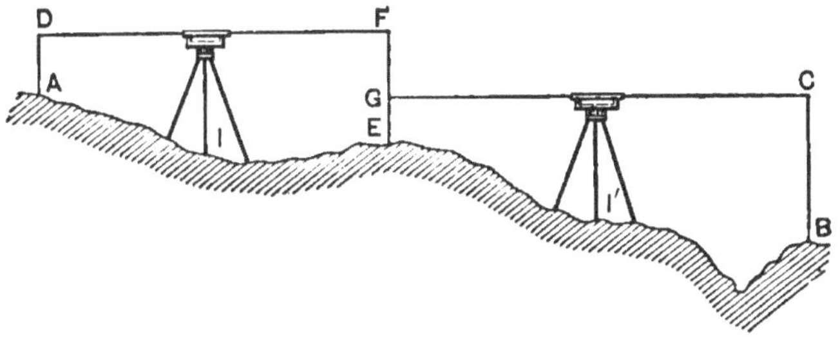

FIG. 11.

auxiliary point or points must be used and readings taken on these points in pairs. Thus in Fig. 11 suppose the difference of elevation of *A* and *B* required:

With the instrument at *I* read on *A* and some intermediate point *E*. Considering the backsights as plus and foresights as minus, the difference of elevation of *A* and *E* is $AD - FE$.

Again, with the instrument at I' the difference of elevation of *E* and *B* is $GE - CB$. The sum of these differences equals the difference of elevation of *A* and *B*, and may be written $(AD + GE) - (EF + CB)$, or, in general, *the sum of the backsights less the sum of the foresights equals the difference of elevation.*

C. *Field-work.*

53. A Datum is a level surface so taken that it shall lie below the lowest point likely to be reached by the profile, to which the surface elevations are referred. It is often spoken of as the datum-line or datum-plane, and is the zero of elevations.

54. A Bench-mark is a permanent mark, such as a copper or other bolt let into the top of a solidly fixed stone, whose height above the datum is known; it may be simply a mark on a stone, or a tack driven into the projecting root of a tree, upon which the rod may be read. In any case it must be so situated that it cannot change its elevation nor is likely to be disturbed within the time for which it is intended to be used as a standard of reference.

The elevation should be marked on some object adjacent to the bench, with the letters B. M. indicating the nature of the point.

55. The Field-work consists in finding the elevation of a number of points on the line established by transit party sufficient to give, when plotted, a fairly correct outline of the surface as seen in profile.

A bench-mark is taken at the beginning of the line, and its distance above mean sea-level or other datum is known or assumed. The level is set with one pair of leveling screws in the line to be run (in order that any change in the position of the bubble may be easily corrected), and the rod is read on the bench. This reading plus the elevation of bench gives the height of instrument (*H. I.*) above the datum.

Readings are taken at every hundred feet along the line, or oftener if the surface changes greatly, until a point is reached beyond which it is desired to move the level. A peg is driven firmly into the ground and the rod read on this; the height of instrument less the rod reading will give its elevation, as it will for the intermediate points. This point is a temporary bench and is called a **turning-point.** It should be marked by a guard-stake if it is desired to use it again. The instrument is now carried beyond the turning-point, set up, and the whole process repeated. Benches and turning-points should be read to hundredths or thousandths of a foot, intermediate points to tenths. Turning-points are marked ⊙ or T. P. in the notes, and their positions, as also the bench-marks, noted by both leveler and rodman in their note-books.

56. The Level Notes may be kept in any convenient form that is easily understood. The following is used more extensively, perhaps, than any other:

Sta.	B. S.	H. I.	F. S.	Elev.	Remarks.
B. M.	5.613	205.613	....	200.0	B. M. on root of L. O. tree 60′ to right of line.
0	....	....	2.3	203.3	
1	....	....	0.8	204.8	
2	....	....	5.7	199.9	
3	....	....	7.8	197.8	
4	....	....	9.9	195.7	
⊙	1.120	196.310	10.423	195.190	On peg at 4 + 30′ – 20′ to left of line, by small P. O. tree.
5	....	....	6.3	190.0	
6	....	....	4.5	191.8	

Here the elevation of the datum was taken 200.00 feet below the first bench-mark. The instrument was set up near Station 2, and a reading of 5.613 taken on the bench; this was written in the *B. S.* column, and when added to the elevation of the bench gives the height of instrument, 205.613. A reading of 2.3 was taken on Sta. 0, recorded in the *F.S.* column, and when subtracted from the *H.I.* yields an elevation of 203.3. The elevations of other points were determined in the same way. A little beyond Station 4 the rodman drove a peg and held the rod on it, yielding a reading of 10.423 and an elevation of 195.190. The instrument was then moved to a point near Station 7 and a reading of 1.120 taken on the peg; this added to 195.190 made the new *H. I.* 196.310, and the process continued with this *H. I.*

In most cases it will be sufficient to read benches and turning-points to hundredths and intermediate points to tenths.

It will be seen from the notes that any error in a turning-point causes the same error in all succeeding points. To guard against this the rodman is required to keep a "peg-book," in which the heights of instrument and elevations of turning-points are recorded, and which must check with the leveler's record.

57. Wind and sunshine affect the accuracy of the work with the level, as is also the case with the transit. For very great accuracy a calm, cloudy day is the best, but the railroad engineer cannot always choose the best times for his work, and must take such precautions as may be possible while he exercises the greatest care to prevent and detect errors. The adjustments should be tested at least once a week, even when the greatest care has been taken, for unequal expansion and other causes may conspire to cause them to change.

By making foresights and backsights to turning-points about equal the error due to curvature will be eliminated; the readings of rodman at these points should also be checked. The rodman should hold his rod vertical, which is sometimes accomplished by means of a level attached to rod; or the leveler can tell by his vertical wire when the rod is in the same vertical plane with the instrument, and by causing the rodman to wave his rod back and forth slowly, after clamping the target, he can tell if the horizontal wire just bisects the target at its highest position.

58. The Rod should be graduated to feet and tenths, reading by target at turning-points and benches; intermediate readings are made by the leveler at his instrument. Strength and durability are essential qualities. The Philadelphia rod seems to answer the purpose as well as any other now manufactured; the Troy rod may be used in the same manner as the Philadelphia rod, but is lighter and less able to stand rough usage.

Article 5. The Topographic Party.

59. The Topographic Party follows the level and secures all the data necessary for making an accurate contour-map of a strip of country extending as far each side of the preliminary as may be needed for the intelligent projection of the location-line. This distance may vary from 50 to 300 or 400 feet, its width depending on the difficulties to be encountered and the degree of precision with which the preliminary approximates to the final location-line. The lateral slope of surface is obtained at the stations of preliminary by means of the hand-level and tape, by the slope-level or clinometer, by cross section rods, or by the transit and stadia. Strictly speaking the topography includes all the surface features, but for railroad work the surface elevations, streams, and nature of surface are the most important; it may be necessary to note the positions of roads, buildings, etc., and should always be done when practicable without undue loss of time. A pocket-compass will be of use in observing the bearings of lines.

60. There are two methods of recording the data obtained; one by means of notes and sketches in a book, the other by drawing the contours directly on the field-sheet as the data are obtained. Station elevations can be taken direct from the leveler's notes, and constitute the base on which the contour elevations rest.

Suppose the hand-level to be used and the notes kept in a book, to be afterwards transferred to the map. Starting with the known center elevation, the topographer notes the height of his eye above the ground and calculates the height of center above or below the next contour; from this the reading of the rod when held on this contour is found, being the height of station above contour plus the height of eye. He directs the slopeman in or out on a line at right angles to preliminary until this reading is given by the hand-level; the distance out is then measured and recorded, just as in setting slope-stakes, and the slopeman directed into position on the next contour, in the same manner.

Thus if 5-foot contour-intervals are employed, and the station elevation is 321.6 feet and the height of eye 5.3 feet, we shall have for the reading at the 320-foot contour $5.3 + (321.6 - 320) = 6.9$. Motion the slopeman down the slope until his rod reads 6.9 and measure the distance out, suppose 21 feet. The 315-foot contour will be 5 feet lower, giving a reading of 11.9, which may be found in like manner at, say, 80 feet out. As the rod reads only to about 12 feet the topographer must move out to this last point, and with the reading $5.3 + 5 = 10.3$ find the 310-foot contour in the same way. On the up-hill side the 325-foot contour will be found with a reading of $5.3 - (325 - 321.6) = 1.9$ feet, and other contours in like manner.

The notes may be written thus

Sta.	Left.	Center Elev.	Right.
824	$\frac{305}{193'}\ \frac{310}{125'}\ \frac{315}{80'},\ \frac{320}{21}$	321.6	$\frac{325}{27'},\ \frac{330}{56'}\ \frac{335}{80'}\ \frac{340}{112}$

The number above the line is the contour elevation, the number below its distance out from center.

If preferred the elevation can be taken at regular distances out and recorded as above; the position of the contour will then be found by interpolation when mapping the work.

61. If the topography is to be plotted in as the work progresses the topographer must have a light drawing-board with a pocket and flap on back for holding the sheets on which the transit-line has been plotted the night before; the station elevations are marked on the line and the contour positions spotted in as obtained by slopemen, after which the contours are sketched in. Points where contours cross transit-line are found in the same manner as side points. The size of the sheets will depend on the taste of topographer and size of drawing-board; 17×24 to 19×28 inches are good sizes.

The topographer will soon learn to guess at the position his contours will occupy at the next station ahead, and will sketch them in lightly, to be erased and corrected when necessary. It is often sufficient to take lateral readings at every second or third station.

62. If the **Slope-level** is used, the inclination of the surface is obtained; then by the use of a scale constructed to show the horizontal distance apart of contours, for the given contour-interval, for slopes varying from 1° to 20°, the position of contours can at once be spotted on the map. Wellington recommends the use of the altazimuth as permitting the employment of either method at will—the altazimuth being merely a hand-level with a clinometer attached.

63. Cross-section Rods are measuring-rods 10 or 12 feet long carrying a level-bubble. By placing one end at the center, bringing the rod horizontal, and noting the height of the end of rod on the down-hill side, the slope may readily be obtained and the contours worked in as before. For very rough, broken ground this method may be preferable to either of the others.

64. If the **Transit and Stadia** are employed, very elaborate topography may be taken with very little field-work, but the observations require considerable reduction. With a suitable topographic protractor and the slide-rule mentioned in **33**, the large number of points that may be obtained from each setting of the transit may be readily plotted and their elevations marked on the plot, after which the contour-lines can be worked in, and other features mapped. For small vertical angles no horizontal reduction is needed.

While not generally favored by railroad engineers in the past, this method is probably the most rapid and economical of any so far employed in topographic work.

Article 6. Preliminary Estimates.

65. After completing the field-work of the preliminary survey the party is usually disbanded, only the transitman, leveler, and topographer being retained to assist the chief of party to complete the map, profile, and estimate of cost.

66. The Map may be drawn to any suitable scale, but less than 400 feet to the inch is not to be recommended where it must be used in projecting location. The transit-line is laid down first and the topography worked in afterwards from the field-map or topographer's notes. If it is wanted on a continuous sheet, the transit-line must first be drawn on a succession of small sheets, which are added as the plotting progresses, a new sheet being slipped under the edge of the preceding and tacked down when required. The overlapping edge is marked by a number of short lines extending over onto the sheet beneath, to enable one to replace in the proper position. When the line has been plotted the sheets are pasted together and the whole shifted so as to bring the transit-line over the continuous sheet. Angular points are then pricked through and the line drawn on the continuous sheet. Ordinarily it will answer to have the map drawn on a succession of small sheets, to be joined together as required.

The plotting had best be done by bearings, though it may be done from the deflection angles, provided care is used to check frequently by bearings. Otherwise an error in one angle will throw all the remaining portion of the line out of position.

If more than one preliminary was run, they should all be shown on the same sheet whenever possible.

67. The Profile will be drawn by the leveler on profile-paper, and shows a developed vertical projection of the line. The scale will depend on the paper used. There are three scales in general use, styled respectively Plates "A," "B," and "C." There is also a metric profile-paper. Plate "A" has the vertical exaggerated 20 to 1 as compared with the horizontal and is the best to use where much rockwork is expected. The vertical exaggeration of Plate "B" is less than of Plate "A"; this plate is most used for ordinary earthwork.

A strip of color laid on below the surface-line, and fading out at the lower edge, adds greatly to the appearance of the profile. The tentative grade-line and points of change should be drawn in red.

68. Preliminary Estimates of quantities are made by assuming a grade-line and drawing it on the profile; then the cuts and fills are taken from the profile, and the corresponding quantities obtained from Table XIX for the base the road is intended to have when completed. The nature of the work, whether ordinary earth or rock, can, of course, be only roughly estimated.

Bridging is estimated from the profile where piling or framed bents may be used, but where piers and long spans are needed special surveys with soundings are required, Culverts, drains, cattle-guards, cross-ties, and rails for main line and sidings, switch stands, buildings, right of way, clearing, and other factors entering into the question of cost must all be considered and allowed for in making up the estimate.

Engineering expenses and unforeseen outlays that are sure to arise should have a liberal allowance.

69. The Report of the chief of party should set forth the advantages and probable cost of each of the several lines run when there is more than one. On this report frequently depends whether or not the line is to be located, and it should be clear and exhaustive, though plainly and concisely worded. The map and profile form an integral part of the report and show from what data the estimates were derived.

CHAPTER III.

LOCATION.

ARTICLE 7. PROJECTING LOCATION.

70. After the preliminary has been mapped and the topography worked in, the engineer proceeds to make a paper location for his guidance in the field. The solution of the varied and complex problems that confront him are more or less interdependent. The guiding principle, applicable to all departments of engineering, that *the best structure is that which for the least cost best answers the purpose for which it was intended*, should control, even though the resulting structure be inferior, in point of scientific design, to some other. The best road as regards construction and grades may be a failure because of excessive first cost, while the cheapest construction will entail such heavy operating expenses that it may be equally unprofitable. The alignment must be as free from curves as possible, while heavy grades are at the same time excluded; these two requirements conflict and must be as well adjusted as possible. The amount of earthwork, of bridging and other structures must be kept down to the lowest limits.

71. Starting at the summit of the most difficult portion of the route, assume a starting-point and elevation; with the dividers set at such a distance to the scale of the map as will give a fall of one contour-space—or half-space—for the assumed grade, step down the slope in such a way that the dividers fall each time on the next lower contour, or half-space, according to the fall assumed in setting dividers. If curve compensation is allowed, the dividers must be reset for each curve, for the same fall, since the grade will be slackened on curves. The points at which the dividers fall are lightly spotted on the map and connected by a **grade contour**, which represents the surface-line having the required gradient. This line will be too broken to be used as a location-line, so we have then to draw on the map a succession of curves and tangents that will approximate sufficiently close to it, at the same time that a proper balance is maintained between earthwork and curvature.

Having lightly plotted the proposed line, the elevations are transferred to profile-paper, thus giving a profile of the line. With a fine thread stretched along the profile, to represent the grade-line, adjust the cuts and fills to suit the nature of the work. In general, fills are cheaper than cuts both in construction and maintenance; and especially is this true where a shallow surface layer of earth is underlaid by rock. It may happen that the material from excavation must be used in embankment, when the cuts and fills must be made to balance by shifting the grade-line until this appears to be the case on the profile.

At the stream crossings the grade-line must be kept safely above high-water mark, so that sufficient waterway is provided, and allowance made therefor.

After locating the most difficult portions pass on to the easier work, returning later on to study the effect this will have on the part first located. It may be necessary to go over the projection several times before you can be reasonably sure that the best location has been projected; even then the study of the line in the field will cause many of the details to be altered, sometimes materially.

Long grades are to be preferred to short ones, but questions of economy may necessitate the latter in order to lighten work; care must be taken that the grades are not so badly "chopped" that they interfere with the easy riding of the train.

In projecting the line it will generally be best to strike the curves first and draw the tangents afterwards, though it sometimes happens that long tangents will control the curves; when this is the case the tangents are drawn to intersection and the curves afterwards put in.

When transition-curves are employed, a slight offset should be made at the beginning and end of curves to allow for their insertion in the field. These offsets will be so small that it is useless to attempt to show them to scale.

72. A Curve-protractor will be of material assistance in finding the degree of curve required to unite two tangents that have been laid down on the map. It consists of a transparent, semicircular protractor having a series of curves from 30′ up to 8°

plainly cut upon it. The curves are on both sides, those on the reverse side having their concavities turned in an opposite sense from those on the face. The scale is usually 400 feet to the inch, and in any case the map and protractor must be drawn to the same scale. Sometimes a set of cardboard or hard-rubber curves are used, but they are inferior to the curve-protractor. To use it, simply prolong tangents to intersection and then place the protractor so that the curve admitting of the best grade is tangent to the two straight lines. Mark the points of tangency, which will be the beginning and end of curve. When the curve is required to pass through a given point the proper curve may be immediately found by trial, whereas the calculations would require some little time.

Reversed curves should never be allowed on main lines. Sufficient tangent should be interposed to allow space for easing off the superelevation of outside rails, or for the insertion of transition-curves when these are to be employed.

73. The Field Corps is substantially that required on the preliminary survey, and the methods of work pretty much the same, except that curves must now be run in, and this necessitates more clearing. If first and second location-lines are to be run (and it is real economy to run both), it will not be necessary to have the stationing continuous on the first, so the pluses arising from "backing up" need only be noted and eliminated when the final location-line is run. If transition-curves are to be inserted, they need not be run the first time, the proper offset being made at the *P.T.* or *P.C.* of the circular curves, which latter are to be run.

On the final location-line the stationing must be continuous, beginning with zero. The stakes are marked as on the preliminary survey, and all hubs that are likely to be used again must be referenced in, the reference-hubs being set well out of the way of disturbance by the plow or scraper.

The leveler should make bench-marks every 1000 or 2000 feet, to be used in running check-levels and in giving grades later on.

From the paper location the notes should be made up in the office, to serve as a guide in the field; however, no attempt should be made to adhere rigidly to them, since slight errors in the mapping will affect the projected line, while in the field the line may be shifted here and there so as to fit the ground more snugly and accord more closely with what the nature of the earthwork demands.

The highest skill of the engineer is required to secure the best location-line, and he should have all the time he needs. Undue haste on location—as on reconnoissance and preliminary—is almost sure to result in increased cost of construction.

Article 8. Simple Curves.

A. Definitions and Formulas.

74. The Circular Curves that are usually employed to unite straight reaches of the railroad may be simple, compound, or reversed. The use of reversed curves should, however, be limited to turnouts and cross-overs.

a. **A Simple Curve** is the arc of a circle.

b. **A Compound Curve** consists of two simple curves, of different radii, both on the same side of a common tangent.

c. **A Reversed Curve** is made up of two curves of contrary flexure having the same or different radii, and a common tangent.

d. **The Point of Curve** (*P. C.*) is the end of tangent and beginning of curve, as at *A*, Fig. 12.

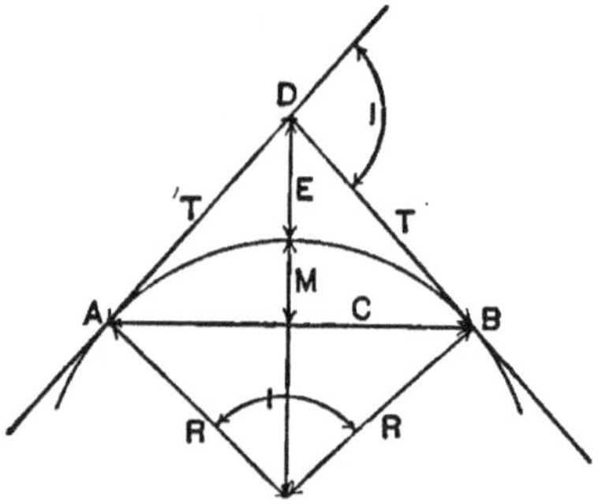

FIG. 12.

e. **The Point of Tangent** (*P. T.*) is the end of curve and beginning of tangent, as at *B* of Fig. 12.

f. **The Point of Intersection** (*P.I.*) is the point where the tangent at the *P.C.* and *P.T.* intersect when produced. (*D* of Fig. 12.)

g. **The Intersection Angle** (*I*) is the angle at the *P.I.* between the tangents meeting there, and equals the angle at the center.

h. **The Tangent Distance** (*T*) is the length of the produced tangent measured from the *P. C.* or *P. T.* to the *P.I.* The term *tangent* is applied to any straight portion of the line, but the letter T will be used to designate the produced portion only.

i. **The Mid-ordinate** (M) is the portion of the radius intercepted between the arc and chord when it cuts the chord at its middle point.

j. **The External** (E) is the part of the radius produced to the *P.I.*, intercepted between curve and the *P.I.*

k. **The Long Chord** (*L.C.*) is the chord joining the *P.C.* and *P.T.* Frequently the term is applied to any chord longer than the unit chord.

l. **The Radius** will be denoted by R.

m. **The Point of Compound Curve** (*P.C.C.*) is the point of common tangency of the two branches of a compound curve. (See Fig. 13.)

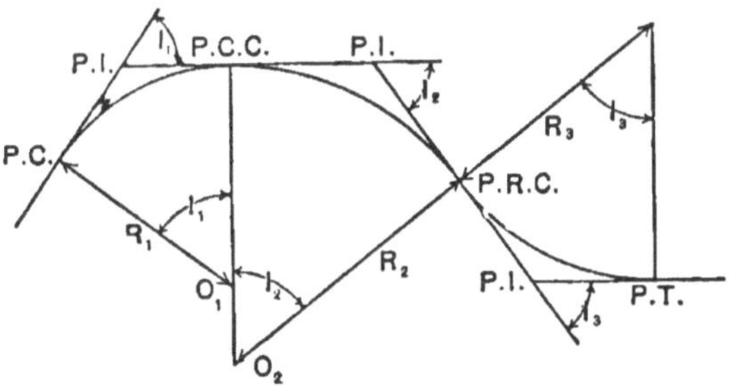

FIG. 13.

n. **The Point of Reversed Curve** (*P.R.C.*) is the point of common tangency of the two branches of a reversed curve.

o. **The Degree of Curve** (D) is the angle at the center subtended by the unit chord. In the United States this chord is 100 feet, in England 66 feet, and where the metric system is employed it is taken at 20 meters. Any convenient chord length may be taken, but for uniformity American engineers have adopted the chord of 100 feet, and unless otherwise stated it is always so understood when we speak of the degree of curve.

Half the degree of curve is called the **deflection-angle**, since it is the angle to be deflected from the tangent to the chord.

If there were any practical method of measuring *around the curve* instead of along the chord, an accurate and convenient ratio for expressing the radius in terms of the degree would be had. Thus if D is the angle at the center subtended by the *arc* of unit length, we have, where a is this unit arc,

$$2\pi R = a \cdot \frac{360}{D}.$$

Hence $$R = \frac{a}{2\pi} \cdot \frac{360}{D}. \quad . \quad . \quad . \quad . \quad . \quad . \quad . \quad . \quad (11)$$

When a equals 100 ft. this becomes

$$R = \frac{100}{2\pi} \cdot \frac{360}{D}. \quad . \quad . \quad . \quad . \quad . \quad . \quad . \quad . \quad (11')$$

R varies inversely as D, so that knowing the radius for a 1° curve, we should have only to divide this by D to get the radius for a $D°$ curve.

Since the chord is employed instead of the arc, we determine R by means of the following problem·

75. Given the Chord C, and Degree of Curve D, to Find the Radius R.

In Fig. 14, AB is the chord C, OE a perpendicular from the center upon AB

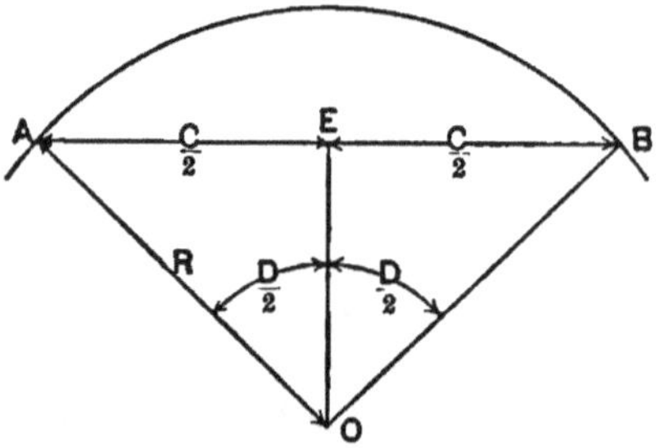

FIG. 14.

From the right triangle AEO we have

$$R \sin \tfrac{1}{2}D = \tfrac{1}{2}C.$$

Whence $$R = \frac{\frac{1}{2}C}{\sin \frac{1}{2}D} = \tfrac{1}{2}C \operatorname{cosec} \tfrac{1}{2}D. \quad . \quad . \quad . \quad . \quad (12)$$

When C is 100 ft.,

$$R = \frac{50}{\sin \frac{1}{2}D} = 50 \operatorname{cosec} \tfrac{1}{2}D. \quad . \quad . \quad . \quad . \quad (12')$$

Comparing results given by formula (12′) with those given by (11′), we have for a few curves:

Degree of Curve.	R by (12′).	R by (11′).	Difference.
1............	5729.65	5729.58	0.07
2............	2864.93	2864.79	0.14
3............	1910.08	1909.86	0.22
5............	1146.28	1145.92	0.36
7............	819.02	818.51	0.51
10............	573.69	572.96	0.73
14............	410.28	409.26	1.02
20............	287.94	286.48	1.46

The difference is seen to be about one half a foot for a 7° curve, one foot for a 14° curve, and one and one-half feet for a 20° curve.

Up to a 7° curve the difference is inconsiderable, and we may stake out curves with 100-foot chords. From 7 to 14 degrees 50-foot chords may be used. Therefore

$$R = \frac{25}{\sin \frac{1}{4}D} = 25 \operatorname{cosec} \tfrac{1}{4}D. \quad . \quad . \quad . \quad . \quad . \quad (12'a)$$

For curves from 14° to 28° we should use 25-foot chords, for which

$$R = \frac{12.5}{\sin \frac{1}{8}D} = 12.5 \operatorname{cosec} \tfrac{1}{8}D. \quad . \quad . \quad . \quad . \quad (12'b)$$

Above 28° shorter chords—say 10 feet—should be used, if the curve cannot be struck from the center. In this case

$$R = \frac{5}{\sin \frac{1}{20}D} = 5 \operatorname{cosec} \tfrac{1}{20}D. \quad . \quad . \quad . \quad . \quad (12'c)$$

Table I of radii was computed by formulas (12′), (12′a), and (12′b).

In practice it is customary to take the radius of a 1° curve as 5730 feet and to assume the radii to vary inversely as the degree; thus for a 4° curve the radius would be $R = \frac{5730}{4} = 1432.5$ feet, while by Table I it is 1432.69 feet—a difference of only .19 foot; for a 12° curve $R = \frac{5730}{12} = 477.5$ feet, while by Table I it is 477.68 feet. The effect of taking 5730 instead of 5729.65 for the radius of a 1° curve is to reduce the error resulting from the assumption that R equals 5730 divided by the degree of curve.

76. The Length of Curve (L) is found by dividing the angle at the center (which equals the intersection angle) by the degree of curve, the result being in chains and decimals of a chain. The number of $P.C. + L$ will give the station number of $P.T.$

EXAMPLE.—The $P.C.$ of a 4° curve having $I = 26° 30'$ is at sta. $104 + 12.5$. Find L and the number of the $P.T.$ Here

$$L = \frac{26.5}{4} = 6.625 \text{ chains.}$$

$104.125 + 6.625 = 110.75$; hence the number of P. T. is $110 + 75$.

77. Use of the Table of Functions of a One-degree Curve.—In the location of railway curves geometrical accuracy will frequently be of less importance than rapidity of field-work, so long as errors are kept within certain limits.

On tangents slight errors of alignment may readily be detected by the unaided eye, but on curves these are not so apparent. Moreover it is not likely that the trackmen will keep them up in the exact position of their location.

To simplify and shorten the field computations engineers make use of a table of functions of a 1° curve, and assume these functions for other curves to vary inversely as their degree, or directly as their radii. Table IX gives values of the tangent distances, long chords, mid-ordinates, and externals for a 1° curve, the radius of which is taken as 5730 feet. To find these functions for other curves, divide the tabular values by the degree of curve. The error resulting from this assumption will, in any practical case, amount to no more than a few tenths or hundredths of a foot.

Table IX may also be used as a metric curve table, the tabular values being taken as meters instead of feet. If the unit metric chord is 20 meters long, this may be taken as one fifth of the tabular unit chord; so to use the table multiply the metric degree by 5 and enter the table with the result as a value of D.

For instance, a 2° metric curve having $I = 40°$ would have a mid-ordinate equal to $\frac{345.6}{2 \times 5} = 34.56$ meters.

For the approximate radius of a metric curve divide 5730 by 5 times the degree. Thus a 4° metric curve would have $R = \frac{5730}{4 \times 5}$

= 286.5 meters. For the exact radius make use of formula (12). Thus for a 4° curve having 20-meter chords $R = \frac{10}{\sin 2^\circ} = 286.54$ meters, a difference of only .04 meters.

If a metric curve is to be retraced with a 100-ft. chain, we convert the metric degree to the degree referred to 100-ft. chords by the relation that a 100-ft. chain = 1.524 chains of 20 meters each; a 20-meter chain = 65.618 ft.; one foot = 0.3048 meters; one meter = 3.2809 ft.

It will sometimes be a sufficiently close approximation to take the 20 meter chain as two thirds of a 100-ft. chain; this will make the metric curve nearly two thirds of the degree the same curve would have when laid out with a 100-ft. chain, and the curve with 100-ft. chords nearly three halves of the degree as laid out with the 20-meter chain. Thus a 4° metric curve would be equivalent to a 6° curve laid out with a 100-ft. chain.

In the problems that follow two methods of solution will be given when practicable—the first being rigid, while the second is based on the use of Table IX. To shorten the formulas the subscript 1 will be written after the letters *T*, *L. C.*, *M*, and *E* when these are the functions of a 1° curve. Thus $T_1 \measuredangle 28^\circ$ means the tangent distance for a 1° curve when $I = 28^\circ$, $L.C._1 \measuredangle 16^\circ$ the long chord for a 1° curve when $I = 16^\circ$, etc.

78. Tables of Natural and Logarithmic Circular Functions.—Many engineers prefer to work altogether by tables of natural sines, cosines, etc., and time may often be saved by their use. Nevertheless logarithmic tables are of frequent advantage, even in the field, and the more important ones, such as the logarithmic sines, cosines, tangents, and cotangents, together with the logarithms of numbers, are given in the back of the book along with the tables of natural functions.

79. Given *R* and *C* to Find *D*.

From equation (12),

$$\sin \tfrac{1}{2}D = \frac{\tfrac{1}{2}C}{R}. \quad . \quad . \quad . \quad . \quad . \quad . \quad . \quad . \quad (13)$$

80. Given *I* and *R* (or *D*) to Find *T*.

If *D* is given, find *R* by (12'); then in Fig. 15 from triangle *OAB* we get

By Table IX. Find the tabular value of T for the given angle I; then

$$T = \frac{T_1}{D}. \qquad (14a)$$

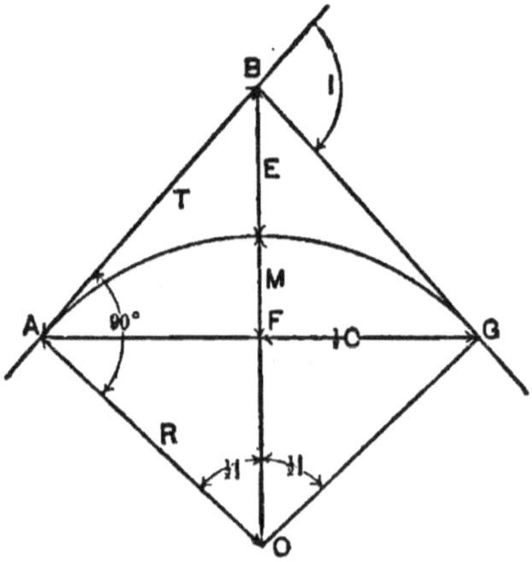

Fig. 15.

Example.—$I = 35° 40'$, $D = 4°$; required T.
By (14), $T = 1432.69 \tan 17° 50' = 460.91$ feet.
By (14a), $T = \frac{1843.4}{4} = 460.85$ feet, a result differing from the value found by the rigid method by only 0.06 foot.

81. Given I and T to Find R or D

From (14),

$$R = \frac{T}{\tan \frac{1}{2}I} = T \cot \tfrac{1}{2}I. \qquad (15)$$

Then by Table I the degree may be found.

By Table IX.

$$D = \frac{T_1}{T}. \qquad (15a)$$

82. Given I and D to Find the Long Chord $L.C.$

First find R by (12) or (12'), or by Table I; then from the triangle OAF of Fig. 15,

$$AF = R \sin \tfrac{1}{2}I.$$
$$\therefore AG = 2AF = L.C. = 2R \sin \tfrac{1}{2}I. \qquad (16)$$

By Table IX.—Find the tabular $L.C.$ for the given angle I; then

$$L.C. = \frac{L.C._1}{D}. \quad . \quad . \quad . \quad . \quad . \quad . \quad . \quad . \quad (16a)$$

83. Given the Radius R and any Chord C to Find the Ordinate to the Curve at any Point.

First Method.—In Fig. 16 let HE be the chord C; $HK = a$ and $KE = b$, the segments into which it is divided by the ordi-

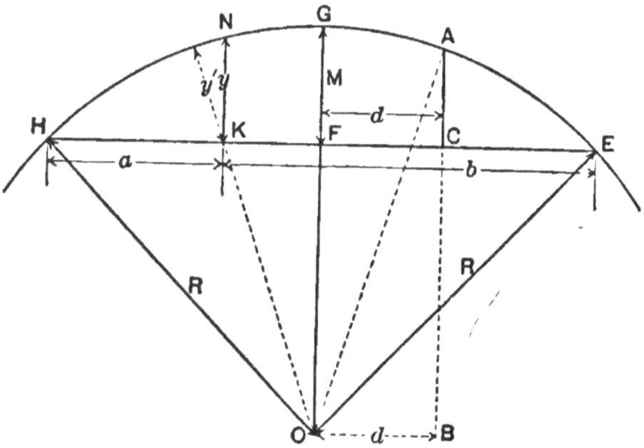

Fig. 16.

nate y. Draw the radius through K; call the portion between chord and curve y'. By geometry,

$$(2R - y')y' = ab,$$

from which

$$y' = \frac{ab}{2R - y'}.$$

But y' is small compared with $2R$, and hence we write

$$y' = \frac{ab}{2R}. \quad . \quad . \quad . \quad . \quad . \quad . \quad . \quad . \quad . \quad . \quad (a)$$

Now y does not differ sensibly from y' in the cases met with in practice, so we write

$$y = \frac{ab}{2R}. \quad . \quad . \quad . \quad . \quad . \quad . \quad . \quad . \quad . \quad . \quad (b)$$

If we write $R = \frac{5730}{D}$, formula (b) becomes

$$y = \frac{abD}{2 \times 5730}. \quad . \quad . \quad . \quad . \quad . \quad . \quad . \quad . \quad (c)$$

Let $\frac{a}{100} = m$, $\frac{b}{100} = n$, and substitute in (c), giving

$$y = \frac{10000}{11460} mnD = 0.873mnD,$$

or very nearly

$$y = \tfrac{7}{8}mnD. \quad . \quad . \quad . \quad . \quad . \quad . \quad . \quad . \quad (17)$$

y is given in feet when m and n are in chains and decimals of a chain.

At the mid-point F, $m = n$, and $y = M$.

$$\therefore\ M = \tfrac{7}{8}n^2D. \quad . \quad . \quad . \quad . \quad . \quad . \quad . \quad . \quad (18)$$

CAUTION.—Formulas (17) and (18), while very convenient for field use in passing obstructions, are liable to error when very long chords or large values of D are used, since they give results that are too small.

If we write the arcs HN, NE for a and b, we shall get results that are too large, yet about as near the true values as by taking m and n to be the segments of the chord. To illustrate we will find a few values of M and compare with the true values taken from Table V.

Degree of Curve.	Length of Arc.	Mid-ord. by $M=\frac{7}{8}(HF)^2D$.	Mid-ord. by $M=\frac{7}{8}(HG)^2D$.	Mid-ord. by Table V.
2.......	2 stations.	1.75	1.75	1.75
2.......	6 "	15.69	15.75	15.69
5.......	2 "	4.37	4.38	4.36
5.......	6 "	38.51	39.38	39.06
8.......	2 "	6.96	7.00	6.97
8.......	4 "	27.29	28.00	27.75
8.......	5 "	42.02	43.75	43.20
8.......	6 "	59.43	63.00	61.93

From this it appears we may use formula (18)—and (17) as well—taking either the segments of the arc or chord for curves not exceeding 4° with arcs up to 600 ft.; for curves from 4° to 6°

they may be used up to 500-ft. arcs, while for curves between 6° and 8° not more than 400 feet of arc may be taken.

Second Method.—First determine the mid-ordinate. In triangle OEF,

$$OF = \sqrt{R^2 - \tfrac{1}{4}C^2};$$

then

$$M = FG = R - \sqrt{R^2 - \tfrac{1}{4}C^2}. \quad . \quad . \quad . \quad . \quad . \quad . \quad (19)$$

To find ordinate AC distant d from the mid-point of EH, draw $OB = d$ parallel to HE; draw AB at right angles to HE. Then

$$BA = \sqrt{R^2 - d^2}.$$

Therefore

$$CA = y = \sqrt{R^2 - d^2} - \sqrt{R^2 - \tfrac{1}{4}C^2}. \quad . \quad . \quad . \quad (20)$$

Third Method.—If the chord C is short, we may regard the arc as an arc of a parabola, for which it is known that ordinates vary as the product of the segments into which they divide the chord. The mid-ordinate being known, we have

$$\frac{y}{M} = \frac{ab}{(\tfrac{1}{2}C)^2}$$

$$\therefore\ y = 4\frac{abM}{C^2}. \quad . \quad . \quad . \quad . \quad . \quad . \quad . \quad . \quad . \quad (21)$$

From formula (b) we have for $y = M$, $a = b = \tfrac{1}{2}C$,

$$M = \frac{(\tfrac{1}{2}C)^2}{2R} = \frac{C^2}{8R}. \quad . \quad . \quad . \quad . \quad . \quad . \quad . \quad (22)$$

The mid-ordinate for any other chord C' is

$$M_1 = \frac{C'^2}{8R}.$$

Hence

$$\frac{M_1}{M} = \frac{C'^2}{C^2}.$$

$$\therefore\ M_1 = M\left(\frac{C'}{C}\right)^2. \quad . \quad . \quad . \quad . \quad . \quad . \quad . \quad . \quad . \quad (23)$$

If $C' = \tfrac{1}{2}C$, this gives

$$M_1 = \tfrac{1}{4}M. \quad . \quad . \quad . \quad . \quad . \quad . \quad . \quad . \quad . \quad . \quad (23')$$

This last relation affords an easy method of staking out a curve when the mid-ordinate of a given chord has been determined. First erect the ordinate M at the mid-point of the chord; then join the ends of chord with the extremity of the ordinate just measured; the lengths of these chords do not differ much from $\frac{1}{2}C$; at their mid-points erect ordinates equal to $\frac{1}{4}M$, giving points on the curve. Proceed in like manner for other points until a sufficient number have been located.

84. Given R and I to Find the External E.

In Fig. 17 $E = GB = OB - OG$.

But $OB = R \sec \frac{1}{2}I$ and $OG = R$.

$$\therefore\ E = R(\sec \tfrac{1}{2}I - 1) = R \operatorname{ex\,sec} \tfrac{1}{2}I. \quad . \quad . \quad . \qquad (24)$$

BY TABLE IX.—Find E for a 1° curve for an intersection angle I; then

$$E = \frac{E_1}{D}. \quad . \quad . \quad . \quad . \quad . \quad . \quad . \quad . \quad . \qquad (24a)$$

85. Given T and I to Find E.

In Fig. 17 draw BC perpendicular to AB, and produce AG to

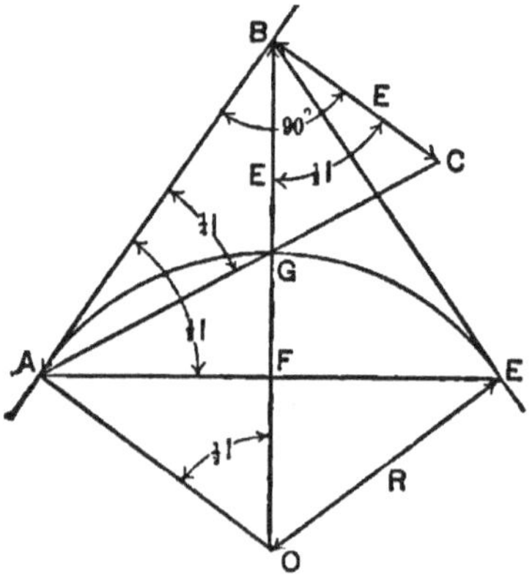

FIG. 17.

intersect BC at C. BC is parallel to AO, and the triangles AGO and GBC are similar; hence $BC = BG = E$. In the right triangle ABC, angle $BAC = \frac{1}{2}BAF = \frac{1}{4}I$. Therefore

$$E = T \tan \tfrac{1}{4}I. \quad . \quad . \quad . \quad . \quad . \quad . \quad . \quad . \qquad (25)$$

EXERCISE.—Derive equation (25) from (24).

86. Given M and I to Find E.

From trigonometry,

$$\sec \tfrac{1}{2}I = \frac{1}{\cos \tfrac{1}{2}I}.$$

Insert this in (24) and we get

$$E = R\frac{1 - \cos \tfrac{1}{2}I}{\cos \tfrac{1}{2}I}. \quad \ldots\ldots\ldots \quad (a)$$

But from Fig. 17, $M = R(1 - \cos \tfrac{1}{2}I)$. Substitute in (a):

$$E = \frac{M}{\cos \tfrac{1}{2}I} = M \sec \tfrac{1}{2}I. \quad \ldots\ldots\ldots \quad (26)$$

87. Given E and I to Find R.

From (24),

$$R = \frac{E}{\sec \tfrac{1}{2}I - 1} = \frac{E}{\text{ex sec } \tfrac{1}{2}I} = E\frac{\cos \tfrac{1}{2}I}{\text{vers } \tfrac{1}{2}I}. \quad \ldots \quad (27)$$

88. Given I and E to Find T.

From (25),

$$T = \frac{E}{\tan \tfrac{1}{4}I} = E \cot \tfrac{1}{4}I. \quad \ldots\ldots\ldots \quad (28$$

89. Given the Chord C and Degree of Curve D to Find the Chord Deflection Offset d.

In Fig. 18 extend EA to H, making $AH = EA = AB$; join

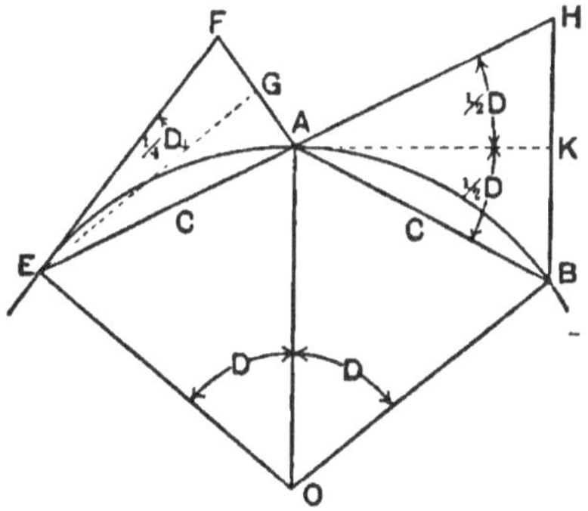

FIG. 18.

H and B and draw AK to the mid-point of HB. Then

$$HK = KB = C \sin \tfrac{1}{2}D.$$

$$\therefore\ d = HB = 2C \sin \tfrac{1}{2}D. \quad \ldots\ldots\ldots \quad (29)$$

When $C = 100'$,

$$d = 200 \sin \tfrac{1}{2}D. \qquad (29')$$

If we write $\sin \frac{1}{2}D = \frac{\frac{1}{2}C}{R}$ from (12) in formula (29), there results

$$d = \frac{C^2}{R}. \qquad (30)$$

For curves up to 7°, $C = 100'$; hence

$$d = \frac{10000}{R}. \qquad (30')$$

For curves from 7° to 14°, $C = 50'$; therefore

$$d = \frac{2500}{R}. \qquad (30'')$$

For R write $\frac{5730}{D}$, and (30'), for $C = 100$, becomes

$$d = \frac{10000}{5730}D = 1.745D; \qquad (31)$$

and for $C = 50$, (30'') becomes

$$d = \frac{2500}{5730}D = .4363D = .873 . \frac{D}{2}. \qquad (31')$$

EXAMPLE.—Find d for a 6° curve, $C = 100$ feet.

By (29'), $d = 200 \times 0.05234 = 10.47$ feet.

By (30'), $d = \frac{10000}{955.4} = 10.47$ feet.

By (31), $d = 1.745 \times 6 = 10.47$ feet.

90. Given the Chord C and Degree of Curve D to Find the Tangential Deflection Offset t.

In Fig. 18 make EF (tangent at E) equal to EA, and join F with A. Draw EG to the mid-point of FA. Angle $AEG = GEF = \frac{1}{4}D$; hence, from the figure,

$$AG = GF = C \sin \tfrac{1}{4}D.$$

$$\therefore t = 2C \sin \tfrac{1}{4}D. \qquad (32)$$

When $C = 100$ feet,

$$t = 200 \sin \tfrac{1}{4}D. \quad . \quad . \quad . \quad . \quad . \quad . \quad . \quad . \quad . \quad . \quad (32')$$

Since $\frac{1}{4}D$ is small, we may write, without material error, $\sin \frac{1}{4}D = \frac{1}{2} \sin \frac{1}{2}D$; then, writing $\sin \frac{1}{2}D = \frac{\frac{1}{2}C}{R}$, as in **89**, we get

$$t = \frac{C^2}{2R}. \quad . \quad . \quad . \quad . \quad . \quad . \quad . \quad . \quad . \quad . \quad . \quad . \quad (33)$$

Making $C = 100$ ft. and writing $R = \frac{5730}{D}$ gives

$$t = \frac{10000}{2 \times 5730} D = 0.873D. \quad . \quad . \quad . \quad . \quad (33')$$

When $C = 50$ feet, (33) yields

$$t = .218D = .436 \times \frac{D}{2}. \quad . \quad . \quad . \quad . \quad . \quad (33'')$$

EXAMPLE.—Find t for a 6° curve, $C = 100$ ft.

By (32') $\qquad t = 200 \sin 1° \, 30' = 5.24$ ft.

By (33'), $\qquad = .873 \times 6 = 5.24$ ft.

91. To Find the Subtangential Deflection Offset t' for a Subchord C'

FIRST METHOD.—By formula (13) find the angle at the center subtended by the subchord C'; call this angle D'. From (32),

$$t' = 2C' \sin \tfrac{1}{4}D'. \quad . \quad . \quad . \quad . \quad . \quad . \quad . \quad . \quad (34)$$

SECOND METHOD.—In Fig. 19, with E as center strike the arcs FG and AH, taking $EF = C'$ and $EA = C$; prolong EG to B. Now assuming that the chords C' and C are proportional to their central angles we have

$$\frac{AB}{C'} = \frac{t}{C}. \quad . \quad . \quad . \quad (a)$$

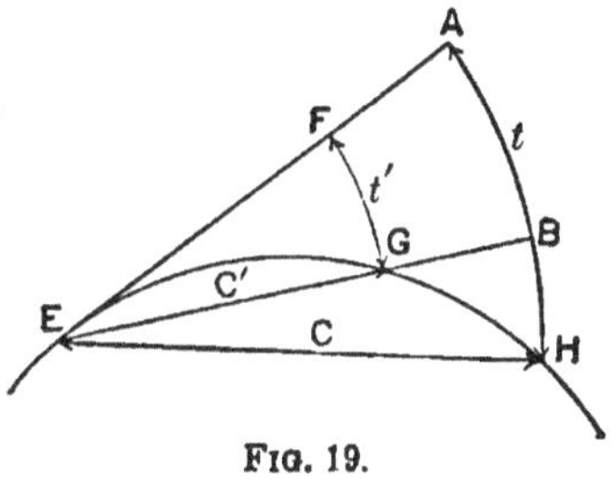

FIG. 19.

From the similar sectors EFG and EAB, since $EB = C$,

Multiplying (*a*) and (*b*) together, term by term,

$$\frac{C}{C'} = \frac{t}{t'} \cdot \frac{C'}{C}.$$

Whence

$$t' = t\left(\frac{C'}{C}\right)^2 \quad . \quad . \quad . \quad . \quad . \quad . \quad . \quad . \quad (35)$$

EXAMPLE.—Find t' for a 7° curve when $C = 60$ ft.

Here $D' = \frac{60}{100} \times 7°$ (very nearly) $= 4°\ 12'$.

By (34), $t' = 2 \times 60 \times 0.01832 = 2.20$ ft.

By (32)' $t = 6.11$ ft.

By (35), $t' = 6.11 \times \left(\frac{60}{100}\right)^2 = 2.20$ ft.

92. To Find the Tangent Offset z.

In Fig. 20, $EB = z$ is the required offset. Let $AE = n$ chains $= 100n$ feet. $AE = FB$, the half-chord having the mid-ordinate $AF = EB$; hence we have, by formula (18),

$$z = \tfrac{7}{8}n^2D. \quad . \quad . \quad . \quad (36)$$

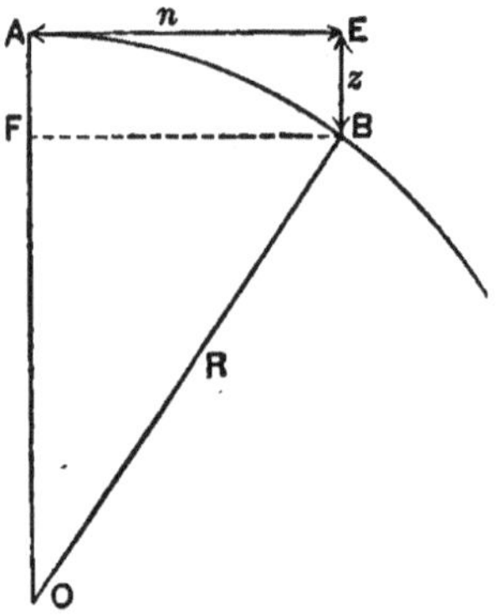

FIG. 20.

In this formula we may take n to be either the length of AE or the arc AB, in chains. If taken equal to AE the offsets will be slightly too small, while if taken equal to AB they will be a little too large. The use of the formula is limited to small values of n and D, as was pointed out in **83**. (See CAUTION.)

Formula (36) is easy of application and of frequent use in locating curves by offsets from the tangents. For curves up to 4° n may be as great as 3, but for sharper curves it should be less.

EXAMPLE.—Find six offsets to a 4° curve at points 50 ft. apart, measured around the curve.

By successive applications of (36) we have

$$\text{for } n = \tfrac{1}{2},\quad z = \tfrac{7}{8} \times \tfrac{1}{4} \times 4 = 0.88 \text{ feet}$$
$$n = 1,\quad z = \tfrac{7}{8} \times 1 \times 4 = 3.50 \text{ ''}$$
$$n = \tfrac{3}{2},\quad z = \tfrac{7}{8} \times \tfrac{9}{4} \times 4 = 7.88 \text{ ''}$$
$$n = 2,\quad z = \tfrac{7}{8} \times 4 \times 4 = 14.00 \text{ ''}$$
$$n = \tfrac{5}{2},\quad z = \tfrac{7}{8} \times \tfrac{25}{4} \times 4 = 21.88 \text{ ''}$$
$$n = 3,\quad z = \tfrac{7}{8} \times 9 \times 4 = 31.50 \text{ ''}$$

The last value of z is in error by about 0.2 ft., but for setting stakes on construction this difference is not material so long as the alignment beyond this point does not depend on it. In setting track-centers the completed road-bed is available and the stakes may be set with the transit, in the usual way.

93. Difference in Length of a Circular Arc and its Long Chord.

First Method.—Let the central angle be a degrees. By (13), $\sin \tfrac{1}{2}a^\circ = \dfrac{c}{2R}$

Changing degrees to circular measure, a (in π meas.) $= \dfrac{\pi a^\circ}{180} = \dfrac{a^\circ}{57.3}$. The length of arc is $Ra = R\dfrac{a^\circ}{57.3}$. Then

$$\text{Arc} - \text{chord} = R\frac{a^2}{57.3} - c. \quad . \quad . \quad . \quad . \quad . \quad (37)$$

Second Method.—An easy approximation may be found as follows:

Referring to Fig. 17, $AE = c$, $GF = M$. Let $AG = b = \dfrac{c}{2} + x$. From the right triangle AFG

$$\left(\frac{c}{2} + x\right)^2 = \frac{c^2}{4} + M^2.$$

Neglecting the x in denominator as small compared with c gives

$$x = \frac{M^2}{c}. \quad . \quad . \quad . \quad . \quad . \quad . \quad . \quad . \quad . \quad . \qquad (b)$$

Then will
$$2b - c = 2x = \frac{2M^2}{c}. \quad . \quad . \quad . \quad . \quad . \quad . \qquad (38)$$

From Huygens' approximation to the length of a circular arc (see Williamson's Differential Calculus, p. 66), arc $= \frac{8b - c}{3}$. Therefore

$$\text{Arc} - \text{chord} = \frac{8b - c}{3} - c = \tfrac{4}{3}(2b - c). \quad . \quad . \qquad (c)$$

Inserting the value of $2b - c$ from (38) gives

$$\text{Arc} - \text{chord} = \frac{8M}{3c}. \quad . \quad . \quad . \quad . \quad . \quad . \quad . \quad . \quad . \quad . \quad . \qquad (d)$$

When the arc is not very great we may write $c = 100n_1$, where n_1 is the number of chains contained in the arc AE. From (18), remembering that $n_1 = 2n$,

$$M = 0.218n_1^2D.$$

Inserting these values of c and M in (d),

$$\text{Arc} - \text{chord} = \frac{8}{3}\,\frac{(.218)^2n_1^4D^2}{100n_1} = \frac{1}{800}n_1^3D^2, \text{ nearly.} \quad . \quad . \qquad (39)$$

Example.—Find the difference in length of arc and chord of a 4° curve when $n_1 = 6$ stations.

The central angle is $4 \times 6 = 24°$; then, from Table IV, $c = 595.74$.

By (37),

$$\text{Arc} - \text{chord} = 1432.7 \times \frac{24}{57.3} - 595.74 = 4.34 \text{ ft.}$$

By (39),

$$\text{Arc} - \text{chord} = \frac{6 \times 6 \times 6 \times 4 \times 4}{800} = 4.32 \text{ ft}$$

Remark.—Formula (38) is interesting as showing what a comparatively small increase in length of line is caused by a considerable lateral deflection in alignment. For instance, a lateral deflection of 2000 feet is made at the mid-point of a line 40,000 feet long; what will be the increase in length?

By (38) the increase is $\frac{2(2000)^2}{40,000} = 200$ feet, giving for the increased length 40,200 feet.

B. Locating Simple Curves.

94. To Locate a Curve with the Chain by Offsets from Chords Produced.

In Fig. 21 let the *P.C.* fall at *B*. If *BC* is a full chain, prolong

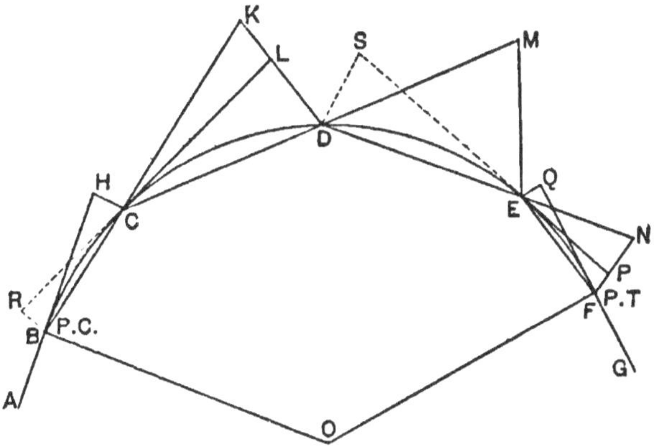

FIG. 21.

the tangent *AB* to *H*, making $BH = BC$; *HC* will equal *t*, which may be calculated by (32′) or (33′). With *B* as center, strike an arc with radius *BH*, and with *H* as center and *t* as radius strike an arc; at *C*, where these arcs intersect, set a stake. Produce *BC* to *K*, making $CK = BC = CD$; strike the arc *KD* from *C* as center; make the chord $KD = d$, calculated from (29′), (30′), or (31). Set a stake at *D* and proceed in like manner for the other points until the *P.T.* is reached, where *FP* is made equal to *t*.

Usually the *P.C.* does not fall at a full station; then $HC = t'$, which may be found by (34) or (35). Using this value of t', we locate *C* as above. At *B* make $BR = t'$, and prolong *RC* to *L*; make $LD = t$ and set a stake at *D*. *EM* will equal *d*, and may be located as before.

We may regard *KD* as equal to $KL + t$, and, finding, *KL*, measure KD and set D without locating R. To do this we have the similar triangles BHC and CKL, from which

$$\frac{KL}{CK} = \frac{t'}{BC},$$

and therefore, since $KC = CD$,

$$KL = t'\frac{CD}{BC}.$$

In like manner at F we have

$$PN = t\frac{EF}{ED}, \quad \text{and} \quad FP = t_1'$$

hence

$$NF = PN + t_1'.$$

Make $EQ = t_1'$, prolong QF, and we have the tangent at F.

EXAMPLE.—Given the $P.C.$ of a 5° curve at 106 + 20 and the angle of intersection 22°, to locate the curve.

Here $$L = \frac{22}{5} = 4.4 \text{ stations.}$$

Therefore the number of the $P.T.$ is

$$106.20 + 4.4 = \text{sta. } 110 + 60.$$

BC in this case is 80 ft., and by (33′)

$$t = 0.873 \times 5 = 4.37 \text{ ft.}$$

By (35), $$t' = 4.37 \times \left(\frac{80}{100}\right)^2 = 2.80 \text{ ft.}$$

Set off $HC = 2.80$ ft., and at D make

$$KD = 2.80 \times \frac{100}{80} + 4.37 = 7.87 \text{ ft.}$$

At E make $ME = d = 8.72$ by (31). This will be at sta. 109; at 110 set a stake by offsetting 8.72 ft. The last chord is 60 long, and hence the offset

$$NF = 4.37 \times \frac{60}{100} + 4.37 \times \left(\frac{60}{100}\right)^2 = 2.62 + 1.57 = 4.19 \text{ ft.}$$

Make $EQ = 1.57$ ft., and prolong QF, the terminal tangent.

95. To Locate a D Degree Curve by Offsets from Tangent. Let AM, Fig. 22, be tangent at A, and E, F, G, etc., points on the curve. The offsets BE, CF, etc., may be found from formula (36),

$$z = \tfrac{7}{8}n^2D,$$

either by taking equal intervals, AB, BC, CM along the tangent or by taking E, F, G, etc., at regular stations around the curve and using the arc length instead of the tangent.

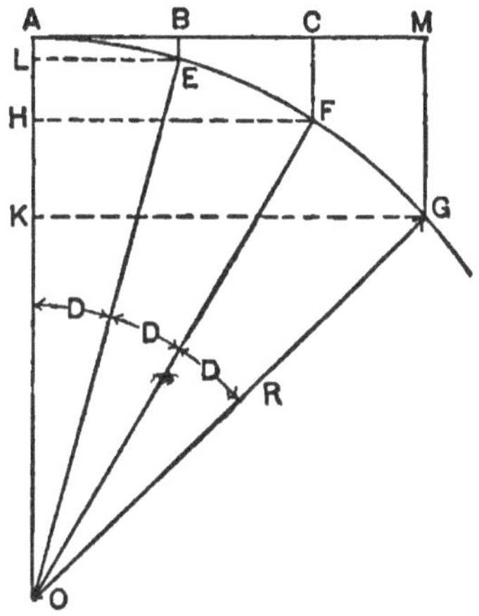

FIG. 22.

When the arc AG is large, or strict accuracy is required, we proceed to find the offsets at regular stations and the lengths of AB, AC, etc. First find R from (12) or (12′); then from triangle OEL,

$$BE = AL = R(1 - \cos D) = R \text{ vers } D,$$

$$AB = LE = R \sin D.$$

In like manner

$$CF = AH = R(1 - \cos 2D) = R \text{ vers } 2D,$$
$$AC = HF = R \sin 2D,$$

and so on for any number of stations.

Should A fall at a plus station, we first find the angle D_1 at the center, then

$$BE = R \text{ vers } D_1,$$
$$AB = R \sin D_1,$$
$$CF = R \text{ vers } (D_1 + D),$$
$$AC = R \sin (D_1 + D),$$
$$\text{etc.} = \quad \text{etc.}$$

The ordinates BE, CF, etc., are evidently equal to the mid-ordinates for long chords $2LE$, $2HF$, etc.; hence we can, if A, E, F, and G, fall at full stations, take them direct from Table V; then take the long chords from Table IV and dividing these by 2, get the required coordinates.

EXAMPLE.—Locate three stations of a 4° curve by offsets every 50 ft. on *curve.*

Referring to Table V, the required offsets are 0.87, 3.49, 7.85, 13.94, 21.77, and 31.31. By Table IV the distances measured along tangent are 50.0, 99.94, 149.76, 199.39, 248.78, and 297.87. With these values we can set out the curve either way from *A*.

Had we used formula (36) we should have had for the values of the offsets 0.87, 3.50, 7.88, 14.00, 21.87, and 31.50.

96. To Locate a Curve by Offsets from a given Long Chord.

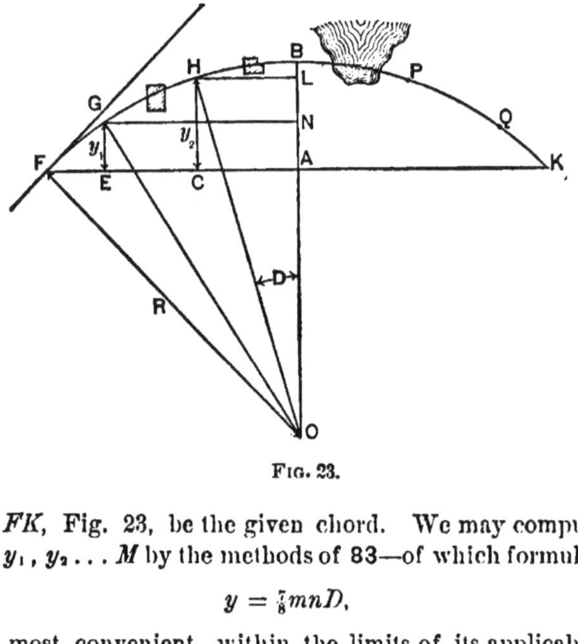

FIG. 23.

Let *FK*, Fig. 23, be the given chord. We may compute the offsets y_1, y_2 . . . *M* by the methods of 83—of which formula (17),

$$y = \tfrac{7}{8}mnD,$$

is the most convenient, within the limits of its applicability—and setting off these ordinates, locate the curve.

Or we may set off the mid-ordinate $M = R$ vers FOA at *A*, and at *C* set off $y_2 = M - R$ vers D, making

$$AC = HL = R \sin D.$$

GE will be

$$y_1 = M - R \text{ vers } 2D, \quad \text{and} \quad AE = R \sin 2D.$$

ANOTHER METHOD is to find the angle *KOF* at the center, and by Table IX determine $BA = M$; then by Tables V and IV determine BL, BN, LH, and NG. Then $HC = M - BL$, which set off at C, and other points in like manner.

EXAMPLE.—Given the $P.C.$ of a 4° curve at station 160 + 75, the angle between tangent and chord = 9°, required the offsets necessary to locate the curve.

Here $$I = 2 \times 9 = 18°.$$

$$\therefore \quad L = \frac{18}{4} = 4.50 \text{ stations.}$$

Hence the $P.T.$ falls at 160.75 + 4.50 = sta. 165 + 25. The mid-point on curve B falls at sta. 163. By Table IX,

$$M = \frac{70.54}{4} = 17.64 \text{ ft.}$$

By Table V the mid-ordinate for two stations of a 4° curve is

$$BL = 3.49.$$

Hence $$HC = 17.64 - 3.49 = 14.15.$$

By Table IV, $$HL = AC = 99.94 \text{ ft.}$$

Measure $AC = 99.94$ ft., and set off $CH = 14.15$ ft., and drive a stake at H. In like manner find

$$GE = 3.70 \quad \text{and} \quad AE = 199.39 \text{ ft.}$$

The points P and Q are also located by means of the coordinates just determined.

If B had fallen at an odd station, the curve could have been located in the same manner, H and P being 100 ft. from B, G and Q 200, etc.

97. To Locate a Curve with Transit and Chain when the Degree D or Radius R is Known.

If R is given, determine D by (13); then, since the angle in the circumference of a circle is half the angle at the center subtended by the same chord, we may locate points on the curve by successive deflections from the tangent.

In Fig. 24 let the $P.C.$ be at A, at which point set the transit, and with the vernier-plates clamped at zero place the telescope in tangent either by sighting the $P.I.$ or by backsighting to some point in the tangent. Deflect from the tangent half the angle at the center for the sub-chord or chord, and direct the head chainman into line while the rear chainman holds his end of the chain at the transit, the chain being kept taut. The stakeman drives a stake at the point where the head chainman's flag rested, and the rear chainman advances to this point. Deflect $\frac{1}{2}D$ from the chord AB just run, and while the rear chainman holds his end of the chain at B direct the head chainman into line at C. Other points are located by deflecting an additional $\frac{1}{2}D$ for each chord length measured, until a point E is reached to which it is desirable to

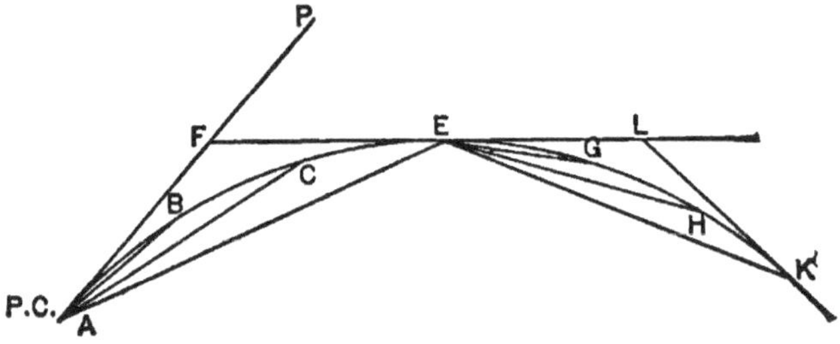

FIG. 24.

move the transit. The angle FAE should not exceed about 15°. Move the transit to E, backsight to A, and deflect $FEA = EAF$, when the telescope will be in tangent, and the curve can be continued until it is again necessary to move the transit. At the *P.T.* put the telescope in tangent by backsighting to the point last occupied by transit and deflecting the tangential angle as at E. The line may now be continued.

98. The Index-angle is read on the vernier-plate, and is the angle between the tangent to the curve at the *P.C.* and any other line passing through a point on the curve when the telescope is directed along this line. It is most frequently taken as the angle between the initial and any subsequent tangent to the curve. Thus at E the index-angle equals $EFP = 2FAE$. At any point on the curve the index-reading in tangent may be found by the following rule, which may be easily deduced from a figure:

From double the index-angle that fixed the point subtract the index-angle in tangent at the last point; the remainder is the index-angle required.

99. Subdeflection-angles may be found by (13) rigidly, or approximately (and with sufficient accuracy except when D is very large) by assuming the central angles to be proportional to their chords. Thus on a 4° curve the central angle for a sub-chord of 25 ft. would be 1°, and the subdeflection-angle 30′.

EXAMPLE.—Locate a 4° curve to left when the *P.C.* is at sta. 81 + 25 and $I = 32° 36'$.

Here $$L = \frac{32.6}{4} = 8.15 \text{ chains.}$$

Hence the *P.T.* will fall at 81.25 + 8.15 = sta. 89 + 40. The first sub-chord is 75 ft. long, and the first deflection-angle will be found by (12).

$$\sin \tfrac{1}{2}\delta = \frac{37.5}{1432.7} = 0.02617$$

$$\therefore \tfrac{1}{2}\delta = 1° 30'.$$

By the approximate rule, since $\frac{1}{2}D = 2°$,

$$\frac{\frac{1}{2}\delta}{2} = \frac{75}{100},$$

whence $\frac{1}{2}\delta = 2 \times \frac{3}{4} = 1° 30'$ as before.

With transit at *P.C.* deflect 1° 30′ from tangent, measure 75 feet, and set sta. 82. Then a deflection of 3° 30′ will determine 83, 5° 30′ sta. 84, 7° 30′ sta. 85. Now remove transit to 85, and with vernier at 7° 30′ backsight to 81 + 25. Reverse telescope and set vernier at 15° 00′, when the telescope will be in tangent. An index angle of 17° will fix 86, and so on.

The last chord will be only 40 feet long, for which the sub-deflection-angle is $\frac{40}{100}$ of 2°, that is, 48′. The index-angle fixing the *P.T.* is therefore 23° 48′.

To get in tangent at 89 + 40 backsight to sta. 85, with vernier at 23° 48′; then by the rule of **98** the index-reading is $(23° 48') \times 2 - 15° = 32° 36' = I$. Set the vernier at this reading and run tangent.

CAUTION.—It is not good practice to set more than 4 or 5 stations on curve from any one point. MR. SHUNK gives the limiting angle to be deflected from tangent as 20°, and says 15° should rarely be exceeded. (*Field Engineer*, p. 82.)

100. The Transit Notes may be conveniently kept in the form below, which shows the notes for the last example.

When possible the tangents should be run to intersection, the angle *I* measured, and the tangent distance calculated. Then

Station.		Deflection-angle.	Index-reading.	Index in Tangent.	Calculated Course	Magnetic Course.	Remarks.
90							
+40	⊙ *P.T.*	0° 48′	23° 48′	32° 36′	N 27° 36′ E	N 27° 30′ E	
89			23° 0′				
88			21° 0′				
87	.		19° 0′				
86			17° 0′				
85	⊙		7° 30′	15° 0′			
84			5° 30′				
83		2° 0′	3° 30′				
82		1° 30′	1° 30′				4° C.L.; *P.I.* set.
+25	⊙ *P.C.* 4°C.L.	0° 0′	0° 0′	0° 0′			*I* = 32° 36′; *T* = 418.9 ft.
81					N 60° 12′ E	N 60° 10′ E	

measure along tangents and set *P.C.* and *P.T.* from the *P.I.* When the curve is run in, the position of the *P.T.* thus found should agree with the one set from the *P.I.* If the error is greater than the circumstances of the case permit, the curve must be rerun and tangents remeasured.

101. Another Form of Notes, and in some respects a better one than the above, is given below. The index-readings are computed as though the entire curve were run from the *P.C.* The notes for the last example would appear as below:

Station.		Deflection-angle.	Index-reading.	Total Angle.	Calculated Course.	Magnetic Course.	Remarks.
90							
+40	⊙ *P.T.*	0° 48′	16° 18′	32° 36′	N 27° 30′ E	N 27° 30′ E	
89			15° 30′				
88			13° 30′				
87			11° 30′				
86			9° 30′				
85	⊙		7° 30′				
84			5° 30′				
83		2° 0′	3° 30′				
82		1° 30′	1° 30′				4° curve left;
+25	⊙ *P.C.* 4°C.L.	0° 0′	0° 0′				*P.I.* set. *I*=32° 16′; *T* = 418.9 ft.
81					N 60° 12′ E	N 60° 10′ N	

The computations are all made before beginning the work, and the notes have the advantage of permitting the tracing of the curve either way from the instrument without additional computations. Suppose the transitman to have run the curve from the *P.C.* to sta. 85, to which point he removes the instrument. He there sets the vernier at 0°—the angle on limb when telescope was in tangent at the *P.C.*—then sighting the *P.C.* he reverses the telescope and deflects to 9° 30′, which will fix sta. 86. Had the tangent at 85 been desired, a reading of 7° 30′—the angle that located that point—would have put the telescope in the plane desired. A reading of 11° 30′ fixes 87, and so on to the *P.T.* Removing to the *P.T.*, the plates are clamped at 7° 30′, and a backsight to sta. 85 taken; then deflecting to 16° 18′, the telescope is in tangent at the *P.T.* Had it been desirable to set 84 from 85, a reading of 5° 30′ would fix that point; others may be found in the same manner.

Any convenient form of notes, which are intelligible to another engineer who may have to retrace the curve, may be used, but it is desirable that some general form should be employed. Either of the preceding forms seems to meet ordinary requirements.

C. Obstacles.

102. To Pass an Obstacle on a Curve.

FIRST. *Suppose the obstacle to be one obstructing vision at one station only.*

In Fig. 25 suppose transit set at *A*, and *B* and *C* located from that point, but the next full station, *H*, to be invisible from *A*.

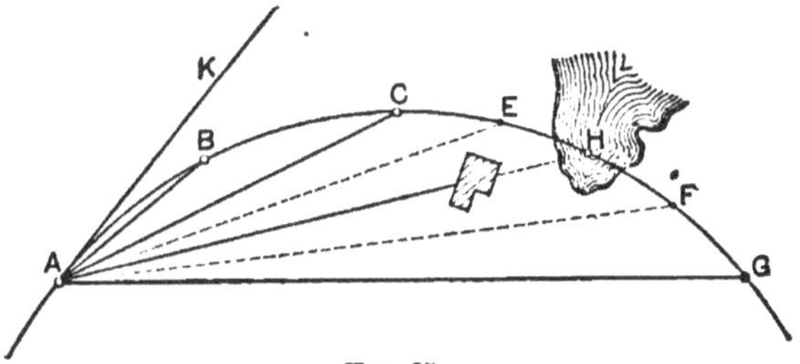

FIG. 25.

Set a plus station at *E*, as near the obstruction as may be convenient, then set *F* 100 feet from *E*. Next make $FG = 100 - CE$, and locate *G* with the corresponding deflection-angle. Other stakes may be set beyond *G*, or the transit may be removed to that point and the curve beyond traced.

SECOND. *Suppose the line of sight obscured for more than one station, as in Fig. 26.*

If transit is at A, deflect an angle HAB that will clear all obstructions, and at the same time cause B to fall at a full station. Then by Table IV, Table IX, or by formula (16) calculate the long chord AB; measure AB and move transit to B; then deflect

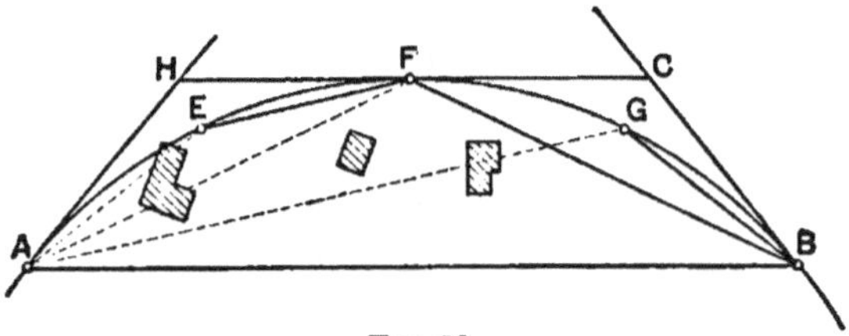

FIG. 26.

the angle $ABC = BAH$ when the telescope will be in tangent. The curve may now be run both ways from B.

If it happen that some stations, as E and F in the figure, are still invisible, they may be located by offsets from chord or tangent.

EXAMPLE.—Let the curve be a 3° curve to right; angle $HAB = 7° 30'$, the deflection-angle for 5 stations. By Table IV the long chord is 498.63 feet, which can now be measured and a hub set at B; then making angle $CBA = 7° 30'$, the telescope will be in tangent and the curve can be traced either way.

103. To Locate a Curve when the *P.C.* is Inaccessible.

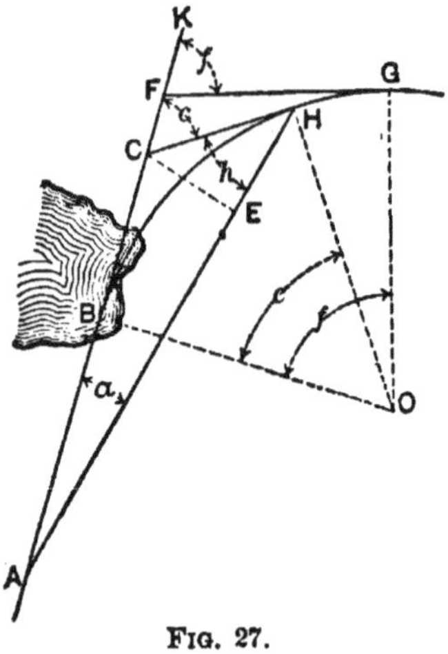

FIG. 27.

In Fig. 27 let the *P.C.* at B be inaccessible; it is desired to reach a point H on accessible ground.

FIRST METHOD. — Assume a point H on the curve such that a line AH from an accessible point A, on tangent, will clear the obstacle; for convenience H should be at a full station. The arc BH and central angle, which equals HCF, are then known. Calculate $BC = T$ by (14) or (14a); then since AB is known, $AC, = AB + BC$, is known.

Now in triangle ACH, from trigonometry,

$$\frac{\tan \frac{1}{2}(h - a)}{\tan \frac{1}{2}(h + a)} = \frac{AC - CH}{AC + CH}$$

But $(h + a) = c$; hence

$$\tan \tfrac{1}{2}(h - a) = \frac{AC - CH}{AC + CH} \tan \tfrac{1}{2}c. \quad . \quad . \quad . \quad (40)$$

Then $\frac{1}{2}(h + a) + \frac{1}{2}(h - a) = h$, the larger angle, and $\frac{1}{2}(h + a) - \frac{1}{2}(h - a) = a$, the smaller angle. AH may be found by the law of sines, or by drawing CE perpendicular to AH, when

$$AH = AC \cos a + CH \cos h. \quad . \quad . \quad . \quad . \quad (41)$$

EXAMPLE.—The $P.C.$ of a 4° curve is at sta. 141 + 25, and it is desired to reach the point H from sta. 139 on tangent.

Suppose H be assumed to fall at sta. 147; the curve length is $L = 147 - 141.25 = 5.75$ chains. Then angle $c = 5.75 \times 4 = 23° 0'$. By Table IX the tangent distance for a 1° curve is $T_1 \measuredangle 23° = 1165.8$ ft.

By (14a), $$T = \frac{1165.8}{4} = 291.45 \text{ ft.}$$

Now $$AC = 291.45 + 225 = 516.45 \text{ ft.},$$
and

$$AC + CH = 516.45 + 291.45 = 807.90,$$

while

$$AC - CH = 225 \text{ ft.};$$

hence, by (40),

$$\tan \tfrac{1}{2}(h - a) = \frac{225}{807.9} \times 0.20345 = 0.05666 = \tan 3° 15'.$$

Therefore

$$h = 11° 30' + 3° 15' = 14° 45',$$

and

$$a = 11° 30' - 3° 15' = 8° 15'.$$

By (41),

$$AH = 516.45 \times 0.98965 + 291.45 \times 0.96705 = 793.0 \text{ ft.}$$

At A deflect 8° 15′ from tangent, measure 793.0 ft. and set a hub; move to this point, backsight to A and deflect 14° 45′ into tangent, then trace in the curve.

SECOND METHOD.—If F, any assumed point in tangent, is visible from A, AF may be measured by some indirect method; then $AF - AB = T$. The tangent for a 1° curve having same intersection-angle, KFG, is $T_1 = T \times D$; find this value of T_1 in Table IX and take out the corresponding value of I. With transit at F deflect the angle KFG, measure $FG = FB = T$, and set hub at G. The station number of G is found by dividing the central angle, $= KFG$, by the degree of curve D. Move to G and trace the curve.

EXAMPLE.—Let AF measure 490.5 ft. from sta. 139 of the last example. Then $AB = 225$ ft., and $BF = 490.5 - 225 = 265.5$ ft. $265.5 \times 4 = 1062$ ft., which by Table IX is the value of T_1 for $I = 21°$. Set transit at F, deflect 21°, and measure $FG = 265.5$ ft.

$$L = \frac{21}{4} = 5.25 \text{ chains;}$$

hence G will fall at $141.25 + 5.25 =$ sta. $146 + 50$. Move to G and run the curve both ways.

THIRD METHOD.—In Fig. 28 let the inaccessible $P.C.$ be at B, and let it be required to reach E from a point C on the curve prolonged backwards from B.

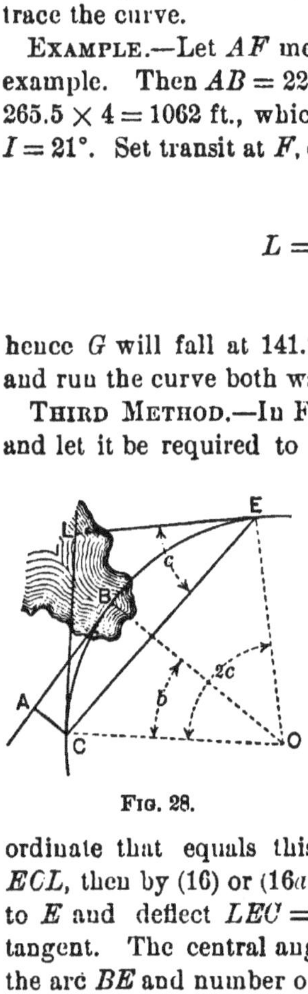

FIG. 28.

At a given point A on tangent calculate the tangent offset by (36) or the methods of **95**, then set this off at right angles to AB; set the transit at C and turn off $ACL = 90° - COB$, when the telescope will be in tangent at C. COB may be found from Table IX by multiplying AC by the degree of curve and taking half the intersection-angle corresponding to the mid-ordinate that equals this product. Now deflect and measure ECL, then by (16) or (16a) calculate CE, which measure. Move to E and deflect $LEC = ECL$ and the telescope will be in tangent. The central angle $BOE = 2LEC - BOC$, from which the arc BE and number of sta. E may be found.

EXAMPLE.—Take the same example as in the last two cases. A is at sta. 139, B at $141 + 25$; hence $AB = 2.25$ stations.

By (36), $\quad z = AC = \frac{7}{8} \times (2.25)^2 \times 4 = 17.72$ ft.

Or by Table IX the angle corresponding to the long chord $(2 \times 2.25) \times 4 = 1800$ ft. is 18° 4′, for which the mid-ordinate is 71.06 ft. For our 4° curve the mid-ordinate will be $\frac{71.06}{4} = 17.77$ ft., which equals AC and agrees closely enough with the value for z above.

Make angle $BAC = 90°$, and measure $AC = 17.72$ ft. Move to C and sight to A, then make angle $ACL = 90° - (9° 2') = 80° 58'$. Suppose an angle $LCE = 16° 1'$ to clear the obstacle. By formula (16),

$$CE = 2R \sin (16° 1') = 2 \times 1432.7 \times 0.27592 = 790.6 \text{ ft.}$$

Measure along CE 790.6 ft. and set a hub; move to E and run the curve.

CE might have been found by means of Table IX, for the long chord of a 1° curve having $I = 2LCE = 32° 2'$ is 3162.0 ft.; divide this by 4 and there results $CE = 790.5$ ft.

104. To Pass to Tangent when the *P.T.* is Inaccessible. This is just the reverse of the preceding problem, and may be accomplished by reversing the processes described above.

When the *P.T.*, however, falls in or beyond a river or lake obstructing the ordinary methods of indirect measurement, the case merits a special solution.

First Method.—In Fig. 29 let the transit be at A, and B the *P.T.* From the known station numbers of A and B the length of curve and angle I may be found; then, by (14), $AC = R \tan \frac{1}{2}I$, or, by (14a), $AC = \frac{T_1 \measuredangle I°}{D}$.

Move to C and deflect the angle I; set a stake F, and one at some other accessible point E; measure angle $ECF = c$. Move to F and measure the angle EFC and the side EF; then in triangle ECF angle $e = 180° - (c + f)$; by trigonometry

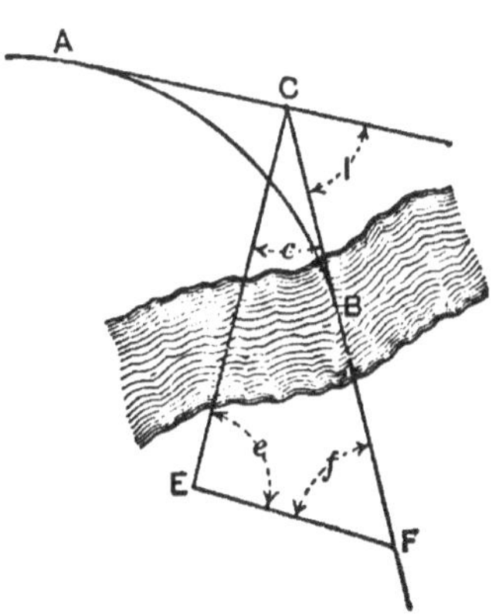

Fig. 29.

Since $BC = AC$, there results $BF = CF - AC$; and as the station number at B is known, that at F becomes known, and the line may be continued.

If B is not the *P.T.*, measure back the distance FB, set transit at B, and continue the curve.

EXAMPLE.—Let the *P.T.* of a 2° *C.L.* fall at sta. 205 + 50—an inaccessible point; suppose A at sta. 200, angle $c = 40°$, $f = 80°$, $EF = 310$ ft.

Here $\qquad I = 5.50 \times 2 = 11°\ 0'$, and $\quad e = 60°$.

By (14*a*), $\qquad T = \dfrac{551.74}{2} = 275.87$ ft.

From (42), applying logarithms,

$$\log CF = 2.49136 + 9.93753 - 9.80807 = 2.6082.$$

Whence $CF = 417.7$ ft. Then $BF = 417.7 - 275.87 = 141.8$ ft.; therefore the number of F will be 206 + 91.8.

SECOND METHOD.—In Fig. 30, with the transit at any point A on the curve, assume a long chord AB and calculate the angle CAB; deflect this angle from the tangent AC, and set a point E beyond obstruction; set also a stake at C in tangent.

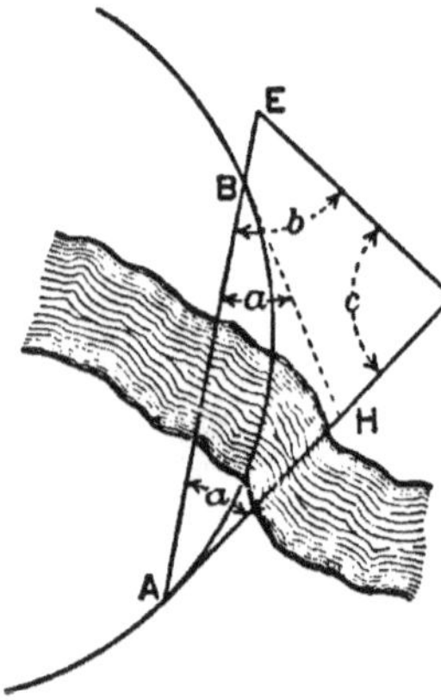

FIG. 30.

Move to E and measure AEC and side EC. Compute AE from the triangle AEC. If this is greater or less than the length of the long chord AB, take their difference BE and set a hub at B. With the transit at B trace out the curve.

EXAMPLE.—Given A at sta. 210 of a 3° C. L., angle $a = 12°$, $b = 92°$, $EC = 181$ ft. Then $c = 76°$, and by solving the triangle AEC, $AE = 844.7$ ft. By Table IX the long chord of a 1° curve for $I = 24°$ is 2382.6 ft.; therefore $AB = \dfrac{2382.6}{3} = 794.2$ ft. Now will $EB = 844.7 - 794.2 = 60.5$ ft., which is the distance along EA that transit must be moved *back* from E.

105. Given the Perpendicular p from a Point to a Tangent, to Find the Point on Tangent at which to Begin a Curve of Given Radius which will Pass through the Given Point.

FIRST SOLUTION.—In Fig. 31 let P be the point, BP the perpendicular. We have to find $BA = x$.

From P draw PC parallel to AB; then in triangle OPC

$$R^2 = x^2 + (R - p)^2.$$

From which

$$x = \sqrt{2Rp - p^2}. \quad . \quad (43)$$

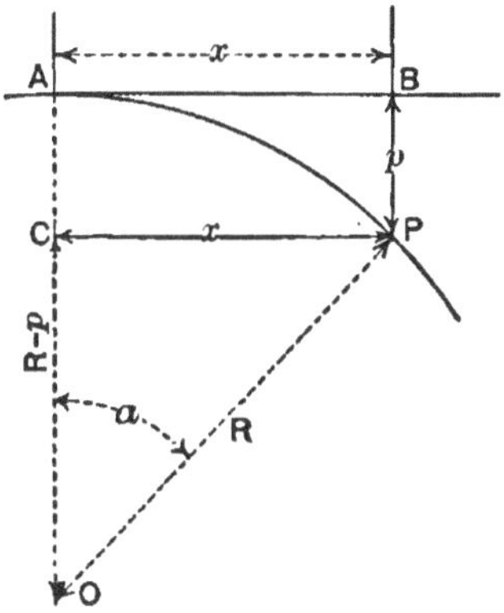

FIG. 31.

SECOND SOLUTION.—Consider $p = AC$ as the mid-ordinate for a long chord $= 2x$; then $p \times D$ = the mid-ordinate for a 1° curve for a central angle equal $2a$. The corresponding long chord may be taken from Table IX. Then

$$x = \frac{1}{2}\frac{L.C.}{D}. \quad . \quad . \quad . \quad . \quad . \quad . \quad . \quad . \quad . \quad . \quad . \quad . \quad (43a)$$

EXAMPLE.—Given $p = 30$ ft., $D = 4°$ ($R = 1432.7$), to find x.

By (43), $\quad x = \sqrt{85,962 - 900} = 291.65$ feet.

By the second method,

$$30 \times 4 = 120,$$

the mid-ordinate for a 1° curve corresponding to an angle of 23° 29′, for which the long chord is 2332.6. Now, by (43a),

$$x = \frac{1}{2} \times \frac{2332.6}{4} = 291.6 \text{ feet.}$$

106. In Fig. 31, Given x and p to Find the Radius of a Curve Tangent to AB at A and Passing through P.

From (43), $$R = \frac{x^2 + p^2}{2p}. \quad . \quad . \quad . \quad . \quad . \quad . \quad . \quad . \quad (44)$$

107. Given the Location of a Point *P* referred to the *P. I.* to Find the Radius of a Curve through *P* which will Unite the Given Tangents.

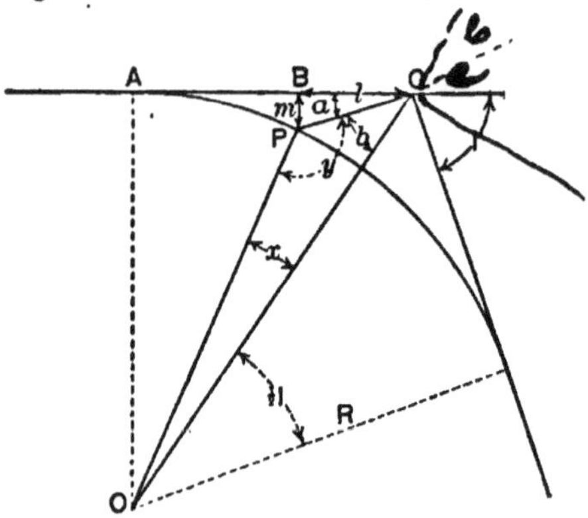

FIG. 32.

In Fig. 32 suppose $BC = l$, $BP = m$ known, and angle a calculated; or PC and a may be measured on the field.

From triangle CAO,

$$b = 90° - (a + \tfrac{1}{2}I), \quad \text{and} \quad CO = R \sec \tfrac{1}{2}I.$$

Now from triangle PCO,

$$\sin y = \frac{CO}{PO} \sin b.$$

Inserting values of PO and CO,

$$\sin y = \frac{R \sec \frac{1}{2}I}{R} . \sin b = \sec \tfrac{1}{2}I . \sin b = \frac{\sin b}{\cos \frac{1}{2}I}, \quad . \quad (45)$$

an equation from which the unknown R has disappeared. Next, from the same triangle, since $x = 180° - (b + y)$,

$$R = \frac{\sin b}{\sin x} . BC. \quad . \quad . \quad . \quad . \quad . \quad . \quad . \quad . \quad . \quad (46)$$

When $I = 90°$, it can easily be shown that

$$R = l + m + \sqrt{2lm}. \quad . \quad . \quad . \quad . \quad . \quad . \quad (47)$$

108. To Locate a Tangent to a Curve from an Outside Point.

FIRST METHOD.—In Fig. 33 let P be the point and AHB the

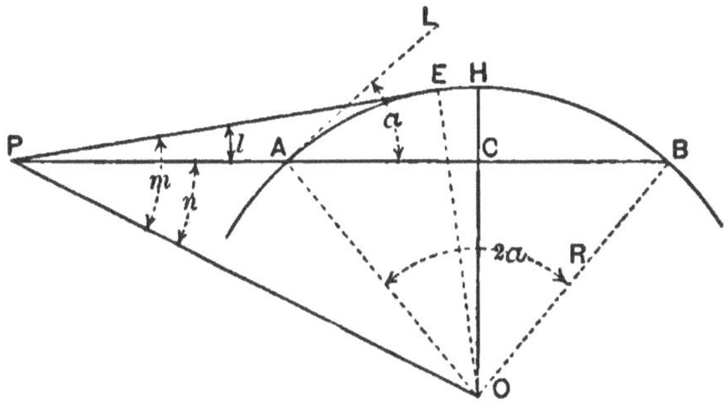

FIG. 33.

curve. Run a trial-line PA cutting the curve in A and B. Measure PA and AB; or measure PA and angle a between the chord AB and tangent AL. Then

$$AB = 2AC = 2R \sin a,$$
$$OC = R \cos a.$$

By geometry, $PE = \sqrt{PA \times PB}$, PE being the required tangent. From the figure,

$$\tan n = \frac{CO}{CP}$$
$$\tan m = \frac{R}{PE}.$$

At P deflect the angle $l = m - n$ from PA and run the tangent.

SECOND METHOD.—In Table IX find the long chord for a central angle $2a$; then

$$AB = 2AC = \frac{L.C.}{D},$$

$$CH = \frac{M_1}{D},$$

and $$CO = R - CH.$$

We may now proceed as before.

109. To Run a Tangent to Two Located Curves of Contrary Flexure.

First Case.—In Fig. 34 let FK and LE be the curves, and $KL = p$ measured on the ground.

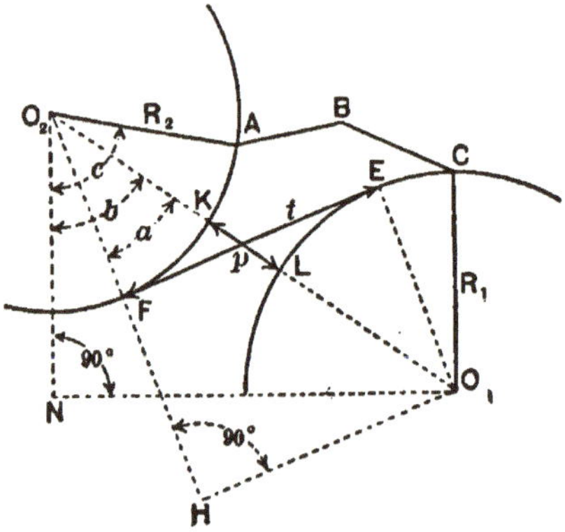

Fig. 34.

Let $FE = t$ be the required tangent.

Draw O_1H parallel and O_2H perpendicular to FE; from the triangle O_1HO_2, since $FH = R_1$,

$$(R_1 + R_2 + p)^2 = (R_1 + R_2)^2 + t^2 ;$$

whence

$$t = \sqrt{2(R_1 + R_2)p + p^2}. \quad . \quad . \quad . \quad (48)$$

Also,

$$\cos a = \frac{R_1 + R_2}{R_1 + R_2 + p}. \quad . \quad . \quad . \quad . \quad . \quad (49)$$

The arcs FK and LE may be found from the angle a and the known curvatures, after which the points F and E may be set.

If t is given and p required, it may easily be found from (48).

Second Case. *p not known.*

Set the transit at a point A on one curve and note the bearing of the tangent to the curve at that point (see Fig. 34); the bearing of the radius O_2A differs from this by 90°. Run a line ABC of one or more courses to intersect the other curve at C. Note the bearings and lengths of these courses and the bearing in tangent at C, from which calculate the bearing of CO_1. R_1 and R_2 being known, the latitudes and departures are next calculated. Let O_2N

be the sum of the northings or southings, O_1N the sum of the eastings or westings; from the triangle O_1O_2N,

$$\tan b = \frac{O_1N}{O_2N},$$

and $$O_1O_2 = \sqrt{\overline{O_1N}^2 + \overline{O_2N}^2}.$$

As before, FE is the required tangent and O_2H perpendicular, while O_1H is parallel thereto.

$$\cos a = \frac{R_1 + R_2}{O_1O_2}.$$

Angle $FO_2N = b - a$ is the bearing of O_2F, while $AO_2F = c - b + a$ is the angle of retreat from the known point A to F, where the tangent may be run. The length of $t = O_1H$ is

$$t = OO_1 \sin a.$$

D. Change of Location.

110. To Locate a Curve Parallel to a Given Curve.

Let p be the perpendicular between parallel tangents, and suppose ABC located (see Fig. 35). If there are no restrictions as to the position of the points E, F, and G on the second curve, we may calculate the new degree of curve D_1 for a radius $R_1 = R + p$, by (13), and trace the curve from any point, as E. Thus

$$\sin \tfrac{1}{2}D_1 = \frac{50}{R_1} = \frac{50}{R + p}.$$

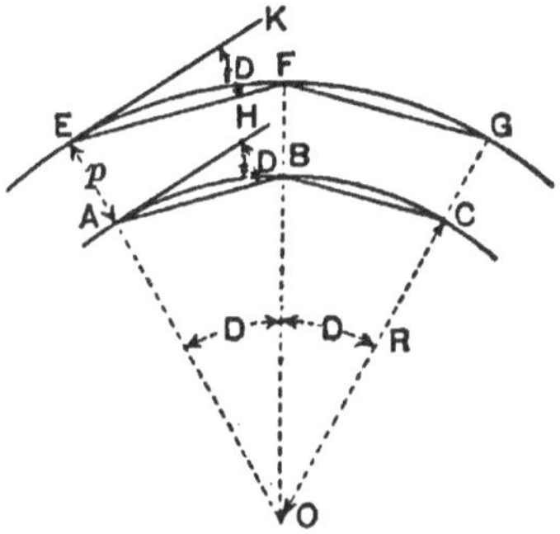

FIG. 35.

If, however, points on the radii through A, B, and C are wanted, they are gotten by using the same degree of curve D and computing the length of chord FE. From similar triangles,

$$\frac{EF}{R_1} = \frac{AB}{R} = \frac{100}{R},$$

whence

$$EF = 100\frac{R_1}{R} = 100\frac{R+p}{R} = 100\left(1+\frac{p}{R}\right). \quad . \quad . \quad (50)$$

Had EFG been the located curve, with radius R, we should have had

$$AB = 100\frac{R-p}{R}. \quad . \quad . \quad . \quad . \quad . \quad . \quad . \quad . \quad . \quad . \quad . \quad . \quad (51)$$

111. To Change the *P.C.* of a Located Curve so that *P.T.* will Fall in a Given Tangent Parallel to Terminal Tangent of Located Curve.

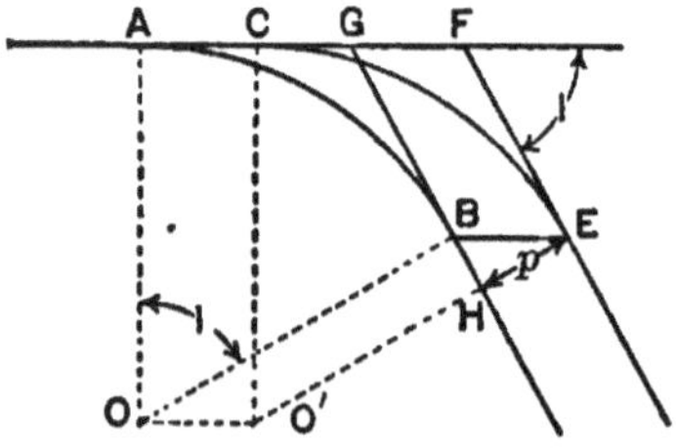

FIG. 36.

Let AB, Fig. 36, be the located curve; FE, the tangent in which the *P.T.* must fall.

Let the distance between tangents be $HE = p$.

Draw BE and OO' parallel to AF; evidently $AC = OO' = BE$, O' being the new position of center.

In triangle BEH,

$$BE = AC = \frac{p}{\sin I} = p \operatorname{cosec} I. \quad . \quad . \quad . \quad . \quad . \quad (52)$$

Set the new *P.C.* by measurement from A, and run the curve CE. Any system of straight lines and curves may be treated as above, provided I is the angle between initial and terminal tangents and p as before.

EXAMPLE.—A located 2° 30′ curve, having $I = 25°$, ends in a tangent 25 ft. outside of desired tangent. Find the change in position of *P.C.*

By (52), $\quad AC = 25 \times 2.36620 = 59.16$ ft.

112. To Find the Change in Radius and Position of *P.C.* if *P.T.* is Required to fall on the same Radial Line but on a Tangent distant *p* from, and parallel to, Terminal Tangent to Located Curve.

In Fig. 37 let AB be the located and CE the required curve. Draw the parallel chords AB and CE. Draw CH and BF perpendicular to AB. The angles $FBE = CAH = \frac{1}{2}I$.

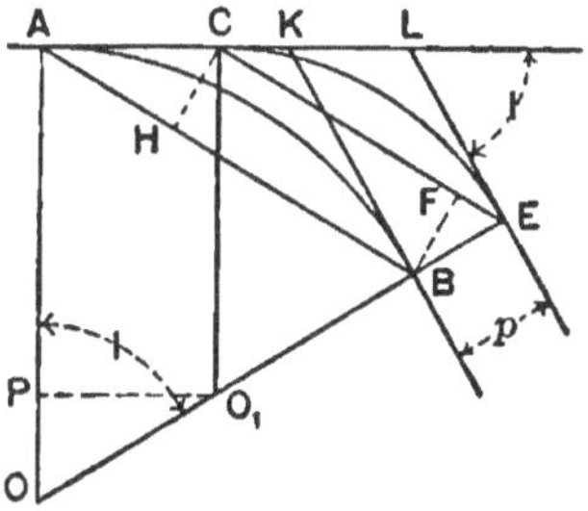

FIG. 37.

From the figure,

$$CH = AC \sin \tfrac{1}{2}I,$$

$$BF = BE \cos \tfrac{1}{2}I = p \cos \tfrac{1}{2}I.$$

Equating,

$$AC \sin \tfrac{1}{2}I = p \cos \tfrac{1}{2}I,$$

whence

$$AC = p \cot \tfrac{1}{2}I. \quad . \quad . \quad . \quad . \quad . \quad . \quad . \quad . \quad (53)$$

In the triangle OPO_1, $O_1P = AC$, $OP = R - R_1$, and

$$(R - R_1) \tan I = AC = p \cot \tfrac{1}{2}I,$$

or

$$R - R_1 = AC \cot I = p \cot \tfrac{1}{2}I \,.\, \cot I.$$

Therefore

$$R_1 = R - AC \cot I = R - p \cot \tfrac{1}{2}I \,.\, \cot I. \quad . \quad . \quad (54)$$

From trigonometry,

$$\cot \tfrac{1}{2}I = \frac{\sin I}{1 - \cos I} \quad \text{and} \quad \cot I = \frac{\cos I}{\sin I}.$$

Inserting these values in (54) gives

$$R_1 = R - p \,.\, \frac{\sin I}{1 - \cos I} \cdot \frac{\cos I}{\sin I} = R - p \frac{\cos I}{1 - \cos I} = R - p \frac{\cos I}{\text{vers } I}.$$

From trigonometry, $$\text{ex sec } I = \frac{\text{vers } I}{\cos I}.$$

EXAMPLE.—A 2° 30′ curve strikes 25 ft. inside a tangent in which the $P.T.$ must fall. Find the necessary change in radius and position of $P.C.$ when $I = 25°$.

By (53) the change in $P.C.$ is

$$AC = 25 \times 4.51071 = 112.77 \text{ ft.}$$

By (54′), $$R_1 = 2292.01 - \frac{25}{.10338} = 2050.38 \text{ ft.}$$

By Table I we find this to be the radius of a 2° 47′ 41″ curve.

113. Given a Located Curve uniting Two Tangents to Find the Change in Position of $P.C.$ or in Radius for a Given Change in the Intersection-angle.

FIRST CASE.—*Radius unchanged.*

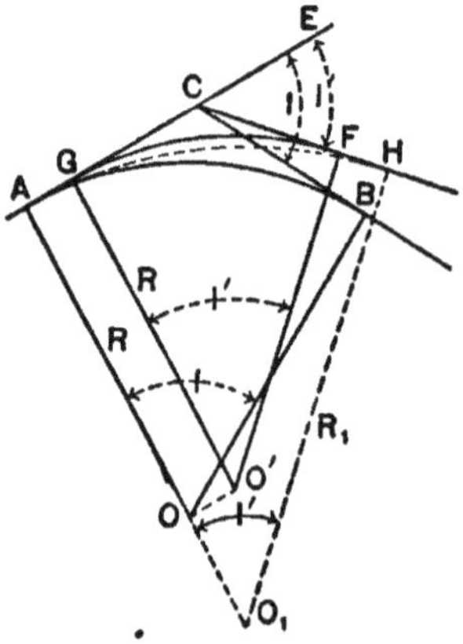

FIG. 38.

In Fig. 38 let $BCE = I$ be the original intersection-angle, $FCE = I'$ the new angle. From the figure,

$$AG = AC - GC,$$

or

$$AG = R(\tan \tfrac{1}{2}I - \tan \tfrac{1}{2}I'). \quad (55)$$

BY TABLE IX.—From the table, for angle I,

$$T = \frac{T_1 \measuredangle I°}{D}$$

For I'

$$T' = \frac{T_1 \cdot \measuredangle I'°}{D}.$$

Then $$AG = T - T'.$$

SECOND CASE.—*P.C. unchanged.*

Here the tangent T for the two curves is the same, and therefore

$$R_1 \tan \tfrac{1}{2}I' = R \tan \tfrac{1}{2}I;$$

BY TABLE IX,

$$T = \frac{T_1 \measuredangle I^\circ}{D} = \frac{T_1 \measuredangle I'^\circ}{D_1}.$$

whence

$$D_1 = \frac{T_1 \measuredangle I'^\circ}{T_1 \measuredangle I^\circ} \cdot D = \frac{T_1 \measuredangle I'^\circ}{T}.$$

114. To Find the Change in R and $P.C.$ for a Given Change in I, the $P.T.$ remaining unchanged

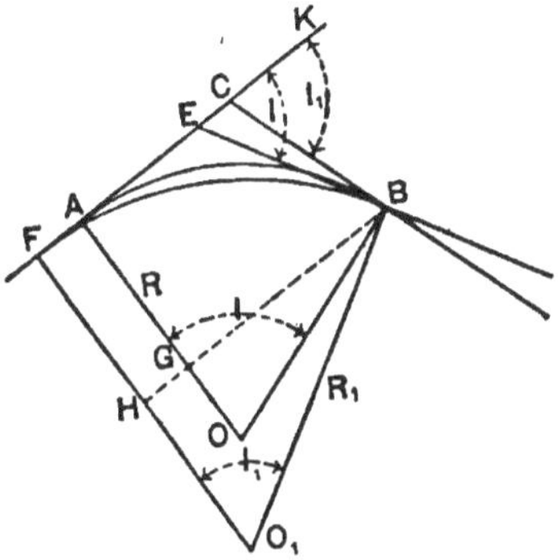

FIG. 39.

In Fig. 39, from the triangles OBG and O_1BH,

$$OG = R \cos I$$

and $$O_1H = R_1 \cos I_1.$$

Now $GA = HF$; hence

$$R_1 - R_1 \cos I_1 = R - R \cos I.$$

Whence

$$R_1 = R\frac{1 - \cos I}{1 - \cos I_1} = R\frac{\text{vers } I}{\text{vers } I_1}. \quad . \quad . \quad . \quad . \quad . \quad (57)$$

Also, $$FA = HG = BH - BG.$$

Inserting values of BH and BG, there results

$$FA = R_1 \sin I_1 - R \sin I. \quad . \quad . \quad . \quad . \quad . \quad (58)$$

115. Given a Located Curve to Find the Change in R for a Given Change in T, I remaining unchanged.

In Fig. 40, from the triangles OAC and O_1EC, since $EA = EC - AC$,

$$R_1 \tan \tfrac{1}{2}I - R \tan \tfrac{1}{2}I = EA = T' - T.$$

Whence $$R_1 = R + (T' - T) \cot \tfrac{1}{2}I. \quad . \quad . \quad . \quad . \quad . \quad (59)$$

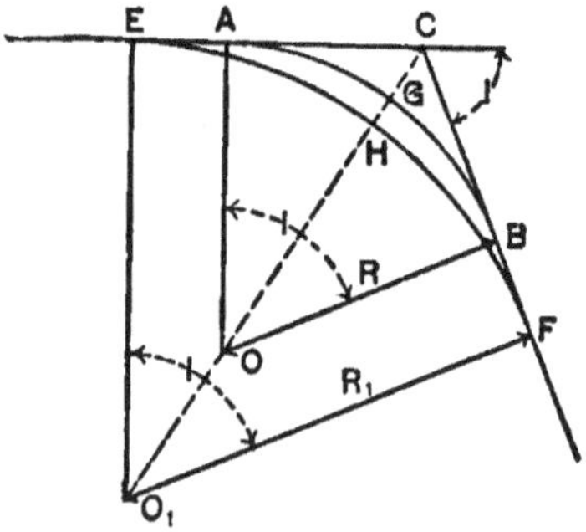

FIG. 40.

BY TABLE IX.—EA being known, $T' = T + EA$. Then, by (15*a*),

$$D_1 = \frac{T_1 \measuredangle I^\circ}{T'}.$$

If the change in vertex of curve is wanted, there results, from (25),

$$E = CG = T \tan \tfrac{1}{4}I, \qquad E' = CH = T' \tan \tfrac{1}{4}I.$$

Therefore $$GH = E' - E = (T' - T) \tan \tfrac{1}{4}I. \quad . \quad . \quad . \quad (60)$$

GH can be found from Table IX after finding D_1 as above.

If R_1 is given and EA wanted, (59) yields

$$EA = T' - T = (R_1 - R) \tan \tfrac{1}{2}I.$$

116. To Find the Radius of a Curve having the Same *P.C.* as a Given Curve, but ending in a Parallel Tangent.

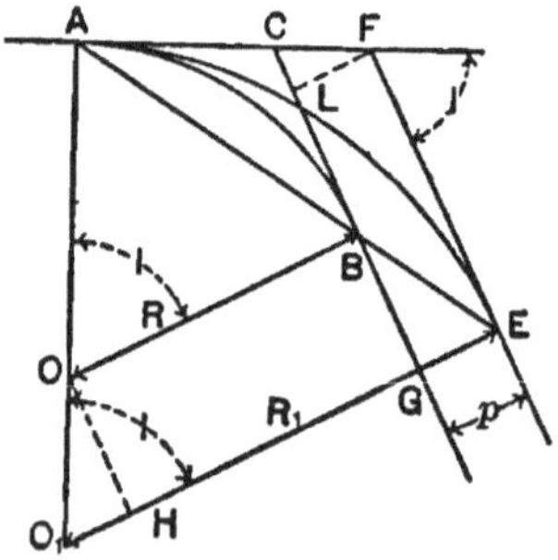

FIG. 41.

In Fig. 41 let the perpendicular distance between tangents be p, and AB be the located curve; $AO_1 = R_1$ is required.

FIRST METHOD.—Draw OH at right angles to O_1E; then

$$O_1E = O_1H + HG + GE,$$

or

$$R_1 = (R_1 - R) \cos I + R + p.$$

From which $$R_1 = R + \frac{p}{1 - \cos I} = R + \frac{p}{\text{vers } I}. \quad . \quad . \quad . \quad (61)$$

SECOND METHOD.—A, B, and E lie on the same straight line, since I is the same for both curves. In triangle BGE angle $EBG = \frac{1}{2}I$, and

$$BE = \frac{p}{\sin \frac{1}{2}I} = p \operatorname{cosec} \tfrac{1}{2}I.$$

From Table IX, $$AB = \frac{L.C._1 \measuredangle I^\circ}{D}.$$

$AE = AB + BE$ is the long chord for curve of degree D_1; therefore

$$D_1 = \frac{L.C._1 \measuredangle I^\circ}{AE}.$$

If desired, R may be found by (12′) or Table I.

THIRD METHOD.—Draw FL parallel to O_1E; then

$$CF = \frac{p}{\sin I} = p \operatorname{cosec} I.$$

From Table IX, $$AC = \frac{T_1 \measuredangle I^\circ}{D}.$$

$AF = AC + CF$, the tangent distance for second curve; hence

$$D_1 = \frac{T_1 \measuredangle I^\circ}{AF}.$$

REMARK.—If transit is set up at B, it will be well to set E by measurement from B, to serve as a check when the curve is run in from A.

ARTICLE 9. COMPOUND CURVES.

A. Location Problems.

117. Given Two Unequal Tangents, their Intersection-angle, and One Radius, to Find the Other Radius of a Compound Curve uniting Tangents.

In Fig. 42, $AH = T_1$ and $BH = T_2$ are the known tangents, $AO_1 = R_1$ the known radius. $BO_2 = R_2$ and the angles I_1 and I_2 must be found before curve can be located.

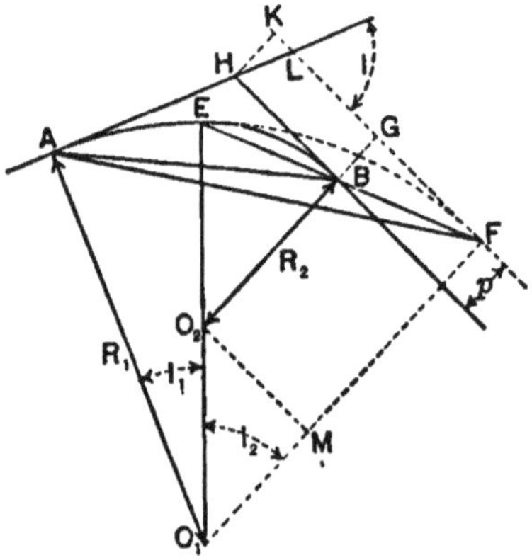

FIG. 42.

Extend first branch to F, so that tangent FL is parallel to BH.

Draw HK and BG perpendicular to FL; draw FB and extend to E; it will pass through the $P.C.C.$, because the central angles EO_1F and EO_2B are equal. Then

$$T_0 = AL = R_1 \tan \tfrac{1}{2}I.$$

In triangle LHK, since $LH = T_0 - T_1$,

$$s = KL = (T_0 - T_1) \cos I,$$

$$p = HK = BG = (T_0 - T_1) \sin I.$$

Now in triangle BGF angle $BFG = \tfrac{1}{2}I_2$, and

$$l = FG = T_0 + s - T_2,$$

$$\tan \tfrac{1}{2} I_2 = \frac{p}{l} . \quad . \quad . \quad . \quad . \quad . \quad . \quad . \quad . \quad (62)$$

Draw O_2M parallel to FL; then

$$(R_1 - R_2) \sin I_2 = l,$$

whence

$$R_2 = R_1 - \frac{l}{\sin I_2} = R_1 - l \operatorname{cosec} I_2. \quad . \quad . \quad (63)$$

Had R_1 been required, the equation would have been

$$R_1 = R_2 + l \operatorname{cosec} I_2.$$

Evidently, $\quad I_1 = I - I_2.$

In the field the points E and B may be located by running in the curve from A as starting-point, or run the chord

$$AF = 2R_1 \sin \tfrac{1}{2} I$$

from A, and at F deflect angle $AFB = \frac{1}{2}I - \frac{1}{2}I_2 = \frac{1}{2}I_1$, measure $FB = l \sec \frac{1}{2}I_2$ and $BE = 2R_2 \sin \frac{1}{2}I_2$.

EXAMPLE.—A 2° curve has the *P.C.* at sta. 110, $T_1 = 590$ ft., $T_2 = 511.8$ ft., $I = 30° \ 50'$. Locate the curve.

By Table IX, $\quad T_0 = 1580/2 = 790$ ft.

By formulas above,

$$s = 200 \times 0.85866 = 171.73 \text{ ft.},$$

$$p = 200 \times 0.51254 = 102.51 \text{ ft.},$$

$$l = 790 + 171.73 - 511.8 = 449.93,$$

$$\tan \tfrac{1}{2} I_2 = \frac{102.51}{449.97} = 0.22778 = \tan 25° \ 40'.$$

Then $\quad I_1 = 30° \ 50' - 25° \ 40' = 5° \ 10'.$

$$R_2 = 2864.93 - \frac{449.97}{.43313} = 1833 \text{ feet.}$$

By Table I this is seen to be the radius of a 3° 7½′ curve.

The length of first branch is 258.3 feet, and of the second 821.3 feet; hence the *P.C.C.* falls at 112 + 58.3, while the *P.T.* is at sta. 120 + 79.6.

118. Given the Long Chord from *P.C.* to *P.T.* of a Compound Curve, the Angles it makes with the Tangents and One Radius, to Find the Other Radius and the Central Angles.

In Fig. 42 AB is known, as also the angles $HAB = a$ and $HBA = b$. Two angles and one side of the triangle HAB are known, and the sides $HA = T_1$ and $HB = T_2$ may be found, after which the solution is the same as in the last problem.

A solution may be reached in a different manner. $I = a + b$, $HAF = \frac{1}{2}I = \frac{1}{2}(a + b)$, and $BAF = \frac{1}{2}(a + b) - a = \frac{1}{2}(b - a)$, $AF = 2R_1 \sin \frac{1}{2}I$. In triangle BAF two sides and the included angle are now known, so AF and angle BFA may be found; $GFB = \frac{1}{2}I_2 = \frac{1}{2}I - BFA$.

Then $$EF = 2R_1 \sin \frac{1}{2}I_2,$$

and $$EB = EF - BF \text{ becomes known.}$$

Then $$EB = 2R_2 \sin \frac{1}{2}I_2 = 2R_1 \sin \frac{1}{2}I_2 - BF,$$

whence $$R_2 = R_1 - \frac{BF}{2 \sin \frac{1}{2}I_2} \quad \quad (64)$$

Evidently $I_1 = I - I_2$

119. Given the Radii and Central Angles of a Compound Curve to Find the Tangent Lengths, the Long Chord from *P.C.* to *P.T.*, and the Angles it makes with Tangents.

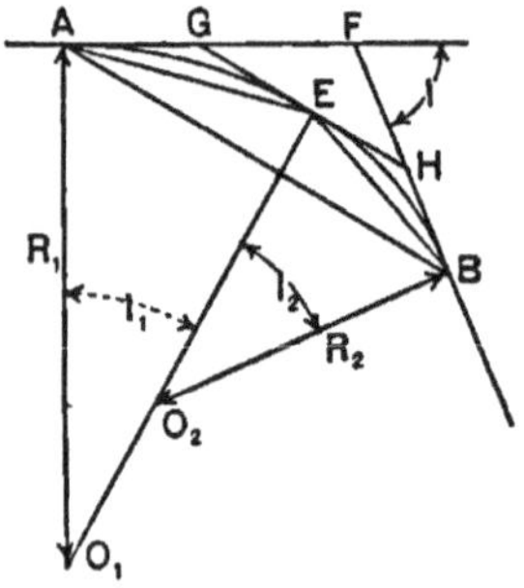

FIG. 43.

In Fig. 43 draw AE and BE from the *P.C.* and *P.T.* to the *P.C.C.*, then calculate AE and BE by (16) or by Table IX. In triangle AEB angle $AEB = 180 - \frac{1}{2}(I_1 + I_2)$. Two sides and the included angle being known, the triangle AEB may be solved for AB and the angles ABE and BAE; then

$$BAF = BAE + \frac{1}{2}I_1,$$

$$ABF = ABE + \frac{1}{2}I_2.$$

The angle AFB of triangle ABF now becomes known and, as

AB is known, the sides $AF = T_1$ and $BF = T_2$ may be computed.

120. Given the Long Chord from $P.C.$ to $P.T.$ of a Compound Curve and the Angles it makes with Tangents to Find the Radii when the Common Tangent is Parallel to Long Chord.

In Fig. 43 let GH be parallel to AB, and $GAB = a$, $HBA = b$ known. Then

$$BAE = EAG = GEA = \tfrac{1}{2}a,$$

and

$$ABE = EBH = HEB = \tfrac{1}{2}b.$$

Also,

$$AEB = 180° - \tfrac{1}{2}(a + b).$$

In triangle AEB, remembering that

$$\sin\,[180 - \tfrac{1}{2}(a + b)] = \sin \tfrac{1}{2}(a + b),$$

$$AE = \frac{AB \sin \tfrac{1}{2}b}{\sin \tfrac{1}{2}(a + b)},$$

and

$$BE = \frac{AB \sin \tfrac{1}{2}a}{\sin \tfrac{1}{2}(a + b)}.$$

Since $AO_1E = a$ and $EO_2B = b$, the radii R_1 and R_2 may be found from formula (16), or (16*a*).

By (16),
$$R_1 = \frac{\tfrac{1}{2}AE}{\sin \tfrac{1}{2}a} = \frac{AB \sin \tfrac{1}{2}b}{2 \sin \tfrac{1}{2}a \,.\, \sin \tfrac{1}{2}(a + b)}. \quad . \quad . \quad . \quad (65)$$

$$R_2 = \frac{\tfrac{1}{2}BE}{\sin \tfrac{1}{2}b} = \frac{AB \sin \tfrac{1}{2}a}{2 \sin \tfrac{1}{2}b \,.\, \sin \tfrac{1}{2}(a + b)}. \quad . \quad . \quad . \quad (66)$$

EXAMPLE.—Required R_1 and R_2, or D_1 and D_2, when $AB = 900$ feet, $a = 12°$, $b = 15°$.

By (65), $R_1 = 2407.0$ ft.

By (66), $R_2 = 1543.7$ ft.

From Table I, $D_1 = 2° \; 22' \; 50''$ and $D_2 = 3° \; 42' \; 44''$.

B. Obstacles.

121. To Locate a Point on the Second Branch of a Compound Curve when the *P.C.C.* is Inaccessible.

Ordinarily the second branch is located by setting transit at the *P.C.C.* and running the curve from that point. An obstacle on either curve may then be passed by the methods given for simple curves.

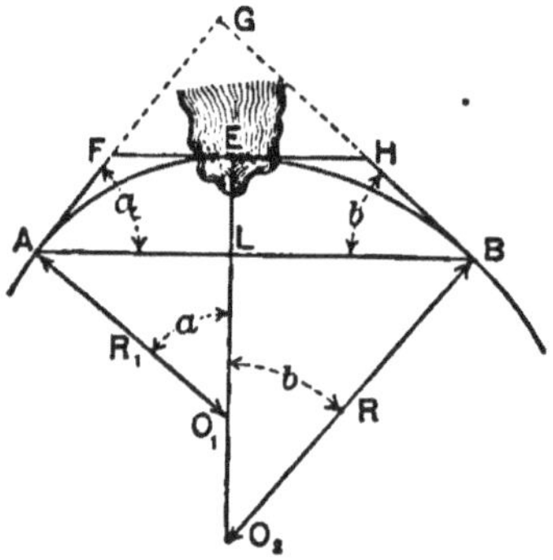

FIG. 44.

When the *P.C.C.* is inaccessible, locate the first branch from the *P.C.* and the second branch from the *P.T.*, if this latter point is known. When this is not the case proceed by one of the following methods:

FIRST. *By means of a long chord.*

In Fig. 44 let *E* be the *P.C.C.*, *A* some known point on first branch, *EF* a tangent at *E*, and *AB* parallel to *FE*. The station numbers of *A* and *E* being known, the arc *AE* and angle *a* are readily found; then

$$EL = R_2 \text{ vers } b = R_1 \text{ vers } a,$$

whence

$$\text{vers } b = \frac{R_1 \text{ vers } a}{R_2}; \quad . \quad . \quad . \quad . \quad . \quad . \quad . \quad . \quad . \quad . \quad (67)$$

next,

$$AB = R_1 \sin a + R_2 \sin b. \quad . \quad . \quad . \quad . \quad . \quad . \quad (68)$$

Deflect $FAB = a$ from tangent at A; measure out AB; set the transit at B and locate the second branch.

BY TABLE IX.—Take the mid-ordinate in table for an intersection-angle $2a$; then

$$EL = \frac{M_1 \measuredangle 2a}{D}.$$

Then $EL \times D_2$ is the mid-ordinate for a 1° curve having $I = 2b$, from which b becomes known. From the table now find AL and LB, the half-chords for angles $2a$ and $2b$, and proceed as before.

SECOND METHOD.—*By means of tangents.*

From Fig. 44, $AF = FE = R_1 \tan \frac{1}{2}a$.

Set transit at F, deflect $GFE = a$, and by some indirect method measure to an accessible point H.

$$EH = FH - FE,$$

and

$$\tan \tfrac{1}{2}b = \frac{EH}{R_2}, \text{ from formula (14).}$$

Angle b is now known and equals GHE, which deflect from EH; then measure $HB = EH$, and with transit at B locate the second branch of curve.

OR BY TABLE IX.—Find $AF = FE$, the tangent distance for $I = a$; then having EH measured, take $T_1 = EH \times D_2$ and find the corresponding angle, which equals b; then proceed to locate curve as above.

EXAMPLE.—Let A be at sta. 126, $P.C.C.$ at $128 + 25$; the degree of first branch 4°, and of second 6°.

By the first method $EL = 17.635$ for $a = 9°$, and $b = 11° 2'$, nearly. $AL = 224.1$ ft., $BL = 182.75$ ft., and therefore $AB = 406.85$ ft. Angle $b = 11° 2'$ corresponds to 183.9 ft. around 6° curve; hence the $P.T.$ number is $130 + 08.9$.

By the second method $AF = 112.74$ ft. Suppose $FH = 264$ ft., then $EH = 151.26$ ft., which multiplied by 6 gives 907.56 ft., corresponding to $I = 18°$. The arc EB is now 300 ft., making B fall at sta. $131 + 25$.

C. Change of Location.

122. Having a Simple Curve Located to Find the $P.C.C.$ so that a Curve of Given Radius shall connect with a Given Tangent Parallel to Tangent to Located Curve.

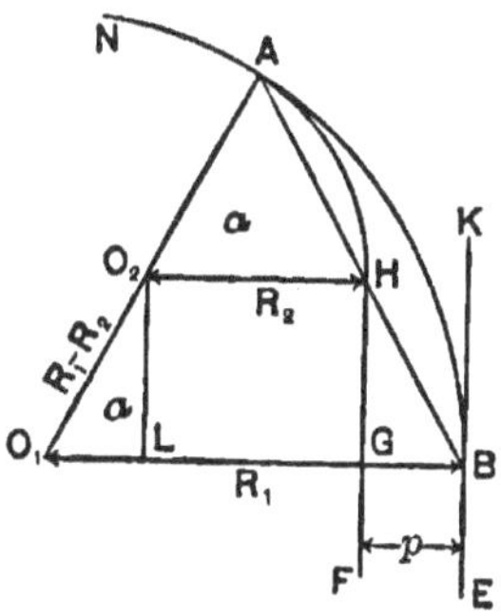

FIG. 45.

Let NAB, Fig. 45, be the located curve, HF the tangent in which the second branch must end. The distance $BG = p$ between tangents is known from measurement. If angle a can be found, the arc BA becomes known and the point A can be located from B. Draw O_2L from the center of second branch perpendicular to O_1B. In triangle O_1O_2L, $O_1O_2 = R_1 - R_2$, and $O_1L = R_1 - (R_2 + p)$; therefore

$$\cos a = \frac{R_1 - R_2 - p}{R_1 - R_2} = 1 - \frac{p}{R_1 - R_2}. \quad . \quad . \quad (69)$$

Then a divided by D_1 gives arc BA.

If desired, BH may be found from the right triangle BHG, in which the side $BG = p$ and angle $GHB = \frac{1}{2}a$ are known—A, H, and B lying in the same straight line; then

$$BH = \frac{p}{\sin \frac{1}{2}a} = p \operatorname{cosec} \tfrac{1}{2}a. \quad . \quad . \quad . \quad . \quad . \quad . \quad (70)$$

Or BA and HA may be found from Table IX, after which

$$BH = BA - HA.$$

EXAMPLE.—A 3° curve ends in a tangent at sta. 160 + 50, 35 ft. outside of desired tangent. Find the point of compounding with a 4° 50′ curve.

From Table I, R for 3° curve equals 1910.08 ft., and for 4° 50′ curve 1185.78 ft.

Then, by (69), $\cos a = 1 - \frac{35}{724.3} = 0.95168.$

From table of cosines angle a is found to be 17° 53′. Dividing this by 3 gives 5.961 stations for the arc BA. Hence the $P.C.C.$ number is 160.50 − 5.961 = sta. 154 + 53.9, and the new $P.T.$ is at sta. 158 + 23.9.

123. Given a Located Compound Curve ending in a Tangent Parallel to, and a Given Distance from, a Tangent in which the Curve is required to end. To Find the Necessary Change in *P.C.C.*

FIRST CASE.—*Terminal branch having shorter radius.*

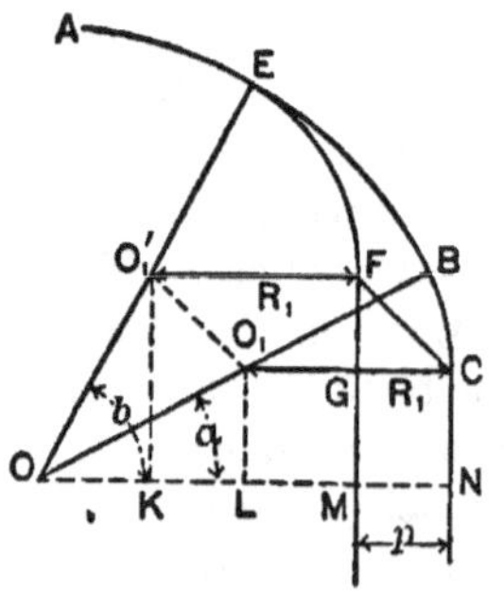

FIG. 46.

In Fig. 46 let ABC be the located curve, AEF the one required; angle $BO_1C = a$ known, and also $MN = p$.

If angle $EOM = b$ can be found, the angle of retreat from B to E will equal $b - a$.

Draw $O_1'K$ and O_1L perpendicular to ON, which is parallel to O_1C.

Then $OK = (R - R_1) \cos b,$

$OL = (R - R_1) \cos a.$

Now $LM = R_1 - KL = R_1 - MN$, from which $KL = MN = p$. Hence

$$(R - R_1) \cos b = (R - R_1) \cos a - p.$$

From which

$$\cos b = \cos a - \frac{p}{R - R_1}. \quad . \quad . \quad . \quad . \quad . \quad . \quad (71)$$

Divide $b - a$ by D, the curvature of first branch, and move back that number of stations from B to the new $P.C.C.$ at E.

Join O_1O_1'; evidently $FC = O_1O_1'$, and angle $KO_1'O_1 = CFG$; $OO_1'O_1 = 90° - \frac{1}{2}(b - a)$, $OO_1'K = 90° - b$. Hence

$$CFG = KO_1'O_1 = [90 - \tfrac{1}{2}(b - a)] - (90 - b) = \tfrac{1}{2}(b + a). \quad (72)$$

From triangle CGF,

$$FC = \frac{p}{\sin \frac{1}{2}(b + a)} = p \operatorname{cosec} \tfrac{1}{2}(b + a). \quad . \quad . \quad (73)$$

Or, from triangle $OO_1'O_1$,

$$FC = O_1'O_1 = 2(R - R_1) \sin \tfrac{1}{2}(b - a).$$

Had AEF been the original curve, b would have been known and a required.

From (71),
$$\cos a = \cos b + \frac{p}{R - R_1}. \quad . \quad . \quad . \quad . \quad . \quad . \quad (74)$$

CF and angle CFM are given by formulas (73) and (72).

EXAMPLE.—A 2° curve compounds with a 4° curve at sta. 82 + 30; $a = 20° 30'$, $p = 40$ feet. Find number of new $P.C.C.$ and distance between $P.T.$s.

From (71), $\cos b = 0.93667 - \dfrac{40}{2864.9 - 1432.7} = 0.90874.$

This yields $b = 24° 20'$, and $b - a = 3° 50'$.

The change in $P.C.C.$ is $\dfrac{3.833}{2} = 1.917$ stations; the $P.C.C.$ number is therefore $82.30 - 1.917 =$ sta. $80 + 38.3$.

By (72), $CFG = \frac{1}{2}(24° 20' + 20° 30') = 22° 25'$.

By (73), $FC = 40 \times 2.62234 = 104.9$ feet.

SECOND CASE.—*The terminal branch having longer radius.* Let CAB, Fig. 47, be the located curve with $P.C.C.$ at A, and let FK be the tangent in which the curve is required to end.

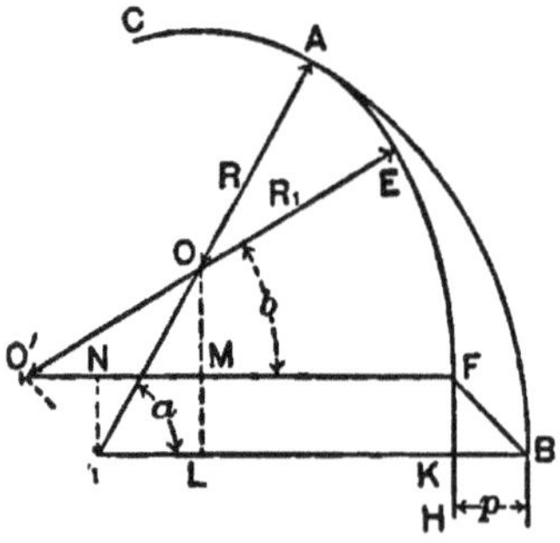

FIG. 47.

The distance $BK = p$, the radii $OA = R$, $O_1A = R_1$, and angle $AO_1B = a$ being known, it will be sufficient to find angle $EO_1'F$ in order to get the angle of advance, $AOE = a - b$. Draw OL and O_1N perpendicular to $O_1'F$ and O_1B. From the triangles $O_1'OM$ and O_1OL,

$$(R_1 - R)\cos b = O_1'N + (R_1 - R)\cos a.$$

But $O_1'N = KB = p$; therefore

$$(R_1 - R)\cos b = p + (R_1 - R)\cos a.$$

Whence
$$\cos b = \cos a + \frac{p}{R_1 - R}. \quad . \quad . \quad . \quad . \quad (75)$$

Then $\frac{a - b}{D}$ will be length of curve from A to E.

$$\text{Angle } KFB = NO_1O_1' = OO_1O_1' - NO_1O.$$

But $OO_1O_1' = 90° - \frac{1}{2}(a - b)$ and $NO_1O = 90 - a$.

$$\therefore \quad KFB = [90° - \tfrac{1}{2}(a - b)] - [90 - a] = \tfrac{1}{2}(a + b).$$

From triangle KFB,

$$FB = \frac{p}{\sin\frac{1}{2}(a + b)} = p \,.\, \operatorname{cosec}\tfrac{1}{2}(a + b). \quad . \quad . \quad . \quad (76)$$

Or, from triangle O_1OO_1', since $O_1O_1' = FB$,

$$FB = 2(R_1 - R)\sin\tfrac{1}{2}(a - b).$$

If AEF had been the located curve, b would have been given and a required. From formula (75),

$$\cos a = \cos b - \frac{p}{R_1 - R}. \quad . \quad . \quad . \quad . \quad . \quad . \quad (77)$$

EXAMPLE.—A 5° curve compounds at sta. 60 with a 2° curve, and the *P.T.* is at sta. 80. What will be the number of *P.C.C.* if the *P.T.* fall in a tangent 81 feet inside of terminal tangent? Here $a = 40°$.

By (75), $\cos b = 0.76604 + \frac{81}{1719} = 0.81316.$

Hence $b = 35°\ 36'$ and $a - b = 4°\ 24'$, corresponding to 220 feet around the 2° curve. The number of the new *P.C.C.* is therefore 62 + 20,

$$\text{angle } KFB = \tfrac{1}{2}(40°\ 0' + 35°\ 36') = 37°\ 48',$$

and $FB = 81 \times 1.63157 = 132.16$ feet.

124. Given a Located Compound Curve to Find Necessary Change in *P.C.C.* and Radius of Second Branch to make the *P.T.* fall in a Tangent Parallel to First Terminal Tangent and in a Point on the Same Radial Line.

FIRST CASE.—*Second branch having shorter radius.*

In Fig. 48, $OB = R$, $O_1B = R_1$ angle a and $HG = p$ are known. $O_2E = R_2$ and angle b must be found; then $\frac{b-a}{D} = BE$ will be the change in *P.C.C.*

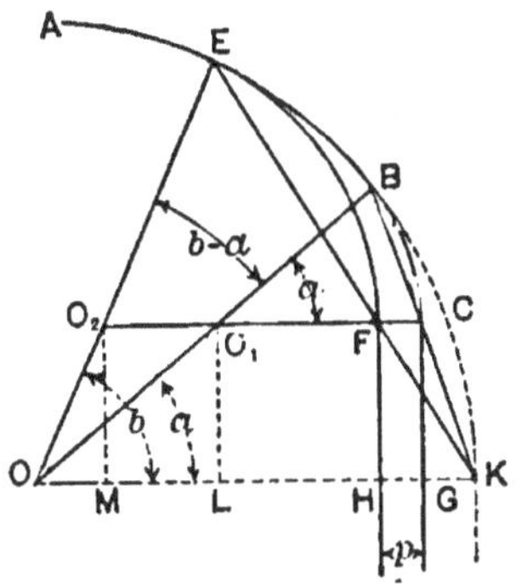

FIG. 48.

Produce first branch to K, where OK is parallel to O_1C. Since $BOK = BO_1C$, B, K, and C lie in the same straight line; and since $EO_2F = EOK$, E, F, and K lie in the same straight line. Therefore

$$KCG = \tfrac{1}{2}a, \quad \text{and} \quad KFH = \tfrac{1}{2}b.$$

From triangles KFH and KCG,

$$\tan \tfrac{1}{2}b = \frac{HK}{FH} = \frac{GK}{GC} + \frac{GH}{FH} = \tan \tfrac{1}{2}a + \frac{p}{FH}$$

But $\qquad FH = O_1L = (R - R_1) \sin a.$

$$\therefore \quad \tan \tfrac{1}{2}b = \tan \tfrac{1}{2}a + \frac{p}{(R - R_1) \sin a}. \quad . \quad . \quad . \quad . \quad . \quad . \quad (78)$$

From triangles OO_1L and OO_2M,

$$(R - R_2) \sin b = (R - R_1) \sin a.$$

When $\qquad R_2 = R - (R - R_1) \dfrac{\sin a}{\sin b}. \quad . \quad . \quad . \quad . \quad . \quad . \quad . \quad (79)$

Had AEF been the first curve located, b and R_2 would be known, a and R_1 required.

From the figure, reasoning as before,

$$\tan \tfrac{1}{2}a = \tan \tfrac{1}{2}b - \frac{p}{(R - R_2) \sin b}, \quad . \quad . \quad . \quad . \quad (80)$$

and

$$R_1 = R - (R - R_2) \frac{\sin b}{\sin a}. \quad . \quad . \quad . \quad . \quad . \quad . \quad (81)$$

SECOND CASE.—*Second branch having longer radius.*

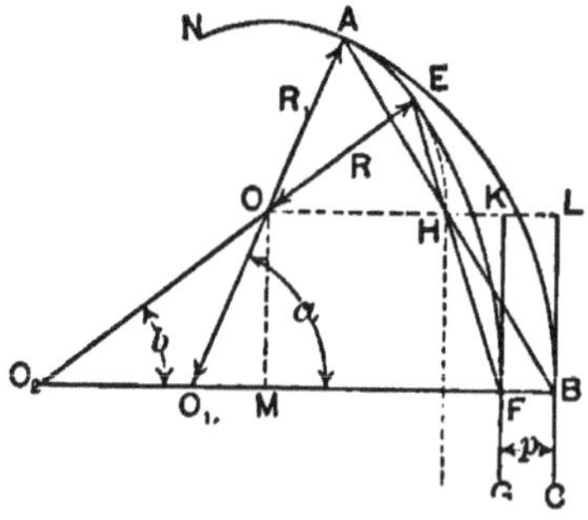

FIG. 49.

In Fig. 49 let AB be the located curve, EF the curve required, $OA = R$, $O_1A = R_1$, $O_2E = R_2$, $FB = p$.

R_2 and angle b are wanted, angle a being known.

We can show, as in first case, that

$$HFK = \tfrac{1}{2}b, \quad HBL = \tfrac{1}{2}a,$$

$$OM = KF = LB = (R_1 - R)\sin a;$$

and hence

$$\tan \tfrac{1}{2}b = \frac{HL}{BL} = \frac{HK}{FK} + \frac{p}{BL}.$$

Or inserting values,

$$\tan \tfrac{1}{2}b = \tan \tfrac{1}{2}a + \frac{p}{(R_1 - R)\sin a}. \quad . \quad . \quad . \quad . \quad . \quad (82)$$

Angle b now becomes known and $\dfrac{a-b}{D} = AE$ in chains, which is the change in position of $P.C.C.$

From triangles OO_1M and OO_2M,

$$(R_2 - R)\sin b = (R_1 - R)\sin a$$

$$\therefore\ R_2 = R + (R_1 - R)\frac{\sin a}{\sin b}. \quad . \quad . \quad . \quad . \quad . \quad (83)$$

Had the new tangent fallen outside the old one, we should have had

$$\tan \tfrac{1}{2}a = \tan \tfrac{1}{2}b - \frac{p}{(R_2 - R)\sin b}, \quad . \quad . \quad . \quad (84)$$

and

$$R_1 = R + (R_2 - R)\frac{\sin b}{\sin a}. \quad . \quad . \quad . \quad . \quad . \quad (85)$$

125. Having a Located Compound Curve, to Find the Change in *P.C.C.* and Radius of Second Branch in order to Cause *P.T.* to Fall at a New Point in Terminal Tangent.

FIRST CASE.—*Second branch having shorter radius.*

In Fig. 50 let NAB be the located curve, and C the point where $P.T.$ is required to fall. Let $BC = k$, $OA = R$, $O_1B = R_1$, and angle $O_1OH = a$ be known; angle b and R_2 are required.

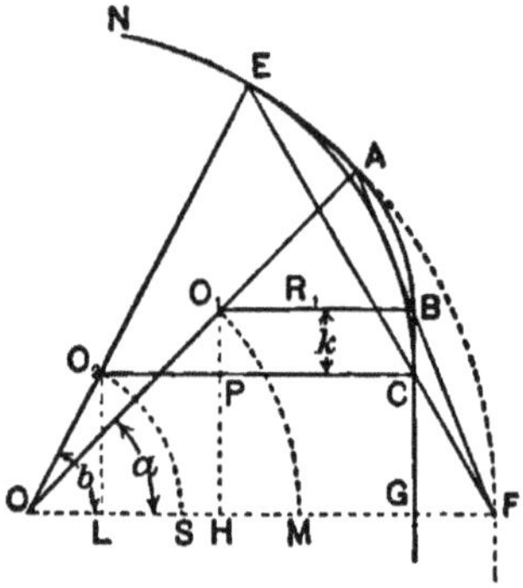

FIG. 50.

Extend first branch to F, making OF parallel to O_1B. A, B, and F lie on a straight line, for angles AO_1B and AOF are equal; likewise E, C, and F lie on the same straight line.

From triangles GBF and GCF,

$$\cot \tfrac{1}{2}b = \frac{GB}{GF} - \frac{CB}{GF} = \cot \tfrac{1}{2}a - \frac{k}{GF}$$

But $GF = HM = (R - R_1)(1 - \cos a) = (R - R_1) \text{ vers } a.$

$$\therefore \cot \tfrac{1}{2}b = \cot \tfrac{1}{2}a - \frac{k}{(R - R_1) \text{ vers } a}. \quad . \quad . \quad . \quad . \quad (86)$$

From triangles OO_1H and OO_2L, since $O_1P = k$,

$$(R - R_2) \sin b = (R - R_1) \sin a - k.$$

Whence

$$R_2 = R + \frac{k - (R - R_1) \sin a}{\sin b}. \quad . \quad . \quad . \quad . \quad (87)$$

Then $b - a$ divided by D gives arc AE. With radius R_2 locate the curve EC from C or E.

Had *NEC* been the located curve, R, R_2, and b would have been known, R_1 and a required. In this case

$$\cot \tfrac{1}{2}a = \cot \tfrac{1}{2}b - \frac{k}{(R - R_2)\ \text{vers}\ b}, \quad . \quad . \quad . \quad . \quad (88)$$

$$R_1 = R - \frac{k + (R - R_2)\sin b}{\sin a}. \quad . \quad . \quad . \quad . \quad (89)$$

SECOND CASE.—*Terminal branch having longer radius*

In Fig. 51 let *NAB* be the located and *NEC* the required curve.

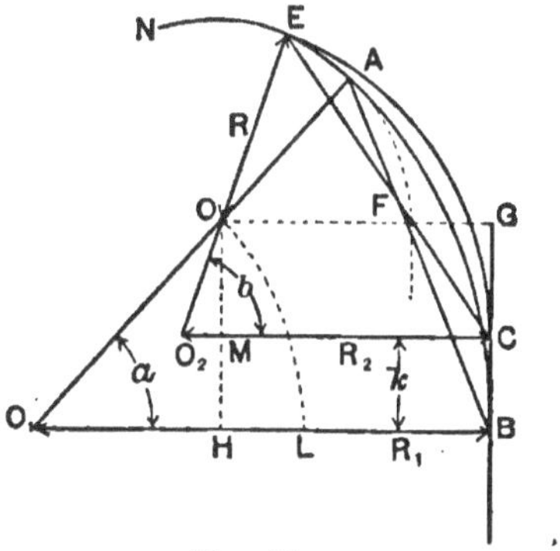

FIG. 51.

Let $CB = k$ be known. Then, as in the first case,

$$\cot \tfrac{1}{2}b = \frac{GC}{FG} = \frac{GB}{FG} - \frac{k}{FG}.$$

$$\therefore \quad \cot \tfrac{1}{2}b = \cot \tfrac{1}{2}a - \frac{k}{(R_1 - R)\ \text{vers}\ a}, \quad . \quad . \quad 90)$$

and $\qquad (R_1 - R)\sin a = (R_2 - R)\sin b + k;$

whence $$R_2 = R + \frac{(R_1 - R)\sin a - k}{\sin b}. \quad . \quad . \quad . \quad (91)$$

Had *NEC* been located and *NAB* required, the equations would have been

$$\cot \tfrac{1}{2}a = \cot \tfrac{1}{2}b + \frac{k}{(R_2 - R) \text{ vers } b}, \quad . \quad . \quad (92)$$

and

$$R_1 = R + \frac{(R_2 - R) \sin b + k}{\sin a}. \quad . \quad . \quad (93)$$

In either of these two cases if k is unknown and the new radius given or assumed, the desired angle and the value of k may be found from the foregoing equations. Or, knowing the new angle, the new radius and value of k may be found from the same equations.

126. To Replace a Curve of Given Radius, which unites Two Tangents with Known Intersection-angle, by a Three-centered Compound Curve.

In Fig. 52 let $OA = R$ be the radius of located curve,

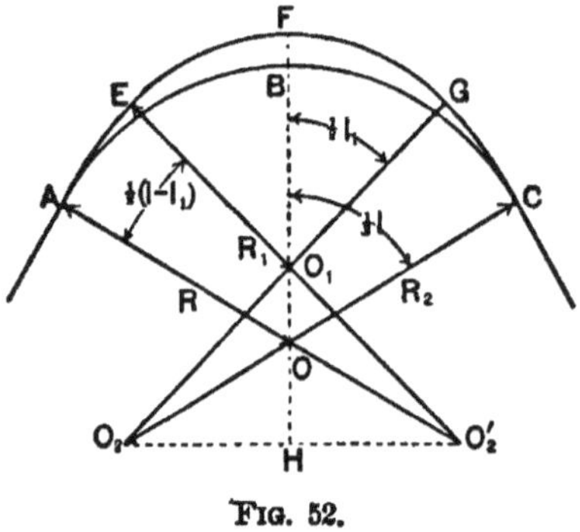

FIG. 52.

$O_2C = O_2'A = R_2$ the radius of terminal portions of the three-centered curve, and the other notation as shown in the figure.

Draw O_2O_2', and draw *FOH* perpendicular thereto. From triangles O_2O_1H and O_2OH,

$$O_2H = (R_2 - R_1) \sin \tfrac{1}{2}I_1 = (R_2 - R) \sin \tfrac{1}{2}I. \quad . \quad . \quad (a)$$

Suppose R_2 and R_1 to be assumed; then equation (a) yields

$$\sin \tfrac{1}{2}I_1 = \frac{R_2 - R}{R_2 - R_1} \sin \tfrac{1}{2}I. \quad . \quad . \quad . \quad . \quad (94)$$

Then $$AO_2'E = CO_2G = \tfrac{1}{2}(I - I_1). \quad . \quad . \quad . \quad (95)$$

Suppose $AO_2'E$, CO_2G, and R_2 to have been assumed. From (95) find I_1; then, from equation (a),

$$R_1 = R_2 - (R_2 - R)\frac{\sin \frac{1}{2}I}{\sin \frac{1}{2}I_1}. \quad . \quad . \quad . \quad . \quad . \quad (96)$$

EXAMPLE.—Given a 4° curve, $I = 38°$, and the terminal branches composed of a 2° curve for two stations, to find R_1 and D_1 for the central portion.

Here $$I_1 = 38° - 2(2 \times 2)° = 30°.$$

From Table I, $R_2 = 2865$ ft., $R = 1432.7$ ft.

Whence $$R_2 - R = 1432.3 \text{ ft.}$$

$$\begin{aligned} \text{Log } 1432.3 &= 3.15603 \\ \text{“ } \sin 19°\ 0' &= 9.51264 \\ \hline & 2.66867 \\ \text{“ } \sin 15°\ 0' &= 9.41300 \\ \hline \therefore \quad \log 1801.7 &= 3.25567 \end{aligned}$$

Therefore $R_1 = 2865 - 1801.7 = 1063.3$ ft., and, by Table I,

$$D_1 = 5°\ 23'.4, \text{ nearly enough.}$$

127. To Substitute a Curve of Given Radius for a Tangent uniting Two Curves.

In Fig. 53 let the tangent $BC = t$, $OB = R$, $O_1C = R_1$, and $O_2A = R_2$ be known.

Angles a, b, and c must be found in order to substitute curve AE for the system $ABCE$.

Draw OF parallel to BC, then $O_1F = R_1 - R$, and, from triangle OO_1F,

$$\tan d = \frac{t}{R_1 - R}, \quad . \quad . \quad . \quad . \quad . \quad . \quad . \quad . \quad . \quad (97)$$

$$OO_1 = \frac{t}{\sin d} = t \,.\, \text{cosec } d = \sqrt{(R_1 - R)^2 + t^2}. \quad . \quad (98)$$

Now in triangle OO_1O_2 three sides are known and the angles c and e may be computed. Thus if s is the half-sum of the sides,

$$\cos \tfrac{1}{2}c = \sqrt{\frac{s(s - OO_1)}{OO_2 \times O_2O_1}}.$$

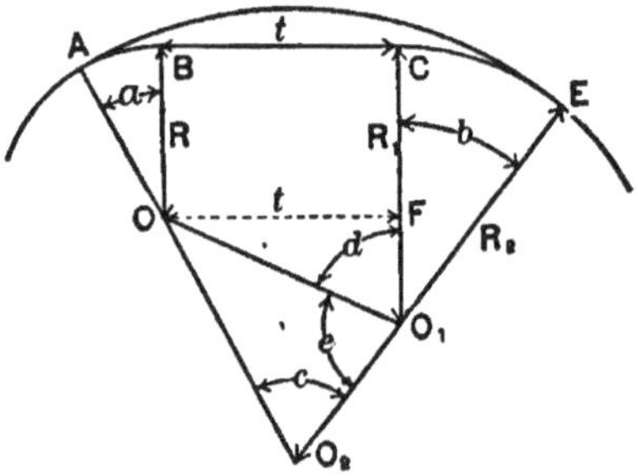

FIG. 53.

Angle e may be found in like manner, then $b = 180° - (e + d)$, and $a = c - b$.

Points A and E may now be located and the curve traced.

EXAMPLE.—A 3° and a 5° curve are united by a tangent 500 feet long. Replace by a 2° curve.

Here $R_1 - R = 1910 - 1146 = 764$ feet.

By (97), $\tan d = \dfrac{500}{764} = 0.65444 = \tan 33° \, 12'$

By (98), $OO_1 = 913.1$ feet.

In triangle OO_1O_2, $OO_1 = 913.1$, $O_1O_2 = 954.9$, and $OO_2 = 1718.7$ feet. Solving for e and c,

$$e = 133° \, 36', \quad c = 23° \, 0'. \quad \text{Then} \quad b = 13° \, 12', \quad a = 9° \, 56'.$$

ARTICLE 10. TRACK PROBLEMS.

128. Reversed Curves should never be employed on main lines because of the shock due to sudden reversal of curvature and superelevation of outside rail. A short tangent should be interposed between the two curves, which may ordinarily be done by changing the end-points of the curve, or slightly altering the radius. If, however, transition curves are employed to ease off both curves, there would seem to be no objection to the use of curves of contrary flexure, provided the track may be kept always in perfect condition. In yards, crossovers, and where connection is made with existing track, reversed curves may be employed, and are often imperative.

129. Having a Located Curve Intersected by a Straight Line, to Connect them by Another Curve.

Either the radius of the joining curve may be given, or else the point on first curve at which the junction must be made. The angle between a tangent to located curve at the point of meeting and the straight line must be measured. Four possible cases occur.

FIRST CASE.—*Joining curve tangent to located curve internally and on same side of cutting line as center.*

In Fig. 54 let GF be joining curve, with center O_1 and radius R_1. Let radius of located curve $OF = R$. Draw O_1G and OH perpendicular to the cutting line produced, and O_1K parallel to AH. If R_1 is known, we must determine angle b, a having been

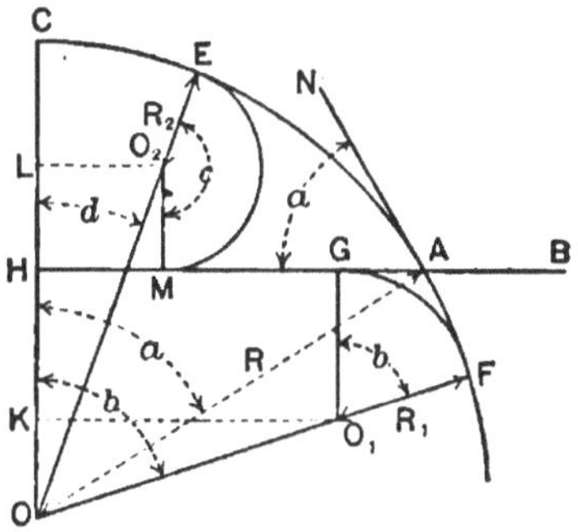

FIG. 54.

measured; then $b - a$ gives the length of arc from A to F where the $P.C.C.$ is to be located. In the triangle KOO_1 we have

$$OK = OH - R_1 \quad \text{and} \quad OO_1 = R - R_1.$$

$$\therefore \quad \cos b = \frac{R \cos a - R_1}{R - R_1}. \qquad (99)$$

Then $\dfrac{b - a}{D} = \text{arc } AF.$

Had F been given, we should have $b = a + AOF$, and, from (99),

$$R_1 = \frac{R (\cos a - \cos b)}{1 - \cos b} = \frac{R (\cos a - \cos b)}{\text{vers } b}. \qquad (100)$$

EXAMPLE.—A 1° curve is cut by a tangent that makes an angle of 64° 32′ with tangent to curve. Unite by means of a 4° curve.

By (99), $\cos b = 0.24000 = \cos 76° \; 07'$, and therefore $b - a = 11° \; 35'$, making AF, of figure, 11.58 stations.

SECOND CASE. — *Joining curve tangent internally to located curve but on opposite side of cutting line from center of located curve.*

In Fig. 54 let arc ME, with center O_2 and radius R_2, be the joining curve. From the figure,

$$\cos d = \frac{R \cos a + R_2}{R - R_2}. \quad . \quad . \quad . \quad . \quad . \quad (101)$$

Then arc $AE = a - d$ divided by D, and $c = 180° - d$.

Had the point E been given and R_2 required, it would have been, from (101)

$$R_2 = \frac{R(\cos d - \cos a)}{1 + \cos d}. \quad . \quad . \quad . \quad . \quad . \quad (102)$$

EXAMPLE.—Take the same example as in first case. Here,

By (101), $\cos d = 0.9068 = \cos 24° \; 56'$.

Then $64° \; 32' - 24° \; 56' = 39° \; 36'$,

equivalent to 39.600 stations around curve from A to E.

THIRD CASE.—*Joining curve tangent externally to located curve, with center on same side of cutting line.*

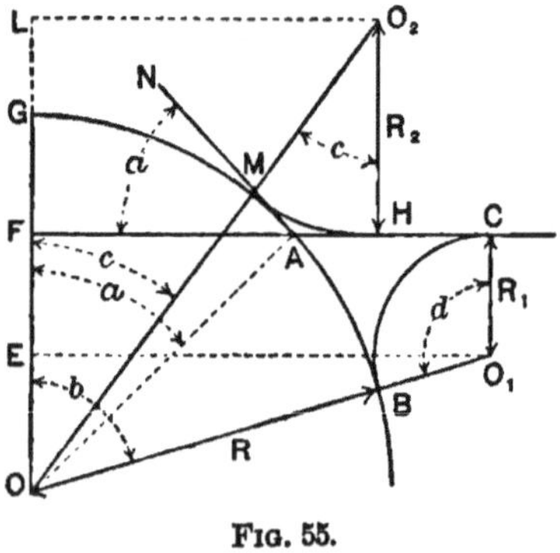

FIG. 55.

In Fig. 55 let arc BC, with center O_1 and radius R_1, be the joining curve. Draw O_1E parallel to CF, and O_1C and OF perpendicular thereto.

From the figure,

$$(R + R_1) \cos b = R \cos a - R_1;$$

$$\therefore \cos b = \frac{R \cos a - R_1}{R + R_1}. \quad . \quad . \quad . \quad . \quad . \quad (103)$$

Then $d = 180 - b$, and $AOB = b - a$. The curve may now be traced on the ground.

If AC is wanted, we have $AC = (R + R_1) \sin b - R \sin a$.

If the point B is fixed and R_1 required, there results, from (103),

$$R_1 = \frac{R (\cos a - \cos b)}{1 + \cos b}. \quad . \quad . \quad . \quad . \quad . \quad (104)$$

EXAMPLE.—Take the example given for the first and second cases

By (103),

$$\cos b = \frac{5730 \times 0.43 - 1432.5}{5730 + 1432.5} = 0.144 = \cos 81° \; 44'$$

$b - a = 81° \; 44' - 64° \; 32' = 17° \; 12'$, equivalent to 17.2 stations on located curve from A to B. Angle $d = 180° - 81° \; 44' = 98° \; 16'$, equivalent to 24.567 stations from B to C on the 4° curve.

FOURTH CASE.—*Joining curve tangent externally to located curve, with center on opposite side of cutting line.*

Let O_2, Fig. 55, be center of joining curve, R_2 its radius. From the figure,

$$(R + R_2) \cos c = R \cos a + R_2.$$

$$\therefore \cos c = \frac{R \cos a + R_2}{R + R_2}. \quad . \quad . \quad . \quad . \quad . \quad . \quad . \quad . \quad . \quad . \quad (105)$$

If M is fixed and R_2 required, (105) yields

$$R_2 = \frac{R(\cos c - \cos a)}{1 - \cos c} = \frac{R(\cos c - \cos a)}{\text{versin } c}. \quad . \quad (106)$$

EXAMPLE.—Take same example as in preceding cases.

By (105), $\qquad \cos c = 0.54403 = \cos 57° \; 02'$.

Then $\qquad a - c = 64° \; 32' - 57° \; 2' = 7° \; 30'$,

calling for a distance of 7.50 stations from A to M around 1° curve. From M to H on 4° curve is 14.258 stations.

130. To Locate a Y

A Y is made up of a system of tracks so arranged as to admit of turning an entire train. Three of the most used arrangements are given below.

FIRST CASE.—*One branch of Y a straight line.*

This is only the special case of the last problem in which the cutting line becomes tangent to both curves. In Fig. 56, if any

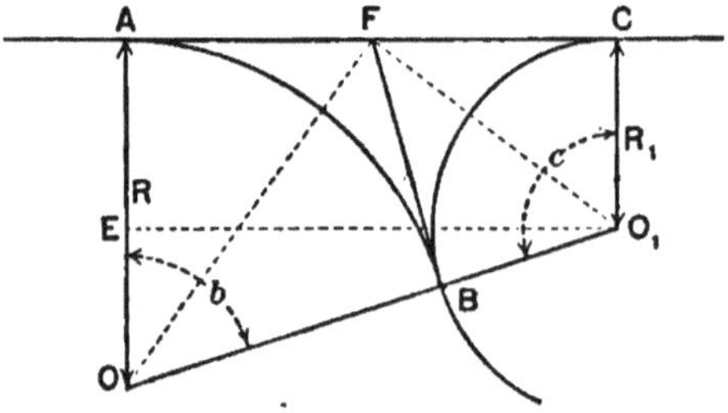

FIG. 56.

one of the points A, B, or C is given, the others may be located by finding the angles c and b. Draw O_1E parallel to CA; then in triangle OO_1E

$$(R + R_1) \cos b = R - R_1.$$

$$\therefore \cos b = \frac{R - R_1}{R + R_1}. \quad . \quad . \quad . \quad . \quad . \quad (107)$$

This follows at once from (103) by making angle $a = 0$. Then angle $c = 180 - b$. If AB were a located curve and the point B given, formula (107) would furnish us a value for R_1.

Another solution is to produce the tangent at B to cut AC at F; then $AF = FC = BF$. Join F with O and O_1; it can easily be seen that angle $OFO_1 = 90°$, and, by geometry,

$$BF = \sqrt{R \times R_1}. \quad . \quad . \quad . \quad . \quad . \quad . \quad (108)$$

Therefore $$\tan \tfrac{1}{2}b = \frac{BF}{R} = \sqrt{\frac{R_1}{R}}, \quad . \quad . \quad . \quad . \quad . \quad . \quad (109)$$

and $$\tan \tfrac{1}{2}c = \frac{BF}{R_1} = \sqrt{\frac{R}{R_1}}. \quad . \quad . \quad . \quad . \quad . \quad . \quad (110)$$

EXAMPLE.—Let AB be a 3° curve, BC a 6° curve, the point A at station 180.

By (107),

$$\cos b = \frac{1910.1 - 955.4}{1910.1 + 955.4} = 0.33317 = \cos 70° 32'.$$

The number of B is 180 + 23.511 = 203 + 51.1. Angle c = 109° 28′, equivalent to 18.244 stations on the 6° curve.

SECOND CASE.—*The three branches curved and convex towards each other.*

Given the three radii and any one of the points A, B, or C, Fig. 57, we have only to find the angles at the center, then divide these angles by the degrees of the respective curves to get their lengths and locate the three branches.

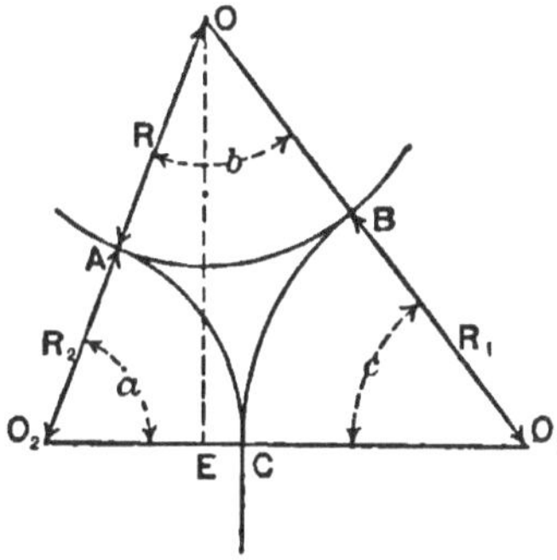

FIG. 57.

In the triangle OO_1O_2, letting $OO_1 = l$, $O_1O_2 = m$, $OO_2 = n$, $s = \frac{1}{2}(l + m + n) = R + R_1 + R_2$, we shall have, by trigonometry,

$$\cos \tfrac{1}{2}a = \sqrt{\frac{s(s - l)}{m \, . \, n}} = \sqrt{\frac{(R + R_1 + R_2)R_2}{(R + R_2)(R_1 + R_2)}}. \quad . \quad (111)$$

Angles b and c may be found in like manner.

The angles may be found otherwise by letting fall a perpendicular from one vertex upon the opposite side, as OE perpendicular to O_1O_2. Then from the relation

$$O_1O_2 : O_1O + OO_2 = OO_1 - OO_2 : O_1E - O_2E$$

determine O_2E and O_1E; then the right triangles O_2OE and O_1OE yield values of cosine a and cosine c, after which b may readily be obtained.

THIRD CASE.—*One branch concave to the other two.*

In Fig. 58 the triangle OO_1O_2 may be solved for the angles at O, O_1, and O_2; for if the radii are given, the sides $OO_1 = R - R_1$, $OO_2 = R - R_2$, and $O_1O_2 = R_1 + R_2$ are known and the solution is the same as for second case. Then b is the central angle for curve AB, $a' = 180 - a$, the central angle for AC, and $c' = 180 - c$, the central angle for curve BC.

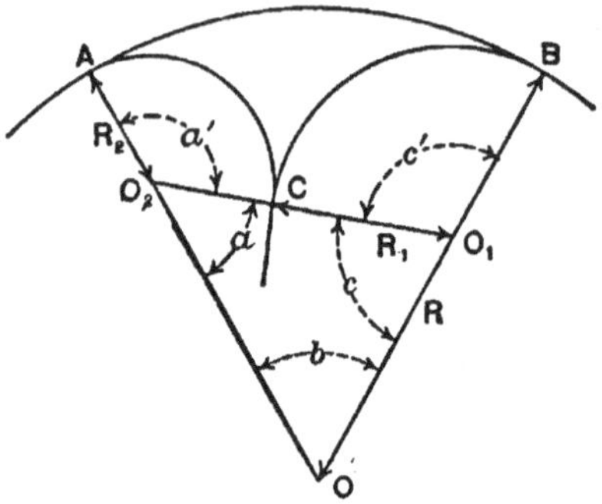

FIG. 58.

EXAMPLE.—If A is at sta. 820 on the 1° curve AB, AC an 8° curve, connect with a 6° curve CB. Here we have

$$O_2O = 5730 - 717 = 5013, \quad O_1O = 5730 - 955 = 4775,$$

and
$$O_2O_1 = 955 + 717 = 1672.$$

Solving this triangle, we get $c = 88°\ 20'$, $b = 19°\ 28'$, and $a = 72°\ 12'$. The number of B is therefore $820 + 19.467 = 839 + 46.7$; the length of CB is $\frac{91.667}{6} = 15.278$ stations, and of AC is $\frac{107.8}{8} = 13.475$ stations.

131. To Locate a Reversed Curve between Parallel Tangents.

FIRST CASE.—*Radii equal.*

(a) The equal radii R and distance p between tangents known. In Fig. 59 draw OE parallel to AG to meet O_1B produced. From triangle OEO_1,

$$\cos a = \frac{2R - p}{2R} = 1 - \frac{p}{2R}, \quad . \quad . \quad . \quad . \quad (112)$$

and
$$OE = 2R \sin a. \quad . \quad . \quad . \quad . \quad . \quad . \quad . \quad . \quad . \quad (113)$$

From triangle ABG,

$$AB = \frac{p}{\sin \frac{1}{2}a} = p \operatorname{cosec} \tfrac{1}{2}a = \sqrt{OE^2 + p^2}. \quad . \quad (114)$$

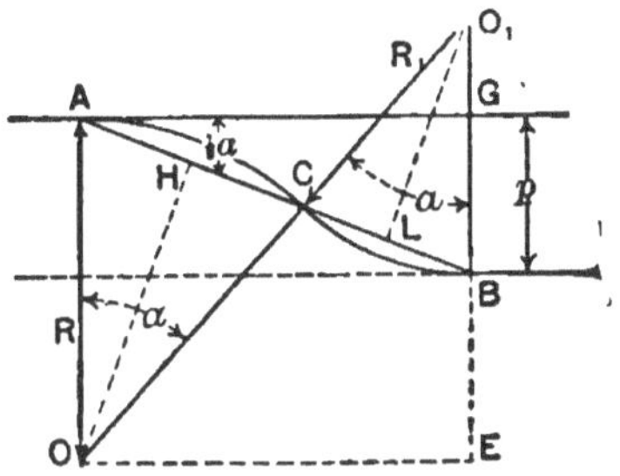

FIG. 59.

(*b*) AG and p known, R required.

Here $AB = \sqrt{AG^2 + p^2} = k$. Draw OH to the mid-point of AC. Triangles AOH and ABG are similar and $AH = \frac{1}{4}k$. Therefore

$$\frac{R}{\frac{1}{4}k} = \frac{k}{p},$$

whence

$$R = \frac{k^2}{4p}. \quad . \quad . \quad . \quad . \quad . \quad . \quad . \quad . \quad (115)$$

EXAMPLE.—Connect two parallel tracks, 30 ft. c. to c. by a 7° reversed curve. From Table I, $R = 819$ feet, and, by (112),

$$\cos a = 1 - \frac{30}{1638} = 0.98167 = \cos 10° \ 59'.$$

By (113), $OE = 1638 \times .19052 = 312.1$ feet.

By (114), $AB = \sqrt{(312.1)^2 + (30)^2} = 313.5$ feet.

If $p = 30$, $OE = 312.1$, or $AB = 313.5$ had been given, we should have had, by (115)

$$R = \frac{(313.5)^2}{120} = 819 \text{ feet.}$$

SECOND CASE.—*Radii unequal.*

(*a*) Suppose the radii $R = OA$ and $R_1 = O_1B$ (Fig. 59) to be known. We must find central angle a and $AB = k$. From the triangle OO_1E,

$$\cos a = \frac{R + R_1 - p}{R + R_1} = 1 - \frac{p}{R + R_1}. \quad . \quad . \quad (116)$$

Then AB will be given by (114).

(*b*) Suppose $AB = k$, p and R known, to find R_1 and angle a. Triangle ABG yields

$$\sin \tfrac{1}{2}a = \frac{p}{k}. \quad . \quad . \quad . \quad . \quad . \quad . \quad . \quad . \quad . \quad . \quad (117)$$

O_1LB is similar to AGB. Hence

$$\frac{R_1}{LB} = \frac{k}{p}.$$

But $AC = 2R \sin \frac{1}{2}a$, and $LB = \frac{1}{2}(k - AC) = \frac{1}{2}C_1$. Inserting this value of LB and solving for R_1,

$$R_1 = \frac{kC_1}{2p}. \quad . \quad . \quad . \quad . \quad . \quad . \quad . \quad . \quad . \quad (118)$$

From similar triangles,

$$\frac{R_1}{R} = \frac{C_1}{k - C_1}.$$

Inserting the value of $C_1 = \frac{2pR_1}{k}$ from (118) and solving for R_1, we get

$$R_1 = \frac{k^2}{2p} - R. \quad . \quad . \quad . \quad . \quad . \quad . \quad . \quad (119)$$

EXAMPLE.—$AB = 300'$, $p = 30'$, $R = 819$ ft., to find angle a and R_1.

By (117), $\sin \frac{1}{2}a = \frac{30}{300} = 0.10000 = \sin 5° 44'$.

Therefore angle $a = 11° 28'$.

By (119), $R_1 = \frac{(300)^2}{60} - 819 = 681$ ft., an $8° 25'$ curve.

132. To Connect Two Parallel Tracks by a Crossover composed of two D° Curves with a Given Length of Tangent between Points of Contrary Flexure.

In Fig. 60 let $AFGB$ be the required crossover, $FG = l$, $EB = p$, and $OA = O'B = R$ known; angle a and $AE = x$ are required.

Draw OM parallel to AE to meet $O'B$ produced; draw also OC parallel and equal to FG; join O and O'. From triangle $OO'C$,

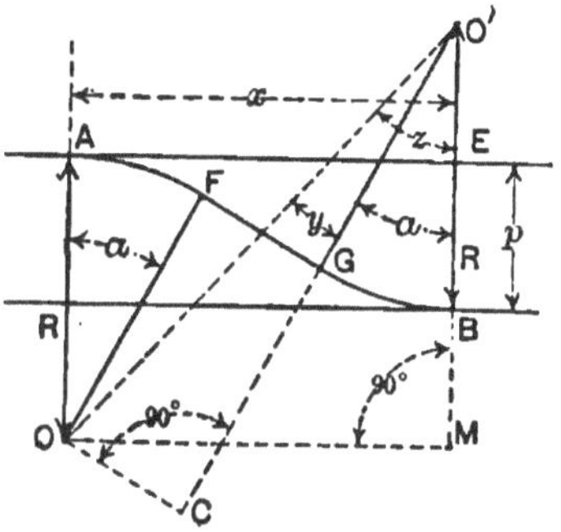

FIG. 60.

$$\tan y = \frac{l}{2R}, \quad . \quad . \quad . \quad . \quad (120)$$

$$OO' = \frac{2R}{\cos y} = 2R \sec y = \sqrt{4R^2 + l^2}. \quad . \quad . \quad (121)$$

Then in triangle $OO'M$,

$$\cos z = \frac{O'M}{OO'} = \frac{2R - p}{2R \sec y} = \left(1 - \frac{p}{2R}\right) \cos y. \quad . \quad (122)$$

Now knowing y and z,

$$a = z - y. \quad . \quad . \quad . \quad . \quad . \quad . \quad . \quad . \quad . \quad . \quad . \quad . \quad (123)$$

Next, $$x = OM = OO' \sin z = 2R \sec y \sin z. \quad . \quad (124)$$

EXAMPLE.—Given $D = 7° 30'$, $p = 62$ ft., $l = 100$ ft., to locate crossover when A is at sta. $86 + 20$.

By (120),

$$\log \tan y = 2 - 3.18441 = 8.81559 = \log \tan 3° 44'.$$

By (121),

$$\log OO' = 3.18441 - 9.99908 = 3.18533 = \log 1532.$$

By (122),

$$\log \cos z = 3.16643 - 3.18533 = 9.98110 = \log \cos 16° 47'.$$

By (123),

$$a = 16° 47' - 3° 44' = 13° 3'.$$

By (124),

$$\log x = 3.18533 + 9.46053 = 2.64586 = \log 442.4.$$

133. To Find the Radius of the Reversed Curve *AFE*, Fig. 61, Given Angles I and I', and $BC = k$.

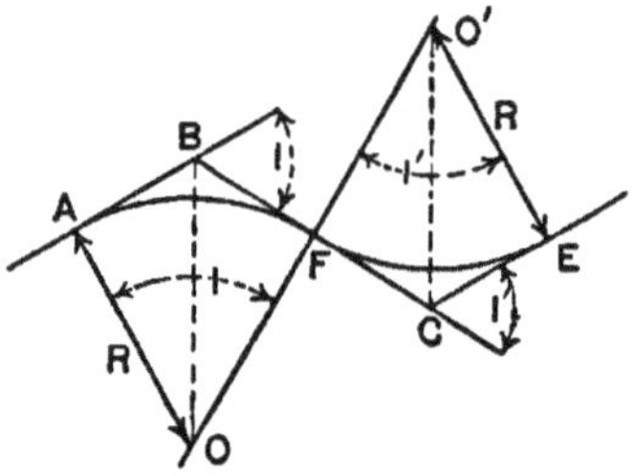

Fig. 61.

From the figure,

$$R \tan \tfrac{1}{2}I = BF,$$

$$R \tan \tfrac{1}{2}I' = CF.$$

Adding,

$$R(\tan \tfrac{1}{2}I + \tan \tfrac{1}{2}I') = BC = k.$$

Whence

$$R = \frac{k}{\tan \frac{1}{2}I + \tan \frac{1}{2}I'}. \quad \dots \dots \quad (125)$$

Example.—Given $I = 10°$, $I' = 20°$, $BC = 700$ feet, to find R.

By (125), $R = \dfrac{700}{0.08749 + 0.17633} = 2653$ ft., a 2° 9¼′ curve.

134. To Locate a Reversed Curve between Fixed Points. In Fig. 62 let $AB = k$, and angles I and I' be known. We

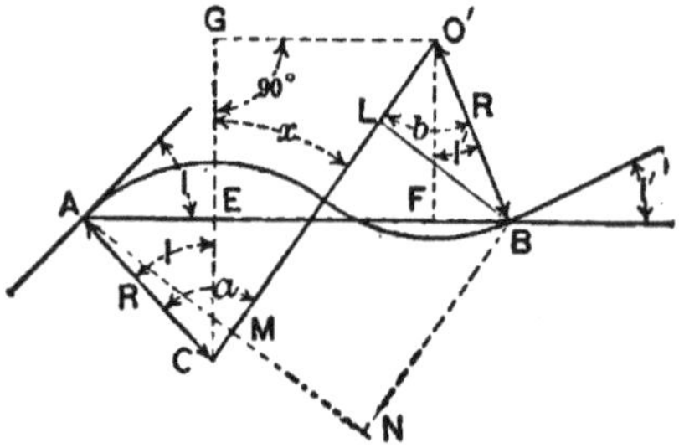

Fig. 62.

have to find R and the angles a and b.

Draw $O'G$ parallel to, and OG and $O'F$ perpendicular to, AB. Angle $AOG = I$ and $BO'F = I'$. Then $OE = R \cos I$ and $O'F = R \cos I'$. Hence

$$G = R(\cos I + \cos I').$$

In triangle $OO'G$, $OO' = 2R$. Therefore

$$\cos x = \frac{R(\cos I + \cos I')}{2R} = \frac{\cos I + \cos I'}{2}, \quad . \quad (126)$$

an expression from which R has disappeared.

We now have $a = I + x$ and $b = I' + x$.

To find R we have $AE + EF + FB = k$,

or $R \sin I + 2R \sin x + R \sin I' = k$.

Whence $$R = \frac{k}{\sin I + \sin I' + 2 \sin x} \quad . \quad . \quad . \quad (127)$$

Another expression for R can be found by drawing AN and BL perpendicular to OO', and BN parallel thereto. Then, since $\measuredangle BAN = x$,

$$R \sin a + R \sin b = k \cos x.$$

$$\therefore \quad R = \frac{k \cos x}{\sin a + \sin b}. \quad . \quad . \quad . \quad . \quad . \quad . \quad (128)$$

EXAMPLE.—Take the example of the last problem,

$$k = 700, \quad I = 10°, \quad I' = 20°.$$

By (126),

$$\cos x = \tfrac{1}{2}(0.98481 + 0.93969) = 0.96225 = \cos 15° \ 48'.$$

We now have $a = 25° \ 48'$ and $b = 35° \ 48'$.

By (128), $R = \dfrac{700 \times 0.96225}{0.43523 + 0.58496} = 660.2$ ft., an 8° 41′ curve.

135. To Connect Two Divergent Tangents by a Reversed Curve.

FIRST CASE.—*Advancing towards the P.I.*

Given the radii R and R_1, the angle I and $AC = k$, to find the angles a and b (Fig. 63).

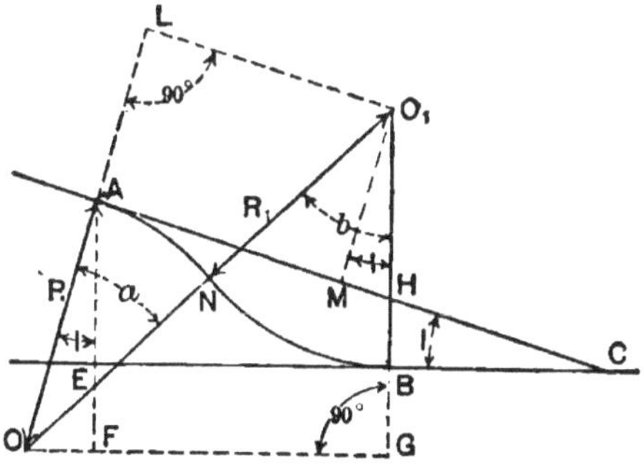

FIG. 63.

Draw OG parallel to the tangent BC to meet O_1B produced.

Then $EF = BG = AF - AE$.

Therefore $BG = R \cos I - k \sin I$.

From triangle OO_1G,

$$\cos b = \frac{R_1 + BG}{R + R_1} = \frac{R_1 + R\cos I - k\sin I}{R + R_1}. \quad . \quad (129)$$

Then $a = MO_1N = b - I$, O_1M being parallel to OA.

SECOND CASE.—*Receding from the P.I.*

In Fig. 63 we have $BC = k_1$, angle I, R, and R_1 given, to find angles a and b.

Produce OA to meet O_1L drawn parallel to CA. AL equals $O_1M = O_1H\cos I$.

$$O_1H = R_1 - HB = R_1 - k_1\tan I.$$

$$\therefore\ AL = O_1M = (R_1 - k_1\tan I)\cos I.$$

Hence

$$OL = R + (R_1 - k_1\tan I)\cos I = R + R_1\cos I - k_1\sin I.$$

From triangle OO_1L,

$$\cos a = \frac{OL}{OO_1} = \frac{R + R_1\cos I - k_1\sin I}{R + R_1}. \quad . \quad . \quad (130)$$

Evidently, $b = a + I.$

136. To Change the *P.R.C.* so that Second Branch of Curve shall End in a Tangent Parallel to Terminal Tangent and Distant *p* therefrom.

In Fig. 64 let MAB be the located curve, $EN = p$. We must

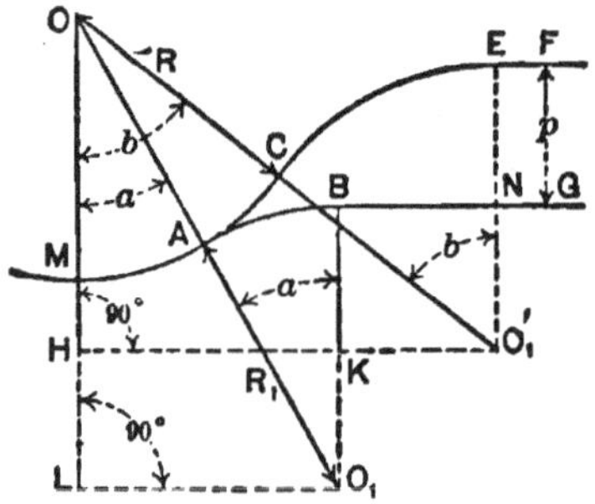

FIG. 64.

determine the angle COA, after which the desired curve ACE may be located.

Draw HO_1' and LO_1 parallel to EF and NG.

$$HL = O_1K = p.$$

From triangles $OO_1'H$ and OO_1L,

$$(R + R_1) \cos b = (R + R_1) \cos a - p.$$

$$\therefore \cos b = \cos a - \frac{p}{R + R_1}. \quad . \quad . \quad . \quad . \quad (131)$$

Angle $AOC = b - a$.

137. To Find the Radius of a Curved Track.

Measure any chord $AB = 2l$, and mid-ordinate $CE = M$.

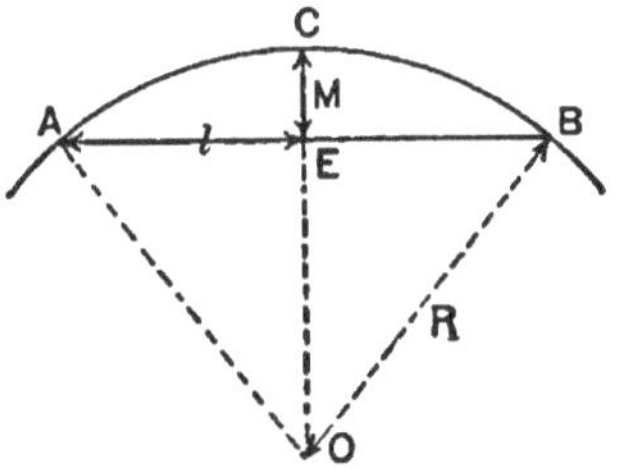

Fig. 65.

Then in the right triangle OAE (Fig. 65),

$$R^2 - (R - M)^2 = l^2.$$

$$\therefore R = \frac{l^2 + M^2}{2M}. \quad . \quad . \quad . \quad . \quad . \quad . \quad . \quad . \quad . \quad (132)$$

CHAPTER IV

TRANSITION-CURVES.

ARTICLE 11.—THEORY OF THE TRANSITION-CURVE.

138. Elevation of Outer Rail on Curves.—To counteract the effect of centrifugal force on curves the outer rail must be elevated above the inner one. It is shown in mechanics that the centrifugal force is

$$F = \frac{Wv^2}{32.16R},$$

where W is the weight, v the velocity in feet per second, 32.16 an average value of the acceleration of gravity in feet per second per second, and R the radius in feet.

In Fig. 66 let the vertical HL represent W, the horizontal KH the centrifugal force, AB the plane of the rails, and $CB = e$ the superelevation of outer rail. From similar triangles,

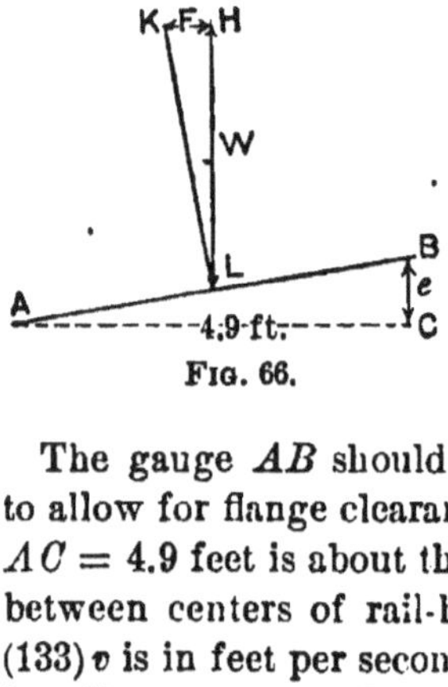

FIG. 66.

$$F = W\frac{e}{AC}.$$

Equate this value of F to that given above and solve for e, giving

$$e = \frac{ACv^2}{32.16R}. \quad . \quad . \quad (133)$$

The gauge AB should be greater on curves than on tangents to allow for flange clearance and the effect of a rigid wheel-base. $AC = 4.9$ feet is about the right value for the horizontal distance between centers of rail-heads for standard gauge. In formula (133) v is in feet per second, but the train velocity is usually given in miles per hour. Let V = velocity in miles per hour, then the velocity in feet per second will be $v = \frac{22}{15}V$. Inserting these values in (133) gives

$$e = \frac{4.9 \times 484 V^2}{32.16 \times 225 R} = \frac{V^2}{3R}, \text{ nearly.} \quad . \quad . \quad . \quad (134)$$

This elevation will be required from the *P.C.* to the *P.T.*, but obviously it cannot be introduced suddenly, so that for easy riding the rate of increase of e should be uniform. From (134) it is seen that e varies inversely with R, which requires that when $e = 0$, $R =$ infinity. Hence R must decrease from infinity to the radius of the circular curve, while e increases from 0 to its maximum value.

139. The True Transition-curve should satisfy formula (134), but so far no such curve has been found that will at the same time admit of the same ease of location as the simple circular curve. According to Rankine the first use of any other than the circular curve was made by Gravatt about 1828 or 1829, the curve employed being the curve of sines. Another method described by Rankine is attributed to William Froude about 1842; this curve was worked up in the *Engineering News* by A. M. Wellington in 1890. Other approximations are the *Railroad Spiral,* developed by W. H. Searles in 1882, and the cubic parabola, described by C. D. Jameson and E. W. Crellin in the *Railroad and Engineering Journal,* 1889.

In 1880 Ellis Holbrook described in the *Railroad Gazette* the true transition-curve applicable to small angles and short lengths of the curve. In 1893 C. L. Crandall published formulæ and tables applicable to large central angles for both the offset and deflection methods.

140. The Notation here employed will be explained with reference to Fig. 67. The curve $CBB'C'$ is the circular curve offset at C and C' from the tangents by the amounts CH and $C'H'$. AGB and $B'G'A'$ are the transition-curves. A is the *P.T.C.*, or point of transition-curve, C the *P.C.*, B the *P.C.*$_1$, B' the *P.T.C.*$_1$, C' the *P.T.*, and A' the *P.T.*$_1$. The co-ordinates of G are $AH = x'$, $HG = y'$; of C, x' and $HC = F$; of B, $AM = x_1$ and $MB = y_1$. The length of curve from *P.T.C.* to any point P is l, and the whole length from *P.T.C.* to *P.C.*$_1$ is l_1.

141. Equation of Transition-curve.—Since the rate of change of e must be uniform, (134) may be written

$$e = kl = \frac{V^2}{3\rho}, \quad \ldots \ldots \quad (135)$$

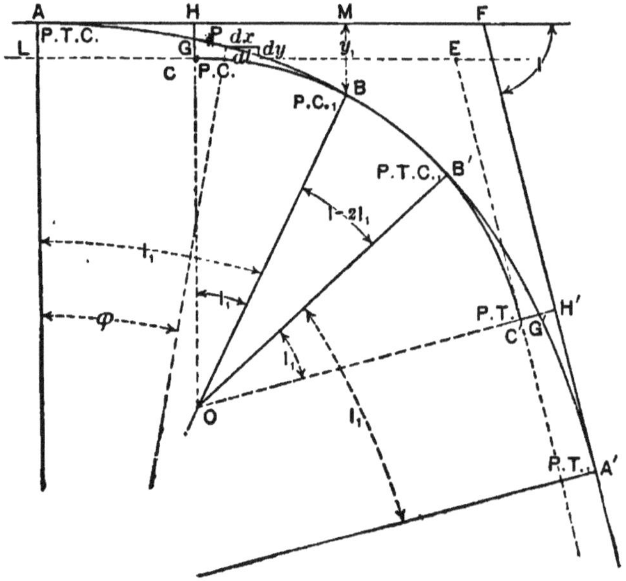

FIG. 67.

in which k is the rate of rise of outer rail along curve, and ρ the varying radius of curvature. From the calculus $\rho d\phi = dl$, whence

$$\rho = \frac{dl}{d\phi}. \quad \ldots \ldots \quad (136)$$

Insert this in (135) and solve for $d\phi$.

$$d\phi = \frac{3k}{V^2} l dl = 2m l dl. \quad \ldots \ldots \quad (137)$$

$2m$ is dependent upon V and k, and is constant for any one curve.

Integrating (137),

$$\phi = ml^2, \quad \ldots \ldots \quad (138)$$

the constant of integration being zero, for l is zero when ϕ is zero.

From the elementary triangle drawn at the point P of transition-curve, Fig. 67, dl being tangent,

$$\frac{dy}{dl} = \sin\phi.$$

Expanding $\sin\phi$ by trigonometry,

$$dy = dl\left(\phi - \frac{\phi^3}{3!} + \frac{\phi^5}{5!} - \frac{\phi^7}{7!} + \ldots\right),$$

in which $3! = 3 \times 2 \times 1$, $5! = 5 \times 4 \times 3 \times 2 \times 1$, etc.

Substituting for ϕ its value from (138),

$$dy = dl\left(ml^2 - \frac{m^3l^6}{6} + \frac{m^5l^{10}}{120} - \frac{m^7l^{14}}{5040} + \ldots\right).$$

Integrating,

$$y = \left(\frac{ml^3}{3} - \frac{m^3l^7}{42} + \frac{m^5l^{11}}{1320} - \frac{m^7l^{15}}{75600} + \ldots\right). \quad . \quad . \quad (139)$$

But $ml^2 = \phi$, where ϕ is in circular measure. To obtain ϕ in degrees, $\phi = \phi^\circ\frac{\pi}{180} = \frac{\phi^\circ}{57.3}$. Inserting this in (139),

$$y = l\left(\frac{\phi^\circ}{171.89} - \frac{\phi^{\circ 3}}{79 \times 10^5} + \frac{\phi^{\circ 5}}{8151 \times 10^8} - \frac{\phi^{\circ 7}}{153245 \times 10^{12}} + \ldots\right),$$

$$\text{or} \qquad y = lC, \quad . \quad . \quad . \quad . \quad . \quad . \quad . \quad . \quad (140)$$

in which C may be found from Table XIV with ϕ° as argument. Interpolation must be resorted to for values of ϕ not given in the table, or y computed by the formula.

From the elementary triangle at P, Fig. 67

$$\frac{dx}{dl} = \cos\phi.$$

Expanding by trigonometry,

Substituting ml^2 for ϕ and integrating,

$$x = l\left(1 - \frac{m^2l^4}{10} + \frac{m^4l^8}{216} - \frac{m^6l^{12}}{9360} + \ldots\right). \quad . \quad . \quad . \quad . \quad (141)$$

Replacing ml^2 by ϕ reduced to degrees,

$$x = l\left(1 - \frac{\phi^{\circ 2}}{32828} + \frac{\phi^{\circ 4}}{2328 \times 10^6} - \frac{\phi^{\circ 6}}{33114 \times 10^{10}} + \ldots\right)$$

or $\qquad x = l - lE. \quad . \quad . \quad . \quad . \quad . \quad . \quad . \quad (142)$

E varies with ϕ°, and may be taken from Table XIV with ϕ° as argument.

142. The Transition-curve Angle I_1 is the value ϕ assumes at the $P.C._1$. From (138),

$$I_1 = ml_1^2. \quad . \quad . \quad . \quad . \quad . \quad . \quad . \quad . \quad (143)$$

From (137) and (136),

$$\frac{dl}{d\phi} = \frac{1}{2ml} = \rho.$$

At the $P.C._1$ $\rho = R$ and may be taken equal to $\frac{5730}{D^\circ}$, so that

$$\frac{1}{2ml_1} = R = \frac{5730}{D^\circ},$$

whence

$$m = \frac{1}{2l_1R} = \frac{D^\circ}{11460l_1}. \quad . \quad . \quad . \quad . \quad . \quad (144)$$

This value of m in (143) gives

$$I_1 = \frac{l_1}{2R} = \frac{D^\circ l_1}{11460}. \quad . \quad . \quad . \quad . \quad . \quad . \quad (145)$$

Reducing this to circular measure by writing $I_1 = I_1^\circ \frac{\pi}{180} = \frac{I_1^\circ}{57.30}$ gives

$$I_1^\circ = 28.65\frac{l_1}{R} = \frac{D^\circ l_1}{200}. \quad . \quad . \quad . \quad . \quad . \quad (146)$$

143. The Coördinates of any point on the curve are given by (140) and (142). The length of the transition-curve being known or assumed, y_1 and x_1 (the coördinates of the $P.C._1$) may be found from these equations by the help of Table XIV; the coördinates of the $P.C.$ (see Fig. 67) will be

$$F = y_1 - R(1 - \cos I_1) = y_1 - R \text{ vers } I_1, \quad . \quad . \quad (147)$$
$$x' = x_1 - R \sin I_1. \quad . \quad . \quad . \quad . \quad . \quad . \quad . \quad . \quad . \quad . \quad . \quad (148)$$

144. Deflection-angles.—With the transit at the $P.T.C.$ (or $P.T._1$ in backing up) the tangent of deflection-angles may be found from the relation $\tan \delta = \frac{y}{x}$. Dividing (139) by (141),

$$\tan \delta = \frac{ml^2}{3} + .009523m^3l^6 + .000167m^5l^{10} + \ldots . \quad (149)$$

From trigonometry the expansion of the angle in terms of its tangent is

$$\delta = \tan \delta - \tfrac{1}{3} \tan^3 \delta + \tfrac{1}{5} \tan^5 \delta - \text{etc.} \quad . \quad . \quad . \quad (a)$$

In (149) write $ml^2 = \phi$ and substitute in (a):

$$\delta = \frac{\phi}{3} - .002823\phi^3 - .000068\phi^5. \quad . \quad . \quad . \quad . \quad (150)$$

From (138) and (143),

$$\frac{\phi}{I_1} = \frac{ml^2}{ml_1^2} = \frac{l^2}{l_1^2} = n^2, \quad . \quad . \quad . \quad . \quad . \quad . \quad . \quad (b)$$

in which $\frac{l}{l_1} = n$. From (b), $\phi = I_1 n^2$, and this in (150) gives

$$\delta = \frac{I_1}{3}n^2 - .002823I_1^3n^6 - .000068I_1^5n^{10}. \quad . \quad . \quad . \quad (c)$$

Both δ and I_1 are in circular measure; to reduce to degrees multiply by $\frac{\pi}{180}$. This gives, neglecting terms involving higher powers of I_1 than the third,

$$\delta^\circ = \frac{I_1^\circ}{3} n^2 - .00000086 \, I_1^3 n^6. \quad . \quad . \quad . \quad . \quad . \quad . \quad (151)$$

The second term is quite small, and in most cases may be entirely neglected in practice.

With the instrument at any intermediate point $x''y''$ the deflection-angle for any point xy, measured from initial tangent, will be

$$\tan \delta = \frac{y - y''}{x - x''} = \tfrac{1}{3}(ml^2 + ml''^2 + mll'') + \tfrac{1}{105}(m^3l^6 + m^3l''^6)$$

$$+ \tfrac{3}{70}(m^3l^5l'' + m^3ll''^5) + \tfrac{8}{105}(m^3l^4l''^2 + m^3l^2l''^4 + m^3l^3l''^3) + \ldots, \quad (152)$$

in which powers of ml^2 higher than the third have been neglected.

Substitute the value of $\tan \delta$ from (152) in (a), write $ml^2 = \phi = I_1n^2$, $ml''^2 = \phi'' = I_1n''^2$, by ($b$), and reduce circular measure to degrees, giving

$$\delta_0 = \frac{I_1^\circ}{3}(n^2 + n''^2 + nn'') - \text{a small correction.} \quad (153)$$

For instrument at $P.T.C.$, $n'' = 0$; then (153) yields

$$(\delta_0^\circ) = \frac{I_1^\circ}{3}n^2 - \text{correction},$$

or

$$(\delta_0^\circ) = \frac{I_1^\circ}{3}A_0 - B_0. \quad . \quad . \quad . \quad . \quad . \quad . \quad . \quad . \quad . \quad (154)$$

(154) is the same as (151), as it should be.

For the transit at the quarter-point of transition-curve $n'' = \frac{l''}{l_1} = \frac{\frac{1}{4}l_1}{l_1} = \frac{1}{4}$; then (153) yields

$$(\delta_{\frac{1}{4}}^\circ) = \frac{I_1^\circ}{3}(n^2 + \tfrac{1}{16} + \tfrac{1}{4}n) - \text{correction},$$

or

$$(\delta_{\frac{1}{4}}^\circ) = \frac{I_1^\circ}{3}A_{\frac{1}{4}} - B_{\frac{1}{4}}. \quad . \quad . \quad . \quad . \quad . \quad . \quad . \quad . \quad . \quad (155)$$

For transit at mid-point of transition-curve $n'' = \frac{1}{2}$, and, from (153),

$$(\delta_{\frac{1}{2}}^\circ) = \frac{I_1^\circ}{3}(n^2 + \tfrac{1}{4} + \tfrac{1}{2}n) - \text{correction},$$

or

$$(\delta_{\frac{1}{2}}^\circ) = \frac{I_1^\circ}{3}A_{\frac{1}{2}} - B_{\frac{1}{2}}. \quad . \quad . \quad . \quad . \quad . \quad . \quad . \quad . \quad . \quad (156)$$

For transit at three-quarter point $n'' = \frac{3}{4}$ and

$$(\delta_{\frac{3}{4}}^\circ) = \frac{I_1^\circ}{3}(n^2 + \tfrac{9}{16} + \tfrac{3}{4}n) - \text{correction},$$

or

$$(\delta_{\frac{3}{4}}^\circ) = \frac{I_1^\circ}{3}A_{\frac{3}{4}} - B_{\frac{3}{4}}. \quad . \quad . \quad . \quad . \quad . \quad . \quad . \quad . \quad . \quad . \quad (157)$$

For transit at $P.C._1$ $n'' = 1$ and

$$(\delta_1^\circ) = \frac{I_1^\circ}{3}(n^2 + 1 + n) - \text{correction},$$

or

$$\delta_1^\circ) = \frac{I_1^\circ}{3}A_1 - B_1. \quad . \quad . \quad . \quad . \quad . \quad . \quad . \quad . \quad . \quad . \quad (158$$

With the transit at the $P.T.C._1$ it will frequently be most convenient to measure the deflections from the tangent to the circular curve at that point. Sometimes this will also be the case for the transit at the $P.C._1$.

By reference to Fig. 68 it will be seen that for the transit at B

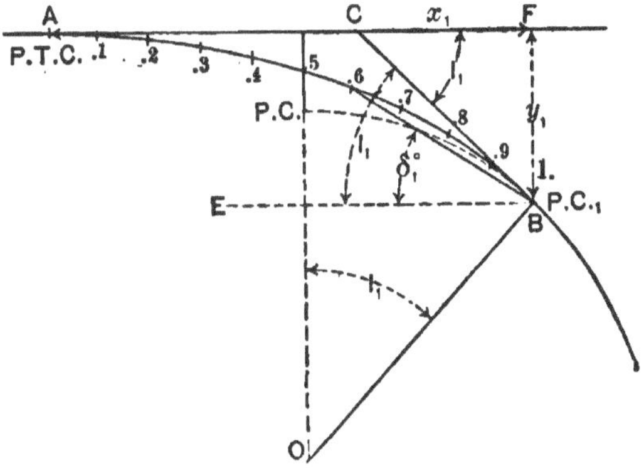

FIG. 68.

the deflection from the tangent BC which serves to fix any point on the curve, as .6, is given by the equation

$$(\delta_c^\circ)_{.6} = I_1^\circ - (\delta_1^\circ)_{.6} = I_1^\circ - \left(\frac{I_1^\circ}{3}A_1 - B_1\right)_{.6},$$

or, in general,

$$(\delta_c^\circ) = \frac{I_1^\circ}{3}A_c + B_1. \quad . \quad . \quad . \quad . \quad . \quad . \quad . \quad . \quad . \quad . \quad . \quad (159)$$

Table XV gives the values of A and B for the five positions of instrument for which equations (154) to (159), inclusive, were deduced. The value of A must be multiplied by $\frac{I_1^\circ}{3}$, but B is taken direct from the table in thousandths of a degree.

If deflection-angles are wanted for other positions of the instrument, or for other points on the curve, they may be computed from equation (153).

145. Tables.—Three tables are given for use with transition-curves.

Table XIV was computed for use with formulas (140) and (142) in determining C and E; ϕ being assumed and C and E computed.

Table XV gives A and B for computing the deflection-angles by (154), (155), (156), (157), (158), and (159) for 20 equidistant stations on the transition-curve. For points not given in the table A and B must be interpolated. Linear interpolation will suffice in most cases, though when I_1° is quite large second differences may be preferable for A. B is given in the table in thousandths of a degree.

Table XVI was calculated by assuming l_1 in lengths varying by increments of 20 feet, then computing I_1° by (146), y_1 by (139), x_1 by (141), F by (147), and x' by (148). y_1 and x_1 will also be given more directly by (140) and (142) with the aid of Table XIV.

The excess in length of transition-curve, measured from $P.T.C.$ to the point on offset at $P.C.$, over x' is tabulated as e; l' is found by trial such that when inserted in (141) or (142) the same value of x' will be obtained as in (148). This may be done by assuming l' a little less than $\frac{l_1}{2}$, then computing x'. More than two trials will rarely be needed to find a sufficiently close value of l'; then $e = l' - x'$. y' is found by (139) after finding l', or ϕ' may be found from (b) of **144**, and used in (140) in connection with Table XIV. $l_1 - l'$ is the length from G (Fig. 67) to the $P.C._1$; the difference in length between this and the length of circular curve from $P.C.$ to $P.C._1$ is tabulated as e'; that is, $e' = (l_1 - l') - \text{arc}$. Then $e + e' = l_1 - (x' + \text{circular arc})$.

For values of l_1 intermediate between those given in the table linear interpolation will suffice, though second differences may be used for F and y_1 if preferred.

146. To Unite the Two Branches of a Compound Curve by a Transition-curve.

The same objections hold to compound curves as to simple curves uniting with a tangent; i.e., where there is a sudden change of curvature there should be a sudden change of super-elevation of outer rail, which of course is not allowable. Instead of compounding the curves, we may offset them at the *P.C.C.* and unite them by means of a portion of a transition-curve tangent to each of the simple curves.

In Fig. 69 *AB* and *CELM* are the simple curves that are to be united by the transition-curve *ANE*. Extend the transition-curve

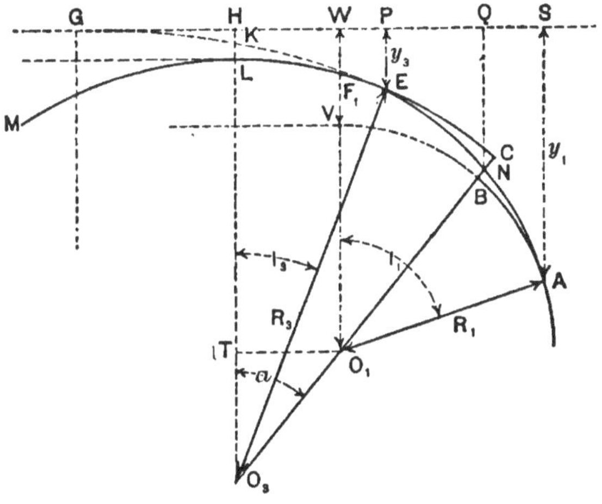

FIG. 69.

to *G*, where its radius of curvature becomes infinite, and let *GS* be its tangent. Call the length of transition-curve from *G* to *A* l_1, from *G* to *E* l_3, and from *E* to *A* l_2. *E* and *A* are points of tangency of simple and transition curves. Then $l_2 = l_1 - l_3$. The coördinates of *A* are $GS = x_1$, $SA = y_1$; and of V (WV perpendicular to GS), $GW = x_1'$, $WV = F_1$; of E, $GP = x_3$, $EP = y_3$; of L (LH perpendicular to GS), $GH = x_3'$, $HL = F_3$. Let $BC = F_2$.

The radius of curvature of transition-curve is inversely proportional to its length from *G*; hence the curvature is proportional to the length of curve; therefore $l_3 : l_1 = D_3 : D_1$, whence

$$l_3 = l_1 \frac{D_3}{D_1}. \qquad (160)$$

Then $$l_2 = l_1 - l_3 = l_1\left(1 - \frac{D_3}{D_1}\right) = l_1\frac{D_1 - D_3}{D_1}. \quad . \quad . \qquad (161)$$

By (138) or (143), $$\frac{I_3}{I_1} = \left(\frac{l_3}{l_1}\right)^2.$$

Equating the value of I_3 from this equation to that resulting from (146) gives

$$I_3 = I_1\left(\frac{l_3}{l_1}\right)^2 = \frac{l_3 D_3^\circ}{200}. \quad . \quad . \quad . \quad . \quad . \quad . \qquad (162)$$

$WV = F_1$ and $HL = F_3$ may be taken from Table XVI with l_1 and l_3 as arguments. Then $O_1W = R_1 + F_1$ $O_3H = R_3 + F_3$. Draw O_1T parallel to GS, then $O_1T = WH$; hence

$$O_3T = (R_3 + F_3) - (R_1 + F_1),$$

and $$O_1T = x_1' - x_3'.$$

Therefore

$$\tan a = \frac{x_1' - x_3'}{(R_3 + F_3) - (R_1 + F_1)}, \quad . \quad . \quad . \qquad (163)$$

$$O_1O_3 = (x_1' - x_3') \operatorname{cosec} a = \sqrt{\overline{O_1T}^2 + \overline{TO_3}^2}. \quad . \qquad (164)$$

Then $$CB = CO_3 - BO_3,$$

or $$F_2 = R_3 - (R_1 + O_1O_3). \quad . \quad . \quad . \quad . \quad . \qquad (165)$$

The lengths of AB and CE are

$$AB = \frac{I_1^\circ - a^\circ}{D_1}100, \quad . \quad . \quad . \quad . \quad . \quad . \quad . \quad . \qquad (166)$$

$$CE = \frac{a^\circ - I_3^\circ}{D_3}. \quad . \quad . \quad . \quad . \quad . \quad . \quad . \quad . \quad . \qquad (167)$$

The excess of transition-curve length over $AB + CE$ is

$$e_2 = l_2 - \left(\frac{I_1^\circ - a}{D_1} + \frac{a - I_1^\circ}{D_3}\right)100. \quad . \quad . \quad . \qquad (168)$$

If AB and CE are quite sharp, we must take account of the arc excess, so that we have then

$$e_2 = l_2 - \left[\left(\frac{I_1^\circ - a}{D_1} + \frac{a - I_3^\circ}{D_3}\right) 100 + \text{arc excess}\right]. \quad (168')$$

The arc excess may be taken from the second column of Table IV, which gives the arc length for one station; this multiplied by the number of stations gives the curve length, which may replace the values within the brackets in (168′).

147. Length of Transition-curve to be Taken.—In practice the rate of change of superelevation of outer rail may vary from $\frac{1}{1200}$ to $\frac{1}{400}$. Call the rate k; then evidently kl_1 must equal the superelevation of outer rail for circular curve; or, by (135),

$$kl_1 = \frac{V^2}{3R}.$$

Writing $R = \frac{5730}{D}$, and solving for l_1

$$l_1 = \frac{V^2D}{17190k}. \quad . \quad . \quad . \quad . \quad . \quad . \quad . \quad (169)$$

For $k = \frac{1}{1200}$, $\quad l_1 = 0.07\,V^2D. \quad . \quad . \quad . \quad . \quad . \quad . \quad . \quad (169')$

For $k = \frac{1}{600}$, $\quad l_1 = 0.035\,V^2D. \quad . \quad . \quad . \quad . \quad . \quad . \quad . \quad (169'')$

For $k = \frac{1}{400}$, $\quad l_1 = 0.023\,V^2D. \quad . \quad . \quad . \quad . \quad . \quad . \quad . \quad (169''')$

The following table gives values of l_1 in feet per degree of circular curve for a few values of V and k.

k	30 Miles per Hour.	35 Miles per Hour	40 Miles per Hour.	45 Miles per Hour.	50 Miles per Hour.	55 Miles per Hour.
$\frac{1}{1200}$	63	86	112	142	175	212
$\frac{1}{600}$	32	43	56	71	83	106
$\frac{1}{400}$	21	29	37	47	58	71

When only a short tangent intervenes between two curves shorter transition-curves must be taken, requiring larger values of k, so that overlapping may be prevented.

For illustration suppose a 5° curve to be eased off with a transition-curve, the highest train-speed being 45 miles per hour and $k = \frac{1}{600}$. By the table the value of l_1 will be $71 \times 5 = 355$ feet, so that we should probably take a 360-ft. transition-curve, requiring an offset of 4.7 feet by Table XVI.

ARTICLE 12 —FIELD-WORK.

A Field Formulas.

148. For the cases most frequently presenting themselves in practice the foregoing formulas may be simplified so as to admit of the rapid location of points on the transition-curve with all the accuracy needed on location, though it is best to use the exact formulas and tables in setting track-centers on the finished roadbed. When the transition-curve angle is quite large it will be better to use the accurate methods on location also, but for the more common cases the following formulas will answer.

149. Simplified Formulas.—In (139) and (140) neglect, as small, all the terms following the first, giving

$$y = \frac{ml^3}{3} = \frac{\phi}{3}l = .005818l\phi^\circ. \qquad (170)$$

In (141) and (142) retain only the first two terms

$$x = l\left(1 - \frac{m^2l^4}{10}\right) = l\left(1 - \frac{\phi^2}{10}\right) = l - .00003l\phi^{\circ 2}, \qquad (171)$$

in which the last term is small for short transition-curves and may often be neglected, x being taken equal to l.

The values of m and I_1 remain as before:

$$m = \frac{1}{2Rl_1} = \frac{D}{11460l_1}, \qquad (144)$$

$$I_1^\circ = 28.65\frac{l_1}{R} = \frac{l_1D}{200}. \qquad (146)$$

In (147) expand cos I_1, giving

$$F = y_1 - R\left(\frac{I_1^2}{2} - \frac{I_1^4}{24} + \frac{I_1^6}{720} - \dots\right).$$

But $R = \frac{l_1}{2I_1}$, by (145). Substitute this for R and neglect all but the first two terms:

$$F = y_1 - \frac{l_1 I_1}{4}.$$

But $l_1 I_1 = 3y_1$, by (170), since $\phi = I_1$ and $y = y_1$ when $l = l_1$; hence

$$F = y_1 - \tfrac{3}{4}y_1 = \tfrac{1}{4}y_1 \quad \dots \dots \quad (172)$$

Likewise expanding sin I_1 in (148),

$$x' = x_1 - R\left(I_1 - \frac{I_1^3}{6} + \frac{I_1^5}{120} - \dots\right);$$

and writing $R = \frac{l_1}{2I_1}$ as above,

$$x' = x_1 - \frac{l_1}{2} + \frac{l_1 I_1^2}{12} - \dots$$

But $x_1 = l_1 - \frac{l_1 I_1^2}{10}$, by (171).

$$\therefore\ x' = l_1 - \frac{l_1 I_1^2}{10} + \frac{l_1 I_1^2}{12} - \frac{l_1}{2} = \frac{l_1}{2} \text{ (nearly)}. \quad \dots \quad (173)$$

By (170),

$$y' = \frac{m}{3}\left(\frac{l_1}{2}\right)^3 = \frac{m}{3} \times \frac{l_1^3}{8} = \frac{y_1}{8} = \frac{F}{2}. \quad \dots \dots \quad (174)$$

In (154), (155), (156), (157), (158), and (159) neglect the correction; then

$$(\delta_0^\circ) = \frac{I_1^\circ}{3} A_0, \quad \dots \dots \quad (175)$$

$$(\delta_{\frac{1}{4}}^\circ) = \frac{I_1^\circ}{3} A_{\frac{1}{4}}, \quad \dots \dots \quad (176)$$

$$(\delta_{\frac{1}{2}}{}^\circ) = \frac{I_1{}^\circ}{3} A_{\frac{1}{2}}, \quad \ldots \ldots \quad (177)$$

$$(\delta_{\frac{3}{4}}{}^\circ) = \frac{I_1{}^\circ}{3} A_{\frac{3}{4}}, \quad \ldots \ldots \quad (178)$$

$$(\delta_4{}^\circ) = \frac{I_1{}^\circ}{3} A_1. \quad \ldots \ldots \quad (179)$$

$$(\delta_c{}^\circ) = \frac{I_1{}^\circ}{3} A_c. \quad \ldots \ldots \quad (180)$$

150. Offsets.—Formula (170) shows that offsets from transition-curve to tangent vary as the cube of the distance from the *P.T.C.*, and it can be shown that offsets from the circular curve to transition-curve follow the same law, reckoning from the *P.C.*$_1$.

Formula (36) may be written

$$z = f(l^2 D), \quad \ldots \ldots \quad (a)$$

in which D is the degree of curve if offset is from tangent, and the difference of degrees if offset is between two curves having a common point of tangency, l being reckoned from the tangent-point.

From (136) and (137),

$$\frac{dl}{d\phi} = \rho = \frac{1}{2ml},$$

and the degree of transition-curve at any point is

$$D_t = \frac{5730}{\rho} = 11460ml = cl. \quad \ldots \ldots \quad (181)$$

Formula (181) shows that the degree of curvature of transition-curve at any point is a function of its length. If the D in (a) is the difference between degrees of circular and transition curves, it will equal $D_1 - D_t$, which is also a function of the length; so in (a) write $D = f(l)$, giving

$$z = f'(l^3), \quad \ldots \ldots \quad (182)$$

which shows that the offset between circular and transition curves varies as the cube of the distance from *P.C.*$_1$. The offset at the *P.C.* is known, being half of F, and may therefore be found for other points; thus midway between $P.C.$ and $P.C._1$ it will be one eighth of its value at $P.C.$, or $\frac{1}{16}F$

151. Compound Curves.—By trial it has been found that e_2 —see formulas (168) and (168′)—equals $e + e'$ from Table XVI when the table is entered with $D = D_1 - D_2$ and l_2 as arguments, up to about $D_1 l_1 = 8000$, which covers all cases in ordinary railroad practice; so we write

$$e_2 = e + e'. \quad . \quad . \quad . \quad . \quad . \quad . \quad (183)$$

The distance AB on sharper curve is found by trial to equal $\frac{1}{2}l_2$ up to about $D_1 l_1 = 4000$, which answers for the ordinary cases arising in practice. When $D_1 l_1$ is greater than this AB (of Fig. 69) must be found from (166).

The point N can be taken midway between C and B, for the radius of transition-curve decreases uniformly from R_2 to R_1 and may be here taken as their mean; hence the offset $BN = NC = \frac{1}{2}F_2$. Other offsets may be found from the relation given by (182). Thus the offset midway between A and B will be

$$z = \frac{F_2}{2}\left(\frac{\frac{1}{4}l_2}{\frac{1}{2}l_2}\right)^3 = \frac{F_2}{16}.$$

Other offsets may be obtained if desired.

B. Setting Out Transition-curves.

152. In first locating the line it will be sufficient to simply offset the curve at the $P.C.$ the amount required for the transition-curve; then, with the transit over this point, bring the telescope parallel to the tangent from which offset was taken and run the circular curve to the $P.T.$, where another offset is made and a tangent parallel to the terminal tangent of circular curve can be run out. The amount to offset will be governed by the length of transition-curve, or if the offset is fixed it governs the length of curve. Either l_1 or F being given, the other may be taken from Table XVI.

153. Location by Offsets.—If the offset is given, l_1 can be taken from Table XVI, interpolating if necessary. At the $P.C.$ bisect the offset, and set a stake at that point; then measure back along tangent the distance x', which for most cases may be taken as $\frac{1}{2}l_1$

(see formula (173)), and set a stake, marking it $P.T.C.$ From the $P.C.$ measure forward around the circular curve a distance equal to $\frac{I_1^\circ}{D}$, which approximately equals $\frac{1}{2}l_1$. Set a stake marked $P.C._1$. At the quarter-point offset from tangent an amount equal to $\frac{1}{16}F$, for, by (170), the offsets are proportional to the cube of the distance from $P.T.C.$, so that

$$y = \frac{F(\frac{1}{4}l_1)^3}{2(\frac{1}{2}l_1)^3} = \frac{1}{16}F.$$

At the three-quarter point offset the same amount from circular curve. If the transition-curve is not over 400 feet long, these are all the points need; if longer, other offsets are similarly found.

EXAMPLE.—At sta. 412 an offset of 4.2 feet was made from a tangent to a 5° curve. Required the data for a transition-curve to connect tangent and circular curve.

By Table XVI it is seen that a 340-ft. transition-curve is required. From the table it is seen that $x' = 169.9$ ft., $I_1^\circ = 8.5°$, and excess of curve over tangent is .02 ft., which we neglect as small. Drive a stake 2.1 ft. from offset hub and mark it 412; measure back along tangent 169.9 ft. to 410 + 30.1, and drive a stake marked $P.T.C.$ Measure forward around circular curve $\frac{8.5}{5} = 1.70$ chains $= 170$ ft., and set a stake marked $P.C._1$ at sta. 413 + 70.1

The approximate offsets are:

At mid-point,	sta. 412,	$t = \frac{4.2}{2}$	$= 2.1$
" one-eighth points,	stas. 410 + 72.6, 413 + 27.6,	$t = 2.1 \times (\frac{1}{4})^3$	$= 0.033$
" quarter-points,	stas. 411 + 15.1, 412 + 85.1,	$t = 0.033 \times 2^3$	$= 0.26$
" three-eighths points,	stas. 411 + 57.6, 412 + 42.6,	$t = 0.033 \times 3^3$	$= 0.89$

Stakes at the one-eighth and three-eighths points were not needed, but were worked out for illustration.

154. Location by Deflections.—The number of chord-lengths being taken as an aliquot part of 20, the deflection-angles for the transit at any one of five positions may be taken from Table XV by multiplying the tabular values of A by $\frac{I_1^\circ}{3}$, I_1° being found from Table XVI or formula (146). If the number of chords is not an aliquot part of 20, or if the transit is at some point other than one of the five for which Table XV was calculated, then the deflection-angles must be computed by (153). The curve is then run out in the usual way.

When I_1 is not more than 15 or 20 degrees the curve may be run from the $P.T.C.$ or $P.T.C._1$ by neglecting the correction B as small. Even when I_1 is greater than 20° the correction may be neglected, provided half the transition-curve is run from the $P.T.C.$ and the remainder with the transit at the mid-point, the telescope being first placed parallel to original tangent.

EXAMPLE.—Take the example of the last section: $l_1 = 340$ ft., $F = 4.2$ ft., $D = 5°$. By formula (146), $I_1 = \frac{340 \times 5}{200} = 8.5°$, the same as given by Table XVI. Then $\frac{I_1^\circ}{3} = 2.833°$. Divide l_1 into 5 parts of 68 ft. each, which will be the chord-length to be used. From Table XV for transit at $P.T.C.$ the deflections will be:

For sta. 410 + 30.1, $P.T.C.$, $(\delta_0^\circ)_0 = 0.$
" " 410 + 98.1, $(\delta_0^\circ)_{.2} = 2.833 \times .04 = 0.1133 = 0°\ 6.8'$.
" " 411 + 66.1, $(\delta_0^\circ)_{.4} = 2.833 \times .16 = 0.4533 = 0°\ 27.2'$.
" " 412 + 34.1, $(\delta_0^\circ)_{.6} = 2.833 \times .36 = 1°\ 1.2'$.
" " 413 + 02.1, $(\delta_0^\circ)_{.8} = 2.833 \times .64 = 1°\ 48.8'$
" " 413 + 70.1, $(\delta_0^\circ)_1 = 2.833 \times 1 = 2°\ 50'$.

Having set out the transition-curve, move to $P.C._1$ at sta. 413 + 70.1, backsight to $P.T.C._1$, and deflect $I_1^\circ - (\delta_0^\circ)_1 = 8°\ 30' - 2°\ 50 = 5°\ 40'$, and run out the circular curve to the $P.T.C._1$, which suppose to fall at sta. 420. Set the transit at this point, and cause the vernier to read zero when the telescope is in tangent to circular curve. The deflections taken from Table XV will now be:

For sta. 420 + 68, $(\delta_c^\circ)_{.8} = 2.833 \times .56 = 1°\ 35.2'$.
" " 421 + 36, $(\delta_c^\circ)_{.6} = 2.833 \times 1.04 = 2°\ 56.8'$.
" " 422 + 04, $(\delta_c^\circ)_{.4} = 2.833 \times 1.44 = 4°\ 4.8'$.
" " 422 + 72, $(\delta_c^\circ)_{.2} = 2.833 \times 1.76 = 4°\ 59.2'$.
" " 423 + 40, $(\delta_c^\circ)_0 = 2.833 \times 2 = 5°\ 40'$.

Set transit at 423 + 40, the $P.T._1$, backsight to 420 and deflect 8° 30′ − 5° 40′ = 2° 50′, when the telescope will be in tangent.

155. Form of Transit Notes.—The following will illustrate a form of notes that will be found to answer.

Let the *P.C.* of a 4° curve be at sta. 160 + 50, and a 200-ft. transition-curve be employed. Let the intersection-angle I be 20°. By Table XVI, $F = 1.16$ ft., $I_1° = 4°$, $x' = 100$ ft., so that *P.T.C.* is at 159 + 50. Take four 50-ft. stations on transition-curve and determine the deflection-angles as in the last section.

Sta.		Deflection-angle.	Central angle.	Calculated Course.	Magnetic Course.	Remarks.
167						
+50 ⊙	$P.T._1$	2° 40′	4° 0′			
166		2° 15′				
+50		1° 40′				
165		0° 55′				
+50 ⊙	$P.T.C._1$	6° 0′	12° 0′			
164		5° 0′				
163		3° 0′				
162		1° 0′				
+50 ⊙	$P.C._1$, 4 C.L.	1° 20′	4° 0′			Set ver. at 2° 40′,
161		0° 45′				B.S. to 159 + 50, and
+50		0° 20′				deflect to 0°. Run
160		0° 5′				circular curve.
+50 ⊙	*P.T.C.*	0° 0′				$l_1 = 200$. $F = 1.16$.
159						
158						

The length of circular curve was $\frac{20 - 2 \times 4}{4} = 3$ stations, since the central angle was 12°. With transit at 161 + 50 set the vernier to 2° 40′, backsight to 159 + 50, and deflect into tangent with the vernier reading zero. With the transit at 164 + 50 cause the vernier to read zero when the transit is in the tangent to circular curve, and run the last transition-curve by deflections from this tangent. With the transit at 166 + 50 backsight to 164 + 50 and deflect 4° − 2° 40′ = 1° 20′, when the telescope will be in tangent and the line may be continued.

ARTICLE 13. TRANSITION-CURVE PROBLEMS.

156. To Find the Tangent Distance and External when Transition-curves are Employed, Offsets Equal.

In Fig. 70 let AB be the circular curve, EF and GH the transition-curves. Let $EK = HK = T_1$ be the tangent distances, and $NK = E_1$ the external required. Let $LK = T'$. Draw PV perpendicular to LK; then in triangle PVK, $VK = F \tan \frac{1}{2}I$; $LK = AP + VK$ is now known, or

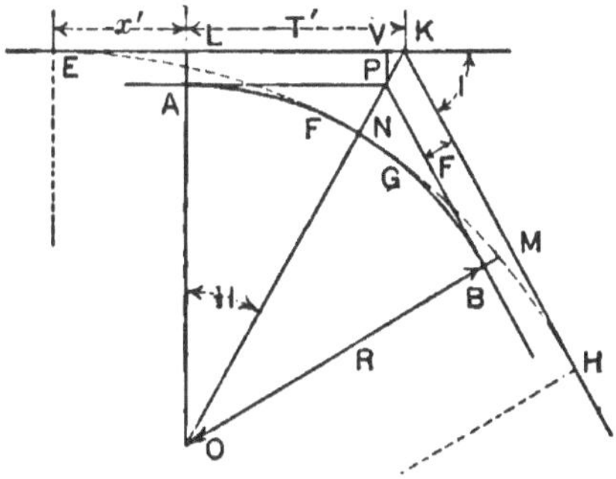

FIG. 70.

$$T' = T + F \tan \tfrac{1}{2}I. \quad . \quad . \quad . \quad . \quad (184)$$

Hence

$$T_1 = x' + T' = x' + T + F \tan \tfrac{1}{2}I. \quad . \quad . \quad . \quad . \quad (185)$$

In triangle PVK, $PK = F \sec \frac{1}{2}I$, so that, letting $PN = E$,

$$E_1 = E + F \sec \tfrac{1}{2}I. \quad . \quad . \quad . \quad . \quad (186)$$

EXAMPLE.—Two tangents intersect at sta. 91 + 37 8; required the tangent and external when $F = 2.62$ ft., $I = 26° 30'$, $D = 4°$. By Table XVI, $l_1 = 300$, $x' = 149.9$. From Table IX, $T = \frac{1349.2}{4} = 337.3$. Then, by (185),

$$T_1 = 149.9 + 337.3 + 2.62 \tan 13° 15' = 487.8 \text{ ft.}$$

The station number of $P.T.C.$ will now be 91.378 − 4.878 = 86 + 50.

By (186),

$$E_1 = \frac{156.7}{4} + 2.62 \sec 13° 15' = 41.87 \text{ ft.}$$

Table XVI gives $I_1° = 6°$; hence the circular curve will cover 26° 30′ − 2 × 6° = 14° 30′, or 3.625 stations, so that the number of the $P.T._1$ will be 86.50 + (2 × 3.00 + 3.625) = 96 + 12.5.

157. Tangent Distance, Offsets Unequal.

In Fig. 71, O, N, and K do not lie in the same straight line.

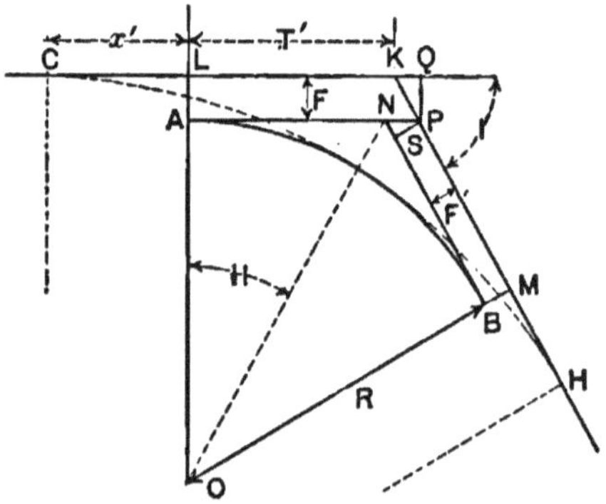

FIG. 71.

Draw PS perpendicular to NB, PQ perpendicular to LK. Let $LA = F$, $MB = F'$.

$$T' = LK = AN + NP - KQ,$$

or

$$T' = T + F' \operatorname{cosec} I - F \cot I; \quad . \quad . \quad . \quad . \quad . \quad (187)$$

$$T_1 = x' + T' = x' + T + F' \operatorname{cosec} I - F \cot I; \quad (188)$$

$$T'' = MK = T - F' \cot I + F \operatorname{cosec} I; \quad . \quad . \quad . \quad (189)$$

$$T_2 = x'' + T - F' \cot I + F \operatorname{cosec} I. \quad . \quad . \quad . \quad (190)$$

EXAMPLE.—Two tangents intersect at sta. 820 and are to be united by a 6° curve having $F = 1.75$, $F' = 2.95$, and $I = 31° 48'$.

By Table IX, $T = \dfrac{1632.3}{6} = 272.05$ ft.

By Table XVI, $l_1 = 200$, $l_2 = 260$, $x' = 100$, $x'' = 129.9$.

By (188),

$$T_1 = 100 + 272.05 + 2.95 \times \operatorname{cosec} 31° 48' - 1.75 \cot 31° 48' = 374.8.$$

By (190),

$$T_2 = 129.9 + 272.05 - 2.95 \cot 31° 48' + 1.75 \operatorname{cosec} 31° 48' = 400.5.$$

158. To Insert Transition-curves without Changing the Position of the Vertex, *B*

In Fig. 72, *ABC* is the located curve, *FGHK* the curve after

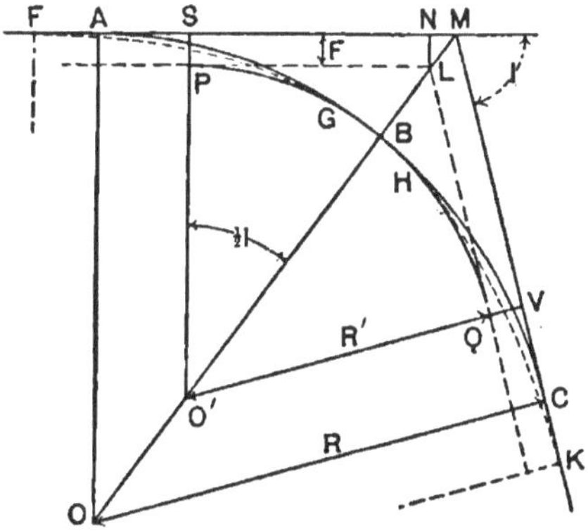

FIG. 72.

inserting transition-curve. The radius of the circular portion has been changed from R to R' in order to make room for the offset $PS = F$. $BM = E$ is the external to located curve, $BL = E'$ the external to circular curve having radius R' and central angle I. In the triangle LNM, $LM = LN \sec \frac{1}{2}I = F \sec \frac{1}{2}I$; hence

$$E' = E - F \sec \tfrac{1}{2}I. \quad . \quad . \quad . \quad . \quad . \quad . \quad . \quad (191)$$

E may be found by (24) or by means of Table IX; then E' becomes known, and from the same table D' is found by dividing the tabular E by E'. D' will be larger than D.

It is sometimes more convenient to assume D' and calculate E' in the same manner as E; then, from (191),

$$F = (E - E') \cos \tfrac{1}{2}I. \quad . \quad . \quad . \quad . \quad . \quad . \quad . \quad (192)$$

If this value of F is too large or too small for the conditions of the problem, a new D' can be assumed and F recomputed.

EXAMPLE.—The *P.C.* of a 5° curve is at sta. 182, and angle $I = 40°$. Compute the data for a new curve to allow for a transition-curve with 1.5 ft. offset.

From Table IX, $E_1 = 367.7$ for $I = 40°$; therefore

$$E = \frac{367.7}{5} = 73.54, \quad F \sec 20° = 1.5 \times 1.0642 = 1.6;$$

then, by (191),

$$E' = 73.54 - 1.60 = 71.94,$$

and

$$D' = \frac{367.7}{71.94} = 5.1113° = 5° \, 6.678', \text{ say } 5° \, 7'.$$

By Table XVI, for $l_1 = 200$, $D = 5° \, 7'$

$$F = 1.45 + \frac{7}{60}(1.75 - 1.45) = 1.485.$$

For $l_1 = 220$,

$$F = 1.76 + \frac{7}{60}(2.11 - 1.76) = 1.842.$$

Then for $F = 1.5$, $D = 5° \, 7'$,

$$l_1 = 200 + 20 \, \frac{1.5 - 1.485}{1.842 - 1.485} = 200.8 \text{ and } x' = 100.4.$$

By (146), $$I_1° = \frac{l_1 D}{200} = \frac{200.8 \times 5.117}{200} = 5.13° = 5° \, 8'.$$

The central angle for circular portion of curve is $40 - 2 \times 5.13 = 29.74°$, equivalent to 581.2 feet around curve.

In Fig. 72, B is at sta. 186 on the 5° curve, and arc $BG = 290.6$ ft. on the 5° 7′ curve. The *P.C.*$_1$ is at $186 - 2.906 =$ sta. $183 + 09.4$, the *P.T.C.* at $183.094 - 2.008 =$ sta. $181 + 08.6$, the *P.T.C.*$_1$ at $188 + 90.6$, and the *P.T.*$_1$ at $190 + 91.4$.

Had D' been assumed equal to 5° 6′ or 5.1° to begin with, we should have had $E' = \dfrac{367.7}{5.1} = 72.10$; then, by (192),

$$F = 1.44 \times .93969 = 1.35 \text{ ft.}$$

l_1 may be found by interpolation from Table XVI as above.

159. To Insert Transition-curves on an Existing Road-bed with the Least Deviation from Old Track.

To satisfy this condition the new track should pass about as far outside the old at the vertex as it does inside at the original *P.C.*; that is, about $\frac{F}{2}$. We shall now have

$$E' = E - F \sec \frac{I}{2} - \frac{F}{2}. \quad . \quad . \quad . \quad . \quad . \quad (193)$$

The remainder of the problem may be solved by **158.**

Transition-curves may be inserted in old track by shifting to suit the existing road-bed, thus adding materially to the safety and easy riding of cars.

160. To Insert Transition-curves at the Ends of a Long Circular Curve without Moving the Central Portion.

In Fig. 73, AC is the circular curve. In order to make room for the offset F the ends must be sharpened by compounding. Let C be the point of compounding, R' the radius of the branch CN, $HN = KB = F$. Let BEG be the transition-curve; the

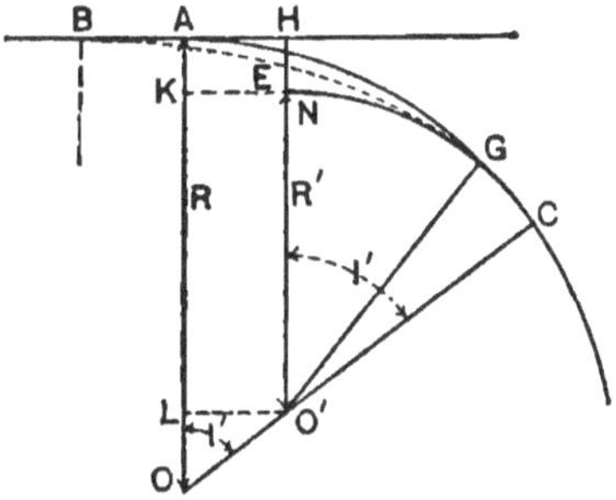

FIG. 73.

closer G comes to C the better, provided the change in radius at C is kept within certain limits. The difference in degrees between the original and the sharpened curve should never exceed 2° and may usually be kept in the neighborhood of 1°.

FIRST METHOD.—Having decided upon the value of F, assume R' so that $D' - D$ is not greater than 2°. Draw $O'L$ parallel to BH; $OO' = R - R'$, and $\cos I' = \frac{OL}{OO'}$, or

$$\cos I' = \frac{R - (R' + F)}{R - R'} = 1 - \frac{F}{R - R'}. \quad . \quad (194)$$

This is the same as (69) in **122**. I' being known, set the transit at C, run out the curve CN, and insert transition-curve in the usual way.

If I' had been assumed in the beginning, R' could be found from (194).

SECOND METHOD.—When the circular curve is flat, and short transition-curves are employed, we may compound the transition-curve with the circular at the $P.C._1$, taking care that the difference of curvatures is not greater than 1° or 2°.

Assume the position of the $P.C._1$ from 100 to 200 feet from the $P.C.$; measure the perpendicular let fall from the $P.C._1$ upon the tangent at the $P.C.$ produced; this will be y_1. The central angle I_1 can be calculated, knowing the length of circular curve from the $P.C.$ to the assumed $P.C._1$, or the angle between tangents may be measured with the transit. The coefficients C and E of (140) and (142) may be found from Table XV with $I_1 = \phi$ as argument; then, from (140) and (142),

$$l_1 = \frac{y_1}{C}, \quad . \quad . \quad . \quad . \quad . \quad . \quad . \quad . \quad . \quad (195)$$

$$x_1 = l_1(1 - E). \quad . \quad . \quad . \quad . \quad . \quad . \quad . \quad (196)$$

Measure back from the foot of the perpendicular let fall from the $P.C._1$ a distance x_1 along tangent, and set the $P.T.C.$

Intermediate points can be located, if needed, by offsets from tangent, computed by (140) or (170); thus at the mid-point the offset is $\frac{1}{16}y_1$.

THIRD METHOD.—From formula (170),

$$l_1 = \frac{y_1}{.005818 I_1^\circ},$$

and, from (36), $\qquad y_1 = \frac{7}{8}n^2 D.$

Therefore $\qquad l_1 = \frac{.875 n^2 D}{.005818 I_1^\circ} = 150\frac{n^2 D}{I_1^\circ}$, nearly.

But $nD = I_1°$; hence

$$l_1 = 150n; \quad . \quad . \quad . \quad . \quad . \quad . \quad . \quad . \quad (197)$$

and as $100n$ is the length of circular curve from $P.C.$ to $P.C._1$, *l_1 is once and a half as great.*

From (146),

$$D' = \frac{200 I_1°}{l_1} = \frac{200 I_1°}{150n} = \frac{200nD}{150n} = \frac{4}{3}D. \quad . \quad . \quad (198)$$

From this equation it is seen that if the break in curvatures is limited to 2°, this method is admissible up to $D = 6°$, independent of the length of transition-curve.

EXAMPLE.—A 4° curve is to have transition-curves inserted at each end; compute the necessary data.

BY FIRST METHOD —Assume a 1.45-ft. offset, and the curvature to be changed from 4° to 5° by compounding. In Table I find $R = 1432.7$, $R' = 1146.3$; then, by (194),

$$\cos I' = 1 - \frac{1.45}{286.4} = .99494 = \cos 5°\ 46'.$$

The length of 5° curve is $\frac{5.767}{5} = 1.153$ stations, and, by Table XVI, $I_1° = 5°$, so that the $P.C._1$ will fall 15.3 ft. back of the $P.C.C.$, while the $P.C.$ will be moved forward $\frac{5.767}{4} - 1.153 =$ .289 stations or 28.9 ft.; the $P.T.C.$ being, by Table XVI, 100 ft. back of the new $P.C.$ will fall $100 - 28.9 = 71.1$ ft. back of old $P.C.$ The transition-curve may now be located in the usual manner.

BY SECOND METHOD.—Assume the $P.C._1$ to fall 150 ft. from the $P.C.$, making $I_1° = 1.5 \times 4 = 6°$. From Table XV, $C =$ .03488, and, by (36), $y_1 = \frac{7}{8}(1.5)^2 \times 4 = 7.875$ ft.

By (195),

$$l_1 = \frac{7.875}{.03488} = 225.8.$$

Now, by (146),

$$6 = \frac{D' \times 225.8}{200},$$

from which $D' = 5.314° = 5° 18.8'$, which differs less than 2° from D

By (196),

$$x_1 = 225.8(1 - .0011) = 225.6 \text{ ft.}$$

To find the position of $P.T.C.$ with reference to the old $P.C.$ consider that the distance from $P.C.$ to foot of perpendicular from the $P.C._1$ is half the chord for angle $2I_1$, and can be taken from Table IX, being equal to $\frac{1}{2} \times \frac{1197.9}{4} = 149.7$. Then $225.6 - 149.7 = 75.9$ feet is the distance from old $P.C.$ back to $P.T.C.$

BY THIRD METHOD.—Assume the $P.C._1$ to be 150 ft. from the old $P.C.$; then, by (197), $l_1 = 225$ ft., and, by (198), the curvature of transition-curve at the $P.C._1$ is $\frac{4}{3} \times 4° = 5° 20'$, giving almost the same results as by the second method. Had we taken the $P.C._1$ 160 ft. from $P.C.$ we should have had $l_1 = 240$, $D' = 5° 20'$; $x_1 = 239.7$, by interpolation from Table XVI; the length along tangent from $P.C.$ to foot of perpendicular from $P.C._1$ 159.9 ft., and therefore $239.7 - 159.9 = 79.8$ ft. as the distance from $P.C.$ to $P.T.C.$

161. To Insert Transition-curves at the *P.C.* and *P.C.C.* of a Compound Curve by Changing the Curvatures of the First Branch.

In Fig. 74 let ABV be the located curve compounding at B. Two cases occur.

FIRST CASE.—*Second branch having shorter radius.*

The offset at $P.C.C.$ must be to outside of located curves; let it be $EB = F_2$ in the figure. Let $CP = F$ be known or assumed.

Draw the tangent BG, and draw EH parallel thereto. Let CE be the changed curve, and CQ parallel to tangent AH. Angle I may be computed from the known station numbers of A and B, or may be measured on the ground. The new tangent distance is $EQ = BG - GK - HQ$ (or LS). From the right triangle GHK, $GK = HK \tan GHK = F_2 \cot I$.

Similarly, $LS = LW \operatorname{cosec} I = F \operatorname{cosec} I$. Therefore

$$T' = EQ = T - F_2 \cot I - F \operatorname{cosec} I. \quad . \quad . \quad (199)$$

T can be found from Table IX or formula (14); then T' is known from (199). The degree of new curve, D', may now be found by means of Table IX, or from Table I by first finding

R' by (15) The transition curve at the $P.C.C.$ may be located by **146** and **151**, while that at the $P.C.$ may be located either by offsets or deflections.

Second Case.—*Second branch having longer radius.*

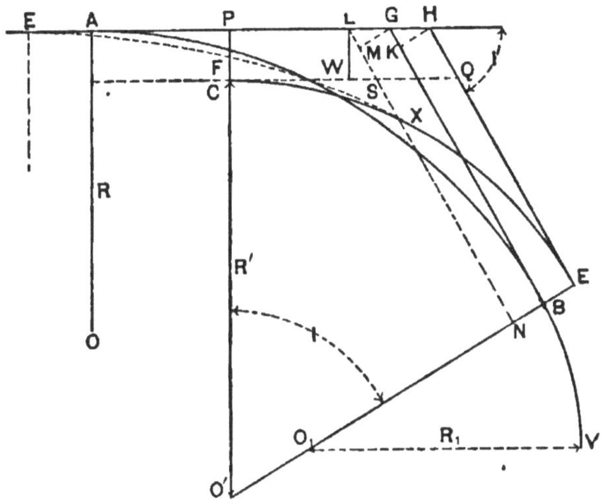

Fig. 74.

In this case the offset must be to inside of curve, and NS is the tangent required. From the figure, letting $NB = F_2$, $NS = T'$,

$$T' = T + F_2 \cot I - F \operatorname{cosec} I. \quad . \quad . \quad . \quad (200)$$

The remainder of the solution is the same as for first case.

Example.—A 5° curve compounds at sta. 280 with a 9° curve; the $P.C.$ is at sta. 272. Required the change in curvature of first branch for an offset of 1.50 ft. at $P.C.C.$ and 2.00 ft. at $P.C.$

Here $I = 8 \times 5° = 40°$, and, by (199),

$$T' = 417.1 - 1.5 \times 1.19175 - 2 \times 1.5557 = 412.2$$

$$D' = \frac{2085.5}{412.2} = 5.06° - 5° \ 3.6'.$$

$D_2 - D' = 9 - 5.06 = 3.94°$ is the difference in curvatures of the two branches of the altered curve. Entering Table XVI with this value for D and $F = 1.5$ ft., we find:

for $l_1 = 220$, $F = 1.06 + .94(1.41 - 1.06) = 1.39$;
" $l_1 = 240$, $F = 1.26 + .94(1.67 - 1.26) = 1.65$.

$$\therefore \text{ for } F = 1.5,\ D = 3.94°,\quad l_1 = 220 + 20\frac{1.5 - 1.39}{1.65 - 1.39} = 228.5$$

Bisect the offset at *P.C.C.*, measure 114.25 ft. along each curve, and set the ends of transition-curve. Midway between these points and *P.C.C.* offset $\frac{1}{16} \times 1.5 = 0.1$ ft.; these are all the points needed.

The length of transition-curve at *P.C.* may be found in like manner, taking $D = 5.06°$ and $F = 2.0$ ft. as arguments in interpolating in Table XVI.

162. To Insert Transition-curves at the Ends of Two Circular Curves of Contrary Flexure united by a Common Tangent. In Fig. 75 let the located line be *ABCE*; the tangent *BC* must be shifted outward at *B* and *C* to the position *HG*, the relative size of offsets being determined by the nature of the ground. The points *B′* and *C′* at which the tangents to circular curves

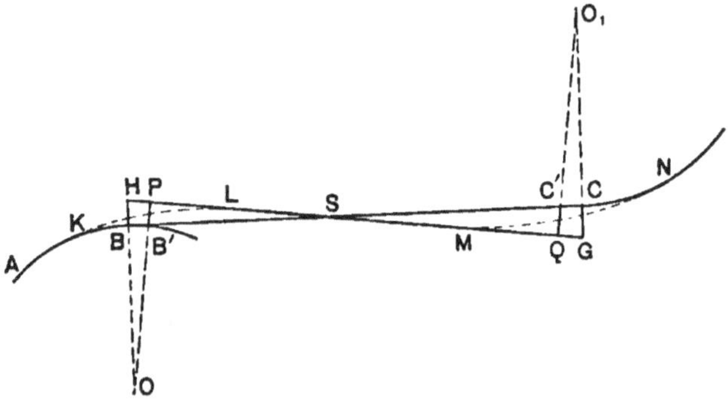

FIG. 75.

will be parallel to *HG* will each move towards *S* a distance due to the increase of central angle, which increase equals $BSH = a$, for which we have

$$\tan a = \frac{BH + CG}{BC}. \quad . \quad . \quad . \quad . \quad . \quad (201)$$

Let the offset at *B′* be *F*, and at *C′*, *F′*. Then

$$F = (R + BH)\cos a - R, \quad . \quad . \quad . \quad . \quad . \quad (202)$$
$$F' = (R' + CG)\cos a - R'. \quad . \quad . \quad . \quad . \quad . \quad (203)$$

F and F', being now known, the transition-curves may be located.

EXAMPLE.—A 6° curve and a 4° curve are united by a tangent 540 ft. long; BH for 6° curve = 4.5 ft.; CG for 4° curve = 3 ft.; B is at sta. 180, C at 185 + 40. Find F and F'.

By (201), $\tan a = \frac{7.5}{540} = .0139 = \tan 0° 48'.$

B will be moved forward $\frac{.8}{6} = .133$ stas. = 13.3 ft. to sta. 180 + 13.3, and C will be moved backwards $\frac{.8}{4} = .2$ stas. or 20 ft. to 185 + 20.

By (202), $F = (955.4 + 4.5)0.99990 - 955.4 = 4.4$ ft.

By (203), $F' = (1432.7 + 3)0.99990 - 1432.7 = 2.86$ ft.

These values call for $l_1 = 317.8$ ft. for 6° curve, and $l_1 = 312.7$ for 4° curve.

REMARK.—It will frequently be found that this problem allows the line to be thrown on better ground. Should the ground require tangent to be shifted inward, the curves must be sharpened by compounding to admit of the necessary offsets.

163. Having Run a Tangent which Falls Outside a Located Curve, to Find the Offset F for a Transition-curve Uniting them.

In Fig. 76 let the tangent be AB; CE the located curve. Set transit at some point C, and bring telescope into tangent to curve. Measure CB and move to B, where angle ABC must be measured; or measure CH perpendicular to AB; then

$$\sin a = \frac{CH}{CB}.$$

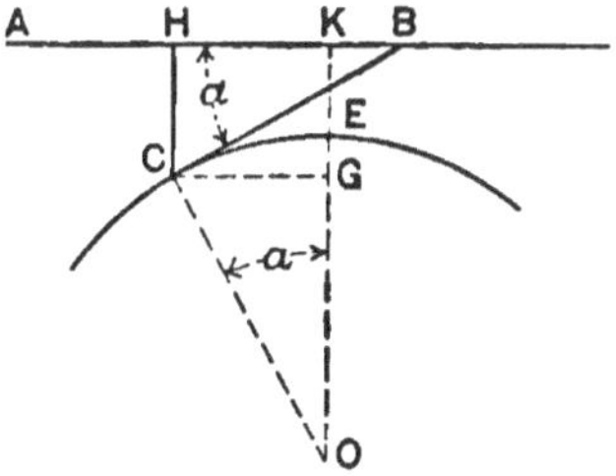

FIG. 76.

Now $EG = R$ vers a; or it is the mid-ordinate for twice a, and may be found from Table IX; then

$$F = CH - EG = CH - R \text{ vers } a. \quad . \quad . \quad . \quad (204)$$

The point E is found from C by the relation $EC = \frac{a}{D}$.

The transition-curve may now be located.

164. Inserting Transition-curves in Old Track.—Sections **159** and **160** afford the means of inserting transition-curves, of which **159** is theoretically the best, though from the amount of track disturbed it may be better to employ **160**. Sometimes the method of **162** may be employed to advantage when the connecting tangent is short. For easing the curves at point of compounding, the method of **161** may be made use of.

The offsets must necessarily be small if the new track is required to occupy the old road-bed. It may be profitable to add to the road-bed when sufficient offset cannot be secured for sharp curves, though ordinarily much good can be accomplished even when the new track is restricted to the old road-bed.

Unless the theoretical *P.C.*, *P.C.C.*, and *P.T.* have been marked by monuments it may be difficult to retrace the old lines. If there is plenty of room, the terminal tangents may be prolonged to intersection and *I* measured, after which the degree of curve may be found by measuring around curve and by approximate measurements of the tangent distances; then one or two assumptions and computations will generally suffice.

In cuts and rough country the curve may be run out by setting transit in center of road-bed and measuring the deflection-angles for a few points around the curve.

After the transition-curves have been inserted permanent monuments should be placed at each end of transition-curve to guide the trackman in keeping up the proper superelevation of outer rail.

165. Remarks on Tabular Interpolations.—The general interpolation formula given in algebra is

$$t = a + pd_1 + \frac{p(p-1)}{2!}d_2 + \frac{p(p-1)(p-2)}{3!}d_3 + \frac{p(p-1)(p-2)(p-3)}{4!}d_4 + \text{etc.},$$

in which t is any term, a the first term taken, p the number of terms from a to t, d_1 the first from a of the first order of differences, d_2 the first of the second order of differences, etc.

In ordinary linear interpolation all terms after the second are neglected; in interpolating by second differences all after the third, etc.

In Tabl XIV linear interpolation will answer for C and ordinarily for E, though second differences may sometimes be needed for the latter.

In Table XV, A is a quadratic function of n, as shown by formula (153), while B is a cubic function of that portion that has been retained. Hence A should be interpolated by second differences, while theoretically B should be interpolated by third differences; but as B is always quite small, its second and third differences will be too small to affect results, and linear interpolations may be made when any are needed.

In Table XVI linear interpolations will generally suffice, though when F and y are large it may be necessary to use second differences.

The examples of **158** and **161** illustrate the method of interpolating in Table XVI for intermediate values of F and D. Values of F were first found for the given degree of curve and assumed values of l_1, so taken that the true l_1 should be between them. From these assumed values of l_1 and F, taken with the required F, the true l_1 was found by linear interpolation.

As an extreme case suppose F and y_1 wanted for an 18° curve when $l_1 = 408$ feet.

First write a few values of y_1 and F so as to obtain the first and second differences.

l_1	y_1	d_1	d_2	F	d_1	d_2
400	81.43			20.63		
		8.08			2.08	
420	89.51		0.35	22.71		0.10
		8.43			2.18	
440	97.94		0.35	24.89		0.10
		8.78			2.28	
460	106.72			27.17		

By the interpolation formula, when $l_1 = 408$,

$$y_1 = 81.43 + \tfrac{8}{20} \times 8.08 + \frac{\frac{8}{20}(\frac{8}{20} - 1)}{2} \times 0.35 = 84.62,$$

$$F = 20.63 + \tfrac{8}{20} \times 2.08 + \frac{\frac{8}{20}(\frac{8}{20} - 1)}{2} \times 0.10 = 21.45.$$

By linear interpolation, $y_1 = 84.66$, $F = 21.46$.

Again, suppose y_1 to be wanted when $l_1 = 430$. By the formula,

$$y_1 = 81.43 + \tfrac{30}{20} \times 8.08 + \frac{\frac{30}{20}(\frac{30}{20} - 1)}{2} \times 0.35 = 93.68,$$

or

$$y_1 = 89.51 + \tfrac{10}{20} \times 8.43 + \frac{\frac{10}{20}(\frac{10}{20} - 1)}{2} \times 0.35 = 93.68.$$

CHAPTER V.

FROGS AND SWITCHES.

ARTICLE 14. TURNOUTS.

A. Turnouts from Straight Lines.

166. A Turnout is a track used in leaving the main line. A **Frog** is placed at the intersection of main and turnout rails.

a. **The Gauge-line** is taken as coinciding with inside face of rail. In making measurements between tracks the distance between corresponding gauge-lines is what is wanted.

b. **The Gauge** of track is the distance between gauge-lines of the rails of that track.

c. **The Point of Switch** is the point at which the turnout curve begins; for a point switch (split switch) this is at the head-block, while with a stub switch it is the length of the switch-rail back of the head-block, which is at the toe of switch.

d. **The Frog-point** is at the intersection of the gauge-lines of intersecting rails, and lies a few inches in front of the blunt point of frog as manufactured.

The angle formed by the intersecting gauge-lines is the **Frog-angle.**

e. **The Frog-number,** N, is the ratio of the axial length to the width of base of frog.

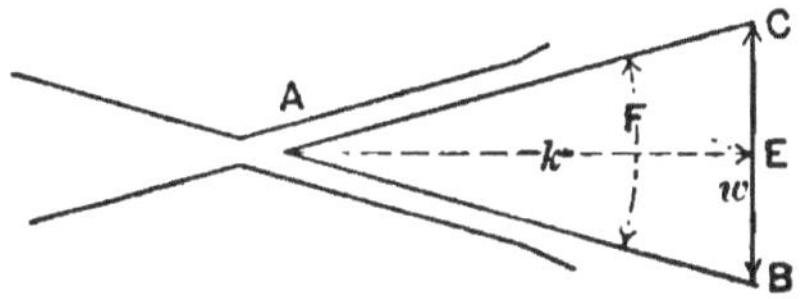

FIG. 77.

In Fig. 77,

$$\frac{AE}{CB} = \frac{k}{w} = N.$$

Letting the frog-angle BAC be F, the figure yields

$$\tan \tfrac{1}{2}F = \frac{\frac{1}{2}w}{k} = \frac{1}{2N}, \quad . \quad . \quad . \quad . \quad . \quad . \quad (205)$$

$$\cot \tfrac{1}{2}F = \frac{k}{\frac{1}{2}w} = 2N. \quad . \quad . \quad . \quad . \quad . \quad . \quad . \quad (206)$$

f. **The Lead,** l, is the distance from point of switch to point of frog, measured along that main rail in which the frog is placed. In Fig. 78, $CB = l$.

g. **The Stub-lead,** $s.l.$, is the distance along main rail from frog-point back to a point where the turnout rail diverges from main rail an amount equal to the throw. In Fig. 78, $KB = s.l. = l -$ length of switch-rail.

h. **The Throw,** t, of switch-rail is the distance the point of a split switch, or toe of stub switch, is moved in opening or closing the switch. A distance of from 5 to 5¾ inches is needed to give necessary clearance for flanges.

k. **The Frog-distance,** $f.d.$, is the length of the chord of outer rail of turnout from the point of a split switch, or toe of stub switch, to the point of frog.

167. Given the Frog-number, N, and the Gauge, g, of a Turnout from a Straight Line, to Find the Lead, l, and Radius, R, of Center Line of Turnout.

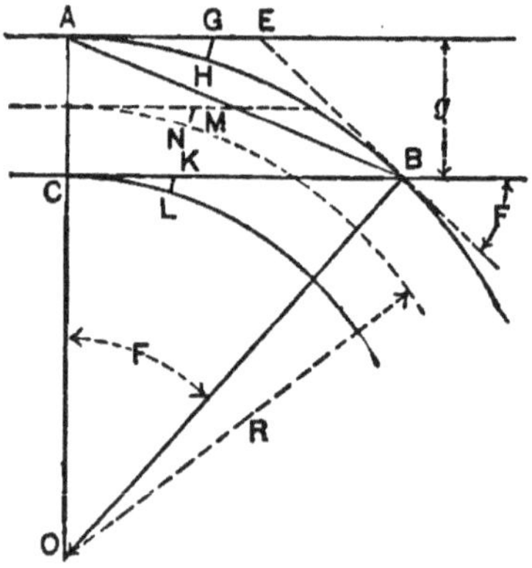

FIG. 78.

In Fig. 78, $AC = g$, $CB = l$, angle $ABC = \frac{1}{2}F$.

From the figure, $l = g \cot \frac{1}{2}F$.

But, (206) $\qquad \cot \frac{1}{2}F = 2N.$

From triangle OBC,

$$(R + \tfrac{1}{2}g)^2 - (R - \tfrac{1}{2}g)^2 = l^2 = 4g^2N^2,$$

whence $$2gR = 4g^2N^2.$$

$$\therefore \quad R = 2gN^2 = lN. \qquad (208)$$

Taking $R = \dfrac{5730}{D}$, inserting in (208), and solving for D,

$$D = \frac{5730}{2gN^2}. \qquad (209)$$

For $g = 4$ ft. $8\frac{1}{2}$ in. these formulas become

$$l = 9.42N \text{ feet}, \qquad (207')$$

$$R = 9.42N^2 \text{ feet}, \qquad (208')$$

$$D = \frac{608}{N^2} \text{ degrees}; \qquad (209')$$

and for $g = 4$ ft. 9 in.,

$$l = 9.5N, \qquad (207'')$$

$$R = 9.5N^2, \qquad (208'')$$

$$D = \frac{603}{N^2}. \qquad (209'')$$

If the frog-distance AB is wanted, we have $AB = \sqrt{l^2 + g^2}$, or

$$f.d. = g\sqrt{4N^2 + 1} = g \operatorname{cosec} \tfrac{1}{2}F. \qquad (210)$$

EXAMPLE.—Find l, R, D, and $f.d.$ for a No. 8 frog and 4.75 feet gauge.

By (207''), $l = 9.5 \times 8 = 76.0$ ft.;

" (208''), $R = 9.5 \times 64 = 608$ ft.;

" (209''), $D = \dfrac{603}{64} = 9° 25'$;

" (210), $f.d. = 4.75\sqrt{257} = 76.14$ ft.

168. Given R (or D) and g, to Find N, l, and F.
From (208) and (209),

$$N = \sqrt{\frac{R}{2g}} = \sqrt{\frac{5730}{2gD}} = \frac{53.52}{\sqrt{gD}}. \quad . \quad . \quad . \quad (211)$$

From (207) and (211),

$$l = 2gN = 2g\sqrt{\frac{R}{2g}} = \sqrt{2gR} = 107\sqrt{\frac{g}{D}}. \quad . \quad (212)$$

From (206) and (211),

$$\cot \tfrac{1}{2}F = 2N = 2\sqrt{\frac{R}{2g}} = \sqrt{\frac{2R}{g}} = \frac{107}{\sqrt{gD}}. \quad . \quad (213)$$

F may also be found from triangle OBC, Fig. 78:

$$\cos F = \frac{R - \frac{1}{2}g}{R + \frac{1}{2}g}. \quad . \quad . \quad . \quad . \quad . \quad (214)$$

169. To Find the Length of Switch-rail, S, when the Frog-number, N, the Throw of Switch, t, and the Gauge, g, are Given.

In Fig. 78, by geometry,

$$HG = \frac{\overline{AG}^2}{2(R + \frac{1}{2}g) + HG}.$$

Neglecting the HG in denominator as small,

$$HG = \frac{\overline{AG}^2}{2(R + \frac{1}{2}g)}.$$

In like manner, $$KL = \frac{\overline{CK}^2}{2(R - \frac{1}{2}g)}$$

Writing $AG = AH = CK = S$, and taking the mean of denominators,

$$t = \frac{S^2}{2R},$$

whence $$S = \sqrt{2Rt} = 2N\sqrt{gt}. \quad . \quad . \quad . \quad . \quad (215)$$

Writing $R = \frac{5730}{D}$,

170. Given the Main Frog-number, N, to Find the Number, N_1, and Lead, l_1, of Crotch-frog for a Turnout from Both Sides of Straight Main Track.

In triangle OCH, Fig. 79, remembering that $R = 2gN^2$,

$$\cos \tfrac{1}{2}F_1 = \frac{R}{R + \frac{1}{2}g} = \frac{4N^2}{4N^2 + 1}. \quad (217)$$

Then, by (206),

$$N_1 = \tfrac{1}{2} \cot \tfrac{1}{2}F_1. \quad . \quad . \quad . \quad (218)$$

From the figure and (205),

$$l_1 = R \tan \tfrac{1}{2}F_1 = \frac{R}{2N_1} = g\frac{N^2}{N_1}; \quad (219)$$

also,

$$l_1 = \sqrt{(R + \tfrac{1}{2}g^2) - R^2} = \sqrt{Rg + \tfrac{1}{4}g^2}$$
$$= g\sqrt{2N^2 + \tfrac{1}{4}}. \quad . \quad . \quad . \quad . \quad (220)$$

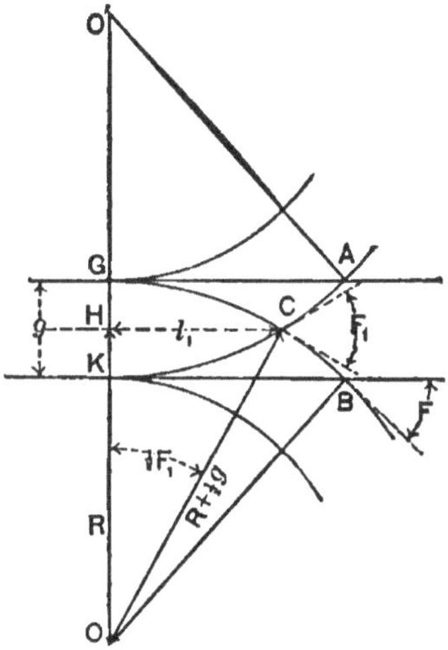

FIG. 79.

Equating these values of l_1 and solving for N_1 gives

$$N_1 = \frac{N^2}{\sqrt{2N^2 + \frac{1}{4}}}. \quad . \quad . \quad . \quad . \quad . \quad (221)$$

If the $\frac{1}{4}$ in denominator be neglected as small compared with $2N^2$, (221) becomes

$$N_1 = \frac{N}{\sqrt{2}} = 0.707N. \quad . \quad . \quad . \quad . \quad . \quad (222)$$

If in (220) we neglect the $\frac{1}{4}$ under radical, there results

$$l_1 = gN\sqrt{2} = 1.414gN = 0.707l. \quad . \quad . \quad . \quad (223)$$

The distance between main and crotch frogs measured along main rail is

$$l - l_1 = 2gN - g\sqrt{2N^2 + \tfrac{1}{4}}, \quad . \quad . \quad . \quad . \quad (224)$$

or, approximately,

$$l - l_1 = 2gN - 1.414gN = 0.586gN = 0.293l. \quad . \quad (225)$$

171. To Find the Radius, R, of Turnout and Lead, l_1, of Crotch-frog in Terms of the Crotch-frog Number, N_1

From (222), $N^2 = 2N_1^2$.

Insert this in (208) and (219), giving

$$R = 2g \cdot 2N_1^2 = 4gN_1^2, \quad . \quad . \quad . \quad . \quad . \quad . \quad (226)$$

$$l_1 = \frac{2gN_1^2}{N_1} = 2gN_1. \quad . \quad . \quad . \quad . \quad . \quad . \quad . \quad (227)$$

REMARK.—In general the frogs kept in stock by manufacturers do not afford suitable combinations of numbers for double turnouts. For instance, the theoretical number of crotch-frog for a number 8 main frog is, by (221) or (222), $N_1 = 5.66$, and we should be compelled to use a number $5\frac{1}{2}$ or 6 for the crotch-frog; this would necessitate a different rate of curvature from crotch to main frog than from head-block to crotch.

172. Given the Numbers of Middle Frog, N_1, and of Main Frogs, N and N', to Find the Radii R_1 from Point of Switch to Crotch-frog, and R and R', from Crotch to Main Frogs.

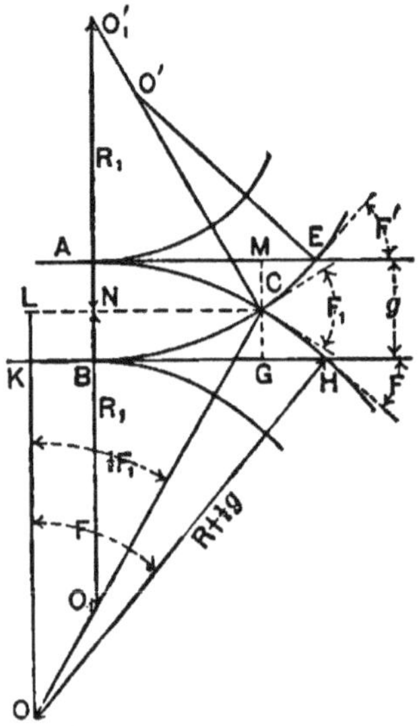

FIG. 80.

In Fig. 80 we have, by (226),

$$O_1N = R_1 = 4gN_1^2,$$

and, by (227),

$$NC = l_1 = 2gN_1.$$

Now if F_1, F, and F' are the angles of the frogs N_1, N, and N', the angle

$$COH = F - \tfrac{1}{2}F_1,$$

and

$$CHG = F - \tfrac{1}{2}(F - \tfrac{1}{2}F_1) = \tfrac{1}{2}(F + \tfrac{1}{2}F_1).$$

Since $CG = \frac{1}{2}g$, the triangle CHG yields

$$GH = \tfrac{1}{2}g \cot \tfrac{1}{2}(F + \tfrac{1}{2}F_1). \quad . \quad (228)$$

But, by trigonometry,

$$\cot(\tfrac{1}{2}F + \tfrac{1}{4}F_1) = \frac{1 - \tan\frac{1}{2}F \cdot \tan\frac{1}{4}F_1}{\tan\frac{1}{2}F + \tan\frac{1}{4}F_1}.$$

Assume $\tan \frac{1}{4}F_1 = \frac{1}{2} \tan \frac{1}{2}F_1$, and write

$$\tan \tfrac{1}{2}F = \frac{1}{2N}, \quad \tfrac{1}{2} \tan \tfrac{1}{2}F_1 = \frac{1}{2} \cdot \frac{1}{2N_1} = \frac{1}{4N_1},$$

and after simplifying and reducing,

$$GH = \frac{2gNN_1}{2N_1 + N} - \frac{g}{4(2N_1 + N)}. \quad . \quad . \quad . \quad (229)$$

The last term is quite small, rarely amounting to as much as one inch, and may be neglected; then

$$GH = \frac{2gNN_1}{2N_1 + N} = \frac{l_1 N}{2N_1 + N} = \frac{lN_1}{2N_1 + N}. \quad . \quad . \quad (230)$$

From the triangles LCO and KHO,

$$(R + \tfrac{1}{2}g) \cos \tfrac{1}{2}F_1 - (R + \tfrac{1}{2}g) \cos F = \tfrac{1}{2}g,$$

whence

$$R + \tfrac{1}{2}g = \frac{g}{2(\cos \frac{1}{2}F_1 - \cos F)}. \quad . \quad . \quad . \quad . \quad (231)$$

In like manner for the curve CE,

$$ME = \tfrac{1}{2}g \cot \tfrac{1}{2}(F' + \tfrac{1}{2}F_1) = \frac{2gN'N_1}{2N_1 + N'}, \quad . \quad . \quad (232)$$

$$R' + \tfrac{1}{2}g = \frac{g}{2(\cos \frac{1}{2}F_1 - \cos F')}. \quad . \quad . \quad . \quad (263)$$

Example.—Given $N_1 = 6$, $N = 8$, and $N' = 9$, to find the lead l_1, the distances GH and ME, and radii R, R_1, and R', g being 4.75 ft.

By (226), $R_1 = 19 \times 6^2 = 684$ ft., an 8° 23′ curve.

By (227), $l_1 = 9.5 \times 6 = 57$ ft.

By (230), $GH = \dfrac{9.5 \times 6 \times 8}{12 + 8} = 22.8$ ft.

By (232), $ME = 24.4$ feet.

By (231), $R = \dfrac{4.75}{2(\cos 4° 46' - \cos 7° 9')} - 2.38 = 547.4$ ft.,

a 10° 28′ curve.

By (233), $R' = \dfrac{4.75}{2(\cos 4° 46' - \cos 6° 22')} - 2.38 = 876.4$ ft.

a 6° 32⅛′ curve.

173. Given the Number, N, of the Two Main Frogs and the Gauge, g, to find the Crotch-frog Number, N_1, its Lead, l_1, and the Radius, R_1, of Curve through Crotch when the Double Turnout is to Same Side of Straight Main Track.

In Fig. 81 the frogs at B and G are of the same number, and may be taken as falling on the same straight line through the center O. Angle $O_1GO = 90° - OGL = F$, and the triangle OO_1G is therefore isosceles; hence

$$O_1G = O_1O = OA - O_1O = \tfrac{1}{2}OA,$$

or
$$R_1 + \tfrac{1}{2}g = \tfrac{1}{2}(R + \tfrac{1}{2}g),$$

whence
$$R_1 = \tfrac{1}{2}R - \tfrac{1}{4}g. \quad \quad (234)$$

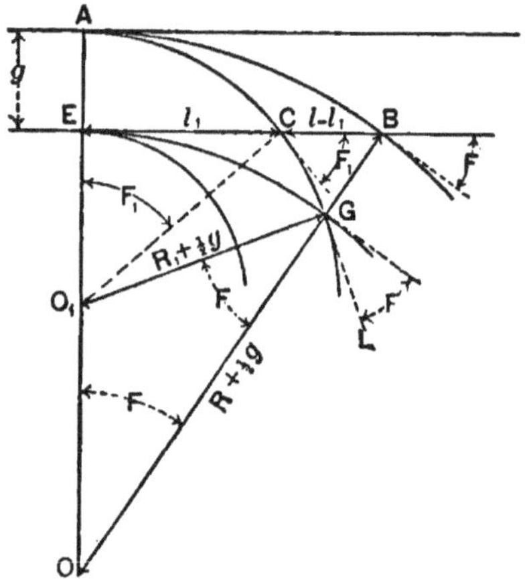

FIG. 81.

Now, by the same reasoning as in **167**, $2gN_1^2 = R_1$, whence

$$N_1 = \sqrt{\frac{R_1}{2g}} = \sqrt{\frac{R}{4g} - \frac{1}{8}}. \quad \quad (235)$$

Neglecting the $\frac{1}{8}$ under radical and writing $R = 2gN^2$ gives

$$N_1 = \frac{N}{\sqrt{2}} = .707N, \quad . \quad . \quad . \quad . \quad . \quad . \quad . \quad . \quad (236)$$

which is identical with (222) for turnouts to opposite sides. For EC and EB, as in **167**, $l_1 = 2gN_1$ and $l = 2gN$. Hence

$$CB = l - l_1 = 2g(N - N_1). \quad . \quad . \quad . \quad . \quad (237)$$

EXAMPLE. — Find N_1, R_1, and $l - l_1$ where $N = 9$ and $g = 4.75$ ft.

By (208″), $R = 9.5 \times 81 = 769.5$ ft., a 7° 27′ curve.
" (234), $R_1 = 384.75 - 1.19 = 383.56$ ft, a 14° 56′ curve.
" (236), $N_1 = .707 \times 9 = 6.36$.
" (237), $CB = l - l_1 = 9.5 \times 2.64 = 25.08$ ft.

REMARK.—It may now be seen that the proper combination of frogs for a double turnout to opposite sides applies also where the turnouts are to same side of straight main line. Also they apply to turnouts from *opposite* sides of curved main line when its radius is not less than that required by main frog for straight track.

174. Given the Number of Main Frogs, N, and of Crotch-frog, N_1, to Find the Radius of Curve between Frog-points of a Double Turnout to Same Side of Straight Track.

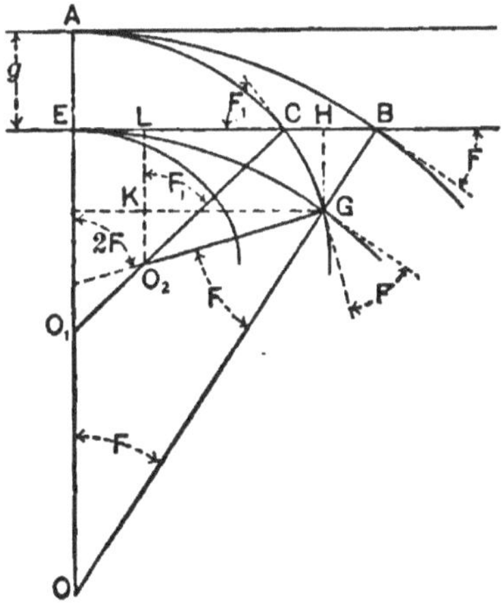

FIG. 82.

In Fig. 82, $O_2G = R_2 + \frac{1}{2}g$, and the chord CG must be determined. The frogs at B and G being of the same number, $O_2GO = GOO_1 = F$ and $CO_1E = F_1$.

Draw GH perpendicular to EB; then in triangle BGH

$$GH = g \cos F.$$

Draw O_2L perpendicular and GK parallel to EB; from triangles O_2GK and O_2CL,

$$(R_2 + \tfrac{1}{2}g)(\cos F_1 - \cos 2F) = KL = GH = g \cos F,$$

whence
$$R_2 + \tfrac{1}{2}g = \frac{g \cos F}{\cos F_1 - \cos 2F}. \quad . \quad . \quad . \quad (238)$$

From triangle O_2CG, since $CO_2G = 2F - F_1$,

$$CG = 2(R_2 + \tfrac{1}{2}g) \sin \tfrac{1}{2}(2F - F_1). \quad . \quad . \quad . \quad (239)$$

EXAMPLE.—Given $N = 8$, $N_1 = 6$, and $g = 4.75$, to locate the turnout.

By (208), $R = 608$ ft.; $R_1 = 342$ ft.
By (238), $R_2 + \tfrac{1}{2}g = 274.5$ ft.
By (239), $CG = 22.8$ ft.

175. Given the Frog-number, N, the Gauge, g, and Distance, p, between Centers, to Unite Main Line with a Parallel Siding when the Reversing-point is at Frog-point.

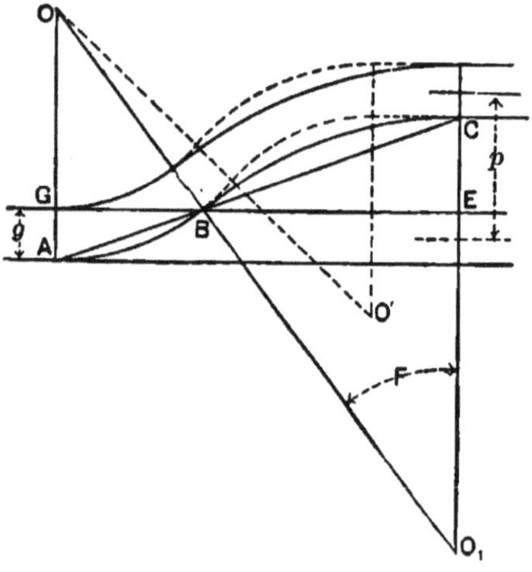

FIG. 83.

In Fig. 83, $BO_1 = R_1 - \tfrac{1}{2}g$ and BE are required.

In triangle BO_1E, $BO_1 = R_1 - \tfrac{1}{2}g$, $EO_1 = R_1 + \tfrac{1}{2}g - p$, and angle $BO_1E = F$. By trigonometry,

$$\frac{(R_1 - \tfrac{1}{2}g) + (R_1 + \tfrac{1}{2}g - p)}{(R_1 - \tfrac{1}{2}g) - (R_1 + \tfrac{1}{2}g - p)} = \frac{\tan \tfrac{1}{2}(180^\circ - F)}{\tan \tfrac{1}{2}F}$$

or
$$\frac{2R_1 - p}{p - g} = \frac{\cot \tfrac{1}{2}F}{\tan \tfrac{1}{2}F} = 4N^2,$$

whence
$$R_1 = 2(p - g)N^2 + \tfrac{1}{2}p, \quad . \quad . \quad . \quad . \quad . \quad (240)$$

$$BE = (R_1 - \tfrac{1}{2}g) \sin F. \quad . \quad . \quad . \quad . \quad . \quad . \quad (241)$$

From the similar triangles ABG and BCE, $\dfrac{BE}{p-g} = \dfrac{l}{g}$, from which

$$BE = \frac{(p - g)l}{g} = \left(\frac{p}{g} - 1\right)l. \quad . \quad . \quad . \quad . \quad (242)$$

Example.—Find R_1 and BE when $N = 8$, $p = 12.35$ ft., $g = 4.75$ ft.

By (240), $R_1 = 15.2 \times 64 + 6.2 = 979$ ft., a 5° 51′ curve.
By (207″), $l = 9.5 \times 8 = 76$ ft.
By (242), $BE = (2.6 - 1) \times 76 = 121.6$ ft.

Remark.—If space requires that the turnout get away from main line more rapidly than by the above method, we can assume the second radius equal to or less than the radius of turnout and find the reversing-point by **131**, and then compute BE.

176. To Lay Out a "Ladder" Track

In yardwork a number of parallel sidings may be conveniently connected with the main line by means of a ladder-track.

In Fig. 84, if the frog-number N and the distance p between center lines of track are given, it is only necessary to determine the distances BC, CE, etc., between frog-points, and BK, CL, etc., between point of switch and point of frog. From triangle BCG,

$$BC = \frac{p}{\sin F} = p \operatorname{cosec} F, \quad . \quad . \quad . \quad . \quad . \quad . \quad . \quad (243)$$

$$BK = BC - KC = p \operatorname{cosec} F - 2gN; \quad . \quad . \quad . \quad (244)$$

or, since $\operatorname{cosec} F = N + \frac{1}{4N}$, (see **186**,)

$$BC = pN + \frac{p}{4N}, \quad \ldots\ldots\ldots\ldots \quad (243')$$

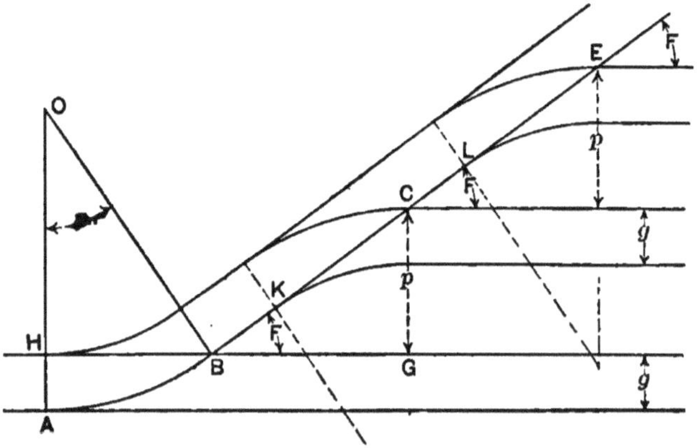

FIG. 84.

$$BK = (p - 2g)N + \frac{p}{4N}. \quad \ldots\ldots\ldots \quad (244')$$

EXAMPLE.—For a No. 8 frog find BC and BK when $p = 12.8$ ft. and $g = 4.75$ ft.

By (243′), $\qquad BC = 12.8 \times 8 + \frac{12.8}{32} = 102.8$ ft.

By (244′), $\qquad BK = 3.3 \times 8 + 0.4 = 26.8$ ft.

B. Turnouts from Curves.

177. Given the Radius of Main Curve, the Frog-number, and the Gauge, to Find the Radius and Lead of Turnout from Concave Side of Main Line.

In Fig. 85, AB is the outer rail of turnout, CB the inner rail of main track. In triangle OAB, since $O_2BA = OAB$,

$$OBA - OAB = F,$$

$$OBA + OAB = 180° - 0,$$

and $\quad OA = R + \frac{1}{2}g, \quad OB = R - \frac{1}{2}g.$

Then, by trigonometry,

$$\frac{(R+\frac{1}{2}g)+(R-\frac{1}{2}g)}{(R+\frac{1}{2}g)-(R-\frac{1}{2}g)}=\frac{\tan\frac{1}{2}(180-\theta)}{\tan\frac{1}{2}F}=\frac{\cot\frac{1}{2}\theta}{\tan\frac{1}{2}F}.$$

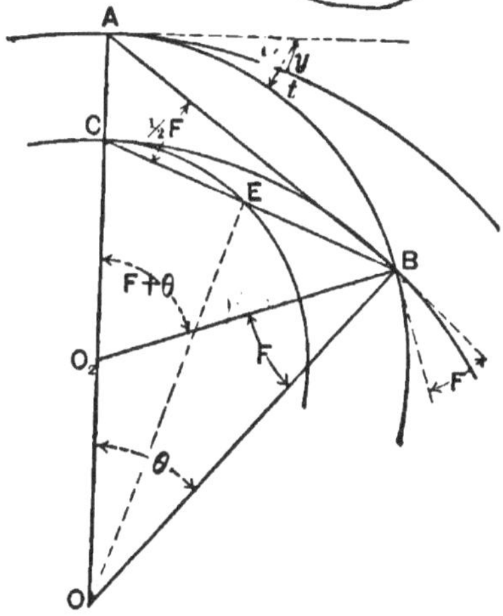

FIG. 85.

Reducing, $$\cot\frac{1}{2}\theta=\frac{2R}{g}\tan\frac{1}{2}F=\frac{R}{gN}. \quad . \quad . \quad . \quad . \quad . \quad . \quad (245)$$

Then $$l=BC=2(R-\frac{1}{2}g)\sin\frac{1}{2}\theta. \quad . \quad . \quad . \quad (246)$$

If the length of AB is wanted, we can show that the angle $ABC=\frac{1}{2}F$; and by solving the triangle ABC, since $ACB=90°+\frac{1}{2}\theta$,

$$AB=\frac{g\cos\frac{1}{2}\theta}{\sin\frac{1}{2}F}. \quad . \quad . \quad . \quad . \quad . \quad . \quad (247)$$

To find R_2, from triangle O_2AB,

$$2(R_2+\frac{1}{2}g)\sin\frac{1}{2}(F+\theta)=AB. \quad . \quad . \quad . \quad . \quad (248)$$

Or, in triangle BO_2C,

$$\frac{(R_2+\frac{1}{2}g)+(R_2-\frac{1}{2}g)}{(R_2+\frac{1}{2}g)-(R_2-\frac{1}{2}g)}=\frac{\tan\frac{1}{2}[180-(F+\theta)]}{\tan\frac{1}{2}F}=\frac{\cot\frac{1}{2}(F+\theta)}{\tan\frac{1}{2}F}.$$

Reducing and solving for R_2,

$$R_2 = \frac{g}{2} \cdot \frac{\cot \frac{1}{2}(F + \theta)}{\tan \frac{1}{2}F} = \frac{g}{2} \cdot \cot \tfrac{1}{2}F . \cot \tfrac{1}{2}(F + \theta). \quad . \quad (249)$$

But, from trigonometry

$$\cot \tfrac{1}{2}(F + \theta) = \cot (\tfrac{1}{2}F + \tfrac{1}{2}\theta) = \frac{1 - \tan \frac{1}{2}F . \tan \frac{1}{2}\theta}{\tan \frac{1}{2}F + \tan \frac{1}{2}\theta}.$$

Substitute this in (249) and write

$$\cot \tfrac{1}{2}F = 2N, \quad \tan \tfrac{1}{2}F = \frac{1}{2N}, \quad \tan \tfrac{1}{2}\theta = \frac{Ng}{R},$$

and reduce; then

$$R_2 = \frac{2gN^2(R - \frac{1}{2}g)}{R + 2gN^2}. \quad . \quad . \quad . \quad . \quad . \quad (250)$$

For $2gN^2$ write R_1, the radius of turnout from straight track, and neglect the $\frac{1}{2}g$ in numerator as small compared with R; then

$$R_2 = \frac{RR_1}{R + R_1}. \quad . \quad . \quad . \quad . \quad . \quad . \quad . \quad . \quad (251)$$

Now write

$$R = \frac{5730}{D}, \quad R_1 = \frac{5730}{D_1}, \quad R_2 = \frac{5730}{D_2}$$

and reduce, yielding

$$D_2 = D + D_1. \quad . \quad . \quad . \quad . \quad . \quad . \quad . \quad . \quad (252)$$

Formula (252) affords an easy method of finding the degree of turnout curve, or, if preferred, the radius may be first found by (251).

Draw OE to the mid-point of CB; OE does not differ greatly from OB or OC; so, if we write $OE = R - \frac{1}{2}g$, there results

$$l = 2(R - \tfrac{1}{2}g) \tan \tfrac{1}{2}\theta = 2(R - \tfrac{1}{2}g)\frac{gN}{R} = 2gN - \frac{g^2N}{R}. \quad (253)$$

The last term is quite small, even in the most extreme case likely to arise in practice; for a turnout from a 6° curve with number 8 frog it amounts to only $2\frac{1}{4}$ inches; neglecting it, we may write, as for straight main track,

$$l = 2gN. \quad . \quad . \quad . \quad . \quad . \quad . \quad . \quad . \quad (254)$$

EXAMPLE.—Turnout from inside of a 4° curve, $N = 8$, $g =$ 4.75 ft.

By (208″), $R_1 = 9.5 \times 64 = 608$ ft., a 9° 26′ curve.

By (252), $D_2 = 4° + 9° 26' = 13° 26'$, for which $R_2 = 426.8$.

By (254), $l = 76$ ft.

178. Given the Frog-number, the Gauge and Radius of Main Curve, to Find the Lead and Radius (or Degree) of Turnout from Convex Side of Main Line.

In triangle AOB of Fig. 86, $A + B = 180° - \theta$,

$$A - B = (180° - O_2AB) - (180° - O_2BA - F) = F.$$

By trigonometry,

$$\frac{(R+\frac{1}{2}g)+(R-\frac{1}{2}g)}{(R+\frac{1}{2}g)-(R-\frac{1}{2}g)} = \frac{\tan\frac{1}{2}(180-\theta)}{\tan\frac{1}{2}F},$$

or

$$\frac{2R}{g} = \frac{\cot\frac{1}{2}\theta}{\tan\frac{1}{2}F}$$

whence

$$\cot\tfrac{1}{2}\theta = \frac{2R}{g}\tan\tfrac{1}{2}F = \frac{R}{gN}. \quad (255)$$

From triangle OCB,

$$l = CB = 2(R + \tfrac{1}{2}g)\sin\tfrac{1}{2}\theta. \quad (256)$$

Assuming $OE = R + \frac{1}{2}g$,

$$l = 2(R + \tfrac{1}{2}g)\tan\tfrac{1}{2}\theta$$

$$2(R+\tfrac{1}{2}g) \cdot \frac{gN}{R} = 2gN + \frac{g^2N}{R}. \quad (257)$$

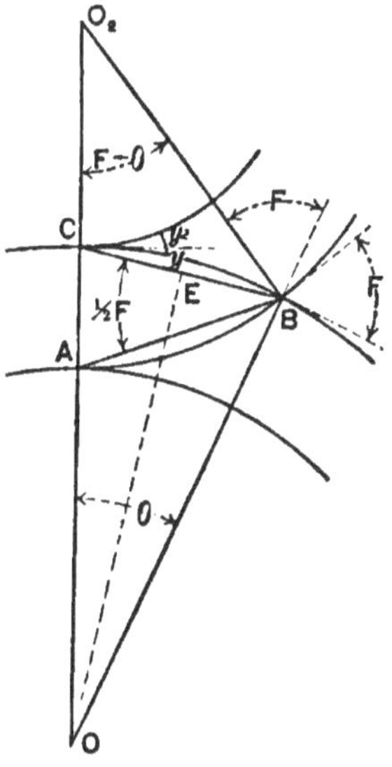

FIG. 86.

Neglecting the last term as small, as in **177**,

$$l = 2gN, \quad . \quad . \quad . \quad . \quad . \quad . \quad . \quad . \quad . \quad (258)$$

which is the same as (207).

In triangle CO_2B, $O_2 = F - \theta$.

$$\therefore \tan\tfrac{1}{2}O_2 = \tan(\tfrac{1}{2}F - \tfrac{1}{2}\theta).$$

We may now follow the same line of reasoning by which (251) was derived, or more simply by assuming the tangent of the difference of two small angles equal to the difference of their tangents; that is, $\tan \frac{1}{2}O_2 = \tan \frac{1}{2}F - \tan \frac{1}{2}O$.

Now it can be easily shown that $\tan \frac{1}{2}O_2 = \frac{gN}{R_2}$; therefore

$$\frac{gN}{R_2} = \frac{1}{2N} - \frac{gN}{R},$$

whence
$$\frac{1}{R_2} = \frac{1}{2gN^2} - \frac{1}{R} = \frac{1}{R_1} - \frac{1}{R},$$

from which
$$R_2 = \frac{RR_1}{R - R_1}. \qquad (259)$$

Write $R_2 = \frac{5730}{D_2}$, $R_1 = \frac{5730}{D_1}$, $R = \frac{5730}{D}$, and solve for D_2.

$$D_2 = D_1 - D, \qquad (260)$$

in which D_1 is the degree of turnout from straight track.

EXAMPLE.—Turnout from outside of a 4° curve, $N = 8$, $g = 4.75$.

By (208″), $R_1 = 9.5 \times 64 = 608$ ft., a 9° 26′ curve.

By (260), $D_2 = 9° 26' - 4° = 5° 26'$, for which $R = 1054.9$.

By (258), $l = 9.5 \times 8 = 76$ ft.

From (255) we have, by inverting,

$$\tan \tfrac{1}{2}\theta = \frac{38}{1432.7} = \tan 1° 31'$$

and, by (256),

$$l = 2870 \times \sin 1° 31' = 75.97 \text{ ft.},$$

a difference of only 0.03 ft. from the value given by (258).

179. To Find Theoretical Length of Switch-rail when the Turnout is from a Curved Track.

A common tangent being drawn at the switch-point, we shall have, as in **169**, for offset from tangent to main curve,

$$y = \frac{S^2}{2R};$$

the offset from tangent to turnout is

$$y_2 = \frac{S^2}{2R_2}.$$

When the turnout is from concave side of main line,

$$y_2 - y = t;$$

therefore $$t = \frac{S^2}{2}\left(\frac{1}{R_2} - \frac{1}{R}\right),$$

whence $$S = \sqrt{\frac{2tRR_2}{R - R_2}}. \quad \ldots \ldots \ldots \quad (261)$$

Writing $R = \frac{5730}{D}$, $R_2 = \frac{5730}{D_2}$, and reducing,

$$S = 107\sqrt{\frac{t}{D_2 - D}} = 107\sqrt{\frac{t}{D_1}}. \quad \ldots \quad (262)$$

When the turnout is from convex side of main line,

$$t = y_2 + y = \frac{S^2}{2}\left(\frac{1}{R_2} + \frac{1}{R}\right),$$

whence $$S = \sqrt{\frac{2tRR_2}{R + R_2}}; \quad \ldots \ldots \ldots \quad (263)$$

from which

$$S = 107\sqrt{\frac{t}{D_2 + D}} = 107\sqrt{\frac{t}{D_1}}. \quad \ldots \quad (264)$$

In (262) and (264) D_1 is the degree of turnout from straight track, and, as these formulas are identical with (216), it is seen that the theoretical length of switch-rail on turnouts from curves is the same as on turnouts from straight line.

Example.—Find S when $t = 0.42$, $N = 8$, $g = 4.75$.

By (208″), $R_1 = 608$ feet, for which $D_1 = 9° 26'$.

By (216), (262), or (264),

$$S = 107\sqrt{\frac{.42}{9.43}} = 22.6 \text{ feet.}$$

180. Given the Distance p between Center Lines of Curved Main Line and Side Track, the Frog-angle, F (or Number, N), and Gauge, g, to Find the Radius and Central Angle of Curve beyond Frog-point.

FIRST CASE.—*Turnout from outside of main line.*

In Fig. 87, O is the center of main curve, O_1 the center of

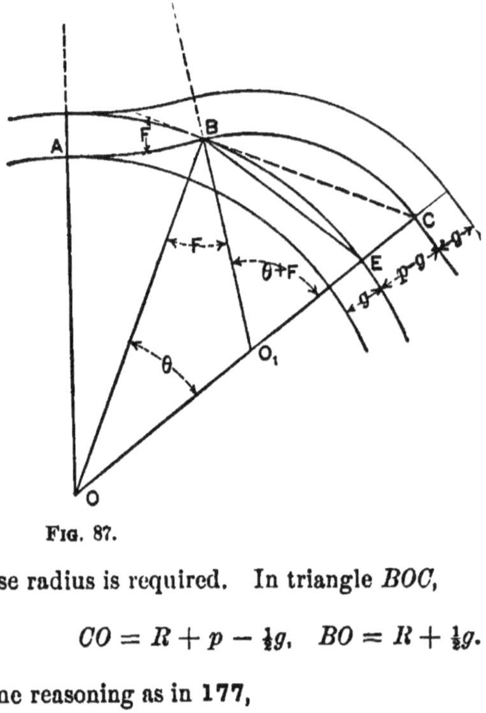

FIG. 87.

curve whose radius is required. In triangle BOC,

$$CO = R + p - \tfrac{1}{2}g, \quad BO = R + \tfrac{1}{2}g.$$

By the same reasoning as in **177**,

$$\cot \tfrac{1}{2}\theta = \frac{2R+p}{p-g}\tan \tfrac{1}{2}F = \frac{2R+p}{2N(p-g)}. \quad . \quad . \quad (265)$$

In triangle OO_1B, $O_1B = R_1 - \tfrac{1}{2}g$; then, by the law of sines,

$$R_1 - \tfrac{1}{2}g = \frac{\sin\theta}{\sin(F+\theta)}(R + \tfrac{1}{2}g). \quad . \quad . \quad . \quad . \quad . \quad . \quad (266)$$

Also, $$BE = 2(R + \tfrac{1}{2}g)\sin\tfrac{1}{2}\theta. \quad . \quad . \quad . \quad . \quad . \quad . \quad . \quad (267)$$

SECOND CASE.—*Turnout from inside of main track.*

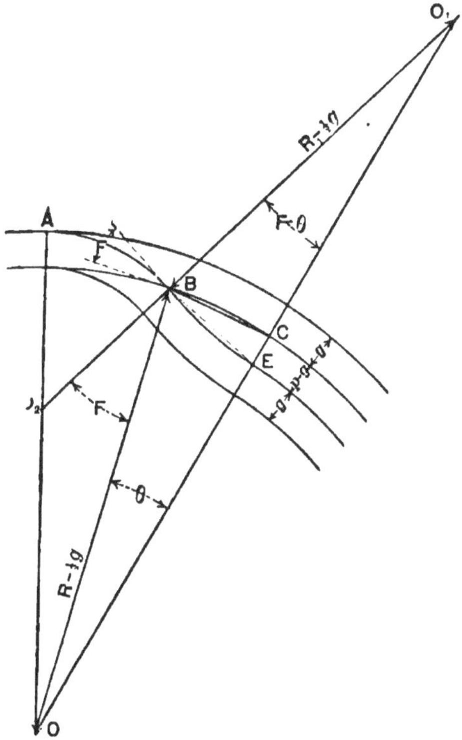

FIG. 88.

In Fig. 88, we have from triangle BOE, reasoning as in **178**,

$$\cot \tfrac{1}{2}\theta = \frac{2R - p}{p - g} \tan \tfrac{1}{2}F = \frac{2R - p}{2N(p - g)}; \quad . \quad . \quad . \quad (268)$$

and from triangle OBO_1,

$$R_1 - \tfrac{1}{2}g = \frac{\sin \theta}{\sin(F - \theta)} (R - \tfrac{1}{2}g). \quad . \quad . \quad . \quad . \quad . \quad . \quad (269)$$

Also, $$BC = 2(R - \tfrac{1}{2}g) \sin \tfrac{1}{2}\theta, \quad . \quad . \quad . \quad . \quad . \quad . \quad . \quad . \quad . \quad (270)$$

and $$BE = 2(R_1 - \tfrac{1}{2}g) \sin \tfrac{1}{2}(F - \theta). \quad . \quad . \quad . \quad . \quad . \quad (271)$$

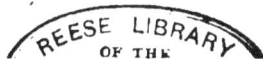

When θ is greater than F, $\sin (F - \theta)$ is negative, and center O_1 falls on same side as O, and

$$R_1 + \tfrac{1}{2}g = \frac{\sin \theta}{\sin (\theta - F)}(R - \tfrac{1}{2}g). \quad . \quad . \quad . \quad . \quad (272)$$

$$BE = 2(R_1 + \tfrac{1}{2}g) \sin \tfrac{1}{2}(\theta - F). \quad . \quad . \quad . \quad (273)$$

C. The Stub Lead.

181. When the frog-number exceeds seven, the length of switch-rail required to give the necessary clearance at heel becomes greater than is allowed in practice. To overcome this difficulty slightly more curvature is given the switch-rail; moreover the physical point of switch is necessarily some distance in advance of the theoretical point. The distance from heel of switch to point of main frog will then be the same as from head-block of stub switch to main-frog point, and is termed the **Stub Lead.** If to this distance the length of switch-rail be added, we get the distance from the head-block of a point switch to the point of main frog, which is the **Short Lead** required in practice.

182. Given the Throw, t, the Gauge, g, and the Frog-number, N, to Find the Stub Lead, $s.l.$

In Fig. 89, KB is the stub lead required; $GN = KL$, the throw.

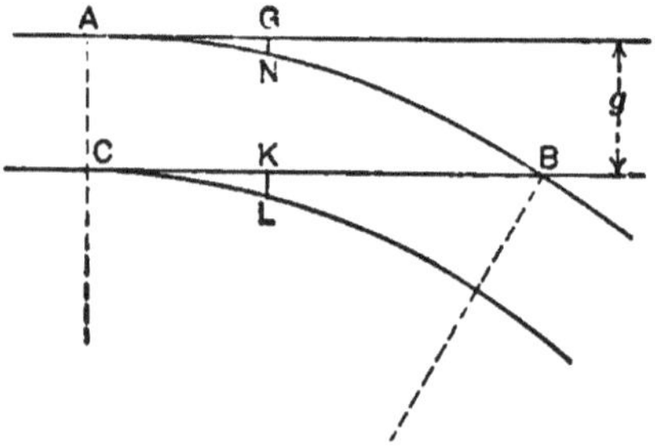

FIG. 89.

From (207), $\qquad l = CB = 2gN,$

and from (215), $\qquad S = CK = 2N\sqrt{gt}.$

From the figure,

$$KB = CB - CK,$$

or $$s.l. = 2N(g - \sqrt{gt}). \quad . \quad . \quad . \quad . \quad . \quad . \quad . \quad (274)$$

Formula (274) may be employed for turnouts from curves as well as straight lines, since it was shown that the formulas from which it was derived may be employed even when the curvature of main track is considerable.

Below is a table of values of $(g - \sqrt{gt})$ for some of the more common values of g and t.

TABLE OF VALUES OF $g - \sqrt{gt}$.

3 Feet Gauge.		4 Feet 8½ Inch Gauge.		4 Feet 9 Inch Gauge.	
Throw.	$g - \sqrt{gt}$.	Throw.	$g - \sqrt{gt}$.	Throw.	$g - \sqrt{gt}$.
Inches.	Feet.	Inches.	Feet.	Inches.	Feet.
3	2.13	5	3.308	5	3.343
3½	2.06	5¼	3.239	5¼	3.275
4	2.00	5½	3.206	5½	3.242

EXAMPLE.—Find the stub lead for $N = 8$, $g = 4.75$ ft., $t = 5$ inches.

From the table, $g - \sqrt{gt} = 3.343$ ft.,

and, by (274), $s.l. = 16 \times 3.343 = 53.49$ ft.

183. The Turnout Table on the next page gives the frog-angles, the radius of center line of turnout from a straight track and its degree, the theoretical lead, the theoretical length of switch-rail for $t = 5$ inches, and the stub lead for certain values of t. The frog-numbers given cover all the usual cases.

Suppose it required to find the short lead for a No. 9 frog and 5-inch throw when the gauge is 4 ft. 9 inches and the length of switch-rail 18 feet. From the table the stub lead is 60.17 feet; hence the short lead is $60.17 + 18 = 78.17$ feet, as against 85.50 ft. for the theoretical lead.

Inspection of the table will show that it makes no very great difference in the tabular quantities whether the gauge be taken as 4 feet 8½ inches or 4 feet 9 inches. However, the numerical coefficients in the formulas involving g are somewhat simpler for the latter value.

TURNOUT TABLE FOR STRAIGHT TRACK.

4 FEET 8½ INCH GAUGE.

Frog No.	Frog Angle.	Theoretical Lead.	Turnout Radius.	Degree of Turnout.	Theoretical Switch-rail for $t = 5$ In.	Stub-lead for a Throw of		
						5 In.	5½ In.	5¾ In.
	° ′	feet	feet	° ′	feet	feet	feet	feet
4	14 15	37.67	150.7	38 2	11.20	26.46	25.91	25.65
5	11 25	47.08	235.4	24 21	14.01	33.08	32.39	32.06
5½	10 23	51.79	284.9	20 7	15.41	36.39	35.63	35.27
6	9 32	56.50	339.0	16 54	16.81	39.70	38.87	38.47
6½	8 48	61.21	397.9	14 24	18.21	43.00	42.11	41.68
7	8 10	65.92	461.4	12 25	19.61	46.31	45.35	44.88
7½	7 38	70.63	529.7	10 49	21.01	49.62	48.59	48.09
8	7 9	75.38	602.7	9 30½	22.41	52.93	51.82	51.30
8½	6 44	80.04	680.4	8 25	23.81	56.24	55.06	54.50
9	6 22	84.75	762.7	7 31	25.21	59.54	58.30	57.71
9½	6 2	89.46	849.8	6 45	26.61	62.85	61.54	60.90
10	5 44	94.17	941.7	6 5	28.01	66.16	64.78	64.12
11	5 12	103.58	1139.4	5 2	30.81	72.78	71.26	70.53
12	4 46	113.00	1356.0	4 18½	33.61	79.39	77.74	76.94
13	4 24	122.42	1591.4	3 36	36.42	86.01	84.21	83.36
14	4 5	131.83	1845.7	3 6	39.22	92.62	90.69	89.77
15	3 49	141.25	2118.7	2 42	42.02	99.24	97.17	96.18

4 FEET 9 INCH GAUGE.

Frog No.	Frog Angle.	Theoretical Lead.	Turnout Radius.	Degree of Turnout.	Theoretical Switch-rail for $t = 5$ In.	Stub-lead for a Throw of		
						5 In.	5½ In.	5¾ In.
	° ′	feet	feet	° ′	feet	feet	feet	feet
4	14 15	38.00	152.0	37 42	11.26	26.74	26.20	25.94
5	11 25	47.50	237.5	24 8	14.07	33.43	32.75	32.42
5½	10 23	52.25	287.4	19 56	15.48	36.77	36.03	35.66
6	9 32	57.00	342 0	16 46	16.88	40.12	39.30	38 90
6½	8 48	61.75	401.4	14 16	18.29	43.46	42.58	42.15
7	8 10	66.50	465.5	12 19	19.70	46.80	45.85	45.39
7½	7 38	71.25	534.4	10 44	21.10	50.15	49.13	48.63
8	7 9	76.00	608.0	9 25½	22.51	53.49	52.40	51.87
8½	6 44	80 75	686.4	8 21	23.92	56.88	55.68	55.11
9	6 22	85.50	769.5	7 27	25.32	60.17	58.95	58.36
9½	6 2	90.25	857.4	6 41	26.73	63.52	62.23	61.60
10	5 44	95.00	950.0	6 2	28.14	66.86	65.50	64.84
11	5 12	104.50	1149.5	4 59	30.95	73.55	72.05	71.32
12	4 46	114.00	1368.0	4 11	33.77	80.23	78.60	77.81
13	4 24	123.50	1605.5	3 34	36.58	86.92	85.15	84 29
14	4 5	133.00	1862.0	3 4½	39.39	93.60	91.70	90.78
15	3 49	142.50	2137.5	2 41	42.21	100.29	98,25	97.26

184. To Stake Out a Turnout.—If the position of head-block is given, fix the frog-point by the foregoing table, remembering that it may be used for turnouts from curves, as well as from straight lines, without material error.

To locate the rail between head-block and point of switch it is sufficient to do so by offsets from main rail. Consider the equation (36) for tangent offsets.

$$z = \tfrac{7}{8}n^2D \text{ for straight main line.}$$
$$z = \tfrac{7}{8}n^2(D_1 \pm D) \text{ for curved main line.}$$

At frog-point $z = g$, and $n = n_1$; hence

$$g = \tfrac{7}{8}n_1{}^2D, \quad \text{or} \quad \tfrac{7}{8}n_1{}^2(D_1 \pm D).$$

At mid-point of curve (practically mid-point of lead), $n = \tfrac{1}{2}n_1$, and

$$z = \tfrac{7}{8} \cdot \tfrac{1}{4}n_1{}^2D, \quad \text{or} \quad \tfrac{7}{8} \cdot \tfrac{1}{4}n_1{}^2(D_1 \pm D) = \tfrac{1}{4}g. \quad . \quad (275)$$

When $n = \tfrac{1}{4}n_1$,

$$z = \tfrac{7}{8} \cdot \tfrac{1}{16}n_1{}^2D, \quad \text{or} \quad \tfrac{7}{8} \cdot \tfrac{1}{16}n_1{}^2(D_1 \pm D) = \tfrac{1}{16}g. \quad . \quad (276)$$

When $n = \tfrac{3}{4}n_1$,

$$z = \tfrac{7}{8} \cdot \tfrac{9}{16}n_1{}^2D, \quad \text{or} \quad \tfrac{7}{8} \cdot \tfrac{9}{16}n^2(D_1 \pm D) = \tfrac{9}{16}g. \quad . \quad (277)$$

These formulas are for the theoretical lead, and afford an easy method of locating the outer rail of turnout with all the accuracy needed in practice.

185. Curving Rails.—In bending rails for curves the proper curvature is determined by measuring the mid-ordinate from a cord held against the inside face of rail-head.

This ordinate may be determined by (18), in which n is the half-length of rail divided by 100. For a 30-ft. rail,

$$M = \tfrac{7}{8}(0.15)^2D = 0.0196D = 0.02D \text{ (nearly)}. \quad . \quad (278)$$

From (209′), $D = \dfrac{608}{N^2}$; and from (209″), $D = \dfrac{603}{N^2}$. Inserting either of these values in (278) gives

$$M = \frac{12}{N^2}, \text{ nearly.} \quad . \quad . \quad . \quad . \quad . \quad . \quad . \quad (279)$$

When the turnout is from a curve compute M from (279), and the mid-ordinate for a rail 30 ft. long on main curve by (278); then the mid-ordinate for turnout rail will be the sum or difference of these values according as the turnout is from concave or convex side of main curve

ARTICLE 15. CROSSOVERS.

186. To Locate a Crossover between Parallel Straight Tracks when the Frog-number, the Distance, p, between Centers, and the Gauge are given, inserting a Tangent between Frog-points.

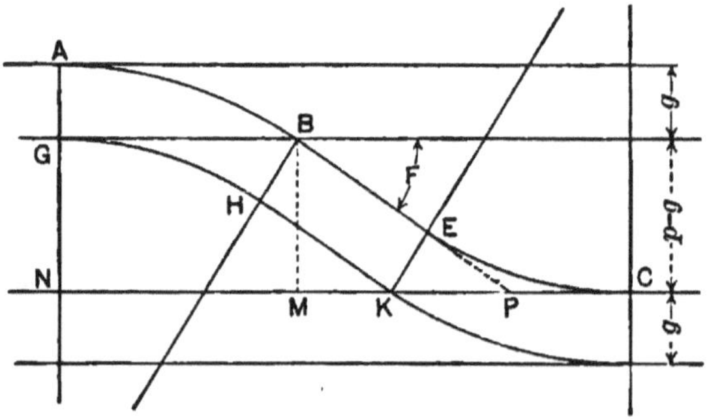

FIG. 90.

In Fig. 90 it is required to find $GB = KC = l$, MK and NC, also $HK = k$.

In the triangle BPM, $BM = p - g$; then

$$BE = k = BP - EP$$

or $$k = (p - g) \operatorname{cosec} F - g \cot F, \quad . \quad . \quad . \quad (280)$$

and $$MK = MP - KP,$$

or $$MK = (p - g) \cot F - g \operatorname{cosec} F. \quad . \quad . \quad . \quad (281)$$

From triangle OBC of Fig. 78,

$$\cos F = \frac{OC}{OB} = \frac{R - \frac{1}{2}g}{R + \frac{1}{2}g} = \frac{2R - g}{2R + g}. \quad . \quad . \quad . \quad (a)$$

In (a) write $R = 2gN^2$ by (208), giving

$$\cos F = \frac{4gN^2 - g}{4gN^2 + g} = \frac{4N^2 - 1}{4N^2 + 1}. \quad . \quad . \quad . \quad . \qquad (b)$$

From Fig 78, triangle OBC,

$$\sin F = \frac{CB}{OB} = \frac{l}{R + \frac{1}{2}g} = \frac{2l}{2R + g} \quad . \quad . \quad . \qquad (c)$$

Writing $l = 2gN$ and $R = 2gN^2$ gives

$$\sin F = \frac{4gN}{4gN^2 + g} = \frac{4N}{4N^2 + 1}. \quad . \quad . \quad . \quad . \qquad (d)$$

From trigonometry, taking the above values of $\sin F$ and $\cos F$,

$$\operatorname{cosec} F = \frac{1}{\sin F} = N + \frac{1}{4N}. \quad . \quad . \quad . \quad . \quad . \qquad (e)$$

$$\cot F = \frac{\cos F}{\sin F} = N - \frac{1}{4N}. \quad . \quad . \quad . \quad . \quad . \qquad (f)$$

Inserting these values in (280) and (281),

$$k = (p - 2g)N + \frac{p}{4N}. \quad . \quad . \quad . \quad . \quad . \quad . \qquad (282)$$

$$MK = (p - 2g)N - \frac{p}{4N}. \quad . \quad . \quad . \quad . \quad . \quad . \qquad (283)$$

By (207), $GB = KC = l = 2gN$; therefore

$$NC = 2l + MK = 4gN + (p - 2g)N - \frac{p}{4N},$$

or

$$NC = (p + 2g)N - \frac{p}{4N} = l + p\left(N - \frac{1}{4N}\right). \quad . \quad . \qquad (284)$$

EXAMPLE.—Find k and MK for a No. 8 frog when $p = 13$ ft. and $g = 4.75$ ft.

By (282), $k = 3.5 \times 8 + 0.4 = 28.4$ feet.
By (283), $MK = 3.5 \times 8 - 0.4 = 27.6$ feet.

187. To Lay Out a Crossover in the Form of a Reversed Curve.

When p is large, or for other reasons it is desirable to get away from main track more rapidly than by the foregoing method, we may lay out the crossover in the form of a reversed curve.

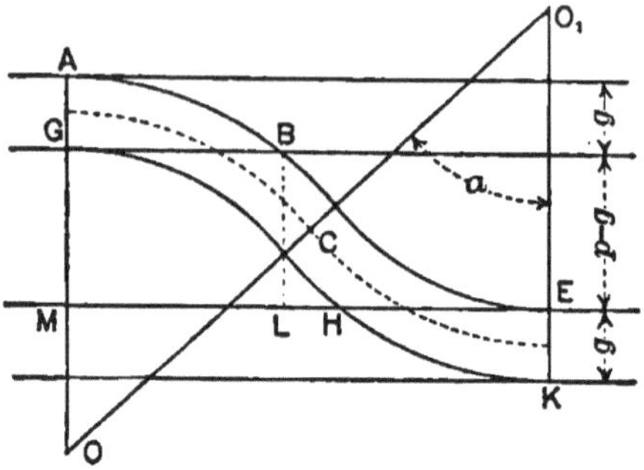

FIG. 91.

In Fig. 91 it is required to find $GB = HE$ and LH.

Find $GB = HE = l$ by (207), and the radius $OC = O_1C$ by (208).

Then, from (113), we have

$$ME = 2R \sin a.$$

The angle a is given by (112). Then

$$LH = 2R \sin a - 2l. \quad . \quad . \quad . \quad . \quad . \quad (285)$$

188. To Lay Out a Crossover when a Fixed Length of Tangent must be Interposed between Points of Reversal of Curvature.

From the given frog-number determine the radius by (208)· then the problem may be solved by **132.**

189. To Lay Out a Crossover in the Form of a Reversed Curve when the Tracks to be Joined are Curved.

In Fig 92 let the notation be as shown. Let $OM = R$, $O_1M = R_1$, $O_2P = O_2C = R_2$.

$$O_1O_2 = R_1 + R_2 = a,$$

$$OO_2 = R + p - R_2 = b,$$

$$OO_1 = R + R_1 = c,$$

$$\tfrac{1}{2}(a + b + c) = s.$$

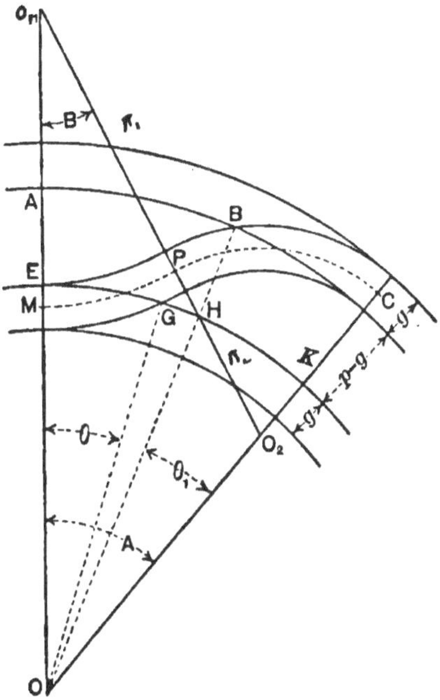

FIG. 92.

Then, from trigonometry,

$$\cos \tfrac{1}{2}A = \sqrt{\frac{s(s-a)}{bc}}, \quad \ldots \ldots \quad (286)$$

$$\cos \tfrac{1}{2}B = \sqrt{\frac{s(s-b)}{ac}}. \quad \ldots \ldots \quad (287)$$

R_1 is determined by **178**, and R_2 by **177**, while R and p are given to begin with.

The angle $(A + B)$ determines the length of arc CP, and angle B the length of arc MP.

The angle θ is given by (255), and θ_1 by (245). Hence angle GOH, which determines the arc GH measured along main track between frog-points, is

$$GOH = A - (\theta + \theta_1).$$

The frog-numbers at G and B need not be equal, only providing that P falls between G and B.

ARTICLE 16. CROSSING-FROGS AND CROSSING-SLIPS.

A. *Crossing-frogs.*

190. When two tracks intersect each other four *crossing-frogs* are required at the intersection of the two sets of rails. The four frogs are sometimes called a set of crossing-frogs.

191. To Find the Length of Rails Intercepted between two Intersecting Straight Tracks when the Angle of Intersection and the Two Gauges are given.

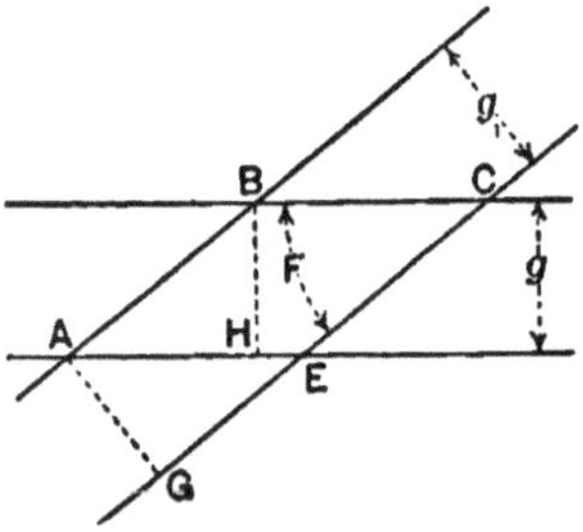

FIG. 93.

In Fig. 93, from triangle ABH,

$$AB = EC = g \operatorname{cosec} F; \quad . \quad (288)$$

and from triangle AEG,

$$AE = BC = g_1 \operatorname{cosec} F. \quad (289)$$

192. Given the Angle of Intersection, a, made by the Center Lines of a Straight and Curved Track, the Gauges g_1 and g, to Find the Angles of the Set of Crossing-frogs.

In Fig. 94, from the triangles OBK and OAH,

$$(R + \tfrac{1}{2}g) \cos F = R \cos a + \tfrac{1}{2}g_1.$$

$$\therefore \cos F = \frac{R \cos a + \frac{1}{2}g_1}{R + \frac{1}{2}g}. \quad . \quad (290)$$

In like manner,

$$\cos F_1 = \frac{R \cos a - \frac{1}{2}g_1}{R + \frac{1}{2}g}. \quad . \quad (291)$$

$$\cos F_2 = \frac{R \cos a - \frac{1}{2}g_1}{R - \frac{1}{2}g}. \quad . \quad (292)$$

$$\cos F_3 = \frac{R \cos a + \frac{1}{2}g_1}{R - \frac{1}{2}g}. \quad . \quad (293)$$

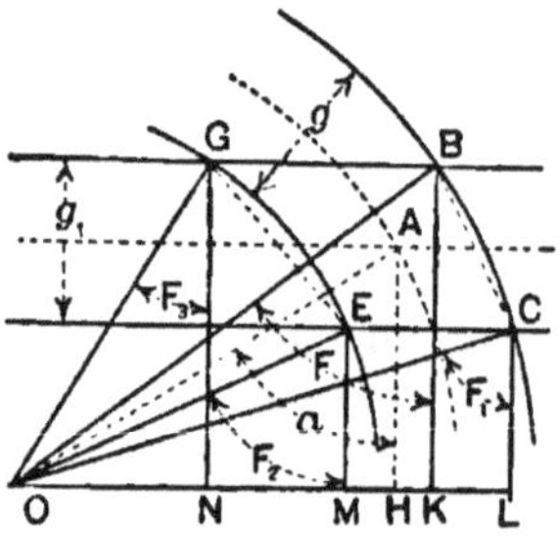

FIG. 94.

From triangle BOC to find the chord BC.

$$BC = 2(R + \tfrac{1}{2}g) \sin \tfrac{1}{2}(F_1 - F). \quad . \quad . \quad . \quad (294)$$

Similarly,

$$GE = 2(R - \tfrac{1}{2}g)\sin\tfrac{1}{2}(F_2 - F_3). \quad . \quad . \quad . \quad . \quad . \quad . \quad . \qquad (295)$$

From triangles EOM and COL, we have

$$EC = ML = (R + \tfrac{1}{2}g)\sin F_1 - (R - \tfrac{1}{2}g)\sin F_2. \quad (296)$$

In like manner,

$$GB = NK = (R + \tfrac{1}{2}g)\sin F - (R - \tfrac{1}{2}g)\sin F_3. \quad (297)$$

193. Given the Angle of Intersection, a, made by the Center Lines of Two Curved Tracks, their Gauges, g and g_1, to Find the Angles of the Crossing-frogs.

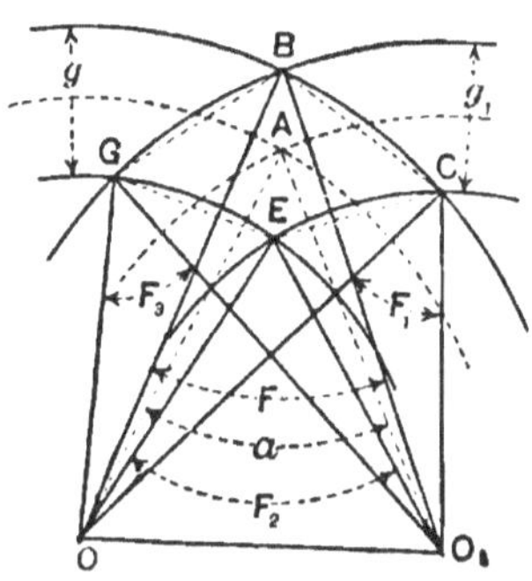

FIG. 95.

In Fig. 95, $OA = R$, $O_1A = R_1$, and angle $OAO_1 = a$ of the triangle OAO_1 are given; whence OO_1 may be determined.

In triangle OBO_1 the side $OB = R + \tfrac{1}{2}g$, $O_1B = R_1 + \tfrac{1}{2}g_1$, and $OO_1 = k$ are known, from which we can determine the angle $OBO_1 = F$.

In like manner from the triangle OCO_1 determine F_1, and from triangle OEO_1 find F_2. F_3 may be found from triangle OGO_1.

To find the chord GB first find angle BO_1O from triangle BO_1O, and angle GO_1O from triangle GO_1O; then

$$GB = 2(R_1 + \tfrac{1}{2}g_1)\sin\tfrac{1}{2}GO_1B. \quad (298)$$

In like manner,

$$EC = 2(R_1 - \tfrac{1}{2}g_1)\sin\tfrac{1}{2}EO_1C, \quad (299)$$

$$BC = 2(R + \tfrac{1}{2}g)\sin\tfrac{1}{2}BOC, \quad . \quad (300)$$

$$GE = 2(R - \tfrac{1}{2}g)\sin\tfrac{1}{2}GOE. \quad . \quad (301)$$

When the tracks intersect, as in Fig. 96, the solution is evidently similar to the foregoing.

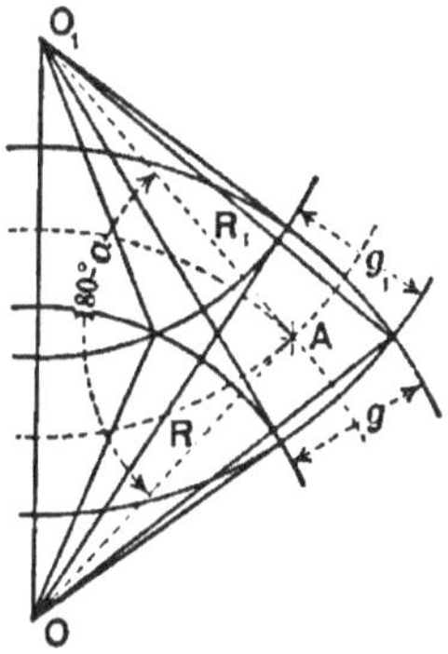

FIG. 96.

B. Crossing-slips.

194. A Crossing-slip is an arrangement of switch-rails, in connection with a set of crossing-frogs, to connect two tracks intersecting at a small angle.

195. Given the Angle of Intersection of Two Straight Tracks, to Find the Length and Radii of Curvature of Slip-rails.

In Fig. 97 determine EA and AB by **191**; then assume GE or BH (according as EA is less or greater than AB) as small as the crossing-frogs will permit. Draw the radii HO and GO; $AH = AG = k$ is the known tangent for the central angle F. Hence

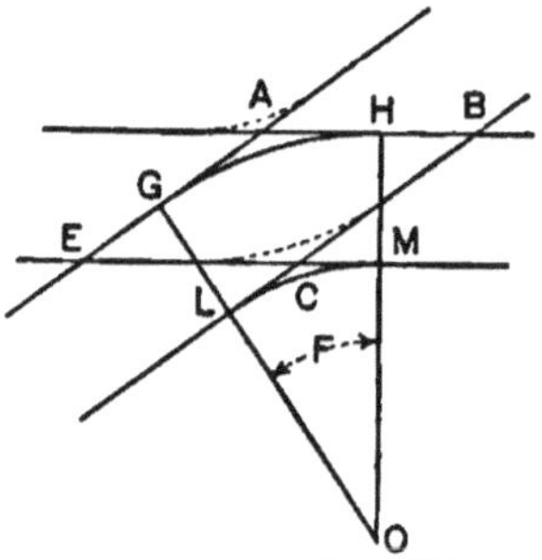

FIG. 97.

$$OG = R + \tfrac{1}{2}g$$
$$= AH \cot \tfrac{1}{2}F = 2kN, \quad (302)$$

$$OL = R - \tfrac{1}{2}g = 2kN - g. \quad . \quad . \quad . \quad . \quad . \quad . \quad . \quad . \quad . \quad . \quad (303)$$

For the theoretical length of rails,

$$GH = (R + \tfrac{1}{2}g) \times \frac{\pi}{180} \times F^\circ = (R + \tfrac{1}{2}g) \times \frac{F^\circ}{57.3}. \quad . \quad . \quad (304)$$

$$LM = (R - \tfrac{1}{2}g) \times \frac{F^\circ}{57.3}. \quad . \quad . \quad . \quad . \quad . \quad (305)$$

196. Given the Angle of Intersection made by the Center Lines of a Straight and a Curved Track, to Find the Radii and Length of Slip-rails.

FIRST CASE.—*Slip-rails inside main curve.*

In Fig. 98 determine the angles F and F_1 at B and C by **192**. Then assume KC as small as constructive reasons will permit. Now

$$\sin \tfrac{1}{2}KOC = \frac{\tfrac{1}{2}KC}{R + \tfrac{1}{2}g}. \quad . \quad . \quad . \quad . \quad . \quad . \quad . \quad . \quad (306)$$

$$b = BOK = (F_1 - F) - KOC, \quad . \quad . \quad . \quad (307)$$

$$c = KO_1H = F + b = F_1 - KOC. \quad . \quad . \quad (308)$$

From **129**, formula (100),

$$R_1 + \tfrac{1}{2}g = O_1H = \frac{(R + \frac{1}{2}g)(\cos F - \cos c)}{1 - \cos c}. \quad . \quad (309)$$

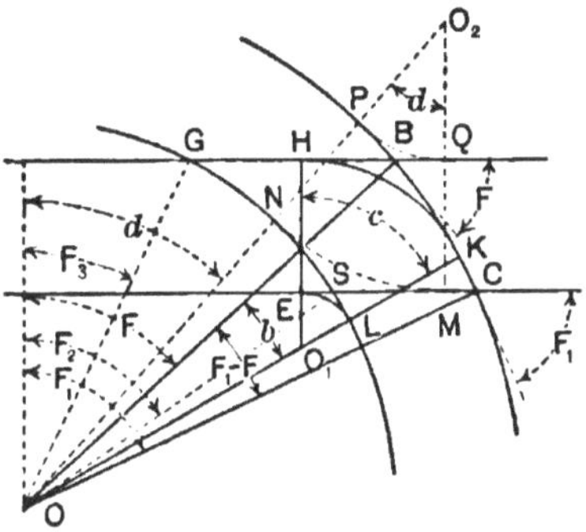

FIG. 98.

For the lengths of slip-rails,

$$HK = (R_1 + \tfrac{1}{2}g) \times \frac{c^\circ}{57.3}. \quad . \quad . \quad . \quad (310)$$

$$EL = (R_1 - \tfrac{1}{2}g) \times \frac{c^\circ}{57.3}. \quad . \quad . \quad . \quad (311)$$

SECOND CASE.—*Slip-rails outside main curve.*

Find the angle F_2 at S, F_3 at G, and GS by the methods of **192**. Assume GN as small as constructive reasons permit, and calculate angle GON, as in (306) Then $NOS = F_2 - F_3 - GON$, $d = F_3 + GON = F_2 - NOS$. By **129**, formula (106),

$$R_2 + \tfrac{1}{2}g = O_2M = \frac{(R - \frac{1}{2}g)(\cos d - \cos F_2)}{1 - \cos d}. \quad . \quad (312)$$

The remainder of the solution is similar to the first case.

197. Given the Angle of Intersection between the Center Lines of Two Curves, to Find Radii and Length of Slip-rails.

FIRST CASE —*Slip-rails on concave side of curves.*

In Fig. 99 take LC as small as constructive reasons permit. Join L with O; then

$$\sin \tfrac{1}{2}LOC = \frac{LC}{2(R + \frac{1}{2}g)}. \quad . \quad . \quad . \quad . \quad . \quad (313)$$

Determine angles COO_1, CO_1O, and side OO_1 by **193**. Make $LM = KO_1$; then

$$MO = (R + \tfrac{1}{2}g) - (R_1 + \tfrac{1}{2}g) = R - R_1,$$

and $$MOO_1 = LOC + COC_1.$$

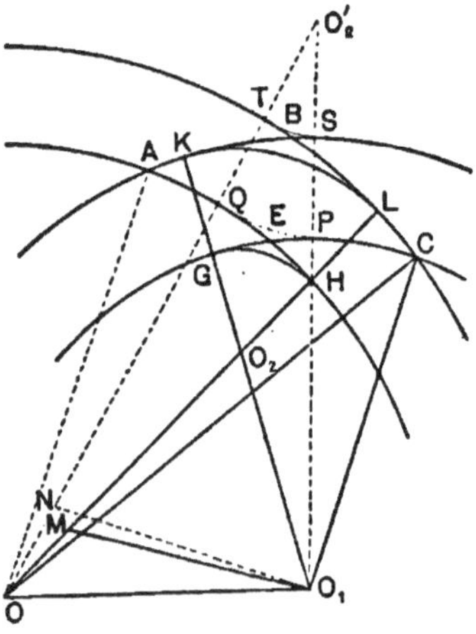

FIG. 99.

In triangle MOO_1 two sides and the included angle are now known, and the triangle may be solved. O_2 is the center of slip-rail curves.

$$O_2MO_1 = O_2O_1M = MOO_1 + MO_1O,$$

and $$MO_2O_1 = 180 - 2O_2MO_1.$$

From the isosceles triangle O_1O_2M, in which O_1M and the three angles are known,

$$MO_2 = \frac{MO_1}{2 \sin \frac{1}{2} MO_2O_1} \quad . \quad . \quad . \quad . \quad . \quad (314)$$

Then

$$R_2 + \tfrac{1}{2}g = O_2L = R_1 + \tfrac{1}{2}g - MO_2, \quad . \quad . \quad . \quad (315)$$

$$R_2 - \tfrac{1}{2}g = O_2H = R_1 - \tfrac{1}{2}g - MO_2. \quad . \quad . \quad . \quad (316)$$

The central angle $KO_2L = MO_2O_1$ being known, GH and KL may be found as in **196**.

SECOND CASE.—*Slip-rails on convex side of curves.*

Let the dotted lines of Fig. 99 represent this case. Assume AQ and compute angle AOQ; produce OQ to O_2', the center of slip-rail curve; make $O_2'N = O_2'O_1$. Reasoning as before, find $O_2'N = O_2'O_1$, after which $O_2'S$, $O_2'P$, and the lengths of TS and QP may be found as in the first case.

Should the curves intersect as in Fig. 96, no difficulty will be found in computing the radii and length of slip-rails by following the methods used above.

These methods furnish the *theoretical length* of slip-rails; but as the theoretical and physical switch-points do not coincide, the actual length will be considerably less.

CHAPTER VI.

CONSTRUCTION.

ARTICLE 17. DEFINITIONS; GENERAL CONSIDERATIONS; VERTICAL CURVES; SUPERELEVATION OF OUTER RAIL.

198. The work of locating the center line having been completed, the field corps is usually disbanded and a new one organized. The **Chief Engineer** still remains in charge, directing the work of construction, passing on bids and estimates, arranging contracts, and attending to such matters of importance as his assistants are unprepared or unauthorized to settle.

199. A Division Engineer is placed in charge of a considerable length of line, made up of several residencies. To him the resident engineers make reports, and from him receive directions and orders relating to construction. These reports will include monthly estimates, which are forwarded to the chief engineer for inspection and approval. Pay-rolls for the men employed are made out in the office of the division engineer, and forwarded to the chief.

200. A Resident Engineer is placed in charge of a few miles of line, called a **Residency**, and has direct charge of the construction. He should have at least two assistants—a rodman and an axeman—and it will be true economy to allow him also an assistant who can take his place at the instrument and assist in superintending construction.

The resident engineer is usually required to set slope-stakes, locate trestles and other bridges, tunnels, culverts, crossings, and other features preceding track-laying, and to make all measurements upon which estimates are based in determining the compensation of the contractor.

201. The Grade-line is determined from the profile, by stretching a fine thread along the paper and so adjusting its position that the proper relation between cut and fill is obtained, at the same time that the maximum gradient is not exceeded. The cuts and fills should be made as small as possible, at the same time that badly broken or chopped grades are avoided.

The Gradient is the rate of change of elevation of grade-line, and is usually expressed in per cents, a 1.2% gradient indicating a rise or fall of 1.2 feet in 100 feet horizontal. When the grade is ascending the gradient is marked plus, and when descending minus.

The word *grade* is frequently used instead of gradient.

The grade-line should be drawn on the profile in red ink, with the points of change marked by a cross or circle, also red. The elevations of these points and the gradients should be written in red above, or below, the grade and surface lines.

The nature of the work and the disposition of material from excavations, and the availability of outside material for embankments will determine whether or not the cuts and fills must balance. If this would necessitate a long haul, it will often be preferable to waste material from excavation and borrow for embankments.

A **Borrow-pit** is an excavation, adjacent to the line, from which material is taken to construct an embankment. It should be separated from the foot of the embankment by a space termed a **Berm**, which should increase with the height of embankment, never falling below a certain minimum width, say six feet. Borrow-pits should be regular in form, with sloping sides and drained so as to prevent water standing in them.

202. A Cross-section is a transverse section taken at each station, and at intermediate points where the longitudinal slope changes considerably, the surface between adjacent cross-sections being approximately such as would be generated by a straight line moving on these end sections as directors.

203. Slope-stakes are set at stations, to mark the points on cross-sections where the side slope meets the ground-surface. On them the cuts or fills are marked on the inside, while the outsides bear the station-numbers.

No slope-stakes are set at the pluses where cross-sections are taken, unless at the top or foot of bank where an opening is left for a bridge or culvert.

The notes are recorded, however, in order that the contents may be correctly calculated.

204. A Grade-point is a point on the intersection of the plane of the road-bed with the ground-surface. If the ground is level transversely, a single stake at the center, marked 0 0, will suffice to locate the point of passage from cut to fill. When the ground is not level transversely, the line of intersection will be oblique to the axis of the road and three grade-stakes are needed, one at the center and one at each side.

If the width of road-bed in excavation differs from the width in embankment, the stake should be set at the edge of the widest base.

205. To Find the Grade-point when the Ground Slopes Uniformly between Stations.

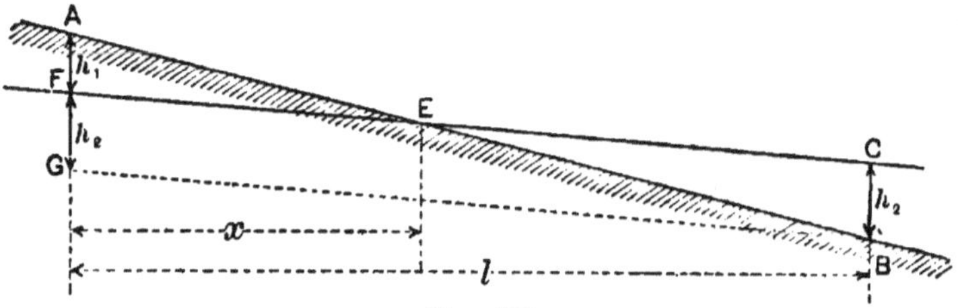

FIG. 100.

In Fig. 100 let AB be the ground-line, FC the grade-line, and E the grade-point. The horizontal distance, x, from A to E is required. Let the cut at A be h_1, the fill at B, h_2, and the length of prismoid l. Draw BG parallel to CF. From the similar triangles AEF and ABG

$$x = \frac{lh_1}{h_1 + h_2}. \qquad (317)$$

If the ground does not slope uniformly, the point E must be found by trial, such that the rod-reading equals the difference between height of instrument and elevation of grade.

206. Vertical Curves.—The angle formed by the junction of two grade-lines should be rounded off either by substituting several small changes for the one large one, or, preferably, by inserting a regular curve. Where the *algebraic difference* of gradients is less than 0.3% no curve will be needed, while for larger differences the length of vertical curve should vary with that difference, unless the circumstances of the case—such as the proximity of other vertical curves, or a bridge—should prescribe its length. In any case the length may be either assumed, or a given rate of change per station fixed upon and the length computed.

The parabola is especially well adapted for vertical curves, because of the ease with which any correction may be found when one is known, since, as will presently be shown, the corrections vary as the square of the distance from the point of tangency. A second property of this curve enables us readily to find the correction at the vertex, or meeting-point of grade-lines.

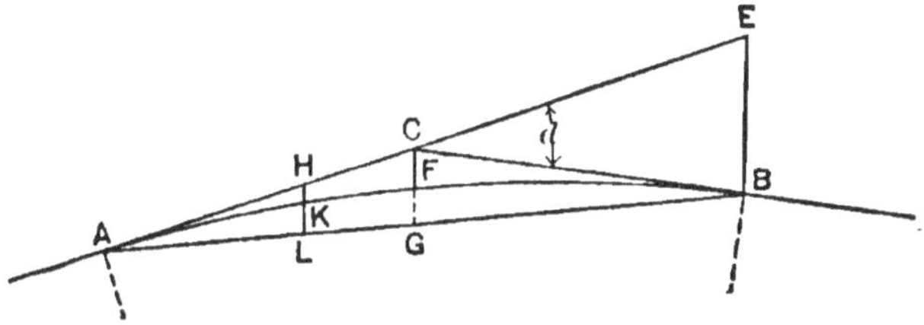

FIG. 101

In Fig. 101 let AC and CB be the intersecting grade-lines, and AFB the curve substituted for them. Produce AC to E to meet a vertical through B. Draw the vertical CG. Then will $CF = FG = m$ by the second property referred to. Since measurements are made horizontally, the similar figures ACG and AEB furnish the relation $CF = \frac{1}{2}CG = \frac{1}{4}EB$. Calling the *algebraic difference* of gradients d, and the length of curve $2l$,

$$m = \frac{1}{4}ld. \quad . \quad . \quad . \quad . \quad . \quad . \quad . \quad . \quad (a)$$

If the rate of change of gradient per station be a, it is evident that

$$l = \frac{d}{2a}. \quad . \quad . \quad . \quad . \quad . \quad . \quad . \quad . \quad (b)$$

The equation of the parabola referred to A as origin may be written

To find the correction $HK = z$ at a distance x from A, we have from the similar triangles AHL and ACG,

$$HL = CG\frac{x}{l} = 2m\frac{x}{l}. \quad . \quad . \quad . \quad . \quad . \quad . \quad . \quad (d)$$

But $KL = y$, and $z = HL - KL$; or, inserting values,

$$z = \frac{2mx}{l} - \left(\frac{2mx}{l} - \frac{mx^2}{l^2}\right),$$

or

$$z = m\frac{x^2}{l^2}. \quad . \quad . \quad . \quad . \quad . \quad . \quad . \quad (318)$$

Insert the value of d from (b) in (a), and the resultant value of m in (318); then

$$z = \frac{l^2a}{2} \times \frac{x^2}{l^2} = \frac{a}{2}x^2. \quad . \quad . \quad . \quad . \quad (318')^*$$

When $x = 1$ station, $z_1 = \frac{1}{2}a$: when $x = 2$ stations, $z_2 = 2a$, etc. It will only be necessary to figure corrections for one-half the curve, as they are the same for corresponding points each side of the vertex. If preferred, however, all corrections may be computed from the first tangent produced.

EXAMPLE.—A $+ 0.9\%$ meets a $- 0.6\%$ grade at sta. 181, the elevation of which is 91.0 ft. Required the corrections, and corrected grade elevations for points 100 ft. apart.

Here the algebraic difference of gradients is $0.9 - (-0.6) = 1.5$ Suppose a be taken as 0.25, or the length of curve as 6 stations.

Formula (a) gives $m = \frac{1}{4} \times 3 \times 1.5 = 1.125$ feet.

At the $P.C.$, sta. 178, $z = 0$; at 179, (318) or (318′) gives $z_1 = 0.125$; at 180, $z_2 = 4 \times 0.125 = .50$. The original and corrected grade elevations are as follows:

Sta.	178	179	180	181	182	183	184
Original elevation.	88.3	89.2	90.1	91.0	90.4	89.8	89.2
Corrections.......	0.0	0.125	0.50	1.125	0.50	0.125	0.0
Corrected elevat'n	88.30	89.075	89.60	89.875	89.90	89.675	89.20

* If a circle be taken as the joining curve we may derive (318′) by finding R in terms of a, then writing $D = 5730 \div R$, and $n = x$, in formula (36).

EXAMPLE 2.—A + 0.3% meets a + 1.1% grade at sta. 312, whose elevation is 155 0 Find corrections.

Take $a = 0.2$ in this case; then $l = \dfrac{+0.3 - (+1.1)}{2 \times .2} = 2$ stations. The corrected grade heights, etc., will be as follows:

Sta.	310	311	312	313	314
Original elevation	154.4	154.7	155.0	156.1	157.2
Corrections	0.0	0.1	0.4	0.1	0.0
Corrected elevation	154.4	154.8	155.4	156.2	157.2

The reason for adding the corrections in this case will be evident from a figure.

The table below gives the corrections in feet, for certain *algebraic differences* of gradients and lengths of curve, at intervals of 50 ft. each way from the vertex. When the difference of gradients is *plus*, the correction must be *subtracted* from the original grade elevation; when the difference is *minus*, the cor-

TABLE OF CORRECTIONS FOR VERTICAL CURVES.

Algebraic Difference of Gradients.	Rate of Change of Grade per Station.	Distance from Vertex in Feet.								
		0	50	100	150	200	250	300	350	400
0.3	0.075	0.15	0.08	0.04	0.01	0				
0 4	.10	.20	.11	.05	.01	0				
0.5	.125	.25	.14	.06	.02	0				
0 6	.15	.30	.17	.08	.02	0				
0.7	.175	.35	.20	.09	.02	0				
0 8	.20	.40	.23	.10	.03	0				
0.9	.225	.45	.25	.11	.03	0				
1.0	.25	.50	.28	.13	.03	0				
1.1	.1833	.83	.57	.37	.21	.09	.02	0		
1.2	.20	.90	.63	.40	.23	.10	.03	0		
1.3	.2167	.98	.68	.44	.24	.11	.03	0		
1.4	.2333	1.05	.73	.47	.26	.12	.03	0		
1.5	.25	1.13	.78	.50	.28	.13	.03	0		
1.6	.2667	1.20	.83	.53	.30	.13	.03	0		
1.7	.2833	1.28	.89	.57	.32	.14	.04	0		
1.8	.30	1.35	.94	.60	.34	.15	.04	0		
1.9	.2375	1.90	1.46	1.07	.74	.48	.27	.12	.03	0
2.0	.25	2.00	1.53	1.13	.78	.50	.28	.13	.03	0
2.1	.2625	2.10	1.61	1.18	.82	.53	.30	.13	.03	0
2.2	.275	2.20	1.68	1.24	.86	.55	.31	.14	.03	0
2.3	.2875	2.30	1.76	1.29	.90	.58	.32	.14	.04	0
2.4	.30	2.40	1.84	1.35	.94	.60	.34	.15	.04	0
2.5	.3125	2.50	1.91	1.41	.97	.63	.35	.16	.04	0
2.6	.325	2.60	1.99	1.46	1.02	.65	.37	.16	.04	0

rection must be *added.* Similar tables for other lengths of curve or differences of gradients may be computed by the engineer, and time in the field saved by their use. In setting grade-stakes, it will be well to set them 50 ft. apart on vertical curves, though to allow for the vertical curve at each regular station will suffice when cross-sectioning.

207. Elevation of Outer Rail on Curves.—In **138** it was shown that the superelevation of outer over inner rail might, for standard gauge track, be given, nearly enough, by the formula

$$e = \frac{V^2}{3R}, \qquad . \quad . \quad . \quad . \quad . \quad . \quad . \quad . \quad (134)$$

in which e is the elevation in feet, and V the velocity in miles per hour. Writing $R = \frac{5730}{D}$ in this formula gives

$$e = 0.000058\, V^2 D. \qquad . \quad . \quad . \quad . \quad . \quad . \quad . \quad (319)$$

The following table has been computed by formula (319).

TABLE OF SUPERELEVATIONS OF OUTER RAIL.

V in Miles per Hour.	Degree of Curve.										
	1°	2°	3°	4°	5°	6°	7°	8°	9°	10°	12°
20	.02	.05	.07	.09	.12	.14	.16	.19	.21	.23	.28
30	.05	.10	.16	.21	.26	.31	.37	.42	.47	.52	.63
40	.09	.19	.28	.37	.46	.56	.65	.74	.84	.93	
50	.15	.29	.44	.58	.73	.87	1.02	1.16			
60	.21	.42	.63	.84	1.04	1.25					

Since grade-stakes are set at the edge of base it will be necessary to determine the difference between these elevations and the elevations of center line. Calling this difference h, the half-base b, and the distance between centers of rail-heads for standard gauge 4.9 feet, we shall have, from similar triangles,

$$h = \frac{b}{4.9}\, e = 0.2be \text{ (nearly)}. \qquad . \quad . \quad . \quad . \quad (320)$$

If the inner rail is required to remain at grade (320) will become

$$h = 0.2be \pm 0.5e, \qquad . \quad . \quad . \quad . \quad . \quad . \quad . \quad . \quad (320')$$

according as the grade-stake is to be set at outer or inner edge of base.

Example.—What will be the value of h when $e = .46$, the base being 14 feet?

By (320), $h = 0.2 \times 7 \times 0.46 = 0.64$ feet. The outside is this much higher than the center, the inside edge this much lower.

The superelevation of outer rail should be computed for the highest speed at which trains are to be run over the curve; the maximum allowed in practice rarely exceeds 8 inches, since a greater elevation would endanger the slow-running freight trains. Even when the theoretical superelevation is given the outer rail, it is more worn than the inner one, either because there are other forces acting, or because of the sliding action of the outer wheel due to imperfect adjustment where the original coning has been destroyed by wear.

Engineers sometimes elevate the outer rail 1 inch per degree up to 3°, and make $e = 3\frac{1}{2}$ inches for a 4° curve, 4 inches for a 5° curve, and $4\frac{1}{2}$ inches for a 6° curve. Still other rules are in use.

If transition-curves are not employed, the difference of elevation is the same from *P.C.* to *P.T.*, fading out to nothing on tangent. The elevation begins on tangent from 50 to 200 feet back of *P.C.*, depending on the amount the outer rail is to be raised.

208. Easing Grades on Curves.—To compensate for the increased resistance due to curvature, it is customary to reduce the grade on curves. This resistance is taken to vary directly as the curvature; a rule often used is to reduce the gradient 0.05 foot per degree of curve

Article 18. Earthwork.

A. Setting Slope-stakes

209. Slope-stakes are set at the points where the side slopes meet the ground-surface, to mark the limits of the excavation or embankment, and to show the constructor what the cut or fill must be. In Fig. 102, *KAE* represents the ground-surface, *HBC* the grade-surface. Let $AB = h$ be the center height. Let $\frac{HL}{KL} = s$ be the side slope, which varies with the nature of the material; for earth-excavation the side slope will average about 1 to 1, so that $s = 1$, while for ordinary earth-embankment it will average about $1\frac{1}{2}$ to 1, so that $s = 1\frac{1}{2}$. The side height for level sections is the same as the center and may be found for any section, so that the distance LH is required. From the equation

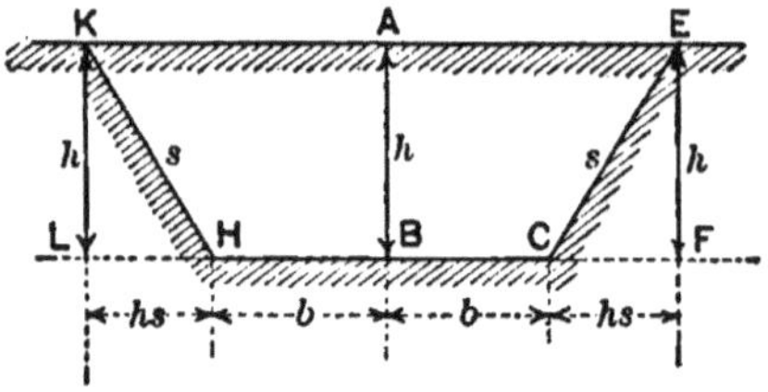

FIG. 102.

defining s, $HL = KLs = hs$. Let the base $HC = 2b$; the "distance out" from center is

$$d.o = BL = b + hs.$$

210. Surface Inclined.—Where the ground slopes transversely the position of the slope-stake cannot be found from the center height unless the slope of the ground-surface, as well as the side slope, is known. The slope-stakes can be most easily and rapidly set by trial.

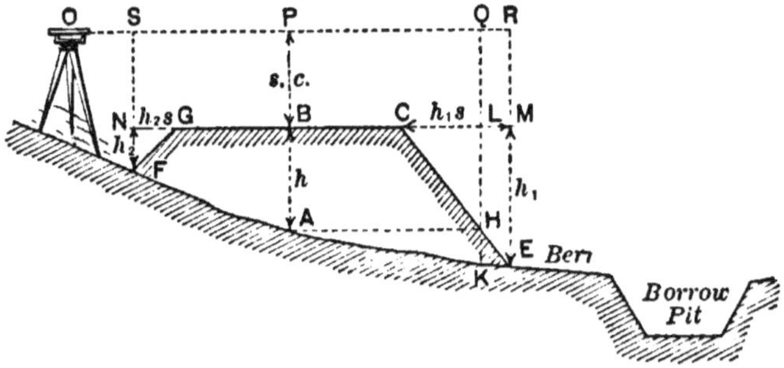

FIG. 103.

In Fig. 103, FAE is the ground-surface, $AB = h$ the fill at the center. We have to find the distances out of E and F from A, and the side heights $ME = h_1$, and $NF = h_2$. Let OP represent the plane of the instrument at a height $H.I.$ above the datum, obtained from the known elevation of a bench or turning-point. PB is the height of the plane of the instrument above the grade. Call PB the **Station Constant** ($s.c.$).

The fill at A will evidently equal the rod reading less the station constant. Mark this on the center stake.

Since the ground slopes downward from A to E, the distance out will be greater than for a level section, while for F, on the higher side, it will be less.

Suppose we take a reading QK at a distance out $= b + hs$; the fill at that point is $LK = QK - QL = r - s.c.$, and the corresponding distance out is $d.o. = b + KL \times s$, which is greater than AH, since LK is greater than h. If now a reading is taken at the distance out $b + s \times LK$, we shall have a fill greater than LK, unless the ground is level from K to E, and therefore $b + s \times LK$, the distance out actually used, will be less than that called for by the reading. However we shall have obtained a closer approximation to the position of E, and by repetitions of this process may come as close to its true position as the conditions require.

The same thing can be accomplished more rapidly by estimating the fall, by the eye, from A to a point $b + h_1 s$ out, then multiply this estimated fall by the slope and add to $b + hs$. Take a reading r at this distance out; then compute the $d.o.$ for the fill $r - s.c.$ and note if this agrees with the actual $d.o.$ If it does not, make a new trial with this reading as a guide.

For ordinary work the actual and computed distance out should be such that if the rod were held at the computed distance the new distance would not differ more than a tenth from that just computed. The stake is then set at the computed distance out. After a little practice it will be found that the second setting of the rod may usually be made to fall as close to the true position as the limit requires.

When the stake is marked and driven the cut or fill at that point and the distance out are recorded in the notes.

As an example suppose $2b = 14$ feet, $s = 1\frac{1}{2}$ (i.e., slope $1\frac{1}{2}$ to 1), $H.I. = 187.3$, grade elevation $= 184.0$. The station constant is $s.c. = 187.3 - 184.0 = 3.3$. Suppose the rod at center to read 8.5; the fill will be $8.5 - 3.3 = 5.2$, which mark on stake as "*F.* 5.2." The distance out, if section were level, would be $d.o. = 7 + 5.2 \times 1.5 = 14.8$; but suppose the ground rises and we estimate the rise as 1 foot, which multiplied by s gives 1.5 feet to be subtracted from 14.8, since this is on the higher side of center for a section in embankment. Let the reading at 13.3 out be 7.7, which gives a fill of $7.7 - 3.3 = 4.4$ feet, calling for a $d.o. = 7 + 4.4 \times 1.5 = 13.6$. This shows we are too far in, but as a reading further out will be less, giving a correspondingly smaller *d o.*, we try a reading at 13.5 feet out. Suppose the reading to be 7.6; the fill will be $7.6 - 3.3 = 4.3$, calling for a distance out of 13.45 feet, which agrees almost exactly with the trial distance. The stake is marked "*F.* 4.3," and the result recorded in the cross-section book.

On the other side of the section suppose we estimate the fall to be 1.5 feet in 15; we should try a reading at $13.8 + 1.5 \times 1.5 = 16.1$, say 16.0 feet. Let this reading be 9.0; the fill will be $9.0 - 3.3 = 5.7$ feet, calling for a *d.o.* $= 7 + 5.7 \times 1.5 = 15.6$, which shows our reading was taken too far out. Try a reading at 15.4, which suppose 8.9; the fill is $8.9 - 3.3 = 5.6$, and the *d.o.* $= 7 + 5.6 \times 1.5 = 15.4$, which agrees exactly with the trial distance.

In excavation the method of proceeding is the same as in embankment, except that s has generally a different value. For solid rock s is usually $\frac{1}{4}$, that is, the slope is taken as $\frac{1}{4}$ to 1; for loose rock, gravel, and ordinary earth the slope may be taken as 1 to 1.

The station constant in cuts is always positive, and the rod reading has to be subtracted from it to obtain the cut. In fills, when the *H.I.* is *greater* than the grade height, the fill equals the difference of the rod reading and the station constant. When the *H.I.* is *less* than the grade height the rod reading plus the *s.c.* gives the fill.

211. The Notes may be kept in the form below, which represents one page of the cross-section book. The cut or fill is written above the line, the distance out below. A plus sign indicates a cut, a minus sign a fill

Sta.	Ground.	Grade.	Left.	Center.	Right.	
161	178.8	184.0	$\frac{-4.4}{13.6}$	−5.2	$\frac{-5.6}{15.4}$	
162	181.6	183.0	$\frac{-0.8}{8.2}$	−1.4	$\frac{-2.0}{10.0}$	
+20	182.2	− 1.0%	$\frac{0.0}{9.0}$	−0.6	$\frac{-1.0}{8.5}$	
+48	182.5		$\frac{+0.9}{9.9}$	0.0	$\frac{-0.4}{7.6}$	
+66	183.5		$\frac{+2.4}{11.4}$	+1.2	$\frac{0.0}{9.0}$	
163	185.0	182.0	$\frac{+4.4}{13.4}$ $\frac{+2.8}{10.0}$	+3.0	$\frac{+4.3}{6.1}$ $\frac{+2.6}{11.6}$	

212. Irregular Sections.—When readings are taken only at the center and sides it is termed a "three-level section." Very irregular ground may require several more readings in order to determine its area; in this case a reading is taken at each change of surface in the section, and the cut or fill, together with the distance out, recorded—the distance being measured from the center to the point where the rod was held in taking the reading.

When the base cuts the ground-surface the section is partly in excavation and partly in embankment, but each side will be staked out in the manner described above. The distance of grade-point from center must be found and recorded.

213. Staking Out Openings.—Where openings are to be left for trestles, culverts, and other structures, stakes must be set to mark the limits of the embankment. Stakes marked T. B. are set at the center and sides to fix the place where the top of bank is to end; other stakes, marked F. S., are set at the foot of slope, the plus at which they fall—together with the distance out from center—being recorded in the note-book. The slope of the toe of dump should be the same as the side slope.

214. Marking Stakes.—All slope and toe stakes that limit excavation or embankment should be driven with tops inclined outward from the center. The cut or fill is marked on inside in plain figures preceded by the letter C. or F. as being more easily understood by the contractor than the plus and minus signs used in the notes. The reverse side should bear the station number.

215. Shrinkage—Growth.—It must be remembered that earthwork in embankment will settle, or *shrink* in volume, even after having been compacted by the feet of the teams during construction. Where the fill is not great, allowance may be made for shrinkage when setting grade-stakes, but in heavy fills allowance should be made when the stakes are set for construction. The proper allowance will vary with the nature of the material, but about 10 per cent will be a fair average. The contract should always specify the amount of shrinkage to be allowed on particular works. If the earth is measured in the borrow-pits, an equivalent allowance should be made, since earth is more compact in embankment than before excavating.

With rock, however, it is found that the volume increases

after excavation, and this increase is termed *growth*. The size of the fragments will determine the growth, which will vary from one half to five eighths of the original volume—the larger the fragments the greater the increase. Little or no allowance need be made for settlement when placed in embankment.

216. Borrow-pits should be regular in form, particularly if the volume of earth moved is to be measured in the borrow-pit. They should be properly drained to prevent water standing in them and should have an ample **berm** between edge of pit and foot of slope, the width of berm increasing with the height of embankment.

B. Areas of Sections.

217. Before the volume of earth in excavation or embankment can be computed the area of each cross-section must be found. To do this divide the section into triangles and trapezoids, find the area of each separately, and take the sum. To shorten the calculations a few simple rules will be deduced.

When the center and side heights are equal we have a **one-level section**; when the center and side heights differ it is a **three-level section**; where the height is found at five places in the section it is a **five-level section**; and so on.

218. To Find the Area of a Three-level Section.

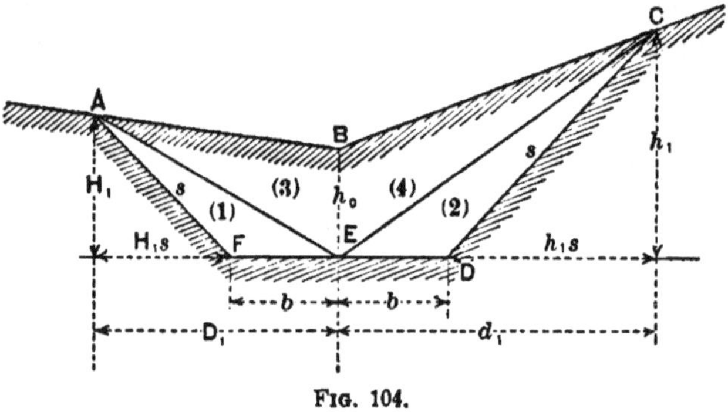

FIG. 104.

In Fig. 104 the area $ABCDF$ is required. Draw EA and EC, dividing the area into four triangles. With the notation of the figure it is seen that triangles (1) and (2) have equal bases b and altitudes H_1 and h_1, while (3) and (4) have the common base h_0 and altitudes $D_1 = b + H_1 s$ and $d_1 = b + h_1 s$. By geometry,

$$\text{Area } (1) + (2) = b\frac{H_1 + h_1}{2},$$

$$\text{Area } (3) + (4) = h_0\frac{D_1 + d_1}{2} = h_0\frac{2b + (H_1 + h_1)s}{2}.$$

The area of the whole section is therefore

$$A = b\frac{H_1 + h_1}{2} + h_0\frac{D_1 + d_1}{2} = b\frac{H_1 + h_1}{2} + h_0\frac{2b + (H_1 + h_1)s}{2}. \quad (321)$$

This formula affords an easy method of obtaining the area of a three-level section, and when written in words becomes the following

RULE.—*Multiply the half-sum of the side heights by the half-base and to this add the product of the center height by the half-sum of the distances out; the result will be the area.*

If the linear measurements are in feet, the area will be in square feet.

When $H_1 = h_1 = h_0$, then $D_1 = d_1 = d = b + h_0 s$ and (321) reduces to

$$A = bh_0 + h_0 d = h_0(b + d) = h_0(2b + h_0 s). \quad . \quad . \quad (322)$$

The section is now a trapezoid or one level section for which (322) may be deduced by the usual rule for the area of a trapezoid.

219. Area of Five-level Section.

Fig. 105 represents this case, where we may evidently divide

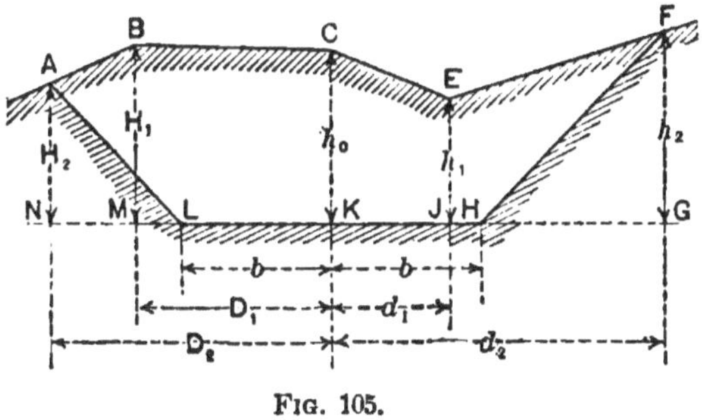

FIG. 105.

the area into triangles and trapezoids, computing the area of each separately and taking their sum for the whole area. The following simpler method may be preferred:

Write the notes as in field-book, except that center height is placed over zero and an additional $\frac{0}{b}$ is written at each end as below.

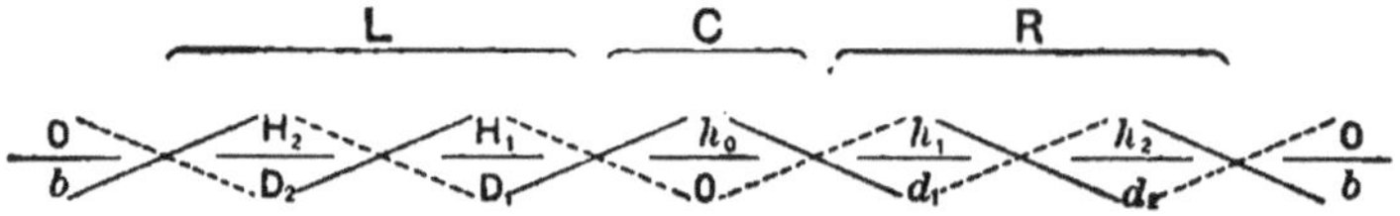

Beginning at the center, multiply heights by distances out in pairs as indicated by the sloping lines, the products of members connected by full lines being plus and of those connected by dotted lines minus. Half the sum of the products will be the area, thus:

$$A = \frac{1}{2}\left[\begin{pmatrix} h_0D_1 + H_1D_2 + H_2b \\ + h_0d_1 + h_1d_2 + h_2b \end{pmatrix} - \begin{pmatrix} 0 + D_1H_2 + 0 \\ 0 + d_1h_2 + 0 \end{pmatrix}\right]. \quad (323)$$

The grouping is symmetrical for areas each side of center, as (323) exhibits; so it will be sufficient to show that the rule is correct for either side. Divide the figure into trapezoids and triangles as shown; then

$$\begin{aligned} \text{Area trapezoid } BCKM &= \tfrac{1}{2}(h_0 + H_1)D_1\,, \\ \text{" \quad " \quad } ABMN &= \tfrac{1}{2}(H_1 + H_2)(D_2 - D_1), \\ \text{" \quad triangle } ALN &= \tfrac{1}{2}H_2(D_2 - b). \end{aligned}$$

The two trapezoids include the triangle; hence the latter must be subtracted from their sum. Doing this and simplifying, we have

$$A_L = \tfrac{1}{2}(h_0D_1 + H_1D_2 - D_1H_2 + H_2b),$$

which is the same as results in (323).

220. General Formula for Areas.—The method of **219** may be applied to any section no matter how irregular. Suppose there have been n levels taken on one side exclusive of the center height; the notes would appear as below:

$$\frac{h_0}{0} \quad \frac{h_1}{d_1} \quad \frac{h_2}{d_2} \quad \cdots \quad \frac{h_{n-1}}{d_{n-1}} \quad \frac{h_n}{d_n} \quad \frac{0}{b}$$

Expanding in the same manner as in **219**,

$$A_R = \tfrac{1}{2}[(h_0d_1 + h_1d_2 + \;..\; h_{n-1}d_n + h_nb) - (d_1h_2 + \;...\; d_{n-1}h_n)]. \quad (324)$$

To show that (324) gives the true area, consider that we have n

trapezoids whose area is positive, and one triangle whose area is negative and equal to $\frac{1}{2} h_n(d_n - b)$.

Writing out the area, we have

$$A_R = \tfrac{1}{2}[(h_0 + h_1)d_1 + (h_1 + h_2)(d_2 - d_1) + \dots (h_{n-1} + h_n)(d_n - d_{n-1}) - h_n(d_n - b)].$$

Performing the indicated operations and simplifying,

$$A_R = \tfrac{1}{2}[(h_0 d_1 + h_1 d_2 + \dots + h_{n-1} d_n + h_n b) - (d_1 h_2 \dots + d_{n-1} h_n),$$

which is the same result obtained in (324).

Evidently n may have any positive integral value.

If preferred, the cross-sections may be plotted on cross-section paper and the area read off by means of a planimeter.

221. Tables of Areas of Level Sections, and the Three-level Correction.—Formula (322) may be employed in computing the areas of level sections for any values of b and s.

Table XVII gives the areas for a few of these values. When many sections are to be figured it will be well for the engineer to compute the necessary tables, provided he is unable to secure published ones for the particular bases and slopes he is working with. It is not within the scope of this volume to give the variety of tables needed; they are published elsewhere.

The area of three-level sections may be found from the areas of level sections by the aid of a suitable correction. Let the height used in entering the tables of level sections be the mean height of the three-level section, $h_m = \frac{H_1 + 2h_0 + h_1}{4}$; the corresponding area, by (322), is

$$A' = h_m(2b + h_m s) = 2bh_m + h_m^2 s. \quad \dots \dots \quad (a)$$

The true area is given by (321):

$$A = b\frac{H_1 + h_1}{2} + h_0 b + h_0 \frac{H_1 + h_1}{2} s$$

$$= 2b\frac{H_1 + 2h_0 + h_1}{4} + 2h_0 \frac{H_1 + 2h_0 + h_1}{4} s - h_0^2 s$$

$$= 2bh_m + 2h_0 h_m s - h_0^2 s. \quad \dots \dots \quad (b)$$

From equations (a) and (b) the correction is

$$c = A' - A = (h_m^2 - 2h_0 h_m + h_0^2)s = (h_m - h_0)^2 s. \quad (325)$$

Table XVIII was computed by (325), and gives values of c which are always positive, and which must be subtracted from the tabular area, found by entering the table with the *mean* height of section, in order to get the true area.

EXAMPLE.—The side heights for a fill having a 14 ft. base and side slopes $1\frac{1}{2}$ to 1 are 8.6 and 16.4 feet, while the center height is 7.7 ft. The mean height is

$$h_m = \frac{8.6 + 2 \times 7.7 + 16.4}{4} = 10.1 \text{ ft.},$$

for which Table XVII gives $A' = 294.4$ sq. ft. For the correction, $h_m - h_0 = 10.1 - 7.7 = 2.4$ ft., for which Table XVIII gives $c = 8.6$ sq. ft.

The true area is now $A = 294.4 - 8.6 = 285.8$ square feet.

C. *Volume of Earthwork.*

222. Cross-sections must be taken at all full stations and at intermediate points, or pluses, where there is a change in longitudinal slope. It will be well in any event to take them so close together that the difference in end heights should not exceed about five feet. The time consumed in making these intermediate measurements will be more than offset by the reliability of the results. A few of the more usual methods of estimating quantities will be given here.

223. Averaging End Areas.—This is the easiest of application and therefore the most generally used, but is open to the objection that it gives inaccurate results. However, when bids are based upon it, both parties to the contract agreeing, it would seem to answer as well as any other method.

If A_1 and A_2 are the end areas and l the length, we shall have

$$V = \frac{A_1 + A_2}{2} l \quad . \quad . \quad . \quad . \quad . \quad . \quad . \quad (326)$$

Stated in words, (326) yields the following

RULE.—*Multiply the half-sum of the end areas by the axial length of prismoid; the result will be the volume.*

If areas are in square feet and length in feet, the volume will be in cubic feet; to reduce to cubic yards divide by 27.

224. Prismoidal Formula.—The parallel sections should be so taken that the surface bounding the volume to be measured may be supposed to be generated by a straight line moving on the bounding lines of the sections as directors in such a manner as to return to its original position. Such a figure is called a *prismoid*.

The height of any section intermediate between the end sections is a function of its distance from either end, and the area of that section will be a quadratic function of its distance from either end.

Now we know from mechanics that Simpson's (Newton's) Rule will hold for any function not higher than the third; so for the mean area this rule yields

$$\text{Mean area} = \frac{A_1 + 4A_m + A_2}{6},$$

where A_1 and A_2 are the end areas, A_m the middle area; hence we have for the volume

$$V = (A_1 + 4A_m + A_2)\frac{l}{6}, \qquad . \quad . \quad . \quad (327)$$

in which l is the axial length of prismoid.

The same result may be obtained geometrically by dividing the prismoid into prisms, wedges, and pyramids, and applying the usual rule for volumes.

Let the end areas be a_1, a_2, a_1', a_2', etc., and the mid-areas a_m, a_m', etc.

For the prism $a_1 = a_2 = a_m$; hence the volume is

$$v = a_1 l = (a_1 + 4a_m + a_2)\frac{l}{6}. \qquad . \quad . \quad . \quad . \quad (a)$$

For the wedge $a_1' = 2a_m'$ and $a_2' = 0$; therefore

$$v' = a_1'\frac{l}{2} = (a_1' + 4a_m' + a_2')\frac{l}{6}. \qquad . \quad . \quad . \quad (b)$$

For the pyramids $a_1'' = 4a_m''$ and $a_2'' = 0$; hence

$$v'' = a_1''\frac{l}{3} = (a_1'' + 4a_m'' + a_2'')\frac{l}{6}. \qquad . \quad . \quad (c)$$

Adding (a), (b), and (c), the total volume is

$$V = v + v' + v''$$

$$= [(a_1+a_1'+a_1'')+4(a_m+a_m'+a_m'')+(a_2+a_2'+a_2'')] \cdot \frac{l}{6}. \quad . \quad (d)$$

But $\qquad a_1 + a_1' + a_1'' = A_1$

$\qquad a_m + a_m' + a_m'' = A_m,$

and $\qquad a_2 + a_2' + a_2'' = A_2;$

therefore $\qquad V = (A_1 + 4A_m + A_2) \cdot \frac{l}{6},$

the same as (327).

Stated in words there results the following

RULE.—*To the sum of the end areas add four times the mid-area, multiply by the length, and divide by 6. The result will be the volume.*

To reduce to cubic yards, divide by 27.

Formula (327) contains three terms, the middle area being derived from the cross-section notes for the end sections at the expense of some little trouble. In the attempt to simplify this formula Dr. George Bruce Halsted in 1881 published a two-term prismoidal formula, giving the volume in terms of one base and a section at two thirds of the length of the prismoid, the formula being

$$V = (A_1 + 3A_{\frac{2}{3}})\frac{l}{4} = (3A_{\frac{1}{3}} + A_2)\frac{l}{4}. \quad . \quad . \quad . \quad (328)$$

In 1894 Professor W. H. Echols showed by the aid of higher mathematics that an indefinite number of two-term formulæ might be derived. The same results were established in 1895 by Professor T. U. Taylor by elementary mathematics.

None of these two-term formulæ have so far been placed in a form suitable for application to earthwork measurement, owing to the difficulty of finding the area of the auxiliary section.

In fact the only objection to the use of (327) is the loss of time required in obtaining the mid-area and the uncertainty as to its accuracy in the case of very irregular sections.

For three-level ground we may construct a section having heights that are means between corresponding end heights, but for very irregular sections there may be uncertainty as to what heights must be averaged to obtain the mid-section heights. For any other than the mid-sections the heights are obtained with more difficulty.

225. Form of Notes.—The record of areas and volumes may be kept in the form below, which represents the cross-section book, with the necessary columns added.

Sta.	Ground	Grade.	L.	C.	R.	End areas.	Mid-areas.	Exc. cu.yds.	Emb. cu.yds.
91	188.5	182.0	$\frac{+6.2}{15.2}$	+6.5	$\frac{+8.9}{17.9}$	+175.53			
			$\frac{+4.1}{13.1}$	+5.3	$\frac{+7.5}{16.5}$		+130.64		
92	185.4	181.0	$\frac{+2.0}{11.0}$	+4.1	$\frac{+6.1}{15.1}$	+89.96		486.4	
			$\frac{+1.0}{10.0}$	+2.1	$\frac{+3.0}{12.0}$		+41.10		
93	180.0	180.0	$\frac{0.0}{9.0}$	0.0	$\frac{0.0}{9.0}$	0.0		158.8	
			$\frac{-2.1}{10.2}$	−3.0	$\frac{-4.3}{13.5}$		−57.95		
94	173.0	179.0	$\frac{-4.2}{13.3}$	−6.0	$\frac{-8.6}{19.9}$	−144.40			232.2

If the method of averaging end areas is employed, the column of mid-areas will not be needed, and may even be omitted when computing by the prismoidal formula. In this case the notes for mid-section and the mid-area should be written in red ink.

An office record should be kept in addition to the record in the cross-section book, to which it will not be necessary to transfer the elevations of ground and grade. If preferred, the areas and volumes may be kept only in the office record, omitting them in the cross-section book.

226. Prismoidal Correction.—The time and labor required to obtain the area of the mid-section makes the use of the prismoidal formula objectionable; for this reason the method of averaging end areas is most often employed. The difference in the two methods will not be great, provided the difference in end heights is not over 3 or 4 ft.; it should never exceed 5 ft.

When the difference exceeds this a considerable error is introduced by the use of (326). It will generally be sufficient to average end areas and then apply a correction if the result must be free from large errors.

(*a*) **Correction for Level Sections.**—Between two level end sections the volume is made up of one prism, one wedge, and two pyramids. For the prism and wedge the true volume is given by averaging end areas, but for the pyramids the error is easily shown to be

$$\frac{2(H_0 - h_0)(H_0 - h_0)s}{2}\left(\frac{l}{2} - \frac{l}{3}\right) \text{ cubic feet,}$$

or

$$C = (H_0 - h_0)^2 s \frac{l}{6 \times 27} \text{ cubic yards.} \quad . \quad . \quad . \quad (329)$$

The table below gives the correction C in cubic yards, computed by (329), for a few values of $H_0 - h_0$, when $s = 1$ and $l = 100$; for any other length and slope multiply by $\frac{ls}{100}$.

TABLE OF PRISMOIDAL CORRECTIONS FOR LEVEL SECTIONS.

$H_0 - h_0$.	.0	.1	.2	.3	.4	.5	.6	.7	.8	.9
0	0.0	0.0	0.0	0.1	0.1	0.2	0.2	0.3	0.4	0.5
1	0.6	0.7	0.9	1.0	1.2	1.4	1.6	1.8	2.0	2.2
2	2.5	2.7	3.0	3.3	3.6	3.9	4.2	4.5	4.8	5.2
3	5.6	5.9	6.3	6.7	7.1	7.6	8.0	8.5	8.9	9.4
4	9.9	10.4	10.9	11.4	12.0	12.5	13.1	13.6	14.2	14.8
5	15.4	16.1	16.7	17.3	18.0	18.7	19.4	20.1	20.8	21.5
6	22.2	23.0	23.7	24.5	25.3	26.1	26.9	27.7	28.5	29.4
7	30.2	31.1	32.0	32.9	33.8	34.7	35.7	36.6	37.6	38.5
8	39.5	40.5	41.5	42.5	43.6	44.6	45.7	46.7	47.8	48.9
9	50.0	51.1	52.2	53.4	54.5	55.7	56.9	58.1	59.3	60.5
10	61.7	63.0	64.2	65.5	66.8	68.1	69.4	70.7	72.0	73.3
11	74.7	76.1	77.4	78.8	80.2	81.6	83.1	84.5	86.0	87.4
12	88.9	90.4	91.9	93.4	94.9	96.5	98.0	99.6	101.1	102.7
13	104.3	105.9	107.6	109.2	110.8	112.5	114.2	115.9	117.6	119.3
14	121.0	122.7	124.5	126.2	128.0	129.8	131.6	133.4	135.2	137.0
15	138.9	140.7	142.6	144.5	146.4	148.3	150.2	152.2	154.1	156.1

(*b*) **Correction for Three-level Sections.**—Formula (329) or the foregoing table may be used in determining the correction for three-level sections when these sections are somewhat similar and the corresponding heights not very different. A general formula may, however, be derived.

Let the center and side heights at one section be H_0, H_1, and H_2, respectively, and let the distance between slope-stakes be $W = D_1 + D_2$; let the corresponding heights at the other end section be h_0, h_1, and h_2 with a distance between slope-stakes of $w = d_1 + d_2$.

By formula (321) the areas will be:

$$\text{Area at first end} = \frac{b}{2}(H_1 + H_2) + \frac{W}{2}H_0, \quad . \quad . \quad . \quad (a)$$

$$\text{Area at second end} = \frac{b}{2}(h_1 + h_2) + \frac{w}{2}h_0, \quad . \quad . \quad . (b)$$

$$4 \times \text{mid-area} = 4\left[\frac{b}{2}\left(\frac{H_1 + h_1}{2} + \frac{H_2 + h_2}{2}\right) + \frac{1}{2}\cdot\frac{W + w}{2}\cdot\frac{H_0 + h_0}{2}\right]$$

$$= b(H_1 + h_1 + H_2 + h_2) + (W + w)\frac{H_0 + h_0}{2}. \quad (c)$$

From (a), (b), and (c) the volume by the prismoidal rule is

$$V = \left[\tfrac{3}{2}b(H_1 + h_1 + H_2 + h_2) + W\left(H_0 + \frac{h_0}{2}\right) + w\left(h_0 + \frac{H_0}{2}\right)\right] \times \frac{l}{6}. \quad (d)$$

From (a) and (b), by averaging end areas,

$$V' = \left[\frac{b}{2}(H_1 + h_1 + H_2 + h_2) + \frac{WH_0}{2} + \frac{wh_0}{2}\right] \times \frac{l}{2}$$

$$= [\tfrac{3}{2}b(H_1 + h_1 + H_2 + h_2) + \tfrac{3}{2}WH_0 + \tfrac{3}{2}wh_0] \times \frac{l}{6}. \quad . \quad . \quad (e)$$

From (d) and (e) the correction is

$$V' - V = \left[\frac{W}{2}(H_0 - h_0) + \frac{w}{2}(h_0 - H_0)\right]\frac{l}{6}$$

$$= (H_0 - h_0)(W - w) \times \frac{l}{12} \text{ cu. ft.} \quad . \quad . \quad (330)$$

The correction in cubic yards is

$$C = (H_0 - h_0)(W - w) \times \frac{l}{12 \times 27}; \quad . \quad . \quad (331)$$

when $l = 100$,

$$C = 0.31(H_0 - h_0)(W - w). \quad . \quad . \quad . \quad . \quad (331')$$

This correction may be either positive or negative.

Example.—Compute the correction for the two prismoids below.

Sta.	L.	C.	R.
160	$\frac{+5.4}{14.4}$	+3.0	$\frac{+6.6}{15.6}$
161	$\frac{+10.8}{19.8}$	+7.0	$\frac{+11.2}{20.2}$
162	$\frac{+5.2}{14.2}$	+9.0	$\frac{+4.8}{13.8}$

For the prismoid between 160 and 161

$$C = (3.0 - 7.0)(30.0 - 40.0) \times \frac{100}{12 \times 27} = +12.3 \text{ cu. yds.}$$

For the prismoid between 161 and 162

$$C = (7.0 - 9.0)(40.0 - 28.0) \times \frac{100}{12 \times 27} = -7.4 \text{ cu. yds.}$$

When the correction C is positive it must be *subtracted from*, and when negative *added to*, the volume as found by averaging end areas; this is evident from (330). C will be negative when the smaller center height and greater width between slope-stakes occur at the same station, as is illustrated in the example.

The general tendency, however, with the method of averaging end areas is to give volumes that are too large, and the error increases with the square of the difference in end heights, as is evident from (329).

227. When the center and sides of roadbed do not pass from cut to fill at the same station we have a volume, part excavation, part embankment, between the same pair of sections, such as is illustrated in Fig. 106.

Between sections ABC and $EDFML$ the solid having bases

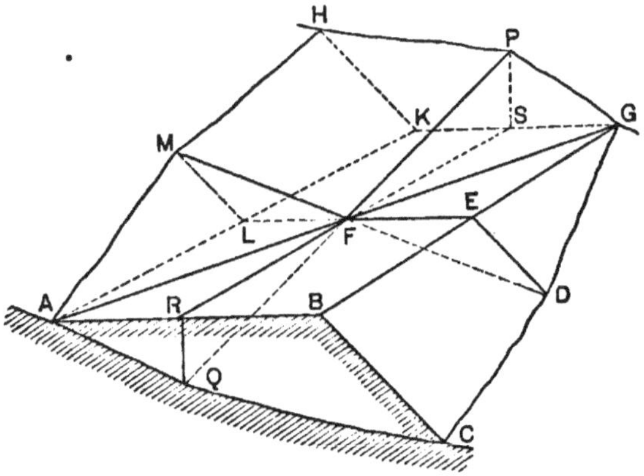

FIG. 106.

ABC and FED is embankment, while the pyramid $FML\text{-}A$ is in excavation.

For such cases as this the method of averaging end areas is most in error, particularly if the ground have a sharp longitudinal as well as transverse slope. Whatever method is employed, the excavations and embankments must be separately computed.

228. Tables of Volumes for Level Sections and Equal End Areas may be used in making preliminary estimates. The average center height for one or more stations is taken from the profile and the volume at once read off from tables, such as Table XIX.

Table XX may be used in finding the volume, after having averaged the end areas, and a correction made by **226** if desired.

229. Side Ditches in cuts have a constant cross-section, and hence a constant volume for each full station. Their contents are separately computed and added after the other computations have been made. They need not be shown in cross-section notes.

230. Earthwork on Curves.—In computing quantities on curves the end sections are assumed to be parallel, and the axial distance between sections taken as the length of the prismoid. If the volume be taken as generated by a moving section, and the center of gravity of this section lie always on a vertical line passing through the axis, this method gives correct results; otherwise not. The result will be too small or too large according as the center of gravity falls without or within the center line of curve.

If the volumes are computed by averaging end areas, it will be a useless refinement to apply a curvature correction; but if the prismoidal formula is employed, and accuracy is desired, it should be applied, especially if the work be in rock.

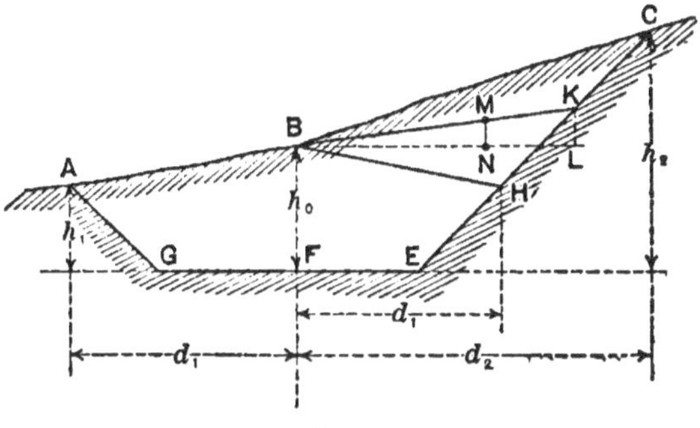

FIG. 107.

To find the curvature correction (*c.c.*) consider Fig. 107, which represents the *mean* section of the prismoid.

The portion *ABHEG* has its center of gravity on the line *BF* (*BH* having the same slope as *BA*); hence the path of its center of gravity will be the same length as the axis of the prismoid, and there will be no error in the computed volume generated by this portion. In the triangle *BCH* draw *BK* to the mid-point of *CH*. The center of gravity of this triangle is at *M*, two thirds of the distance *BK* from *B*. Now, by Guldin's rule (theorem of Pappus) the volume generated equals the area multiplied by the path of the center of gravity, the center of rotation being in the plane of the area.

Draw *BL* horizontal and take *N* on a vertical through *M*; let the angle in degrees at the center be θ.

The volume generated by the triangle *BCH* is

$$V = BCH\frac{\pi\theta^\circ}{180}(R + BN).$$

But the calculated volume is

$$V_0 = BCH \times l = BCH\frac{\pi\theta^\circ}{180}R.$$

Hence the curvature correction will be

$$c.c. = V - V_0 = BCH\frac{\pi\theta^\circ}{180}BN.$$

$$\text{But } BN = \frac{2}{3}BL = \frac{2}{3}\cdot\frac{d_1 + d_2}{2} = \frac{d_1 + d_2}{3}$$

$$\therefore\ c.c. = \frac{\pi}{540}BCH(d_1 + d_2)\theta^\circ = .006BCH(d_1 + d_2)\theta^\circ. \quad (332)$$

When the sections are 100 ft. apart $\theta^\circ = D$ and the correction becomes

$$c.c. = .006BCH(d_1 + d_2)D. \quad . \quad . \quad . \quad . \quad (332')$$

The area of the triangle *BCH* is easily seen to be

$$A = \tfrac{1}{2}[b(h_2 - h_1) + h_0(d_2 - d_1)]. \quad . \quad . \quad (333)$$

If the triangle *BCH* is on the convex side of curve the correction must be added, if on the concave side it must be subtracted.

For light work the correction is small, but for heavy work with steep transverse slope on sharp curves it may be considerable. In practice we may use the mi'   for the mean area without material error.

EXAMPLE.—Find the correction per station on an 8° curve, 28

ft. base, side slopes $1\frac{1}{2}$ to 1, inside height 10 ft., outside height 30 ft., end sections equal.

231. Overhaul.—Contract prices are usually based on a certain maximum length of haul, and all material carried farther than this is termed overhaul, for which the contractor receives extra compensation.

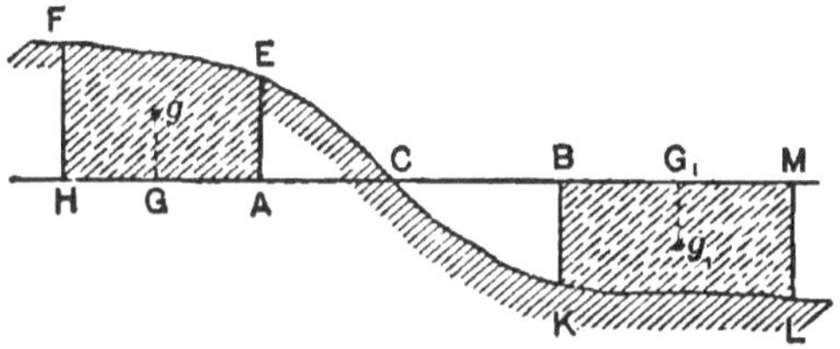

FIG. 108

In Fig. 108 let AB be the length of free haul, the points A and B being fixed on the profile so that the volume ACE equals the volume CBK; this may be done by trial computations, or closely enough in some cases from the profile alone. Let the mass $EFHA$ be removed to $BKLM$, and let the centers of mass in the two positions be at g and g_1 respectively; the length of overhaul to be paid for will be $GG_1 - AB = GA + BG_1$. To find g and g_1 accurately requires that the sum of the moments of the elementary masses equal the moment of the whole mass with respect to any chosen point. It will answer in practice to multiply the volume per station by the distance of its center of mass (found by dividing the station length in the inverse ratio of its end areas) from some selected point, as C, and equate this to the product of the whole mass by the unknown distance of its center of gravity from the same point, then solve for this distance. Indeed, it will answer in most cases to find a point that divides the mass into two equal parts and treat this as the center of gravity; such a point may be readily found by trial.

Sometimes it is specified that the overhaul must be found by finding the distance of the center of gravity of the *whole* mass moved from the center of gravity of the same mass after depositing in embankment and deducting from this the length of free haul, the remainder being called the overhaul. This is the easier method, but requires every yard moved to be carried the entire length of free haul before any overhaul whatever is counted.

EXAMPLE.—Let $AEFH = 5000$ cu. yds., $GA = 200$ ft., G_1B

= 300 ft., and the price paid for overhaul 1 cent per cubic yard per 100 ft.

The additional compensation above the contract price will be

$$\frac{200 + 300}{100} \times 5000 \times .01 = \$250.00.$$

Article 19. Grade and Ballast Stakes, Culverts, Bridges, and Tunnels.

232. Grade and Center Stakes.—After the excavations and embankments have been brought approximately to the level called for on the cross-section stakes, the engineer must go carefully over the road, setting center stakes every hundred feet on tangents and flat curves, and every 50 or 25 feet on sharp curves—the distance between center stakes depending on the sharpness of the curves. On tangents it will be sufficient to drive a grade-stake beside each center stake, so that its top will be at the height to which the finished surface must come, due allowance being made for shrinkage.

On curves grade-stakes must be set at each side a distance equal to the half-base from the center; the proper elevation or depression of these stakes must be found by **207**, formula (320).

The *P.C.*'s and *P.T.*'s are recovered by means of the reference-points set during location.

233. Ballast-stakes are set on the completed sub-grade at the proper width of ballast-base—just as in slope-staking—with their tops at the level of the final grade. They should be set at intervals of 50 ft. on tangents and flat curves, and at 25 ft. on sharp curves.

234. Track Centers are set for the guidance of trackmen as soon as the road-bed is ready to receive the cross-ties and rails.

235. The Opening left for a culvert, drain, or trestle bridge is measured from top of bank to top of bank; the manner in which it should be staked out is described in **213**.

A note of the size of drain and the material of which it is to be built, whether glazed earthenware pipe, box drain, stone culvert, etc., should be made in the note-book opposite the notes for the opening.

After the culvert or drain has been built the earth is filled in over and around it, and face or wing walls built to protect the bank at the points where culvert or drain meets its face.

For trestle bridges it must be remembered that the bank-sills set back from the top of bank a distance sufficient to give firm bearing, usually about 6 ft. for ordinary earth, and allowance made therefor in staking out the opening. The length of opening is designated by the number of bents between bank-sills: thus a 12-bent opening, where the distance between bents is 14 ft., would be $13 \times 14 - 12 = 170$ ft. The bent spacing depends upon the size of timbers available and upon the weight of locomotives to be run over the road.

Whatever the nature of the structure, ample waterway should always be provided for the heaviest storms; failure to do this is the cause of many a costly wreck.

Center stakes are set for each trestle-bent, and if piles are to be driven a stake should mark the position of each pile. If the bridge is not at right angles to the stream it will often be best to set the bents askew, but this should be avoided whenever possible. After the piles have been driven **cut-off levels** are given by the engineer, for which a tack is set in the pile at a definite distance below the point of cut-off, allowance being made for cap, stringer, etc. If the bridge is on a grade, the rate of rise per bent must be figured out and allowed for. On curves the proper superelevation of outer rail must be computed by the method of **207**.

For details of trestles see Foster's *Trestle Bridges*.

236. The Piers and Abutments for truss bridges must be very accurately located, the spacing being done with a steel tape whose constants are known, and the center and limits being marked by stakes. On tangents the centers are easily located and referenced, but on curves this is not so easy, as the center of track cannot be taken as the center of pier on account of the clearance necessary for trains.

Bridges on curves should be avoided whenever possible, but when they cannot be avoided the centers of piers are to be placed at the intersection of pier-axis and "bridge-chord."

In Fig. 109 ABC is the center line of track, AE and CF the pier-axes. At the mid-point of the arc AC the tangent EF, parallel to AC, is drawn; make $AN = NE = CL = LF$, and draw NL, which is the bridge-chord. The points N and L are the centers of the piers

Should L or N be inaccessible, they may be located from a point P on some accessible portion of the curve. To do this take PQ perpendicular to LN, such that

$$PQ = HB - KB = R(\text{vers } b - \tfrac{1}{2} \text{ vers } a); \quad . \quad . \quad (334)$$

then will

$$QL = QK + KL = R(\sin b + \sin a). \quad . \quad . \quad . \quad (335)$$

The manner of building the piers, determining the nature of the foundation, and erecting the bridge come properly within

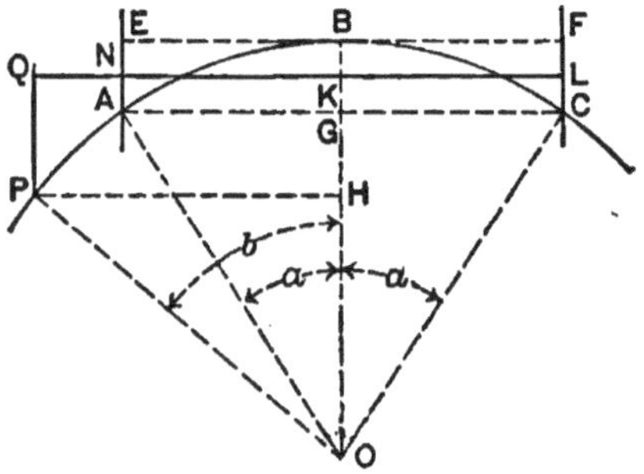

FIG. 109.

the province of the bridge engineer and require too much space to be outlined here. For preliminary estimates it will often be sufficient for the locating engineer to make soundings with gas-pipe in order to determine the depth to a suitable foundation and the nature of the overlying deposits, the core forced up within the pipe serving for the latter purpose.

237. Tunnels, like bridges, require great nicety in the measurements by which they are constructed. The angular measurements should be made with the best available transit in the best possible adjustment, and repetitions and reversals made to eliminate errors as much as possible. Linear measurements should be made with a steel tape the constants of which are known, so that correction may be made for temperature, etc.

If headings are to be driven from the ends and an unobstructed view of the summit is obtainable, a point may be fixed in the same vertical plane as the axis of the road, and will serve for giving the alignments of the headings.

Sometimes several points must be located on the mountain in the plane of the axis, and triangulation resorted to to secure the desired end.

The most accurate work can often be done at night, sightings being made to a plummet-lamp, or in the early morning before the sun's heat has produced great changes in the density of the air.

Within the tunnel, alignment is made by sighting to a plummet-lamp suspended from a plug 'et into the roof. Work is usually carried on from both ends, so that it is necessary to secure accurate alignment. If the entire tunnel is on a tangent, this is not difficult when working from the ends; but when the tunnel is on a curve (the curve falling most often at the ends), or when alignment must be transferred down a shaft, the operation is much more difficult.

The Mont Cenis Tunnel, over seven miles long, was constructed from the ends—one end being on a curve—yet there was no trouble in making a fit where the headings met.

When headings are driven from the foot of a shaft it is necessary to secure a point in the surface on each side of shaft in the plane of tunnel-axis and to transfer these points by plumb-lines to the bottom. By connecting these transferred points the direction of the axis may be secured. In the Hoosac Tunnel a line was transferred down a shaft 1000 ft. deep and carried 2050 ft. with a final error of only nine sixteenths of an inch.

Levels are run over the surface with great care, and may be transferred down a shaft by measuring its depth with a rod or steel tape. The grade of the bottom must be sufficient for drainage.

The dimensions of the tunnel will depend on the height of engines and the purposes for which it was intended.

For detailed information regarding tunnels the student is referred to Drinker's and Sims's books on the subject and to the current engineering journals.

Article 20. Monthly and Final Estimates.

238. Monthly Estimates are made by the engineers in charge of construction about the end of each month, and upon these the division engineer bases his estimate, which he forwards to the chief engineer for approval. The contractor receives his compensation some days later, usually about the 15th or 20th of the month following. Monthly estimates should always be based on actual measurements and never guessed at, particularly if several classifications are to be made. The total quantity of work done or material delivered is to be estimated, then the difference between any estimate and the last preceding one will be the estimate upon which the contractor receives his installments.

239. For Earthwork, measurements (when needed) are only approximate, but it is best to make them with level and tape even for monthly estimates. It will be sufficient to compute volumes by averaging end areas, no attention being paid to the prismoidal correction. Care must be taken, however, that such estimates are not in excess—in fact, it is well to keep slightly within the actual quantities on account of the greater cost and labor required to finish the work, which would make the latter part appear so much less profitable to the contractor as sometimes to induce a disposition to abandon the work before completion.

240. The Classification of Earthwork.—It is customary to group earthwork in excavation, according to the difficulty of removal, into three classes—earth-excavation, loose rock, and solid rock—though other classifications are frequently made.

Earth-excavation includes all earth, sand, loam, and loose stones that can be moved with the plow and scraper.

Loose rock includes all stones and detached boulders less than from 1 to 3 cubic yards in size, and all slate, shale, or cemented gravel requiring the use of the pick and bar, but which may be removed without blasting.

Solid rock includes all boulders above a certain size (usually from 1 to 3 cubic yards, as specified) and all rock masses that cannot be removed without blasting.

The relative prices vary, but a ratio of 1 : 3 : 7 will not be far from an average for the more common conditions arising in railroad work.

The necessity for a correct classification is evident, and the engineer should keep full notes, and make careful measurements whenever a given volume involves more than one class of earthwork. It is customary to specify that his decision is final, and therefore his measurements should be carefully made during the progress of the work, and notes on the nature of material taken at the same time.

A note-book should be kept showing the measurements and amounts of each class of material for each station, together with the date of completion and acceptance.

241. A Progress Profile should accompany the monthly estimate to exhibit graphically the amount of work done during the month, different colors being used for the different months. The final profile should show approximately the progress of the work. The colors may be laid on with a brush, or hatchings made with a pen; in neither case should the color obscure the lines of the profile-paper. A duplicate progress profile should be retained in the division engineer's office; if transparent profile-paper is employed, one may be simply traced through from the other. A further advantage of the transparent paper is that blue-prints of any portion of the profile may be readily made when duplicates are desired, provided the drawings are in black or any color admitting blue printing.

242. Masonry is to be measured in cubic yards, and any material on hand, but not in place, is to be measured and estimated. The classification of masonry must be according to specifications. Foundation-pits for piers or culverts must be measured as soon as completed, and before the masonry has been put in place.

243. Bridges must be estimated by measurement, or by checking up material in place and that on hand but not in place.

For trestle bridges, or foundations requiring piling, the actual number of linear feet below cap must be measured; this necessitates the constant supervision of the engineer or an assistant, sometimes known as a "pile-recorder," whose duty it is to see that all piles come up to specifications and are driven in accordance therewith.

All framing-timber in place, or delivered but not in place, is to be included in the estimate, the amount being obtained by measurement.

Steel spans or trestles are to be estimated, in the same manner as wooden trestles, by checking up or measuring the material on hand and in place.

244. Track Material must be checked up either by the "material clerk" or the engineer in charge of track. Ballasting properly belongs with the graduation, but may be put in place after the rails have been laid; in either case it is estimated in accordance with the specifications.

For preliminary and monthly estimates it will be sufficient to estimate track material by means of tables showing the number of cross-ties for a given spacing and the weight of steel for a given rail section, but before the final estimate is made all material must be measured or counted.

245. Blank Estimate-sheets are sent out from the chief engineer's office to be filled out by the engineers making estimate, who should retain a copy of each estimate rendered. On these sheets should appear the total quantity estimated, the amount of the last preceding estimate, and the estimate for the month, which will be the difference of the other two.

The division engineer's estimate must show not only the quantity of material, but its value in dollars and cents computed from the contract price. The footings of the several columns then serve as a check upon each other.

246. The Monthly Payments are not made for the full amount estimated, but about 15 or 20 per cent is retained until after the final estimate has been made, in order to insure the completion of the work by the contractor, and to be used as a fund from which to withhold the amount of damages provided in the contract for failure to comply with all its provisions.

247. Extras incident to minor changes, or to the protection or drainage of the work, are usually shown on the final estimate, but a better way would be to require the contractor to present his bill for extras at the end of each month, and to incorporate them in the monthly estimate when they are just. The engineer should take measurements upon any extra work at the time of its completion, and should keep a record thereof. If the extras are of a nature not admitting of measurement, he should note the compensation to be allowed at the time the extra work is done.

248. The Final Estimate must include all earthwork moved, all material in bridges, all masonry in foundations, culverts, piers, and tunnels, and all other material supplied or work done in compliance with the contract. The engineer should keep his notes full and complete during the construction of the work, in order to be able to meet the contractor's claims for extras or complaints as to classification. Any items that may have been overlooked in making up the monthly estimates must be included here.

249. Acceptance.—Until the engineer has pronounced the work satisfactory and formally accepted it the contractor is liable for its condition, and must make good all damage caused by accident or storm. The road-bed and track may be accepted without special test; but all spans should be subjected to a specified test-load, under which they must show not more than a certain maximum deflection, so their acceptance will come last.

Sometimes the contract requires a particular structure or class of structures to be maintained in good order for a certain length of time after completion, and a percentage is retained to cover the case.

After final acceptance the work is paid for in accordance with the final estimate.

249. Acceptance.—Until the engineer has pronounced the work satisfactory and formally accepted it the contractor is liable for its condition, and must make good all damage caused by accident or storm. The road bed and track may be accepted without special test; but all spans should be subjected to a specified test load, under which they must show not more than a certain maximum deflection, so their acceptance will come last.

Sometimes the contract requires a particular structure or class of structures to be maintained in good order for a certain length of time after completion, and a percentage is retained to cover the case.

After final acceptance the work is paid for in accordance with the final estimate.

TABLES.

Deg.	Radius.	Deg.	Radius.	Deg.	Radius.	Deg.	Radius.	Deg.	Radius.
0° 0'	Infinite	1° 0'	5729.65	2° 0'	2864.93	3° 0'	1910.08	4° 0'	1432.69
1	343775.	1	5635.72	1	2841.26	1	1899.53	1	1426.74
2	171887.	2	5544.83	2	2817.97	2	1889.09	2	1420.85
3	114592.	3	5456.82	3	2795.06	3	1878.77	3	1415.01
4	85943.7	4	5371.56	4	2772.53	4	1868.56	4	1409.21
5	68754.9	5	5288.92	5	2750.35	5	1858.47	5	1403.46
6	57295.8	6	5208.79	6	2728.52	6	1848.48	6	1397.76
7	49110.7	7	5131.05	7	2707.04	7	1838.59	7	1392.10
8	42971.8	8	5055.59	8	2685.89	8	1828.82	8	1386.49
9	38197.2	9	4982.33	9	2665.08	9	1819.14	9	1380.92
10	34377.5	10	4911.15	10	2644.58	10	1809.57	10	1375.40
11	31252.3	11	4841.98	11	2624.39	11	1800.10	11	1369.92
12	28647.8	12	4774.74	12	2604.51	12	1790.73	12	1364.49
13	26444.2	13	4709.33	13	2584.93	13	1781.45	13	1359.10
14	24555.4	14	4645.69	14	2565.65	14	1772.27	14	1353.75
15	22918.3	15	4583.75	15	2546.64	15	1763.18	15	1348.45
16	21485.9	16	4523.44	16	2527.92	16	1754.19	16	1343.15
17	20222.1	17	4464.70	17	2509.47	17	1745.26	17	1337.65
18	19098.6	18	4407.46	18	2491.29	18	1736.48	18	1332.77
19	18093.4	19	4351.67	19	2473.37	19	1727.75	19	1327.63
20	17188.8	20	4297.28	20	2455.70	20	1719.12	20	1322.53
21	16370.2	21	4244.23	21	2438.29	21	1710.56	21	1317.46
22	15626.1	22	4192.47	22	2421.12	22	1702.10	22	1312.43
23	14946.7	23	4141.96	23	2404.19	23	1693.72	23	1307.45
24	14323.6	24	4092.66	24	2387.50	24	1685.42	24	1302.50
25	13751.0	25	4044.51	25	2371.04	25	1677.20	25	1297.58
26	13222.1	26	3997.49	26	2354.80	26	1669.06	26	1292.71
27	12732.4	27	3951.54	27	2338.78	27	1661.00	27	1287.87
28	12277.7	28	3906.54	28	2322.98	28	1653.01	28	1283.07
29	11854.3	29	3862.74	29	2307.39	29	1645.11	29	1278.30
30	11459.2	30	3819.83	30	2292.01	30	1637.28	30	1273.57
31	11089.6	31	3777.85	31	2276.84	31	1629.52	31	1268.87
32	10743 0	32	3736.79	32	2261.86	32	1621.84	32	1264.21
33	10417.5	33	3696.61	33	2247.08	33	1614.22	33	1259.58
34	10111.1	34	3657.29	34	2232.49	34	1606.68	34	1254.98
35	9822.18	35	3618.80	35	2218.09	35	1599.21	35	1250.42
36	9549.34	36	3581.10	36	2203.87	36	1591.81	36	1245.89
37	9291.29	37	3544.19	37	2189.84	37	1584.48	37	1241.40
38	9046.75	38	3508.02	38	2175.98	38	1577.21	38	1236.94
39	8814.78	39	3472.59	39	2162.30	39	1570.01	39	1232.51
40	8594.42	40	3437.87	40	2148.79	40	1562.88	40	1228.11
41	8384.80	41	3403.83	41	2135.44	41	1555.81	41	1223.74
42	8185.16	42	3370.46	42	2122.26	42	1548.80	42	1219.40
43	7994.81	43	3337.74	43	2109.24	43	1541.86	43	1215.30
44	7813.11	44	3305.65	44	2096.39	44	1534.98	44	1210.82
45	7639.49	45	3274.17	45	2083.68	45	1528.16	45	1206.57
46	7473.42	46	3243.29	46	2071.13	46	1521.40	46	1202.36
47	7314.41	47	3212.98	47	2058.73	47	1514.70	47	1198.17
48	7162.03	48	3183.23	48	2046.48	48	1508.06	48	1194.01
49	7015.87	49	3154.03	49	2034.37	49	1501.48	49	1189.88
50	6875.55	50	3125.36	50	2022.41	50	1494.95	50	1185.78
51	6740.74	51	3097.20	51	2010.59	51	1488.48	51	1181.71
52	6611.12	52	3069.55	52	1998.90	52	1482.07	52	1177.66
53	6486.38	53	3042.39	53	1987.35	53	1475.71	53	1173.65
54	6366.26	54	3015.71	54	1975.93	54	1469.41	54	1169.66
55	6250.51	55	2989.48	55	1964.64	55	1463.16	55	1165.70
56	6138.90	56	2963.71	56	1953.48	56	1456.96	56	1161.76
57	6031.20	57	2938.39	57	1942.44	57	1450.81	57	1157.85
58	5927.22	58	2913.49	58	1931.53	58	1444.72	58	1153.97
59	5826.76	59	2889.01	59	1920.75	59	1438.68	59	1150.11
60	5729.65	60	2864.93	60	1910.08	60	1432.69	60	1146.28

TABLE I.—RADII.

Deg.	Radius.	Deg.	Radius.	Deg.	Radius.	Deg.	Radius.	Deg.	Radius.
5° 0′	1146.28	6° 0′	955.37	7° 0′	818.64	8° 0′	716.34	9° 0′	636.78
1	1142.47	1	952.72	1	816.70	1	714.85	1	635.61
2	1138.69	2	950.09	2	814.76	2	713.37	2	634.44
3	1134.94	3	947.48	3	812.83	3	711.90	3	633.27
4	1131.21	4	944.88	4	810.92	4	710.43	4	632.10
5	1127.50	5	942.29	5	809.01	5	708.96	5	630.94
6	1123.82	6	939.72	6	807.11	6	707.51	6	629.79
7	1120.16	7	937.16	7	805.22	7	706.05	7	628.64
8	1116.52	8	934.62	8	803.34	8	704.60	8	627.49
9	1112.91	9	932.09	9	801.47	9	703.16	9	626.35
10	1109.33	10	929.57	10	799.61	10	701.73	10	625.21
11	1105.76	11	927.07	11	797.75	11	700.30	11	624.08
12	1102.22	12	924.58	12	795.91	12	698.88	12	622.95
13	1098.70	13	922.10	13	794.07	13	697.46	13	621.82
14	1095.20	14	919.64	14	792.24	14	696.05	14	620.70
15	1091.73	15	917.19	15	790.42	15	694.65	15	619.58
16	1088.28	16	914.75	16	788.61	16	693.24	16	618.47
17	1084.85	17	912.33	17	786.80	17	691.85	17	617.36
18	1081.44	18	909.92	18	785.01	18	690.46	18	616.25
19	1078.05	19	907.52	19	783.22	19	689.08	19	615.15
20	1074.68	20	905.13	20	781.44	20	687.70	20	614.05
21	1071.34	21	902.76	21	779.67	21	686.33	21	612.96
22	1068.01	22	900.40	22	777.91	22	684.96	22	611.87
23	1064.71	23	898.05	23	776.15	23	683.60	23	610.78
24	1061.43	24	895.71	24	774.40	24	682.25	24	609.70
25	1058.16	25	893.39	25	772.66	25	680.89	25	608.62
26	1054.92	26	891.08	26	770.93	26	679.55	26	607.55
27	1051.70	27	888.78	27	769.21	27	678.21	27	606.48
28	1048.48	28	886.49	28	767.49	28	676.88	28	605.41
29	1045.31	29	884.21	29	765.78	29	675.54	29	604.35
30	1042.14	30	881.95	30	764.08	30	674.22	30	603.29
31	1039.00	31	879.69	31	762.39	31	672.90	31	602.23
32	1035.87	32	877.45	32	760.70	32	671.59	32	601.18
33	1032.76	33	875.22	33	759.02	33	670.28	33	600.13
34	1029.67	34	873.00	34	757.35	34	668.98	34	599.09
35	1026.60	35	870.80	35	755.69	35	667.68	35	598.04
36	1023.55	36	868.60	36	754.03	36	666.39	36	597.01
37	1020.51	37	866.41	37	752.38	37	665.10	37	595.97
38	1017.49	38	864.24	38	750.74	38	663.82	38	594.94
39	1014.50	39	862.08	39	749.10	39	662.54	39	593.91
40	1011.51	40	859.92	40	747.48	40	661.26	40	592.89
41	1008.55	41	857.78	41	745.86	41	659.99	41	591.87
42	1005.60	42	855.65	42	744.24	42	658.73	42	590.85
43	1002.67	43	853.53	43	742.63	43	657.47	43	589.84
44	999.76	44	851.42	44	741.03	44	656.22	44	588.83
45	996.87	45	849.32	45	739.44	45	654.97	45	587.83
46	993.99	46	847.23	46	737.86	46	653.72	46	586.82
47	991.13	47	845.15	47	736.28	47	652.48	47	585.83
48	988.28	48	843.08	48	734.70	48	651.25	48	584.83
49	985.45	49	841.02	49	733.14	49	650.02	49	583.84
50	982.64	50	838.97	50	731.58	50	648.79	50	582.85
51	979.84	51	836.93	51	730.03	51	647.57	51	581.86
52	977.06	52	834.90	52	728.48	52	646.35	52	580.88
53	974.29	53	832.89	53	726.94	53	645.14	53	579.90
54	971.54	54	830.88	54	725.41	54	643.94	54	578.92
55	968.81	55	828.88	55	723.88	55	642.73	55	577.95
56	966.09	56	826.89	56	722.36	56	641.53	56	576.98
57	963.39	57	824.91	57	720.85	57	640.34	57	576.02
58	960.70	58	822.93	58	719.34	58	639.15	58	575.06
59	958.03	59	820.97	59	717.84	59	637.96	59	574.10
60	955.37	60	819.02	60	716.34	60	636.78	60	573.14

TABLE I.—RADII.

Deg.	Radius.	Deg.	Radius.	Deg.	Radius.	Deg.	Radius.	Deg.	Radius.
10° 0′	573.14	12° 0′	477.68	14° 0′	409.32	16° 0′	358.17	18° 0′	318.39
2	571.24	2	476.36	2	408.35	2	357.43	2	317.80
4	569.35	4	475.05	4	407.38	4	356.69	4	317.22
6	567.47	6	473.74	6	406.42	6	355.95	6	316.63
8	565.60	8	472.44	8	405.46	8	355.21	8	316.05
10	563.75	10	471.15	10	404.51	10	354.48	10	315.47
12	561.91	12	469.86	12	403.56	12	353.75	12	314.89
14	560.08	14	468.58	14	402.61	14	353.03	14	314.32
16	558.26	16	467.31	16	401.67	16	352.30	16	313.75
18	556.45	18	466.04	18	400.74	18	351.58	18	313.18
20	554.66	20	464.78	20	399.80	20	350.86	20	312.61
22	552.88	22	463.53	22	398.88	22	350.15	22	312.04
24	551.11	24	462.29	24	397.95	24	349.44	24	311.47
26	549.35	26	461.05	26	397.03	26	348.72	26	310.91
28	547.60	28	459.82	28	396.13	28	348.02	28	310.35
30	545.87	30	458.59	30	395.21	30	347.32	30	309.79
32	544.14	32	457.38	32	394.30	32	346.62	32	309.23
34	542.42	34	456.16	34	393.40	34	345.93	34	308.68
36	540.72	36	454.96	36	392.50	36	345.23	36	308.13
38	539.03	38	453.76	38	391.61	38	344.54	38	307.58
40	537.34	40	452.57	40	390.72	40	343.85	40	307.03
42	535.67	42	451.38	42	389.83	42	343.16	42	306.48
44	534.01	44	450.20	44	388.95	44	342.48	44	305.93
46	532.36	46	449.02	46	388.07	46	341.80	46	305.39
48	530.71	48	447.86	48	387.20	48	341.12	48	304.85
50	529.08	50	446.69	50	386.33	50	340.45	50	304.31
52	527.46	52	445.54	52	385.47	52	339.78	52	303.77
54	525.85	54	444.39	54	384.60	54	339.11	54	303.24
56	524.25	56	443.24	56	383.75	56	338.44	56	302.70
58	522.65	58	442.11	58	382.89	58	337.77	58	302.17
11° 0′	521.07	13° 0′	440.97	15° 0′	382.04	17° 0′	337.11	19° 0′	301.64
2	519.50	2	439.85	2	381.19	2	336.45	2	301.12
4	517.93	4	438.73	4	380.35	4	335.80	4	300.59
6	516.38	6	437.61	6	379.51	6	335.14	6	300.07
8	514.84	8	436.50	8	378.68	8	334.49	8	299.54
10	513.30	10	435.40	10	377.84	10	333.84	10	299.02
12	511.77	12	434.30	12	377.02	12	333.19	12	298.50
14	510.26	14	433.21	14	376.19	14	332.55	14	297.99
16	508.75	16	432.12	16	375.37	16	331.91	16	297.47
18	507.25	18	431.04	18	374.55	18	331.27	18	296.96
20	505.76	20	429.96	20	373.74	20	330.63	20	296.45
22	504.28	22	428.98	22	372.93	22	330.00	22	295.94
24	502.80	24	427.82	24	372.12	24	329.37	24	295.43
26	501.34	26	426.76	26	371.32	26	328.74	26	294.92
28	499.88	28	425.71	28	370.52	28	328.11	28	294.42
30	498.43	30	424.66	30	369.72	30	327.48	30	293.91
32	496.99	32	423.61	32	368.93	32	326.86	32	293.41
34	495.56	34	422.57	34	368.14	34	326.24	34	292.91
36	494.14	36	421.54	36	367.35	36	325.62	36	292.41
38	492.73	38	420.51	38	366.57	38	325.01	38	291.92
40	491.32	40	419.49	40	365.79	40	324.40	40	291.42
42	489.92	42	418.47	42	365.01	42	323.79	42	290.93
44	488.53	44	417.45	44	364.24	44	323.18	44	290.44
46	487.15	46	416.44	46	363.47	46	322.57	46	289.95
48	485.77	48	415.44	48	362.70	48	321.97	48	289.46
50	484.40	50	414.44	50	361.94	50	321.37	50	288.98
52	483.05	52	413.44	52	361.18	52	320.77	52	288.49
54	481.69	54	412.45	54	360.42	54	320.17	54	288.01
56	480.35	56	411.47	56	359.67	56	319.57	56	287.53
58	479.01	58	410.49	58	358.92	58	318.98	58	287.05
60	477.68	60	409.51	60	358.17	60	318.39	60	286.57

′	0″	10″	15″	20″	30″	40″	45″	50″	′
0	.00000	.00278	.00417	.00556	.00833	.01111	.01250	.01389	0
1	.01667	.01944	.02083	.02222	.02500	.02778	.02917	.03055	1
2	.03333	.03611	.03750	.03889	.04167	.04444	.04583	.04722	2
3	.05000	.05278	.05417	.05556	.05833	.06111	.06250	.06389	3
4	.06667	.06944	.07083	.07222	.07500	.07778	.07917	.08056	4
5	.08333	.08611	.08750	.08889	.09167	.09444	.09583	.09722	5
6	.10000	.10278	.10417	.10556	.10833	.11111	.11250	.11389	6
7	.11667	.11944	.12083	.12222	.12500	.12778	.12917	.13056	7
8	.13333	.13611	.13750	.13889	.14167	.14444	.14583	.14722	8
9	.15000	.15278	.15417	.15556	.15833	.16111	.16250	.16389	9
10	.16667	.16944	.17083	.17222	.17500	.17778	.17917	.18056	10
11	.18333	.18611	.18750	.18889	.19167	.19444	.19583	.19722	11
12	.20000	.20278	.20417	.20556	.20833	.21111	.21250	.21389	12
13	.21667	.21944	.22083	.22222	.22500	.22778	.22917	.23056	13
14	.23333	.23611	.23750	.23889	.24167	.24444	.24583	.24722	14
15	.25000	.25278	.25417	.25556	.25833	.26111	.26250	.26389	15
16	.26667	.26944	.27083	.27222	.27500	.27778	.27917	.28056	16
17	.28333	.28611	.28750	.28889	.29167	.29444	.29583	.29722	17
18	.30000	.30278	.30417	.30556	.30833	.31111	.31250	.31389	18
19	.31667	.31944	.32083	.32222	.32500	.32778	.32917	.33056	19
20	.33333	.33611	.33750	.33889	.34167	.34444	.34583	.34722	20
21	.35000	.35278	.35417	.35556	.35833	.36111	.36250	.36389	21
22	.36667	.36944	.37083	.37222	.37500	.37778	.37917	.38056	22
23	.38333	.38611	.38750	.38889	.39167	.39444	.39583	.39722	23
24	.40000	.40278	.40417	.40556	.40833	.41111	.41250	.41389	24
25	.41667	.41944	.42083	.42222	.42500	.42778	.42917	.43056	25
26	.43333	.43611	.43750	.43889	.44167	.44444	.44583	.44722	26
27	.45000	.45278	.45417	.45556	.45833	.46111	.46250	.46389	27
28	.46667	.46944	.47083	.47222	.47500	.47778	.47917	.48056	28
29	.48333	.48611	.48750	.48889	.49167	.49444	.49583	.49722	29
30	.50000	.50278	.50417	.50556	.50833	.51111	.51250	.51389	30
31	.51667	.51944	.52083	.52222	.52500	.52778	.52917	.53056	31
32	.53333	.53611	.53750	.53889	.54167	.54444	.54583	.54722	32
33	.55000	.55278	.55417	.55556	.55833	.56111	.56250	.56389	33
34	.56667	.56944	.57083	.57222	.57500	.57778	.57917	.58056	34
35	.58333	.58611	.58750	.58889	.59167	.59444	.59583	.59722	35
36	.60000	.60278	.60417	.60556	.60833	.61111	.61250	.61389	36
37	.61667	.61944	.62083	.62222	.62500	.62778	.62917	.63056	37
38	.63333	.63611	.63750	.63889	.64167	.64444	.64583	.64722	38
39	.65000	.65278	.65417	.65556	.65833	.66111	.66250	.66389	39
40	.66667	.66944	.67083	.67222	.67500	.67778	.67917	.68056	40
41	.68333	.68611	.68750	.68889	.69167	.69444	.69583	.69722	41
42	.70000	.70278	.70417	.70556	.70833	.71111	.71250	.71389	42
43	.71667	.71944	.72083	.72222	.72500	.72778	.72917	.73056	43
44	.73333	.73611	.73750	.73889	.74167	.74444	.74583	.74722	44
45	.75000	.75278	.75417	.75556	.75833	.76111	.76250	.76389	45
46	.76667	.76944	.77083	.77222	.77500	.77778	.77917	.78056	46
47	.78333	.78611	.78750	.78889	.79167	.79444	.79583	.79722	47
48	.80000	.80278	.80417	.80556	.80833	.81111	.81250	.81389	48
49	.81667	.81944	.82083	.82222	.82500	.82778	.82917	.83056	49
50	.83333	.83611	.83750	.83889	.84167	.84444	.84583	.84722	50
51	.85000	.85278	.85417	.85556	.85833	86111	.86250	.86389	51
52	.86667	.86944	.87083	.87222	.87500	.87778	.87917	.88056	52
53	.88333	.88611	.88750	.88889	.89167	.89444	.89583	.89722	53
54	.90000	.90278	.90417	.90556	.90833	.91111	.91250	.91389	54
55	.91667	.91944	.92083	.92222	.92500	.92778	.92917	.93056	55
56	.93333	.93611	.93750	.93889	.94167	.94444	.94583	.94722	56
57	.95000	.95278	.95417	.95556	.95833	.96111	.96250	.96389	57
58	.96667	.96944	.97083	.97222	.97500	.97778	.97917	.98056	58
59	.98333	.98611	.98750	.98889	.99167	.99444	.99583	.99722	59
′	0″	10″	15″	20″	30″	40″	45″	50″	′

TABLE III.—TANGENTIAL OFFSETS.

Deg. of Curve.	OFFSETS 100 FEET ALONG CURVE.						
	0°	1°	2°	3°	4°	5°	6°
0′	0.00	0.87	1.75	2.62	3.49	4.36	5.23
2	0.03	0.90	1.77	2.65	3.52	4.39	5.26
4	0.06	0.93	1.80	2.68	3.55	4.42	5.29
6	0.09	0.96	1.83	2.71	3.58	4.45	5.32
8	0.12	0.99	1.86	2.73	3.61	4.48	5.35
10	0.15	1.02	1.89	2.76	3.64	4.51	5.38
12	0.17	1.05	1.92	2.79	3.66	4.54	5.41
14	0.20	1.08	1.95	2.82	3.69	4.57	5.44
16	0.23	1.11	1.98	2.85	3.72	4.59	5.47
18	0.26	1.13	2.01	2.88	3.75	4.62	5.50
20	0.29	1.16	2.04	2.91	3.78	4.65	5.52
22	0.32	1.19	2.07	2.94	3.81	4.68	5.55
24	0.35	1.22	2.09	2.97	3.84	4.71	5.58
26	0.38	1.25	2.12	3.00	3.87	4.74	5.61
28	0.41	1.28	2.15	3.02	3.90	4.77	5.64
30	0.44	1.31	2.18	3.05	3.93	4.80	5.67
32	0.47	1.34	2.21	3.08	3.96	4.83	5.70
34	0.49	1.37	2.24	3.11	3.98	4.86	5.73
36	0.52	1.40	2.27	3.14	4.01	4.89	5.76
38	0.55	1.43	2.30	3.17	4.04	4.91	5.79
40	0.58	1.45	2.33	3.20	4.07	4.94	5.81
42	0.61	1.48	2.36	3.23	4.10	4.97	5.84
44	0.64	1.51	2.39	3.26	4.13	5.00	5.87
46	0.67	1.54	2.41	3.29	4.16	5.03	5.90
48	0.70	1.57	2.44	3.32	4.19	5.06	5.93
50	0.73	1.60	2.47	3.35	4.22	5.09	5.96
52	0.76	1.63	2.50	3.37	4.25	5.12	5.99
54	0.79	1.66	2.53	3.40	4.28	5.15	6.02
56	0.81	1.69	2.56	3.43	4.30	5.18	6.05
58	0.84	1.72	2.59	3.46	4.33	5.20	6.08
60	0.87	1.75	2.62	3.49	4.36	5.23	6.11

Deg. of Curve.	OFFSETS 50 FEET ALONG CURVE.						
	7°	8°	9°	10°	11°	12°	13°
0′	1.53	1.75	1.96	2.18	2.40	2.62	2.84
2	1.53	1.75	1.97	2.19	2.41	2.62	2.84
4	1.54	1.76	1.98	2.20	2.41	2.63	2.85
6	1.55	1.77	1.99	2.20	2.42	2.64	2.86
8	1.56	1.77	1.99	2.21	2.43	2.65	2.86
10	1.56	1.78	2.00	2.22	2.44	2.65	2.87
12	1.57	1.79	2.01	2.23	2.44	2.66	2.88
14	1.58	1.80	2.01	2.23	2.45	2.67	2.89
16	1.59	1.80	2.02	2.24	2.46	2.68	2.89
18	1.59	1.81	2.03	2.25	2.47	2.68	2.90
20	1.60	1.82	2.04	2.25	2.47	2.69	2.91
22	1.61	1.83	2.04	2.26	2.48	2.70	2.92
24	1.61	1.83	2.05	2.27	2.49	2.70	2.92
26	1.62	1.84	2.06	2.28	2.49	2.71	2.93
28	1.63	1.85	2.07	2.28	2.50	2.72	2.94
30	1.64	1.85	2.07	2.29	2.51	2.73	2.94
32	1.64	1.86	2.08	2.30	2.52	2.73	2.95
34	1.65	1.87	2.09	2.31	2.52	2.74	2.96
36	1.66	1.88	2.09	2.31	2.53	2.75	2.97
38	1.67	1.88	2.10	2.32	2.54	2.76	2.97
40	1.67	1.89	2.11	2.33	2.55	2.76	2.98
42	1.68	1.90	2.12	2.33	2.55	2.77	2.99
44	1.69	1.91	2.12	2.34	2.56	2.78	3.00
46	1.69	1.91	2.13	2.35	2.57	2.79	3.00
48	1.70	1.92	2.14	2.36	2.57	2.79	3.01
50	1.71	1.93	2.15	2.36	2.58	2.80	3.02
52	1.72	1.93	2.15	2.37	2.59	2.81	3.02
54	1.72	1.94	2.16	2.38	2.60	2.81	3.03
56	1.73	1.95	2.17	2.39	2.60	2.82	3.04
58	1.74	1.96	2.17	2.39	2.61	2.83	3.05
60	1.75	1.96	2.18	2.40	2.62	2.84	3.05

TABLE IV.—LONG CHORDS.

Degree of Curve.		Actual Arc, One Station.	Long Chords. 2 Stations.	3 Stations.	4 Stations.	5 Stations.	6 Stations.
0°	10′	100.000	200.00	300.00	400.00	500.00	599.99
	20	.000	200.00	300.00	399.99	499.98	599.97
	30	.000	200.00	299.99	399.98	499.96	599.93
	40	.001	200.00	299.99	399.97	499.93	599.88
	50	.001	200.00	299.98	399.95	499.89	599.82
1		100.001	199.99	299.97	399.92	499.85	599.73
	10	.002	199.99	299.96	399.90	499.79	599.64
	20	002	199.99	299.95	399.87	499.73	599.53
	30	.003	199.98	299.93	399.83	499.66	599.40
	40	.003	199.98	299.92	399.79	499.58	599.26
	50	.004	199.97	299.90	399.74	499.49	599.11
2		100.005	199.97	299.88	399.70	499.39	598.93
	10	.006	199.96	299.86	399.64	499.29	598.75
	20	.007	199.96	299.83	399.59	499.17	598.55
	30	.008	199.95	299.81	399.52	499.05	598.34
	40	.009	199.95	299.78	399.46	498.92	598.11
	50	.010	199.94	299.76	399.39	498.78	597.86
3		100.011	199.93	299.73	399.32	498.63	597.60
	10	.013	199.92	299.70	399.24	498.47	597.33
	20	.014	199.92	299.66	399.15	498.31	597.04
	30	.015	199.91	299.63	399.07	498.14	596.74
	40	.017	199.90	299.59	398.98	497.96	596.42
	50	.019	199.89	209.55	398.88	497.77	596.09
4		100.020	199.88	299.51	398.78	497.57	595.74
	10	.022	199.87	299.47	398.68	497.36	595.38
	20	.024	199.86	299.43	398.57	497.15	595.01
	30	.026	199.85	299.38	398.46	496.92	594.62
	40	.028	199.83	299.34	398.34	496.69	594.21
	50	.030	199.82	299.29	398.22	496.45	593.79
5		100.032	199.81	299.24	398.10	496.20	593.36
	10	.034	199.80	299.19	397.97	495.94	592.91
	20	.036	199.78	299.13	397.84	495.68	592.45
	30	.038	199.77	299.08	397.70	495.41	591.97
	40	.041	199.76	209.02	397.56	495.12	591.48
	50	.043	199.74	298.96	397.41	494.83	590.97
6		100.046	199.73	298.90	397.26	494.53	590.45
	10	.048	199.71	298.84	307.11	494.23	589.91
	20	.051	199.70	298.78	396.95	493.91	589.36
	30	.054	199.68	298.71	396.79	493.59	588.80
	40	.056	199.66	298.65	396.62	493.26	588.22
	50	.059	199.64	298.58	396.45	492.92	587.63
7		100.062	199.63	298.51	396.28	492.57	587.02
	10	.016	199.51	298.30	395.91	491.97	586.12
	20	.017	199.49	298.21	395.71	491.60	585.46
	30	.018	199.47	298.13	395.52	491.21	584.80
	40	.018	199.44	298.05	395.32	490.82	584.13
	50	.019	199.42	297.96	395.11	490.42	583.44
8		100.020	199.39	297.87	394.90	490.01	582.72
	10	.021	199.37	297.78	394.69	489.59	582.01
	20	.022	199.34	297.69	394.47	489.16	581.27
	30	.023	199.31	297.60	394.25	488.73	580.52
	40	.024	199.28	297.50	394.02	488.28	579.76
	50	.025	199.26	297.41	393.79	487.83	578.98
9		100.026	199.23	297.31	393.55	487.37	578.18
	10	.027	199.20	297.21	393.31	486.90	577.38
	20	.028	199.17	297.10	393.07	486.43	576.56
	30	.029	199.14	297.00	392.82	485.95	575.73
	40	.030	199.11	296.90	392.57	485.45	574.88
	50	.031	199.08	296.79	392.31	484.95	574.02
10		100.032	199.05	296.68	392.05	484.44	573.14

TABLE IV.—LONG CHORDS.

Degree of Curve.	Actual Arc, One Station.	Long Chords. 1 Station.	2 Stations.	3 Stations.	4 Stations.	5 Stations.
10° 0′	100.032	99.91	199.05	296.68	392.05	484.44
10	.033	99.90	199.02	296.57	391.79	483.92
20	.034	99.90	198.98	296.45	391.51	483.39
30	.035	99.90	198.95	296.33	391.24	482.86
40	.036	99.89	198.92	296.22	390.96	482.32
50	.037	99.89	198.88	296.10	390.68	481.77
11	100.038	99.88	198.85	295.98	390.39	481.21
10	.040	99.88	198.81	295.86	390.10	480.64
20	.041	99.88	198.78	295.74	389.81	480.07
30	.042	99.87	198.74	295.61	389.50	479.48
40	.043	99.87	198.71	295.48	389.20	478.89
50	.044	99.87	198.67	295.35	388.89	478.29
12	100.046	99.86	198.63	295.22	388.58	477.68
10	.047	99.86	198.59	295.09	388.27	477.07
20	.048	99.85	198.55	294.95	387.95	476.44
30	.050	99.85	198.51	294.82	387.62	475.81
40	.051	99.85	198.48	294.68	387.29	475.18
50	.052	99.84	198.43	294.54	386.95	474.52
13	100.054	99.84	198.39	294.40	386.62	473.87
10	.055	99.84	198.35	294.26	386.28	473.20
20	.056	99.83	198.31	294.11	385.93	472.53
30	.058	99.83	198.27	293.96	385.58	471.86
40	.059	99.82	198.23	293.81	385.23	471.17
50	.061	99.82	198.18	293.66	384.87	470.48
14	100.062	99.81	198.14	293.51	384.51	469.77

TABLE V.—MID-ORDINATES TO LONG CHORDS.

Degree of Curve.	1 Station.	2 Stations.	3 Stations.	4 Stations.	5 Stations.	6 Stations.
0° 10′	.04	.15	.33	.58	.91	1.31
20	.07	.29	.65	1.16	1.82	2.62
30	.11	.44	.98	1.75	2.73	3.93
40	.15	.58	1.31	2.33	3.64	5.24
50	.18	.73	1.64	2.91	4.55	6.54
1	.22	.87	1.96	3.49	5.45	7.85
10	.26	1.02	2.29	4.07	6.36	9.16
20	.29	1.16	2.62	4.65	7.27	10.47
30	.33	1.31	2.95	5.24	8.18	11.78
40	.36	1.45	3.27	5.82	9.09	13.08
50	.40	1.60	3.60	6.40	9.99	14.39
2	.44	1.75	3.93	6.98	10.90	15.69
10	.47	1.89	4.25	7.56	11.81	17.00
20	.51	2.04	4.58	8.14	12.72	18.30
30	.55	2.18	4.91	8.72	13.62	19.61
40	.58	2.33	5.23	9.30	14.53	20.91
50	.62	2.47	5.56	9.88	15.44	22.21
3	.65	2.62	5.89	10.46	16.34	23.52
10	.69	2.76	6.22	11.04	17.25	24.82
20	.73	2.91	6.54	11.62	18.15	26.12
30	.76	3.05	6.87	12.20	19.06	27.42
40	.80	3.20	7.20	12.78	19.96	28.71
50	.84	3.35	7.52	13.36	20.86	30.01
4	.87	3.49	7.85	13.94	21.77	31.31

TABLE V.—MID-ORDINATES TO LONG CHORDS.

Degree of Curve.	1 Station.	2 Stations.	3 Stations.	4 Stations.	5 Stations.	6 Stations.
4° 0′	.87	3.49	7.85	13.94	21.77	31.31
10	.91	3.64	8.18	14.52	22.67	32.60
20	.95	3.78	8.50	15.10	23.57	33.90
30	.98	3.93	8.83	15.68	24.47	35.19
40	1.02	4.07	9.15	16.26	25.37	36.48
50	1.05	4 22	9.48	16.84	26.27	37.77
5	1.09	4.36	9.81	17 42	27.17	39.06
10	1.13	4.51	10.13	17.99	28.07	40.35
20	1.16	4.65	10.46	18.57	28.97	41.63
30	1.20	4.80	10.79	19.15	29.87	42.92
40	1.24	4.94	11.11	19.72	30.76	44.20
50	1.27	5.09	11.44	20.30	31.66	45.48
6	1.31	5.23	11.76	20.88	32.55	46.76
10	1.35	5.38	12.09	21.45	33.45	48.04
20	1.38	5.52	12.41	22.03	34.34	49.31
30	1.42	5.67	12.74	22.60	35.23	50.59
40	1.46	5.81	13.06	23.18	36.13	51.86
50	1.49	5.96	13.39	23.75	37.02	53.13
7	1.53	6.11	13.72	24.33	37.91	54.40
10	1.56	6.25	14.03	24.89	38.79	55.64
20	1.60	6.39	14 36	25.46	39.66	56.90
30	1.64	6.54	14.68	26.04	40.55	58.16
40	1.67	6.68	15.01	26.61	41.43	59.41
50	1.71	6.83	15.33	27.18	42.32	60.68
8	1.74	6.97	15.65	27.75	43.20	61.93
10	1.78	7.12	15.98	28.32	44.08	63.18
20	1.82	7.26	16.30	28.89	44.96	64.43
30	1.85	7.41	16.62	29.46	45.84	65.68
40	1.89	7.55	16.95	30.03	46.72	66.92
50	1.93	7.70	17.27	30.60	47.60	68.17
9	1.96	7.84	17.59	31.17	48.47	69.40
10	2.00	7.98	17.92	31.73	49.35	70.64
20	2.04	8.13	18.24	32.30	50.22	71.88
30	2.07	8.27	18.56	32.87	51.09	73.11
40	2.11	8.42	18 88	33.43	51.96	74.34
50	2.14	8.56	19.21	34.00	52.83	75.56
10	2.18	8.71	19.53	34.56	53.70	76.79
10	2 22	8.85	19.85	35.13	54.57	
20	2.25	9.00	20.17	35.69	55.43	
30	2.29	9.14	20.49	36.26	56.29	
40	2.33	9.28	20.82	36.82	57.16	
50	2.36	9.43	21.14	37.38	58.02	
11	2.40	9.57	21.46	37.94	58.88	
10	2.44	9.72	21.78	38.50	59.73	
20	2.47	9.86	22.10	39 06	60.58	
30	2.51	10.01	22.42	39.62	61.44	
40	2.54	10.15	22.74	40.18	62.30	
50	2.58	10.29	23.06	40.74	63.15	
12	2 62	10.44	23.38	41.30	64.00	
10	2.65	10.58	23.70	41.86	64.85	
20	2.69	10.73	24.02	42.41	65.69	
30	2.73	10.87	24.34	42.97	66.54	
40	2.76	11.01	24.66	43.52	67.38	
50	2.80	11.16	24.97	44.08	68.22	
13	2.83	11.30	25.29	44.63	69.06	
10	2.87	11.45	25.61	45.18	69.90	
20	2.91	11.59	25.93	45.73	70.73	
30	2.94	11.73	26.25	46.29	71.57	
40	2.98	11.88	26.57	46.84	72.40	
50	3.02	12.02	26.88	47.39	73.23	
14	3.05	12.16	27.20	47.93	74.06	

N	0	1	2	3	4	5	6	7	8	9
100	00000	00043	00087	00130	00173	00217	00260	00303	00346	00389
1	0432	0475	0518	0561	0604	0647	0689	0732	0775	0817
2	0860	0903	0945	0988	1030	1072	1115	1157	1199	1242
3	1284	1326	1368	1410	1452	1494	1536	1578	1620	1662
4	1703	1745	1787	1828	1870	1912	1953	1995	2036	2078
5	2119	2160	2202	2243	2284	2325	2366	2407	2449	2490
6	2531	2572	2612	2653	2694	2735	2776	2816	2857	2898
7	2938	2979	3019	3060	3100	3141	3181	3222	3262	3302
8	3342	3383	3423	3463	3503	3543	3583	3623	3663	3703
9	3743	3782	3822	3862	3902	3941	3981	4021	4060	4100
110	04139	04179	04218	04258	04297	04336	04376	04415	04454	04493
1	4532	4571	4610	4650	4689	4727	4766	4805	4844	4883
2	4922	4961	4999	5038	5077	5115	5154	5192	5231	5269
3	5308	5346	5385	5423	5461	5500	5538	5576	5614	5652
4	5690	5729	5767	5805	5843	5881	5918	5956	5994	6032
5	6070	6108	6145	6183	6221	6258	6296	6333	6371	6408
6	6446	6483	6521	6558	6595	6633	6670	6707	6744	6781
7	6819	6856	6893	6930	6967	7004	7041	7078	7115	7151
8	7188	7225	7262	7298	7335	7372	7408	7445	7482	7518
9	7555	7591	7628	7664	7700	7737	7773	7809	7846	7882
120	07918	07954	07990	08027	08063	08099	08135	08171	08207	08243
1	8279	8314	8350	8386	8422	8458	8493	8529	8565	8600
2	8636	8672	8707	8743	8778	8814	8849	8884	8920	8955
3	8991	9026	9061	9096	9132	9167	9202	9237	9272	9307
4	9342	9377	9412	9447	9482	9517	9552	9587	9621	9656
5	9691	9726	9760	9795	9830	9864	9899	9934	9968	10003
6	10037	10072	10106	10140	10175	10209	10243	10278	10312	0346
7	0380	0415	0449	0483	0517	0551	0585	0619	0653	0687
8	0721	0755	0789	0823	0857	0890	0924	0958	0992	1025
9	1059	1093	1126	1160	1193	1227	1261	1294	1327	1361
130	11394	11428	11461	11494	11528	11561	11594	11628	11661	11694
1	1727	1760	1793	1826	1860	1893	1926	1959	1992	2024
2	2057	2090	2123	2156	2189	2222	2254	2287	2320	2352
3	2385	2418	2450	2483	2516	2548	2581	2613	2646	2678
4	2710	2743	2775	2808	2840	2872	2905	2937	2969	3001
5	3033	3066	3098	3130	3162	3194	3226	3258	3290	3322
6	3354	3386	3418	3450	3481	3513	3545	3577	3609	3640
7	3672	3704	3735	3767	3799	3830	3862	3893	3925	3956
8	3988	4019	4051	4082	4114	4145	4176	4208	4239	4270
9	4301	4333	4364	4395	4426	4457	4489	4520	4551	4582
140	14613	14644	14675	14706	14737	14768	14799	14829	14860	14891
1	4922	4953	4983	5014	5045	5076	5106	5137	5168	5198
2	5229	5259	5290	5320	5351	5381	5412	5442	5473	5503
3	5534	5564	5594	5625	5655	5685	5715	5746	5776	5806
4	5836	5866	5897	5927	5957	5987	6017	6047	6077	6107
5	6137	6167	6197	6227	6256	6286	6316	6346	6376	6406
6	6435	6465	6495	6524	6554	6584	6613	6643	6673	6702
7	6732	6761	6791	6820	6850	6879	6909	6938	6967	6997
8	7026	7056	7085	7114	7143	7173	7202	7231	7260	7289
9	7319	7348	7377	7406	7435	7464	7493	7522	7551	7580
150	17609	17638	17667	17696	17725	17754	17782	17811	17840	17869

0	1	2	3	4	5	6	7	8	9
17609	17638	17667	17696	17725	17754	17782	17811	17840	17869
7898	7926	7955	7984	8013	8041	8070	8099	8127	8156
8184	8213	8241	8270	8298	8327	8355	8384	8412	8441
8469	8498	8526	8554	8583	8611	8639	8667	8696	8724
8752	8780	8808	8837	8865	8893	8921	8949	8977	9005
9033	9061	9089	9117	9145	9173	9201	9229	9257	9285
9312	9340	9368	9396	9424	9451	9479	9507	9535	9562
9590	9618	9645	9673	9700	9728	9756	9783	9811	9838
9866	9893	9921	9948	9976	20003	20030	20058	20085	20112
20140	20167	20194	20222	20249	0276	0303	0330	0358	0385
20412	20439	20466	20493	20520	20548	20575	20602	20629	20656
0683	0710	0737	0763	0790	0817	0844	0871	0898	0925
0952	0978	1005	1032	1059	1085	1112	1139	1165	1192
1219	1245	1272	1299	1325	1352	1378	1405	1431	1458
1484	1511	1537	1564	1590	1617	1643	1669	1696	1722
1748	1775	1801	1827	1854	1880	1906	1932	1958	1985
2011	2037	2063	2089	2115	2141	2167	2194	2220	2246
2272	2298	2324	2350	2376	2401	2427	2453	2479	2505
2531	2557	2583	2608	2634	2660	2686	2712	2737	2763
2789	2814	2840	2866	2891	2917	2943	2968	2994	3019
23045	23070	23096	23121	23147	23172	23198	23223	23249	23274
3300	3325	3350	3376	3401	3426	3452	3477	3502	3528
3553	3578	3603	3629	3654	3679	3704	3729	3754	3779
3805	3830	3855	3880	3905	3930	3955	3980	4005	4030
4055	4080	4105	4130	4155	4180	4204	4229	4254	4279
4304	4329	4353	4378	4403	4428	4452	4477	4502	4527
4551	4576	4601	4625	4650	4674	4699	4724	4748	4773
4797	4822	4846	4871	4895	4920	4944	4969	4993	5018
5042	5066	5091	5115	5139	5164	5188	5212	5237	5261
5285	5310	5334	5358	5382	5406	5431	5455	5479	5503
25527	25551	25575	25600	25624	25648	25672	25696	25720	25744
5768	5792	5816	5840	5864	5888	5912	5935	5959	5983
6007	6031	6055	6079	6102	6126	6150	6174	6198	6221
6245	6269	6293	6316	6340	6364	6387	6411	6435	6458
6482	6505	6529	6553	6576	6600	6623	6647	6670	6694
6717	6741	6764	6788	6811	6834	6858	6881	6905	6928
6951	6975	6998	7021	7045	7068	7091	7114	7138	7161
7184	7207	7231	7254	7277	7300	7323	7346	7370	7393
7416	7439	7462	7485	7508	7531	7554	7577	7600	7623
7646	7669	7692	7715	7738	7761	7784	7807	7830	7852
27875	27898	27921	27944	27967	27989	28012	28035	28058	28081
8103	8126	8149	8171	8194	8217	8240	8262	8285	8307
8330	8353	8375	8398	8421	8443	8466	8488	8511	8533
8556	8578	8601	8623	8646	8668	8691	8713	8735	8758
8780	8803	8825	8847	8870	8892	8914	8937	8959	8981
9003	9026	9048	9070	9092	9115	9137	9159	9181	9203
9226	9248	9270	9292	9314	9336	9358	9380	9403	9425
9447	9469	9491	9513	9535	9557	9579	9601	9623	9645
9667	9688	9710	9732	9754	9776	9798	9820	9842	9863
9885	9907	9929	9951	9973	9994	30016	30038	30060	30081
30103	30125	30146	30168	30190	30211	30233	30255	30276	30298

0	1	2	3	4	5	6	7	8	9
30103	30125	30146	30168	30190	30211	30233	30255	30276	30298
0320	0341	0363	0384	0406	0428	0449	0471	0492	0514
0535	0557	0578	0600	0621	0643	0664	0685	0707	0728
0750	0771	0792	0814	0835	0856	0878	0899	0920	0942
0963	0984	1006	1027	1048	1069	1091	1112	1133	1154
1175	1197	1218	1239	1260	1281	1302	1323	1345	1366
1387	1408	1429	1450	1471	1492	1513	1534	1555	1576
1597	1618	1639	1660	1681	1702	1723	1744	1765	1785
1806	1827	1848	1869	1890	1911	1931	1952	1973	1994
2015	2035	2056	2077	2098	2118	2139	2160	2181	2201
32222	32243	32263	32284	32305	32325	32346	32366	32387	32408
2428	2449	2469	2490	2510	2531	2552	2572	2593	2613
2634	2654	2675	2695	2715	2736	2756	2777	2797	2818
2838	2858	2879	2899	2919	2940	2960	2980	3001	3021
3041	3062	3082	3102	3122	3143	3163	3183	3203	3224
3244	3264	3284	3304	3325	3345	3365	3385	3405	3425
3445	3465	3486	3506	3526	3546	3566	3586	3606	3626
3646	3666	3686	3706	3726	3746	3766	3786	3806	3826
3846	3866	3885	3905	3925	3945	3965	3985	4005	4025
4044	4064	4084	4104	4124	4143	4163	4183	4203	4223
34242	34262	34282	34301	34321	34341	34361	34380	34400	34420
4439	4459	4479	4498	4518	4537	4557	4577	4596	4616
4635	4655	4674	4694	4713	4733	4753	4772	4792	4811
4830	4850	4869	4889	4908	4928	4947	4967	4986	5005
5025	5044	5064	5083	5102	5122	5141	5160	5180	5199
5218	5238	5257	5276	5295	5315	5334	5353	5372	5392
5411	5430	5449	5468	5488	5507	5526	5545	5564	5583
5603	5622	5641	5660	5679	5698	5717	5736	5755	5774
5793	5813	5832	5851	5870	5889	5908	5927	5946	5965
5984	6003	6021	6040	6059	6078	6097	6116	6135	6154
36173	36192	36211	36229	36248	36267	36286	36305	36324	36342
6361	6380	6399	6418	6436	6455	6474	6493	6511	6530
6549	6568	6586	6605	6624	6642	6661	6680	6698	6717
6736	6754	6773	6791	6810	6829	6847	6866	6884	6903
6922	6940	6959	6977	6996	7014	7033	7051	7070	7088
7107	7125	7144	7162	7181	7199	7218	7236	7254	7273
7291	7310	7328	7346	7365	7383	7401	7420	7438	7457
7475	7493	7511	7530	7548	7566	7585	7603	7621	7639
7658	7676	7694	7712	7731	7749	7767	7785	7803	7822
7840	7858	7876	7894	7912	7931	7949	7967	7985	8003
38021	38039	38057	38075	38093	38112	38130	38148	38166	38184
8202	8220	8238	8256	8274	8292	8310	8328	8346	8364
8382	8399	8417	8435	8453	8471	8489	8507	8525	8543
8561	8578	8596	8614	8632	8650	8668	8686	8703	8721
8739	8757	8775	8792	8810	8828	8846	8863	8881	8899
8917	8934	8952	8970	8987	9005	9023	9041	9058	9076
9094	9111	9129	9146	9164	9182	9199	9217	9235	9252
9270	9287	9305	9322	9340	9358	9375	9393	9410	9428
9445	9463	9480	9498	9515	9533	9550	9568	9585	9602
9620	9637	9655	9672	9690	9707	9724	9742	9759	9777
39794	39811	39829	39846	39863	39881	39898	39915	39933	39950

N	0	1	2	3	4	5	6	7	8	9
250	39794	39811	39829	39846	39863	39881	39898	39915	39933	39950
1	9967	9985	40002	40019	40037	40054	40071	40088	40106	40123
2	40140	40157	0175	0192	0209	0226	0243	0261	0278	0295
3	0312	0329	0346	0364	0381	0398	0415	0432	0449	0466
4	0483	0500	0518	0535	0552	0569	0586	0603	0620	0637
5	0654	0671	0688	0705	0722	0739	0756	0773	0790	0807
6	0824	0841	0858	0875	0892	0909	0926	0943	0960	0976
7	0993	1010	1027	1044	1061	1078	1095	1111	1128	1145
8	1162	1179	1196	1212	1229	1246	1263	1280	1296	1313
9	1330	1347	1363	1380	1397	1414	1430	1447	1464	1481
260	41497	41514	41531	41547	41564	41581	41597	41614	41631	41647
1	1664	1681	1697	1714	1731	1747	1764	1780	1797	1814
2	1830	1847	1863	1880	1896	1913	1929	1946	1963	1979
3	1996	2012	2029	2045	2062	2078	2095	2111	2127	2144
4	2160	2177	2193	2210	2226	2243	2259	2275	2292	2308
5	2325	2341	2357	2374	2390	2406	2423	2439	2455	2472
6	2488	2504	2521	2537	2553	2570	2586	2602	2619	2635
7	2651	2067	2684	2700	2716	2732	2749	2765	2781	2797
8	2813	2830	2846	2862	2878	2894	2911	2927	2943	2959
9	2975	2091	3008	3024	3040	3056	3072	3088	3104	3120
270	43136	43152	43169	43185	43201	43217	43233	43249	43265	43281
1	3297	3313	3329	3345	3361	3377	3393	3409	3425	3441
2	3457	3473	3489	3505	3521	3537	3553	3569	3584	3600
3	3616	3632	3648	3664	3680	3696	3712	3727	3743	3759
4	3775	3791	3807	3823	3838	3854	3870	3886	3902	3917
5	3933	3949	3965	3981	3996	4012	4028	4044	4059	4075
6	4091	4107	4122	4138	4154	4170	4185	4201	4217	4232
7	4248	4264	4279	4295	4311	4326	4342	4358	4373	4389
8	4404	4420	4436	4451	4467	4483	4498	4514	4529	4545
9	4560	4576	4592	4607	4623	4638	4654	4669	4685	4700
280	44716	44731	44747	44762	44778	44793	44809	44824	44840	44855
1	4871	4886	4902	4917	4932	4948	4963	4979	4994	5010
2	5025	5040	5056	5071	5086	5102	5117	5133	5148	5163
3	5179	5194	5209	5225	5240	5255	5271	5286	5301	5317
4	5332	5347	5362	5378	5393	5408	5423	5439	5454	5469
5	5484	5500	5515	5530	5545	5561	5576	5591	5606	5621
6	5637	5652	5667	5682	5697	5712	5728	5743	5758	5773
7	5788	5803	5818	5834	5849	5864	5879	5894	5909	5924
8	5939	5954	5969	5984	6000	6015	6030	6045	6060	6075
9	6090	6105	6120	6135	6150	6165	6180	6195	6210	6225
290	46240	46255	46270	46285	46300	46315	46330	46345	46359	46374
1	6389	6404	6419	6434	6449	6464	6479	6494	6509	6523
2	6538	6553	6568	6583	6598	6613	6627	6642	6657	6672
3	6687	6702	6716	6731	6746	6761	6776	6790	6805	6820
4	6835	6850	6864	6879	6894	6909	6923	6938	6953	6967
5	6982	6997	7012	7026	7041	7056	7070	7085	7100	7114
6	7129	7144	7159	7173	7188	7202	7217	7232	7246	7261
7	7276	7290	7305	7319	7334	7349	7363	7378	7392	7407
8	7422	7436	7451	7465	7480	7494	7509	7524	7538	7553
9	7567	7582	7596	7611	7625	7640	7654	7669	7683	7698
300	47712	47727	47741	47756	47770	47784	47709	47813	47828	47842

0	1	2	3	4	5	6	7	8	9
47712	47727	47741	47756	47770	47784	47799	47813	47828	47842
7857	7871	7885	7900	7914	7929	7943	7958	7972	7986
8001	8015	8029	8044	8058	8073	8087	8101	8116	8130
8144	8159	8173	8187	8202	8216	8230	8244	8259	8273
8287	8302	8316	8330	8344	8359	8373	8387	8401	8416
8430	8444	8458	8473	8487	8501	8515	8530	8544	8558
8572	8586	8601	8615	8629	8643	8657	8671	8686	8700
8714	8728	8742	8756	8770	8785	8799	8813	8827	8841
8855	8869	8883	8897	8911	8926	8940	8954	8968	8982
8996	9010	9024	9038	9052	9066	9080	9094	9108	9122
49136	49150	49164	49178	49192	49206	49220	49234	49248	49262
9276	9290	9304	9318	9332	9346	9360	9374	9388	9402
9415	9429	9443	9457	9471	9485	9499	9513	9527	9541
9554	9568	9582	9596	9610	9624	9638	9651	9665	9679
9693	9707	9721	9734	9748	9762	9776	9790	9803	9817
9831	9845	9859	9872	9886	9900	9914	9927	9941	9955
9969	9982	9996	50010	50024	50037	50051	50065	50079	50092
50106	50120	50133	0147	0161	0174	0188	0202	0215	0229
0243	0256	0270	0284	0297	0311	0325	0338	0352	0365
0379	0393	0406	0420	0433	0447	0461	0474	0488	0501
50515	50529	50542	50556	50569	50583	50596	50610	50623	50637
0651	0664	0678	0691	0705	0718	0732	0745	0759	0772
0786	0799	0813	0826	0840	0853	0866	0880	0893	0907
0920	0934	0947	0961	0974	0987	1001	1014	1028	1041
1055	1068	1081	1095	1108	1121	1135	1148	1162	1175
1188	1202	1215	1228	1242	1255	1268	1282	1295	1308
1322	1335	1348	1362	1375	1388	1402	1415	1428	1441
1455	1468	1481	1495	1508	1521	1534	1548	1561	1574
1587	1601	1614	1627	1640	1654	1667	1680	1693	1706
1720	1733	1746	1759	1772	1786	1799	1812	1825	1838
51851	51865	51878	51891	51904	51917	51930	51943	51957	51970
1983	1996	2009	2022	2035	2048	2061	2075	2088	2101
2114	2127	2140	2153	2166	2179	2192	2205	2218	2231
2244	2257	2270	2284	2297	2310	2323	2336	2349	2362
2375	2388	2401	2414	2427	2440	2453	2466	2479	2492
2504	2517	2530	2543	2556	2569	2582	2595	2608	2621
2634	2647	2660	2673	2686	2699	2711	2724	2737	2750
2763	2776	2789	2802	2815	2827	2840	2853	2866	2879
2892	2905	2917	2930	2943	2956	2969	2982	2994	3007
3020	3033	3046	3058	3071	3084	3097	3110	3122	3135
53148	53161	53173	53186	53199	53212	53224	53237	53250	53263
3275	3288	3301	3314	3326	3339	3352	3364	3377	3390
3403	3415	3428	3441	3453	3466	3479	3491	3504	3517
3529	3542	3555	3567	3580	3593	3605	3618	3631	3643
3656	3668	3681	3694	3706	3719	3732	3744	3757	3769
3782	3794	3807	3820	3832	3845	3857	3870	3882	3895
3908	3920	3933	3945	3958	3970	3983	3995	4008	4020
4033	4045	4058	4070	4083	4095	4108	4120	4133	4145
4158	4170	4183	4195	4208	4220	4233	4245	4258	4270
4283	4295	4307	4320	4332	4345	4357	4370	4382	4394
54407	54419	54432	54444	54456	54469	54481	54494	54506	54518

0	1	2	3	4	5	6	7	8	9
54407	54419	54432	54444	54456	54469	54481	54494	54506	54518
4531	4543	4555	4568	4580	4593	4605	4617	4630	4642
4654	4667	4679	4691	4704	4716	4728	4741	4753	4765
4777	4790	4802	4814	4827	4839	4851	4864	4876	4888
4900	4913	4925	4937	4949	4962	4974	4986	4998	5011
5023	5035	5047	5060	5072	5084	5096	5108	5121	5133
5145	5157	5169	5182	5194	5206	5218	5230	5242	5255
5267	5279	5291	5303	5315	5328	5340	5352	5364	5376
5388	5400	5413	5425	5437	5449	5461	5473	5485	5497
5509	5522	5534	5546	5558	5570	5582	5594	5606	5618
55630	55642	55654	55666	55678	55691	55703	55715	55727	55739
5751	5763	5775	5787	5799	5811	5823	5835	5847	5859
5871	5883	5895	5907	5919	5931	5943	5955	5967	5979
5991	6003	6015	6027	6038	6050	6062	6074	6086	6098
6110	6122	6134	6146	6158	6170	6182	6194	6205	6217
6229	6241	6253	6265	6277	6289	6301	6312	6324	6336
6348	6360	6372	6384	6396	6407	6419	6431	6443	6455
6467	6478	6490	6502	6514	6526	6538	6549	6561	6573
6585	6597	6608	6620	6632	6644	6656	6667	6679	6691
6703	6714	6726	6738	6750	6761	6773	6785	6797	6808
56820	56832	56844	56855	56867	56879	56891	56902	56914	56926
6937	6949	6961	6972	6984	6996	7008	7019	7031	7043
7054	7066	7078	7089	7101	7113	7124	7136	7148	7159
7171	7183	7194	7206	7217	7229	7241	7252	7264	7276
7287	7299	7310	7322	7334	7345	7357	7368	7380	7392
7403	7415	7426	7438	7449	7461	7473	7484	7496	7507
7519	7530	7542	7553	7565	7576	7588	7600	7611	7623
7634	7646	7657	7669	7680	7692	7703	7715	7726	7738
7749	7761	7772	7784	7795	7807	7818	7830	7841	7852
7864	7875	7887	7898	7910	7921	7933	7944	7955	7967
57978	57990	58001	58013	58024	58035	58047	58058	58070	58081
8092	8104	8115	8127	8138	8149	8161	8172	8184	8195
8206	8218	8229	8240	8252	8263	8274	8286	8297	8309
8320	8331	8343	8354	8365	8377	8388	8399	8410	8422
8433	8444	8456	8467	8478	8490	8501	8512	8524	8535
8546	8557	8569	8580	8591	8602	8614	8625	8636	8647
8659	8670	8681	8692	8704	8715	8726	8737	8749	8760
8771	8782	8794	8805	8816	8827	8838	8850	8861	8872
8883	8894	8906	8917	8928	8939	8950	8961	8973	8984
8995	9006	9017	9028	9040	9051	9062	9073	9084	9095
59106	59118	59129	59140	59151	59162	59173	59184	59195	59207
9218	9229	9240	9251	9262	9273	9284	9295	9306	931
9329	9340	9351	9362	9373	9384	9395	9406	9417	94
9439	9450	9461	9472	9483	9494	9506	9517	9528	9
9550	9561	9572	9583	9594	9605	9616	9627	9638	9649
9660	9671	9682	9693	9704	9715	9726	9737	9748	9759
9770	9780	9791	9802	9813	9824	9835	9846	9857	9868
9879	9890	9901	9912	9923	9934	9945	9956	9966	9977
9988	9999	60010	60021	60032	60043	60054	60065	60076	60086
60097	60108	0119	0130	0141	0152	0163	0173	0184	0195
60206	60217	60228	60239	60249	60260	60271	60282	60293	60304

N	0	1	2	3	4	5	6	7	8	9
400	60206	60217	60228	60239	60249	60260	60271	60282	60293	60304
1	0314	0325	0336	0347	0358	0369	0379	0390	0401	0412
2	0423	0433	0444	0455	0466	0477	0487	0498	0509	0520
3	0531	0541	0552	0563	0574	0584	0595	0606	0617	0627
4	0638	0649	0660	0670	0681	0692	0703	0713	0724	0735
5	0746	0756	0767	0778	0788	0799	0810	0821	0831	0842
6	0853	0863	0874	0885	0895	0906	0917	0927	0938	0949
7	0959	0970	0981	0991	1002	1013	1023	1034	1045	1055
8	1066	1077	1087	1098	1109	1119	1130	1140	1151	1162
9	1172	1183	1194	1204	1215	1225	1236	1247	1257	1268
410	61278	61289	61300	61310	61321	61331	61342	61352	61363	61374
1	1384	1395	1405	1416	1426	1437	1448	1458	1469	1479
2	1490	1500	1511	1521	1532	1542	1553	1563	1574	1584
3	1595	1606	1616	1627	1637	1648	1658	1669	1679	1690
4	1700	1711	1721	1731	1742	1752	1763	1773	1784	1794
5	1805	1815	1826	1836	1847	1857	1868	1878	1888	1899
6	1909	1920	1930	1941	1951	1962	1972	1982	1993	2003
7	2014	2024	2034	2045	2055	2066	2076	2086	2097	2107
8	2118	2128	2138	2149	2159	2170	2180	2190	2201	2211
9	2221	2232	2242	2252	2263	2273	2284	2294	2304	2315
420	62325	62335	62346	62356	62366	62377	62387	62397	62408	62418
1	2428	2439	2449	2459	2469	2480	2490	2500	2511	2521
2	2531	2542	2552	2562	2572	2583	2593	2603	2613	2624
3	2634	2644	2655	2665	2675	2685	2696	2706	2716	2726
4	2737	2747	2757	2767	2778	2788	2798	2808	2818	2829
5	2839	2849	2859	2870	2880	2890	2900	2910	2921	2931
6	2941	2951	2961	2972	2982	2992	3002	3012	3022	3033
7	3043	3053	3063	3073	3083	3094	3104	3114	3124	3134
8	3144	3155	3165	3175	3185	3195	3205	3215	3225	3236
9	3246	3256	3266	3276	3286	3296	3306	3317	3327	3337
430	63347	63357	63367	63377	63387	63397	63407	63417	63428	63438
1	3448	3458	3468	3478	3488	3498	3508	3518	3528	3538
2	3548	3558	3568	3579	3589	3599	3609	3619	3629	3639
3	3649	3659	3669	3679	3689	3699	3709	3719	3729	3739
4	3749	3759	3769	3779	3789	3799	3809	3819	3829	3839
5	3849	3859	3869	3879	3889	3899	3909	3919	3929	3939
6	3949	3959	3969	3979	3988	3998	4008	4018	4028	4038
7	4048	4058	4068	4078	4088	4098	4108	4118	4128	4137
8	4147	4157	4167	4177	4187	4197	4207	4217	4227	4237
9	4246	4256	4266	4276	4286	4296	4306	4316	4326	4335
440	64345	64355	64365	64375	64385	64395	64404	64414	64424	64434
1	4444	4454	4464	4473	4483	4493	4503	4513	4523	4532
2	4542	4552	4562	4572	4582	4591	4601	4611	4621	4631
3	4640	4650	4660	4670	4680	4689	4699	4709	4719	4729
4	4738	4748	4758	4768	4777	4787	4797	4807	4816	4826
5	4836	4846	4856	4865	4875	4885	4895	4904	4914	4924
6	4933	4943	4953	4963	4972	4982	4992	5002	5011	5021
7	5031	5040	5050	5060	5070	5079	5089	5099	5108	5118
8	5128	5137	5147	5157	5167	5176	5186	5196	5205	5215
9	5225	5234	5244	5254	5263	5273	5283	5292	5302	5312
450	65321	65331	65341	65350	65360	65369	65379	65389	65398	65408

0	1	2	3	4	5	6	7	8	9
65321	65331	65341	65350	65360	65369	65379	65389	65398	65408
5418	5427	5437	5447	5456	5466	5475	5485	5495	5504
5514	5523	5533	5543	5552	5562	5571	5581	5591	5600
5610	5619	5629	5639	5648	5658	5667	5677	5686	5696
5706	5715	5725	5734	5744	5753	5763	5772	5782	5792
5801	5811	5820	5830	5839	5849	5858	5868	5877	5887
5896	5906	5916	5925	5935	5944	5954	5963	5973	5982
5992	6001	6011	6020	6030	6039	6049	6058	6068	6077
6087	6096	6106	6115	6124	6134	6143	6153	6162	6172
6181	6191	6200	6210	6219	6229	6238	6247	6257	6266
66276	66285	66295	66304	66314	66323	66332	66342	66351	66361
6370	6380	6389	6398	6408	6417	6427	6436	6445	6455
6464	6474	6483	6492	6502	6511	6521	6530	6539	6549
6558	6567	6577	6586	6596	6605	6614	6624	6633	6642
6652	6661	6671	6680	6689	6699	6708	6717	6727	6736
6745	6755	6764	6773	6783	6792	6801	6811	6820	6829
6839	6848	6857	6867	6876	6885	6894	6904	6913	6922
6932	6941	6950	6960	6969	6978	6987	6997	7006	7015
7025	7034	7043	7052	7062	7071	7080	7089	7099	7108
7117	7127	7136	7145	7154	7164	7173	7182	7191	7201
67210	67219	67228	67237	67247	67256	67265	67274	67284	67293
7302	7311	7321	7330	7339	7348	7357	7367	7376	7385
7394	7403	7413	7422	7431	7440	7449	7459	7468	7477
7486	7495	7504	7514	7523	7532	7541	7550	7560	7569
7578	7587	7596	7605	7614	7624	7633	7642	7651	7660
7669	7679	7688	7697	7706	7715	7724	7733	7742	7752
7761	7770	7779	7788	7797	7806	7815	7825	7834	7843
7852	7861	7870	7879	7888	7897	7906	7916	7925	7934
7943	7952	7961	7970	7979	7988	7997	8006	8015	8024
8034	8043	8052	8061	8070	8079	8088	8097	8106	8115
68124	68133	68142	68151	68160	68169	68178	68187	68196	68205
8215	8224	8233	8242	8251	8260	8269	8278	8287	8296
8305	8314	8323	8332	8341	8350	8359	8368	8377	8386
8395	8404	8413	8422	8431	8440	8449	8458	8467	8476
8485	8494	8502	8511	8520	8529	8538	8547	8556	8565
8574	8583	8592	8601	8610	8619	8628	8637	8646	8655
8664	8673	8681	8690	8699	8708	8717	8726	8735	8744
8753	8762	8771	8780	8789	8797	8806	8815	8824	8833
8842	8851	8860	8869	8878	8886	8895	8904	8913	8922
8931	8940	8949	8958	8966	8975	8984	8993	9002	9011
69020	69028	69037	69046	69055	69064	69073	69082	69090	69099
9108	9117	9126	9135	9144	9152	9161	9170	9179	9188
9197	9205	9214	9223	9232	9241	9249	9258	9267	9276
9285	9294	9302	9311	9320	9329	9338	9346	9355	9364
9373	9381	9390	9399	9408	9417	9425	9434	9443	9452
9461	9469	9478	9487	9496	9504	9513	9522	9531	9539
9548	9557	9566	9574	9583	9592	9601	9609	9618	9627
9636	9644	9653	9662	9671	9679	9688	9697	9705	9714
9723	9732	9740	9749	9758	9767	9775	9784	9793	9801
9810	9819	9827	9836	9845	9854	9862	9871	9880	9888
69897	69906	69914	69923	69932	69940	69949	69958	69966	69975

0	1	2	3	4	5	6	7	8	9
69897	69906	69914	69923	69932	69940	69949	69958	69966	69975
9984	9992	70001	70010	70018	70027	70036	70044	70053	70062
70070	70079	0088	0096	0105	0114	0122	0131	0140	0148
0157	0165	0174	0183	0191	0200	0209	0217	0226	0234
0243	0252	0260	0269	0278	0286	0295	0303	0312	0321
0329	0338	0346	0355	0364	0372	0381	0389	0398	0406
0415	0424	0432	0441	0449	0458	0467	0475	0484	0492
0501	0509	0518	0526	0535	0544	0552	0561	0569	0578
0586	0595	0603	0612	0621	0629	0638	0646	0655	0663
0672	0680	0689	0697	0706	0714	0723	0731	0740	0749
70757	70766	70774	70783	70791	70800	70808	70817	70825	70834
0842	0851	0859	0868	0876	0885	0893	0902	0910	0919
0927	0935	0944	0952	0961	0969	0978	0986	0995	1003
1012	1020	1029	1037	1046	1054	1063	1071	1079	1088
1096	1105	1113	1122	1130	1139	1147	1155	1164	1172
1181	1189	1198	1206	1214	1223	1231	1240	1248	1257
1265	1273	1282	1290	1299	1307	1315	1324	1332	1341
1349	1357	1366	1374	1383	1391	1399	1408	1416	1425
1433	1441	1450	1458	1466	1475	1483	1492	1500	1508
1517	1525	1533	1542	1550	1559	1567	1575	1584	1592
71600	71609	71617	71625	71634	71642	71650	71659	71667	71675
1684	1692	1700	1709	1717	1725	1734	1742	1750	1759
1767	1775	1784	1792	1800	1809	1817	1825	1834	1842
1850	1858	1867	1875	1883	1892	1900	1908	1917	1925
1933	1941	1950	1958	1966	1975	1983	1991	1999	2008
2016	2024	2032	2041	2049	2057	2066	2074	2082	2090
2099	2107	2115	2123	2132	2140	2148	2156	2165	2173
2181	2189	2198	2206	2214	2222	2230	2239	2247	2255
2263	2272	2280	2288	2296	2304	2313	2321	2329	2337
2346	2354	2362	2370	2378	2387	2395	2403	2411	2419
72428	72436	72444	72452	72460	72469	72477	72485	72493	72501
2509	2518	2526	2534	2542	2550	2558	2567	2575	2583
2591	2599	2607	2616	2624	2632	2640	2648	2656	2665
2673	2681	2689	2697	2705	2713	2722	2730	2738	2746
2754	2762	2770	2779	2787	2795	2803	2811	2819	2827
2835	2843	2852	2860	2868	2876	2884	2892	2900	2908
2916	2925	2933	2941	2949	2957	2965	2973	2981	2989
2997	3006	3014	3022	3030	3038	3046	3054	3062	3070
3078	3086	3094	3102	3111	3119	3127	3135	3143	3151
3159	3167	3175	3183	3191	3199	3207	3215	3223	3231
73239	73247	73255	73263	73272	73280	73288	73296	73304	73312
3320	3328	3336	3344	3352	3360	3368	3376	3384	3392
3400	3408	3416	3424	3432	3440	3448	3456	3464	3472
3480	3488	3496	3504	3512	3520	3528	3536	3544	3552
3560	3568	3576	3584	3592	3600	3608	3616	3624	3632
3640	3648	3656	3664	3672	3679	3687	3695	3703	3711
3719	3727	3735	3743	3751	3759	3767	3775	3783	3791
3799	3807	3815	3823	3830	3838	3846	3854	3862	3870
3878	3886	3894	3902	3910	3918	3926	3933	3941	3949
3957	3965	3973	3981	3989	3997	4005	4013	4020	4028
74036	74044	74052	74060	74068	74076	74084	74092	74099	74107

0	1	2	3	4	5	6	7	8	9
74036	74044	74052	74060	74068	74076	74084	74092	74099	74107
4115	4123	4131	4139	4147	4155	4162	4170	4178	4186
4194	4202	4210	4218	4225	4233	4241	4249	4257	4265
4273	4280	4288	4296	4304	4312	4320	4327	4335	4343
4351	4359	4367	4374	4382	4390	4398	4406	4414	4421
4429	4437	4445	4453	4461	4468	4476	4484	4492	4500
4507	4515	4523	4531	4539	4547	4554	4562	4570	4578
4586	4593	4601	4609	4617	4624	4632	4640	4648	4656
4663	4671	4679	4687	4695	4702	4710	4718	4726	4733
4741	4749	4757	4764	4772	4780	4788	4796	4803	4811
74819	74827	74834	74842	74850	74858	74865	74873	74881	74889
4896	4904	4912	4920	4927	4935	4943	4950	4958	4966
4974	4981	4989	4997	5005	5012	5020	5028	5035	5043
5051	5059	5066	5074	5082	5089	5097	5105	5113	5120
5128	5136	5143	5151	5159	5166	5174	5182	5189	5197
5205	5213	5220	5228	5236	5243	5251	5259	5266	5274
5282	5289	5297	5305	5312	5320	5328	5335	5343	5351
5358	5366	5374	5381	5389	5397	5404	5412	5420	5427
5435	5442	5450	5458	5465	5473	5481	5488	5496	5504
5511	5519	5526	5534	5542	5549	5557	5565	5572	5580
75587	75595	75603	75610	75618	75626	75633	75641	75648	75656
5664	5671	5679	5686	5694	5702	5709	5717	5724	5732
5740	5747	5755	5762	5770	5778	5785	5793	5800	5808
5815	5823	5831	5838	5846	5853	5861	5868	5876	5884
5891	5899	5906	5914	5921	5929	5937	5944	5952	5959
5967	5974	5982	5989	5997	6005	6012	6020	6027	6035
6042	6050	6057	6065	6072	6080	6087	6095	6103	6110
6118	6125	6133	6140	6148	6155	6163	6170	6178	6185
6193	6200	6208	6215	6223	6230	6238	6245	6253	6260
6268	6275	6283	6290	6298	6305	6313	6320	6328	6335
76343	76350	76358	76365	76373	76380	76388	76395	76403	76410
6418	6425	6433	6440	6448	6455	6462	6470	6477	6485
6492	6500	6507	6515	6522	6530	6537	6545	6552	6559
6567	6574	6582	6589	6597	6604	6612	6619	6626	6634
6641	6649	6656	6664	6671	6678	6686	6693	6701	6708
6716	6723	6730	6738	6745	6753	6760	6768	6775	6782
6790	6797	6805	6812	6819	6827	6834	6842	6849	6856
6864	6871	6879	6886	6893	6901	6908	6916	6923	6930
6938	6945	6953	6960	6967	6975	6982	6989	6997	7004
7012	7019	7026	7034	7041	7048	7056	7063	7070	7078
77085	77093	77100	77107	77115	77122	77129	77137	77144	77151
7159	7166	7173	7181	7188	7195	7203	7210	7217	7225
7232	7240	7247	7254	7262	7269	7276	7283	7291	7298
7305	7313	7320	7327	7335	7342	7349	7357	7364	7371
7379	7386	7393	7401	7408	7415	7422	7430	7437	7444
7452	7459	7466	7474	7481	7488	7495	7503	7510	7517
7525	7532	7539	7546	7554	7561	7568	7576	7583	7590
7597	7605	7612	7619	7627	7634	7641	7648	7656	7663
7670	7677	7685	7692	7699	7706	7714	7721	7728	7735
7743	7750	7757	7764	7772	7779	7786	7793	7801	7808
77815	77822	77830	77837	77844	77851	77859	77866	77873	77880

0	1	2	3	4	5	6	7	8	9
77815	77822	77830	77837	77844	77851	77859	77866	77873	77880
7887	7895	7902	7909	7916	7924	7931	7938	7945	7952
7960	7967	7974	7981	7988	7996	8003	8010	8017	8025
8032	8039	8046	8053	8061	8068	8075	8082	8089	8097
8104	8111	8118	8125	8132	8140	8147	8154	8161	8168
8176	8183	8190	8197	8204	8211	8219	8226	8233	8240
8247	8254	8262	8269	8276	8283	8290	8297	8305	8312
8319	8326	8333	8340	8347	8355	8362	8369	8376	8383
8390	8398	8405	8412	8419	8426	8433	8440	8447	8455
8462	8469	8476	8483	8490	8497	8504	8512	8519	8526
78533	78540	78547	78554	78561	78569	78576	78583	78590	78597
8604	8611	8618	8625	8633	8640	8647	8654	8661	8668
8675	8682	8689	8696	8704	8711	8718	8725	8732	8739
8746	8753	8760	8767	8774	8781	8789	8796	8803	8810
8817	8824	8831	8838	8845	8852	8859	8866	8873	8880
8888	8895	8902	8909	8916	8923	8930	8937	8944	8951
8958	8965	8972	8979	8986	8993	9000	9007	9014	9021
9029	9036	9043	9050	9057	9064	9071	9078	9085	9092
9099	9106	9113	9120	9127	9134	9141	9148	9155	9162
9169	9176	9183	9190	9197	9204	9211	9218	9225	9232
79239	79246	79253	79260	79267	79274	79281	79288	79295	79302
9309	9316	9323	9330	9337	9344	9351	9358	9365	9372
9379	9386	9393	9400	9407	9414	9421	9428	9435	9442
9449	9456	9463	9470	9477	9484	9491	9498	9505	9511
9518	9525	9532	9539	9546	9553	9560	9567	9574	9581
9588	9595	9602	9609	9616	9623	9630	9637	9644	9650
9657	9664	9671	9678	9685	9692	9699	9706	9713	9720
9727	9734	9741	9748	9754	9761	9768	9775	9782	9789
9796	9803	9810	9817	9824	9831	9837	9844	9851	9858
9865	9872	9879	9886	9893	9900	9906	9913	9920	9927
79934	79941	79948	79955	79962	79969	79975	79982	79989	79996
80003	80010	80017	80024	80030	80037	80044	80051	80058	80065
0072	0079	0085	0092	0099	0106	0113	0120	0127	0134
0140	0147	0154	0161	0168	0175	0182	0188	0195	0202
0209	0216	0223	0229	0236	0243	0250	0257	0264	0271
0277	0284	0291	0298	0305	0312	0318	0325	0332	0339
0346	0353	0359	0366	0373	0380	0387	0393	0400	0407
0414	0421	0428	0434	0441	0448	0455	0462	0468	0475
0482	0489	0496	0502	0509	0516	0523	0530	0536	0543
0550	0557	0564	0570	0577	0584	0591	0598	0604	0611
80618	80625	80632	80638	80645	80652	80659	80665	80672	80679
0686	0693	0699	0706	0713	0720	0726	0733	0740	0747
0754	0760	0767	0774	0781	0787	0794	0801	0808	0814
0821	0828	0835	0841	0848	0855	0862	0868	0875	0882
0889	0895	0902	0909	0916	0922	0929	0936	0943	0949
0956	0963	0969	0976	0983	0990	0996	1003	1010	1017
1023	1030	1037	1043	1050	1057	1064	1070	1077	1084
1090	1097	1104	1111	1117	1124	1131	1137	1144	1151
1158	1164	1171	1178	1184	1191	1198	1204	1211	1218
1224	1231	1238	1245	1251	1258	1265	1271	1278	1285
81291	81298	81305	81311	81318	81325	81331	81338	81345	81351

0	1	2	3	4	5	6	7	8	9
81291	81298	81305	81311	81318	81325	81331	81338	81345	81351
1358	1365	1371	1378	1385	1391	1398	1405	1411	1418
1425	1431	1438	1445	1451	1458	1465	1471	1478	1485
1491	1498	1505	1511	1518	1525	1531	1538	1544	1551
1558	1564	1571	1578	1584	1591	1598	1604	1611	1617
1624	1631	1637	1644	1651	1657	1664	1671	1677	1684
1690	1697	1704	1710	1717	1723	1730	1737	1743	1750
1757	1763	1770	1776	1783	1790	1796	1803	1809	1816
1823	1829	1836	1842	1849	1856	1862	1869	1875	1882
1889	1895	1902	1908	1915	1921	1928	1935	1941	1948
81954	81961	81968	81974	81981	81987	81994	82000	82007	82014
2020	2027	2033	2040	2046	2053	2060	2066	2073	2079
2086	2092	2099	2105	2112	2119	2125	2132	2138	2145
2151	2158	2164	2171	2178	2184	2191	2197	2204	2210
2217	2223	2230	2236	2243	2249	2256	2263	2269	2276
2282	2289	2295	2302	2308	2315	2321	2328	2334	2341
2347	2354	2360	2367	2373	2380	2387	2393	2400	2406
2413	2419	2426	2432	2439	2445	2452	2458	2465	2471
2478	2484	2491	2497	2504	2510	2517	2523	2530	2536
2543	2549	2556	2562	2569	2575	2582	2588	2595	2601
82607	82614	82620	82627	82633	82640	82646	82653	82659	82666
2672	2679	2685	2692	2698	2705	2711	2718	2724	2730
2737	2743	2750	2756	2763	2769	2776	2782	2789	2795
2802	2808	2814	2821	2827	2834	2840	2847	2853	2860
2866	2872	2879	2885	2892	2898	2905	2911	2918	2924
2930	2937	2943	2950	2956	2963	2969	2975	2982	2988
2995	3001	3008	3014	3020	3027	3033	3040	3046	3052
3059	3065	3072	3078	3085	3091	3097	3104	3110	3117
3123	3129	3136	3142	3149	3155	3161	3168	3174	3181
3187	3193	3200	3206	3213	3219	3225	3232	3238	3245
83251	83257	83264	83270	83276	83283	83289	83296	83302	83308
3315	3321	3327	3334	3340	3347	3353	3359	3366	3372
3378	3385	3391	3398	3404	3410	3417	3423	3429	3436
3442	3448	3455	3461	3467	3474	3480	3487	3493	3499
3506	3512	3518	3525	3531	3537	3544	3550	3556	3563
3569	3575	3582	3588	3594	3601	3607	3613	3620	3626
3632	3639	3645	3651	3658	3664	3670	3677	3683	3689
3696	3702	3708	3715	3721	3727	3734	3740	3746	3753
3759	3765	3771	3778	3784	3790	3797	3803	3809	3816
3822	3828	3835	3841	3847	3853	3860	3866	3872	3879
83885	83891	83897	83904	83910	83916	83923	83929	83935	83942
3948	3954	3960	3967	3973	3979	3985	3992	3998	4004
4011	4017	4023	4029	4036	4042	4048	4055	4061	4067
4073	4080	4086	4092	4098	4105	4111	4117	4123	4130
4136	4142	4148	4155	4161	4167	4173	4180	4186	4192
4198	4205	4211	4217	4223	4230	4236	4242	4248	4255
4261	4267	4273	4280	4286	4292	4298	4305	4311	4317
4323	4330	4336	4342	4348	4354	4361	4367	4373	4379
4386	4392	4398	4404	4410	4417	4423	4429	4435	4442
4448	4454	4460	4466	4473	4479	4485	4491	4497	4504
84510	84516	84522	84528	84535	84541	84547	84553	84559	84566

N	0	1	2	3	4	5	6	7	8	9
700	84510	84516	84522	84528	84535	84541	84547	84553	84559	84566
1	4572	4578	4584	4590	4597	4603	4609	4615	4621	4628
2	4634	4640	4646	4652	4658	4665	4671	4677	4683	4689
3	4696	4702	4708	4714	4720	4726	4733	4739	4745	4751
4	4757	4763	4770	4776	4782	4788	4794	4800	4807	4813
5	4819	4825	4831	4837	4844	4850	4856	4862	4868	4874
6	4880	4887	4893	4899	4905	4911	4917	4924	4930	4936
7	4942	4948	4954	4960	4967	4973	4979	4985	4991	4997
8	5003	5009	5016	5022	5028	5034	5040	5046	5052	5058
9	5065	5071	5077	5083	5089	5095	5101	5107	5114	5120
710	85126	85132	85138	85144	85150	85156	85163	85169	85175	85181
1	5187	5193	5199	5205	5211	5217	5224	5230	5236	5242
2	5248	5254	5260	5266	5272	5278	5285	5291	5297	5303
3	5309	5315	5321	5327	5333	5339	5345	5352	5358	5364
4	5370	5376	5382	5388	5394	5400	5406	5412	5418	5425
5	5431	5437	5443	5449	5455	5461	5467	5473	5479	5485
6	5491	5497	5503	5509	5516	5522	5528	5534	5540	5546
7	5552	5558	5564	5570	5576	5582	5588	5594	5600	5606
8	5612	5618	5625	5631	5637	5643	5649	5655	5661	5667
9	5673	5679	5685	5691	5697	5703	5709	5715	5721	5727
720	85733	85739	85745	85751	85757	85763	85769	85775	85781	85788
1	5794	5800	5806	5812	5818	5824	5830	5836	5842	5848
2	5854	5860	5866	5872	5878	5884	5890	5896	5902	5908
3	5914	5920	5926	5932	5938	5944	5950	5956	5962	5968
4	5974	5980	5986	5992	5998	6004	6010	6016	6022	6028
5	6034	6040	6046	6052	6058	6064	6070	6076	6082	6088
6	6094	6100	6106	6112	6118	6124	6130	6136	6141	6147
7	6153	6159	6165	6171	6177	6183	6189	6195	6201	6207
8	6213	6219	6225	6231	6237	6243	6249	6255	6261	6267
9	6273	6279	6285	6291	6297	6303	6308	6314	6320	6326
730	86332	86338	86344	86350	86356	86362	86368	86374	86380	86386
1	6392	6398	6404	6410	6415	6421	6427	6433	6439	6445
2	6451	6457	6463	6469	6475	6481	6487	6493	6499	6504
3	6510	6516	6522	6528	6534	6540	6546	6552	6558	6564
4	6570	6576	6581	6587	6593	6599	6605	6611	6617	6623
5	6629	6635	6641	6646	6652	6658	6664	6670	6676	6682
6	6688	6694	6700	6705	6711	6717	6723	6729	6735	6741
7	6747	6753	6759	6764	6770	6776	6782	6788	6794	6800
8	6806	6812	6817	6823	6829	6835	6841	6847	6853	6859
9	6864	6870	6876	6882	6888	6894	6900	6906	6911	6917
740	86923	86929	86935	86941	86947	86953	86958	86964	86970	86976
1	6982	6988	6994	6999	7005	7011	7017	7023	7029	7035
2	7040	7046	7052	7058	7064	7070	7075	7081	7087	7093
3	7099	7105	7111	7116	7122	7128	7134	7140	7146	7151
4	7157	7163	7169	7175	7181	7186	7192	7198	7204	7210
5	7216	7221	7227	7233	7239	7245	7251	7256	7262	7268
6	7274	7280	7286	7291	7297	7303	7309	7315	7320	7326
7	7332	7338	7344	7349	7355	7361	7367	7373	7379	7384
8	7390	7396	7402	7408	7413	7419	7425	7431	7437	7442
9	7448	7454	7460	7466	7471	7477	7483	7489	7495	7500
750	87506	87512	87518	87523	87529	87535	87541	87547	87552	87558

N	0	1	2	3	4	5	6	7	8	9
750	87506	87512	87518	87523	87529	87535	87541	87547	87552	87558
1	7564	7570	7576	7581	7587	7593	7599	7604	7610	7616
2	7622	7628	7633	7639	7645	7651	7656	7662	7668	7674
3	7679	7685	7691	7697	7703	7708	7714	7720	7726	7731
4	7737	7743	7749	7754	7760	7766	7772	7777	7783	7789
5	7795	7800	7806	7812	7818	7823	7829	7835	7841	7846
6	7852	7858	7864	7869	7875	7881	7887	7892	7898	7904
7	7910	7915	7921	7927	7933	7938	7944	7950	7955	7961
8	7967	7973	7978	7984	7990	7996	8001	8007	8013	8018
9	8024	8030	8036	8041	8047	8053	8058	8064	8070	8076
760	88081	88087	88093	88098	88104	88110	88116	88121	88127	88133
1	8138	8144	8150	8156	8161	8167	8173	8178	8184	8190
2	8195	8201	8207	8213	8218	8224	8230	8235	8241	8247
3	8252	8258	8264	8270	8275	8281	8287	8292	8298	8304
4	8309	8315	8321	8326	8332	8338	8343	8349	8355	8360
5	8366	8372	8377	8383	8389	8395	8400	8406	8412	8417
6	8423	8429	8434	8440	8446	8451	8457	8463	8468	8474
7	8480	8485	8491	8497	8502	8508	8513	8519	8525	8530
8	8536	8542	8547	8553	8559	8564	8570	8576	8581	8587
9	8593	8598	8604	8610	8615	8621	8627	8632	8638	8643
770	88649	88655	88660	88666	88672	88677	88683	88689	88694	88700
1	8705	8711	8717	8722	8728	8734	8739	8745	8750	8756
2	8762	8767	8773	8779	8784	8790	8795	8801	8807	8812
3	8818	8824	8829	8835	8840	8846	8852	8857	8863	8868
4	8874	8880	8885	8891	8897	8902	8908	8913	8919	8925
5	8930	8936	8941	8947	8953	8958	8964	8969	8975	8981
6	8986	8992	8997	9003	9009	9014	9020	9025	9031	9037
7	9042	9048	9053	9059	9064	9070	9076	9081	9087	9092
8	9098	9104	9109	9115	9120	9126	9131	9137	9143	9148
9	9154	9159	9165	9170	9176	9182	9187	9193	9198	9204
780	89209	89215	89221	89226	89232	89237	89243	89248	89254	89260
1	9265	9271	9276	9282	9287	9293	9298	9304	9310	9315
2	9321	9326	9332	9337	9343	9348	9354	9360	9365	9371
3	9376	9382	9387	9393	9398	9404	9409	9415	9421	9426
4	9432	9437	9443	9448	9454	9459	9465	9470	9476	9481
5	9487	9492	9498	9504	9509	9515	9520	9526	9531	9537
6	9542	9548	9553	9559	9564	9570	9575	9581	9586	9592
7	9597	9603	9609	9614	9620	9625	9631	9636	9642	9647
8	9653	9658	9664	9669	9675	9680	9686	9691	9697	9702
9	9708	9713	9719	9724	9730	9735	9741	9746	9752	9757
790	89763	89768	89774	89779	89785	89790	89796	89801	89807	89812
1	9818	9823	9829	9834	9840	9845	9851	9856	9862	9867
2	9873	9878	9883	9889	9894	9900	9905	9911	9916	9922
3	9927	9933	9938	9944	9949	9955	9960	9966	9971	9977
4	9982	9988	9993	9998	90004	90009	90015	90020	90026	90031
5	90037	90042	90048	90053	0059	0064	0069	0075	0080	0086
6	0091	0097	0102	0108	0113	0119	0124	0129	0135	0140
7	0146	0151	0157	0162	0168	0173	0179	0184	0189	0195
8	0200	0206	0211	0217	0222	0227	0233	0238	0244	0249
9	0255	0260	0266	0271	0276	0282	0287	0293	0298	0304
800	90309	90314	90320	90325	90331	90336	90342	90347	90352	90358

0	1	2	3	4	5	6	7	8	9
90309	90314	90320	90325	90331	90336	90342	90347	90352	90358
0363	0369	0374	0380	0385	0390	0396	0401	0407	0412
0417	0423	0428	0434	0439	0445	0450	0455	0461	0466
0472	0477	0482	0488	0493	0499	0504	0509	0515	0520
0526	0531	0536	0542	0547	0553	0558	0563	0569	0574
0580	0585	0590	0596	0601	0607	0612	0617	0623	0628
0634	0639	0644	0650	0655	0660	0666	0671	0677	0682
0687	0693	0698	0703	0709	0714	0720	0725	0730	0736
0741	0747	0752	0757	0763	0768	0773	0779	0784	0789
0795	0800	0806	0811	0816	0822	0827	0832	0838	0843
90849	90854	90859	90865	90870	90875	90881	90886	90891	90897
0902	0907	0913	0918	0924	0929	0934	0940	0945	0950
0956	0961	0966	0972	0977	0982	0988	0993	0998	1004
1009	1014	1020	1025	1030	1036	1041	1046	1052	1057
1062	1068	1073	1078	1084	1089	1094	1100	1105	1110
1116	1121	1126	1132	1137	1142	1148	1153	1158	1164
1169	1174	1180	1185	1190	1196	1201	1206	1212	1217
1222	1228	1233	1238	1243	1249	1254	1259	1265	1270
1275	1281	1286	1291	1297	1302	1307	1312	1318	1323
1328	1334	1339	1344	1350	1355	1360	1365	1371	1376
91381	91387	91392	91397	91403	91408	91413	91418	91424	91429
1434	1440	1445	1450	1455	1461	1466	1471	1477	1482
1487	1492	1498	1503	1508	1514	1519	1524	1529	1535
1540	1545	1551	1556	1561	1566	1572	1577	1582	1587
1593	1598	1603	1609	1614	1619	1624	1630	1635	1640
1645	1651	1656	1661	1666	1672	1677	1682	1687	1693
1698	1703	1709	1714	1719	1724	1730	1735	1740	1745
1751	1756	1761	1766	1772	1777	1782	1787	1793	1798
1803	1808	1814	1819	1824	1829	1834	1840	1845	1850
1855	1861	1866	1871	1876	1882	1887	1892	1897	1903
91908	91913	91918	91924	91929	91934	91939	91944	91950	91955
1960	1965	1971	1976	1981	1986	1991	1997	2002	2007
2012	2018	2023	2028	2033	2038	2044	2049	2054	2059
2065	2070	2075	2080	2085	2091	2096	2101	2106	2111
2117	2122	2127	2132	2137	2143	2148	2153	2158	2163
2169	2174	2179	2184	2189	2195	2200	2205	2210	2215
2221	2226	2231	2236	2241	2247	2252	2257	2262	2267
2273	2278	2283	2288	2293	2298	2304	2309	2314	2319
2324	2330	2335	2340	2345	2350	2355	2361	2366	2371
2376	2381	2387	2392	2397	2402	2407	2412	2418	2423
92428	92433	92438	92443	92449	92454	92459	92464	92469	92474
2480	2485	2490	2495	2500	2505	2511	2516	2521	2526
2531	2536	2542	2547	2552	2557	2562	2567	2572	2578
2583	2588	2593	2598	2603	2609	2614	2619	2624	2629
2634	2639	2645	2650	2655	2660	2665	2670	2675	2681
2686	2691	2696	2701	2706	2711	2716	2722	2727	2732
2737	2742	2747	2752	2758	2763	2768	2773	2778	2783
2788	2793	2799	2804	2809	2814	2819	2824	2829	2834
2840	2845	2850	2855	2860	2865	2870	2875	2881	2886
2891	2896	2901	2906	2911	2916	2921	2927	2932	2937
92942	92947	92952	92957	92962	92967	92973	92978	92983	92988

0	1	2	3	4	5	6	7	8	9
92942	92947	92952	92957	92962	92967	92973	92978	92983	92988
2993	2998	3003	3008	3013	3018	3024	3029	3034	3039
3044	3049	3054	3059	3064	3069	3075	3080	3085	3090
3095	3100	3105	3110	3115	3120	3125	3131	3136	3141
3146	3151	3156	3161	3166	3171	3176	3181	3186	3192
3197	3202	3207	3212	3217	3222	3227	3232	3237	3242
3247	3252	3258	3263	3268	3273	3278	3283	3288	3293
3298	3303	3308	3313	3318	3323	3328	3334	3339	3344
3349	3354	3359	3364	3369	3374	3379	3384	3389	3394
3399	3404	3409	3414	3420	3425	3430	3435	3440	3445
93450	93455	93460	93465	93470	93475	93480	93485	93490	93495
3500	3505	3510	3515	3520	3526	3531	3536	3541	3546
3551	3556	3561	3566	3571	3576	3581	3586	3591	3596
3601	3606	3611	3616	3621	3626	3631	3636	3641	3646
3651	3656	3661	3666	3671	3676	3682	3687	3692	3697
3702	3707	3712	3717	3722	3727	3732	3737	3742	3747
3752	3757	3762	3767	3772	3777	3782	3787	3792	3797
3802	3807	3812	3817	3822	3827	3832	3837	3842	3847
3852	3857	3862	3867	3872	3877	3882	3887	3892	3897
3902	3907	3912	3917	3922	3927	3932	3937	3942	3947
93952	93957	93962	93967	93972	93977	93982	93987	93992	93997
4002	4007	4012	4017	4022	4027	4032	4037	4042	4047
4052	4057	4062	4067	4072	4077	4082	4086	4091	4096
4101	4106	4111	4116	4121	4126	4131	4136	4141	4146
4151	4156	4161	4166	4171	4176	4181	4186	4191	4196
4201	4206	4211	4216	4221	4226	4231	4236	4240	4245
4250	4255	4260	4265	4270	4275	4280	4285	4290	4295
4300	4305	4310	4315	4320	4325	4330	4335	4340	4345
4349	4354	4359	4364	4369	4374	4379	4384	4389	4394
4399	4404	4409	4414	4419	4424	4429	4433	4438	4443
94448	94453	94458	94463	94468	94473	94478	94483	94488	94493
4498	4503	4507	4512	4517	4522	4527	4532	4537	4542
4547	4552	4557	4562	4567	4571	4576	4581	4586	4591
4596	4601	4606	4611	4616	4621	4626	4630	4635	4640
4645	4650	4655	4660	4665	4670	4675	4680	4685	4689
4694	4699	4704	4709	4714	4719	4724	4729	4734	4738
4743	4748	4753	4758	4763	4768	4773	4778	4783	4787
4792	4797	4802	4807	4812	4817	4822	4827	4832	4836
4841	4846	4851	4856	4861	4866	4871	4876	4880	4885
4890	4895	4900	4905	4910	4915	4919	4924	4929	4934
94939	94944	94949	94954	94959	94963	94968	94973	94978	94983
4988	4993	4998	5002	5007	5012	5017	5022	5027	5032
5036	5041	5046	5051	5056	5061	5066	5071	5075	5080
5085	5090	5095	5100	5105	5109	5114	5119	5124	5129
5134	5139	5143	5148	5153	5158	5163	5168	5173	5177
5182	5187	5192	5197	5202	5207	5211	5216	5221	5226
5231	5236	5240	5245	5250	5255	5260	5265	5270	5274
5279	5284	5289	5294	5299	5303	5308	5313	5318	5323
5328	5332	5337	5342	5347	5352	5357	5361	5366	5371
5376	5381	5386	5390	5395	5400	5405	5410	5415	5419
95424	95429	95434	95439	95444	95448	95453	95458	95463	95468

0	1	2	3	4	5	6	7	8	9
95424	95429	95434	95439	95444	95448	95453	95458	95463	95468
5472	5477	5482	5487	5492	5497	5501	5506	5511	5516
5521	5525	5530	5535	5540	5545	5550	5554	5559	5564
5569	5574	5578	5583	5588	5593	5598	5602	5607	5612
5617	5622	5626	5631	5636	5641	5646	5650	5655	5660
5665	5670	5674	5679	5684	5689	5694	5698	5703	5708
5713	5718	5722	5727	5732	5737	5742	5746	5751	5756
5761	5766	5770	5775	5780	5785	5789	5794	5799	5804
5809	5813	5818	5823	5828	5832	5837	5842	5847	5852
5856	5861	5866	5871	5875	5880	5885	5890	5895	5899
95904	95909	95914	95918	95923	95928	95933	95938	95942	95947
5952	5957	5961	5966	5971	5976	5980	5985	5990	5995
5999	6004	6009	6014	6019	6023	6028	6033	6038	6042
6047	6052	6057	6061	6066	6071	6076	6080	6085	6090
6095	6099	6104	6109	6114	6118	6123	6128	6133	6137
6142	6147	6152	6156	6161	6166	6171	6175	6180	6185
6190	6194	6199	6204	6209	6213	6218	6223	6227	6232
6237	6242	6246	6251	6256	6261	6265	6270	6275	6280
6284	6289	6294	6298	6303	6308	6313	6317	6322	6327
6332	6336	6341	6346	6350	6355	6360	6365	6369	6374
96379	96384	96388	96393	96398	96402	96407	96412	96417	96421
6426	6431	6435	6440	6445	6450	6454	6459	6464	6468
6473	6478	6483	6487	6492	6497	6501	6506	6511	6515
6520	6525	6530	6534	6539	6544	6548	6553	6558	6562
6567	6572	6577	6581	6586	6591	6595	6600	6605	6609
6614	6619	6624	6628	6633	6638	6642	6647	6652	6656
6661	6666	6670	6675	6680	6685	6689	6694	6699	6703
6708	6713	6717	6722	6727	6731	6736	6741	6745	6750
6755	6759	6764	6769	6774	6778	6783	6788	6792	6797
6802	6806	6811	6816	6820	6825	6830	6834	6839	6844
96848	96853	96858	96862	96867	96872	96876	96881	96886	96890
6895	6900	6904	6909	6914	6918	6923	6928	6932	6937
6942	6946	6951	6956	6960	6965	6970	6974	6979	6984
6988	6993	6997	7002	7007	7011	7016	7021	7025	7030
7035	7039	7044	7049	7053	7058	7063	7067	7072	7077
7081	7086	7090	7095	7100	7104	7109	7114	7118	7123
7128	7132	7137	7142	7146	7151	7155	7160	7165	7169
7174	7179	7183	7188	7192	7197	7202	7206	7211	7216
7220	7225	7230	7234	7239	7243	7248	7253	7257	7262
7267	7271	7276	7280	7285	7290	7294	7299	7304	7308
97313	97317	97322	97327	97331	97336	97340	97345	97350	97354
7359	7364	7368	7373	7377	7382	7387	7391	7396	7400
7405	7410	7414	7419	7424	7428	7433	7437	7442	7447
7451	7456	7460	7465	7470	7474	7479	7483	7488	7493
7497	7502	7506	7511	7516	7520	7525	7529	7534	7539
7543	7548	7552	7557	7562	7566	7571	7575	7580	7585
7589	7594	7598	7603	7607	7612	7617	7621	7626	7630
7635	7640	7644	7649	7653	7658	7663	7667	7672	7676
7681	7685	7690	7695	7699	7704	7708	7713	7717	7722
7727	7731	7736	7740	7745	7749	7754	7759	7763	7768
97772	97777	97782	97786	97791	97795	97800	97804	97809	97813

N	0	1	2	3	4	5	6	7	8	9
950	97772	97777	97782	97786	97791	97795	97800	97804	97809	97813
1	7818	7823	7827	7832	7836	7841	7845	7850	7855	7859
2	7864	7868	7873	7877	7882	7886	7891	7896	7900	7905
3	7909	7914	7918	7923	7928	7932	7937	7941	7946	7950
4	7955	7959	7964	7968	7973	7978	7982	7987	7991	7996
5	8000	8005	8009	8014	8019	8023	8028	8032	8037	8041
6	8046	8050	8055	8059	8064	8068	8073	8078	8082	8087
7	8091	8096	8100	8105	8109	8114	8118	8123	8127	8132
8	8137	8141	8146	8150	8155	8159	8164	8168	8173	8177
9	8182	8186	8191	8195	8200	8204	8209	8214	8218	8223
960	98227	98232	98236	98241	98245	98250	98254	98259	98263	98268
1	8272	8277	8281	8286	8290	8295	8299	8304	8308	8313
2	8318	8322	8327	8331	8336	8340	8345	8349	8354	8358
3	8363	8367	8372	8376	8381	8385	8390	8394	8399	8403
4	8408	8412	8417	8421	8426	8430	8435	8439	8444	8448
5	8453	8457	8462	8466	8471	8475	8480	8484	8489	8493
6	8498	8502	8507	8511	8516	8520	8525	8529	8534	8538
7	8543	8547	8552	8556	8561	8565	8570	8574	8579	8583
8	8588	8592	8597	8601	8605	8610	8614	8619	8623	8628
9	8632	8637	8641	8646	8650	8655	8659	8664	8668	8673
970	98677	98682	98686	98691	98695	98700	98704	98709	98713	98717
1	8722	8726	8731	8735	8740	8744	8749	8753	8758	8762
2	8767	8771	8776	8780	8784	8789	8793	8798	8802	8807
3	8811	8816	8820	8825	8829	8834	8838	8843	8847	8851
4	8856	8860	8865	8869	8874	8878	8883	8887	8892	8896
5	8900	8905	8909	8914	8918	8923	8927	8932	8936	8941
6	8945	8949	8954	8958	8963	8967	8972	8976	8981	8985
7	8989	8994	8998	9003	9007	9012	9016	9021	9025	9029
8	9034	9038	9043	9047	9052	9056	9061	9065	9069	9074
9	9078	9083	9087	9092	9096	9100	9105	9109	9114	9118
980	99123	99127	99131	99136	99140	99145	99149	99154	99158	99162
1	9167	9171	9176	9180	9185	9189	9193	9198	9202	9207
2	9211	9216	9220	9224	9229	9233	9238	9242	9247	9251
3	9255	9260	9264	9269	9273	9277	9282	9286	9291	9295
4	9300	9304	9308	9313	9317	9322	9326	9330	9335	9339
5	9344	9348	9352	9357	9361	9366	9370	9374	9379	9383
6	9388	9392	9396	9401	9405	9410	9414	9419	9423	9427
7	9432	9436	9441	9445	9449	9454	9458	9463	9467	9471
8	9476	9480	9484	9489	9493	9498	9502	9506	9511	9515
9	9520	9524	9528	9533	9537	9542	9546	9550	9555	9559
990	99564	99568	99572	99577	99581	99585	99590	99594	99599	99603
1	9607	9612	9616	9621	9625	9629	9634	9638	9642	9647
2	9651	9656	9660	9664	9669	9673	9677	9682	9686	9691
3	9695	9699	9704	9708	9712	9717	9721	9726	9730	9734
4	9739	9743	9747	9752	9756	9760	9765	9769	9774	9778
5	9782	9787	9791	9795	9800	9804	9808	9813	9817	9822
6	9826	9830	9835	9839	9843	9848	9852	9856	9861	9865
7	9870	9874	9878	9883	9887	9891	9896	9900	9904	9909
8	9913	9917	9922	9926	9930	9935	9939	9944	9948	9952
9	9957	9961	9965	9970	9974	9978	9983	9987	9991	9996
1000	00000	00004	00009	00013	00017	00022	00026	00030	00035	00039

′	0° Sine	0° Cosine	1° Sine	1° Cosine	2° Sine	2° Cosine	′
0	—∞	10.00000	8.24186	9.99993	8.54282	9.99974	60
1	6.46373	00000	24903	99993	54642	99973	59
2	76476	00000	25609	99993	54999	99973	58
3	94085	00000	26304	99993	55354	99972	57
4	7.06579	00000	26988	99992	55705	99972	56
5	16270	00000	27661	99992	56054	99971	55
6	24188	00000	28324	99992	56400	99971	54
7	30882	00000	28977	99992	56743	99970	53
8	36682	00000	29621	99992	57084	99970	52
9	41797	00000	30255	99991	57421	99969	51
10	7.46373	10.00000	8.30879	9.99991	8.57757	9.99969	50
11	50512	00000	31495	99991	58089	99968	49
12	54291	00000	32103	99990	58419	99968	48
13	57767	00000	32702	99990	58747	99967	47
14	60985	00000	33292	99990	59072	99967	46
15	63982	00000	33875	99990	59395	99967	45
16	66784	00000	34450	99989	59715	99966	44
17	69417	9.99999	35018	99989	60033	99966	43
18	71900	99999	35578	99989	60349	99965	42
19	74248	99999	36131	99989	60662	99964	41
20	7.76475	9.99999	8.36678	9.99988	8.60973	9.99964	40
21	78594	99999	37217	99988	61282	99963	39
22	80615	99999	37750	99988	61589	99963	38
23	82545	99999	38276	99987	61894	99962	37
24	84393	99999	38796	99987	62196	99962	36
25	86166	99999	39310	99987	62497	99961	35
26	87870	99999	39818	99986	62795	99961	34
27	89509	99999	40320	99986	63091	99960	33
28	91088	99999	40816	99986	63385	99960	32
29	92612	99998	41307	99985	63678	99959	31
30	7.94084	9.99998	8.41792	9.99985	8.63968	9.99959	30
31	95508	99998	42272	99985	64256	99958	29
32	96887	99998	42746	99984	64543	99958	28
33	98223	99998	43216	99984	64827	99957	27
34	99520	99998	43680	99984	65110	99956	26
35	8.00779	99998	44139	99983	65391	99956	25
36	02002	99998	44594	99983	65670	99955	24
37	03192	99997	45044	99983	65947	99955	23
38	04350	99997	45489	99982	66223	99954	22
39	05478	99997	45930	99982	66497	99954	21
40	8.06578	9.99997	8.46366	9.99982	8.66769	9.99953	20
41	07650	99997	46799	99981	67039	99952	19
42	08696	99997	47226	99981	67308	99952	18
43	09718	99997	47650	99981	67575	99951	17
44	10717	99996	48069	99980	67841	99951	16
45	11693	99996	48485	99980	68104	99950	15
46	12647	99996	48896	99979	68367	99949	14
47	13581	99996	49304	99979	68627	99949	13
48	14495	99996	49708	99979	68886	99948	12
49	15391	99996	50108	99978	69144	99948	11
50	8.16268	9.99995	8.50504	9.99978	8.69400	9.99947	10
51	17128	99995	50897	99977	69654	99946	9
52	17971	99995	51287	99977	69907	99946	8
53	18798	99995	51673	99977	70159	99945	7
54	19610	99995	52055	99976	70409	99944	6
55	20407	99994	52434	99976	70658	99944	5
56	21189	99994	52810	99975	70905	99943	4
57	21958	99994	53183	99975	71151	99942	3
58	22713	99994	53552	99974	71395	99942	2
59	23456	99994	53919	99974	71638	99941	1
60	24186	99993	54282	99974	71880	99940	0
′	89° Cosine	89° Sine	88° Cosine	88° Sine	87° Cosine	87° Sine	′

′	3°		4°		5°		′
	Sine	Cosine	Sine	Cosine	Sine	Cosine	
0	8.71880	9.99940	8.84358	9.99894	8.94030	9.99834	60
1	72120	99940	84539	99893	94174	99833	59
2	72359	99939	84718	99892	94317	99832	58
3	72597	99938	84897	99891	94461	99831	57
4	72834	99938	85075	99891	94603	99830	56
5	73069	99937	85252	99890	94746	99829	55
6	73303	99936	85429	99889	94887	99828	54
7	73535	99936	85605	99888	95029	99827	53
8	73767	99935	85780	99887	95170	99825	52
9	73997	99934	85955	99886	95310	99824	51
10	8.74226	9.99934	8.86128	9.99885	8.95450	9.99823	50
11	74454	99933	86301	99884	95589	99822	49
12	74680	99932	86474	99883	95728	99821	48
13	74906	99932	86645	99882	95867	99820	47
14	75130	99931	86816	99881	96005	99819	46
15	75353	99930	86987	99880	96143	99817	45
16	75575	99929	87156	99879	96280	99816	44
17	75795	99929	87325	99879	96417	99815	43
18	76015	99928	87494	99878	96553	99814	42
19	76234	99927	87661	99877	96689	99813	41
20	8.76451	9.99926	8.87829	9.99876	8.96825	9.99812	40
21	76667	99926	87995	99875	96960	99810	39
22	76883	99925	88161	99874	97095	99809	38
23	77097	99924	88326	99873	97229	99808	37
24	77310	99923	88490	99872	97363	99807	36
25	77522	99923	88654	99871	97496	99806	35
26	77733	99922	88817	99870	97629	99804	34
27	77943	99921	88980	99869	97762	99803	33
28	78152	99920	89142	99868	97894	99802	32
29	78360	99920	89304	99867	98026	99801	31
30	8.78568	9.99919	8.89464	9.99866	8.98157	9.99800	30
31	78774	99918	89625	99865	98288	99798	29
32	78979	99917	89784	99864	98419	99797	28
33	79183	99917	89943	99863	98549	99796	27
34	79386	99916	90102	99862	98679	99795	26
35	79588	99915	90260	99861	98808	99793	25
36	79789	99914	90417	99860	98937	99792	24
37	79990	99913	90574	99859	99066	99791	23
38	80189	99913	90730	99858	99194	99790	22
39	80388	99912	90885	99857	99322	99788	21
40	8.80585	9.99911	8.91040	9.99856	8.99450	9.99787	20
41	80782	99910	91195	99855	99577	99786	19
42	80978	99909	91349	99854	99704	99785	18
43	81173	99909	91502	99853	99830	99783	17
44	81367	99908	91655	99852	99956	99782	16
45	81560	99907	91807	99851	9.00082	99781	15
46	81752	99906	91959	99850	00207	99780	14
47	81944	99905	92110	99848	00332	99778	13
48	82134	99904	92261	99847	00456	99777	12
49	82324	99904	92411	99846	00581	99776	11
50	8.82513	9.99903	8.92561	9.99845	9.00704	9.99775	10
51	82701	99902	92710	99844	00828	99773	9
52	82888	99901	92859	99843	00951	99772	8
53	83075	99900	93007	99842	01074	99771	7
54	83261	99899	93154	99841	01196	99769	6
55	83446	99898	93301	99840	01318	99768	5
56	83630	99898	93448	99839	01440	99767	4
57	83813	99897	93594	99838	01561	99765	3
58	83996	99896	93740	99837	01682	99764	2
59	84177	99895	93885	99836	01803	99763	1
60	84358	99894	94030	99834	01923	99761	0
′	Cosine	Sine	Cosine	Sine	Cosine	Sine	′
	86°		85°		84°		

′	6°		7°		8°		′
	Sine	Cosine	Sine	Cosine	Sine	Cosine	
0	9.01923	9.99761	9.08589	9.99675	9.14356	9.99575	60
1	02043	99760	08692	99674	14445	99574	59
2	02163	99759	08795	99672	14535	99572	58
3	02283	99757	08897	99670	14624	99570	57
4	02402	99756	08999	99669	14714	99568	56
5	02520	99755	09101	99667	14803	99566	55
6	02639	99753	09202	99666	14891	99565	54
7	02757	99752	09304	99664	14980	99563	53
8	02874	99751	09405	99663	15069	99561	52
9	02992	99749	09506	99661	15157	99559	51
10	9.03109	9.99748	9.09606	9.99659	9.15245	9.99557	50
11	03226	99747	09707	99658	15333	99556	49
12	03342	99745	09807	99656	15421	99554	48
13	03458	99744	09907	99655	15508	99552	47
14	03574	99742	10006	99653	15596	99550	46
15	03690	99741	10106	99651	15683	99548	45
16	03805	99740	10205	99650	15770	99546	44
17	03920	99738	10304	99648	15857	99545	43
18	04034	99737	10402	99647	15944	99543	42
19	04149	99736	10501	99645	16030	99541	41
20	9.04262	9.99734	9.10599	9.99643	9.16116	9.99539	40
21	04376	99733	10697	99642	16203	99537	39
22	04490	99731	10795	99640	16289	99535	38
23	04603	99730	10893	99638	16374	99533	37
24	04715	99728	10990	99637	16460	99532	36
25	04828	99727	11087	99635	16545	99530	35
26	04940	99726	11184	99633	16631	99528	34
27	05052	99724	11281	99632	16716	99526	33
28	05164	99723	11377	99630	16801	99524	32
29	05275	99721	11474	99629	16886	99522	31
30	9.05386	9.99720	9.11570	9.99627	9.16970	9.99520	30
31	05497	99718	11666	99625	17055	99518	29
32	05607	99717	11761	99624	17139	99517	28
33	05717	99716	11857	99622	17223	99515	27
34	05827	99714	11952	99620	17307	99513	26
35	05937	99713	12047	99618	17391	99511	25
36	06046	99711	12142	99617	17474	99509	24
37	06155	99710	12236	99615	17558	99507	23
38	06264	99708	12331	99613	17641	99505	22
39	06372	99707	12425	99612	17724	99503	21
40	9.06481	9.99705	9.12519	9.99610	9.17807	9.99501	20
41	06589	99704	12612	99608	17890	99499	19
42	06696	99702	12706	99607	17973	99497	18
43	06804	99701	12799	99605	18055	99495	17
44	06911	99699	12892	99603	18137	99494	16
45	07018	99698	12985	99601	18220	99492	15
46	07124	99696	13078	99600	18302	99490	14
47	07231	99695	13171	99598	18383	99488	13
48	07337	99693	13263	99596	18465	99486	12
49	07442	99692	13355	99595	18547	99484	11
50	9.07548	9.99690	9.13447	9.99593	9.18628	9.99482	10
51	07653	99689	13539	99591	18709	99480	9
52	07758	99687	13630	99589	18790	99478	8
53	07863	99686	13722	99588	18871	99476	7
54	07968	99684	13813	99586	18952	99474	6
55	08072	99683	13904	99584	19033	99472	5
56	08176	99681	13994	99582	19113	99470	4
57	08280	99680	14085	99581	19193	99468	3
58	08383	99678	14175	99579	19273	99466	2
59	08486	99677	14266	99577	19353	99464	1
60	08589	99675	14356	99575	19433	99462	0
′	Cosine	Sine	Cosine	Sine	Cosine	Sine	′
	83°		82°		81°		

′	9° Sine	9° Cosine	10° Sine	10° Cosine	11° Sine	11° Cosine	′
0	9.19433	9.99462	9.23967	9.99335	9.28060	9.99195	60
1	19513	99460	24039	99333	28125	99192	59
2	19592	99458	24110	99331	28190	99190	58
3	19672	99456	24181	99328	28254	99187	57
4	19751	99454	24253	99326	28319	99185	56
5	19830	99452	24324	99324	28384	99182	55
6	19909	99450	24395	99322	28448	99180	54
7	19988	99448	24466	99319	28512	99177	53
8	20067	99446	24536	99317	28577	99175	52
9	20145	99444	24607	99315	28641	99172	51
10	9.20223	9.99442	9.24677	9.99313	9.28705	9.99170	50
11	20302	99440	24748	99310	28769	99167	49
12	20380	99438	24818	99308	28833	99165	48
13	20458	99436	24888	99306	28896	99162	47
14	20535	99434	24958	99304	28960	99160	46
15	20613	99432	25028	99301	29024	99157	45
16	20691	99429	25098	99299	29087	99155	44
17	20768	99427	25168	99297	29150	99152	43
18	20845	99425	25237	99294	29214	99150	42
19	20922	99423	25307	99292	29277	99147	41
20	9.20999	9.99421	9.25376	9.99290	9.29340	9.99145	40
21	21076	99419	25445	99288	29403	99142	39
22	21153	99417	25514	99285	29466	99140	38
23	21229	99415	25583	99283	29529	99137	37
24	21306	99413	25652	99281	29591	99135	36
25	21382	99411	25721	99278	29654	99132	35
26	21458	99409	25790	99276	29716	99130	34
27	21534	99407	25858	99274	29779	99127	33
28	21610	99404	25927	99271	29841	99124	32
29	21685	99402	25995	99269	29903	99122	31
30	9.21761	9.99400	9.26063	9.99267	9.29966	9.99119	30
31	21836	99398	26131	99264	30028	99117	29
32	21912	99396	26199	99262	30090	99114	28
33	21987	99394	26267	99260	30151	99112	27
34	22062	99392	26335	99257	30213	99109	26
35	22137	99390	26403	99255	30275	99106	25
36	22211	99388	26470	99252	30336	99104	24
37	22286	99385	26538	99250	30398	99001	23
38	22361	99383	26605	99248	30459	99099	22
39	22435	99381	26672	99245	30521	99096	21
40	9.22509	9.99379	9.26739	9.99243	9.30582	9.99093	20
41	22583	99377	26806	99241	30643	99091	19
42	22657	99375	26873	99238	30704	99088	18
43	22731	99372	26940	99236	30765	99086	17
44	22805	99370	27007	99233	30826	99083	16
45	22878	99368	27073	99231	30887	99080	15
46	22952	99366	27140	99229	30947	99078	14
47	23025	99364	27206	99226	31008	99075	13
48	23098	99362	27273	99224	31068	99072	12
49	23171	99359	27339	99221	31129	99070	11
50	9.23244	9.99357	9.27405	9.99219	9.31189	9.99067	10
51	23317	99355	27471	99217	31250	99064	9
52	23390	99353	27537	99214	31310	99062	8
53	23462	99351	27602	99212	31370	99059	7
54	23535	99348	27668	99209	31430	99056	6
55	23607	99346	27734	99207	31490	99054	5
56	23679	99344	27799	99204	31549	99051	4
57	23752	99342	27864	99202	31609	99048	3
58	23823	99340	27930	99200	31669	99046	2
59	23895	99337	27995	99197	31728	99043	1
60	23967	99335	28060	99195	31788	99040	0
′	Cosine	Sine	Cosine	Sine	Cosine	Sine	′
	80°		79°		78°		

′	12° Sine	12° Cosine	13° Sine	13° Cosine	14° Sine	14° Cosine	′
0	9.31788	9.99040	9.35209	9.98872	9.38368	9.98690	60
1	31847	99038	35263	98869	38418	98687	59
2	31907	99035	35318	98867	38469	98684	58
3	31966	99032	35373	98864	38519	98681	57
4	32025	99030	35427	98861	38570	98678	56
5	32084	99027	35481	98858	38620	98675	55
6	32143	99024	35536	98855	38670	98671	54
7	32202	99022	35590	98852	38721	98668	53
8	32261	99019	35644	98849	38771	98665	52
9	32319	99016	35698	98846	38821	98662	51
10	9.32378	9.99013	9.35752	9.98843	9.38871	9.98659	50
11	32437	99011	35806	98840	38921	98656	49
12	32495	99008	35860	98837	38971	98652	48
13	32553	99005	35914	98834	39021	98649	47
14	32612	99002	35968	98831	39071	98646	46
15	32670	99000	36022	98828	39121	98643	45
16	32728	98997	36075	98825	39170	98640	44
17	32786	98994	36129	98822	39220	98636	43
18	32844	98991	36182	98819	39270	98633	42
19	32902	98989	36236	98816	39319	98630	41
20	9.32960	9.98986	9.36289	9.98813	9.39369	9.98627	40
21	33018	98983	36342	98810	39418	98623	39
22	33075	98980	36395	98807	39467	98620	38
23	33133	98978	36449	98804	39517	98617	37
24	33190	98975	36502	98801	39566	98614	36
25	33248	98972	36555	98798	39615	98610	35
26	33305	98969	36608	98795	39664	98607	34
27	33362	98967	36660	98792	39713	98604	33
28	33420	98964	36713	98789	39762	98601	32
29	33477	98961	36766	98786	39811	98597	31
30	9.33534	9.98958	9.36819	9.98783	9.39860	9.98594	30
31	33591	98955	36871	98780	39909	98591	29
32	33647	98953	36924	98777	39958	98588	28
33	33704	98950	36976	98774	40006	98584	27
34	33761	98947	37028	98771	40055	98581	26
35	33818	98944	37081	98768	40103	98578	25
36	33874	98941	37133	98765	40152	98574	24
37	33931	98938	37185	98762	40200	98571	23
38	33987	98936	37237	98759	40249	98568	22
39	34043	98933	37289	98756	40297	98565	21
40	9.34100	9.98930	9.37341	9.98753	9.40346	9.98561	20
41	34156	98927	37393	98750	40394	98558	19
42	34212	98924	37445	98746	40442	98555	18
43	34268	98921	37497	98743	40490	98551	17
44	34324	98919	37549	98740	40538	98548	16
45	34380	98916	37600	98737	40586	98545	15
46	34436	98913	37652	98734	40634	98541	14
47	34491	98910	37703	98731	40682	98538	13
48	34547	98907	37755	98728	40730	98535	12
49	34602	98904	37806	98725	40778	98531	11
50	9.34658	9.98901	9.37858	9.98722	9.40825	9.98528	10
51	34713	98898	37909	98719	40873	98525	9
52	34769	98896	37960	98715	40921	98521	8
53	34824	98893	38011	98712	40968	98518	7
54	34879	98890	38062	98709	41016	98515	6
55	34934	98887	38113	98706	41063	98511	5
56	34989	98884	38164	98703	41111	98508	4
57	35044	98881	38215	98700	41158	98505	3
58	35099	98878	38266	98697	41205	98501	2
59	35154	98875	38317	98694	41252	98498	1
60	35209	98872	38368	98690	41300	98494	0
′	Cosine	Sine	Cosine	Sine	Cosine	Sine	′
	77°		76°		75°		

′	15° Sine	15° Cosine	16° Sine	16° Cosine	17° Sine	17° Cosine	′
0	9.41300	9.98494	9.44034	9.98284	9.46594	9.98060	60
1	41347	98491	44078	98281	46635	98056	59
2	41394	98488	44122	98277	46676	98052	58
3	41441	98484	44166	98273	46717	98048	57
4	41488	98481	44210	98270	46758	98044	56
5	41535	98477	44253	98266	46800	98040	55
6	41582	98474	44297	98262	46841	98036	54
7	41628	98471	44341	98259	46882	98032	53
8	41675	98467	44385	98255	46923	98029	52
9	41722	98464	44428	98251	46964	98025	51
10	9.41768	9.98460	9.44472	9.98248	9.47005	9.98021	50
11	41815	98457	44516	98244	47045	98017	49
12	41861	98453	44559	98240	47086	98013	48
13	41908	98450	44602	98237	47127	98009	47
14	41954	98447	44646	98233	47168	98005	46
15	42001	98443	44689	98229	47209	98001	45
16	42047	98440	44733	98226	47249	97997	44
17	42093	98436	44776	98222	47290	97993	43
18	42140	98433	44819	98218	47330	97989	42
19	42186	98429	44862	98215	47371	97986	41
20	9.42232	9.98426	9.44905	9.98211	9.47411	9.97982	40
21	42278	98422	44948	98207	47452	97978	39
22	42324	98419	44992	98204	47492	97974	38
23	42370	98415	45035	98200	47533	97970	37
24	42416	98412	45077	98196	47573	97966	36
25	42461	98409	45120	98192	47613	97962	35
26	42507	98405	45163	98189	47654	97958	34
27	42553	98402	45206	98185	47694	97954	33
28	42599	98398	45249	98181	47734	97950	32
29	42644	98395	45292	98177	47774	97946	31
30	9.42690	9.98391	9.45334	9.98174	9.47814	9.97942	30
31	42735	98388	45377	98170	47854	97938	29
32	42781	98384	45419	98166	47894	97934	28
33	42826	98381	45462	98162	47934	97930	27
34	42872	98377	45504	98159	47974	97926	26
35	42917	98373	45547	98155	48014	97922	25
36	42962	98370	45589	98151	48054	97918	24
37	43008	98366	45632	98147	48094	97914	23
38	43053	98363	45674	98144	48133	97910	22
39	43098	98359	45716	98140	48173	97906	21
40	9.43143	9.98356	9.45758	9.98136	9.48213	9.97902	20
41	43188	98352	45801	98132	48252	97898	19
42	43233	98349	45843	98129	48292	97894	18
43	43278	98345	45885	98125	48332	97890	17
44	43323	98342	45927	98121	48371	97886	16
45	43367	98338	45969	98117	48411	97882	15
46	43412	98334	46011	98113	48450	97878	14
47	43457	98331	46053	98110	48490	97874	13
48	43502	98327	46095	98106	48529	97870	12
49	43546	98324	46136	98102	48568	97866	11
50	9.43591	9.98320	9.46178	9.98098	9.48607	9.97861	10
51	43635	98317	46220	98094	48647	97857	9
52	43680	98313	46262	98090	48686	97853	8
53	43724	98309	46303	98087	48725	97849	7
54	43769	98306	46345	98083	48764	97845	6
55	43813	98302	46386	98079	48803	97841	5
56	43857	98299	46428	98075	48842	97837	4
57	43901	98295	46469	98071	48881	97833	3
58	43946	98291	46511	98067	48920	97829	2
59	43990	98288	46552	98063	48959	97825	1
60	44034	98284	46594	98060	48998	97821	0
′	74° Cosine	74° Sine	73° Cosine	73° Sine	72° Cosine	72° Sine	′

′	18° Sine	18° Cosine	19° Sine	19° Cosine	20° Sine	20° Cosine	′
0	9.48998	9.97821	9.51264	9.97567	9.53405	9.97299	60
1	49037	97817	51301	97563	53440	97294	59
2	49076	97812	51338	97558	53475	97289	58
3	49115	97808	51374	97554	53509	97285	57
4	49153	97804	51411	97550	53544	97280	56
5	49192	97800	51447	97545	53578	97276	55
6	49231	97796	51484	97541	53613	97271	54
7	49269	97792	51520	97536	53647	97266	53
8	49308	97788	51557	97532	53682	97262	52
9	49347	97784	51593	97528	53716	97257	51
10	9.49385	9.97779	9.51629	9.97523	9.53751	9.97252	50
11	49424	97775	51666	97519	53785	97248	49
12	49462	97771	51702	97515	53819	97243	48
13	49500	97767	51738	97510	53854	97238	47
14	49539	97763	51774	97506	53888	97234	46
15	49577	97759	51811	97501	53922	97229	45
16	49615	97754	51847	97497	53957	97224	44
17	49654	97750	51883	97492	53991	97220	43
18	49692	97746	51919	97488	54025	97215	42
19	49730	97742	51955	97484	54059	97210	41
20	9.49768	9.97738	9.51991	9.97479	9.54093	9.97206	40
21	49806	97734	52027	97475	54127	97201	39
22	49844	97729	52063	97470	54161	97196	38
23	49882	97725	52099	97466	54195	97192	37
24	49920	97721	52135	97461	54229	97187	36
25	49958	97717	52171	97457	54263	97182	35
26	49996	97713	52207	97453	54297	97178	34
27	50034	97708	52242	97448	54331	97173	33
28	50072	97704	52278	97444	54365	97168	32
29	50110	97700	52314	97439	54399	97163	31
30	9.50148	9.97696	9.52350	9.97435	9.54433	9.97159	30
31	50185	97691	52385	97430	54466	97154	29
32	50223	97687	52421	97426	54500	97149	28
33	50261	97683	52456	97421	54534	97145	27
34	50298	97679	52492	97417	54567	97140	26
35	50336	97674	52527	97412	54601	97135	25
36	50374	97670	52563	97408	54635	97130	24
37	50411	97666	52598	97403	54668	97126	23
38	50449	97662	52634	97399	54702	97121	22
39	50486	97657	52669	97394	54735	97116	21
40	9.50523	9.97653	9.52705	9.97390	9.54769	9.97111	20
41	50561	97649	52740	97385	54802	97107	19
42	50598	97645	52775	97381	54836	97102	18
43	50635	97640	52811	97376	54869	97097	17
44	50673	97636	52846	97372	54903	97092	16
45	50710	97632	52881	97367	54936	97087	15
46	50747	97628	52916	97363	54969	97083	14
47	50784	97623	52951	97358	55003	97078	13
48	50821	97619	52986	97353	55036	97073	12
49	50858	97615	53021	97349	55069	97068	11
50	9.50896	9.97610	9.53056	9.97344	9.55102	9.97063	10
51	50933	97606	53092	97340	55136	97059	9
52	50970	97602	53126	97335	55169	97054	8
53	51007	97597	53161	97331	55202	97049	7
54	51043	97593	53196	97326	55235	97044	6
55	51080	97589	53231	97322	55268	97039	5
56	51117	97584	53266	97317	55301	97035	4
57	51154	97580	53301	97312	55334	97030	3
58	51191	97576	53336	97308	55367	97025	2
59	51227	97571	53370	97303	55400	97020	1
60	51264	97567	53405	97299	55433	97015	0
′	Cosine	Sine	Cosine	Sine	Cosine	Sine	′
	71°		70°		69°		

′	21° Sine	21° Cosine	22° Sine	22° Cosine	23° Sine	23° Cosine	′
0	9.55433	9.97015	9.57358	9.96717	9.59188	9.96403	60
1	55466	97010	57389	96711	59218	96397	59
2	55499	97005	57420	96706	59247	96392	58
3	55532	97001	57451	96701	59277	96387	57
4	55564	96996	57482	96696	59307	96381	56
5	55597	96991	57514	96691	59336	96376	55
6	55630	96986	57545	96686	59366	96370	54
7	55663	96981	57576	96681	59396	96365	53
8	55695	96976	57607	96676	59425	96360	52
9	55728	96971	57638	96670	59455	96354	51
10	9.55761	9.96966	9.57669	9.96665	9.59484	9.96349	50
11	55793	96962	57700	96660	59514	96343	49
12	55826	96957	57731	96655	59543	96338	48
13	55858	96952	57762	96650	59573	96333	47
14	55891	96947	57793	96645	59602	96327	46
15	55923	96942	57824	96640	59632	96322	45
16	55956	96937	57855	96634	59661	96316	44
17	55988	96932	57885	96629	59690	96311	43
18	56021	96927	57916	96624	59720	96305	42
19	56053	96922	57947	96619	59749	96300	41
20	9.56085	9.96917	9.57978	9.96614	9.59778	9.96294	40
21	56118	96912	58008	96608	59808	96289	39
22	56150	96907	58039	96603	59837	96284	38
23	56182	96903	58070	96598	59866	96278	37
24	56215	96898	58101	96593	59895	96273	36
25	56247	96893	58131	96588	59924	96267	35
26	56279	96888	58162	96582	59954	96262	34
27	56311	96883	58192	96577	59983	96256	33
28	56343	96878	58223	96572	60012	96251	32
29	56375	96873	58253	96567	60041	96245	31
30	9.56408	9.96868	9.58284	9.96562	9.60070	9.96240	30
31	56440	96863	58314	96556	60099	96234	29
32	56472	96858	58345	96551	60128	96229	28
33	56504	96853	58375	96546	60157	96223	27
34	56536	96848	58406	96541	60186	96218	26
35	56568	96843	58436	96535	60215	96212	25
36	56599	96838	58467	96530	60244	96207	24
37	56631	96833	58497	96525	60273	96201	23
38	56663	96828	58527	96520	60302	96196	22
39	56695	96823	58557	96514	60331	96190	21
40	9.56727	9.96818	9.58588	9.96509	9.60359	9.96185	20
41	56759	96813	58618	96504	60388	96179	19
42	56790	96808	58648	96498	60417	96174	18
43	56822	96803	58678	96493	60446	96168	17
44	56854	96798	58709	96488	60474	96162	16
45	56886	96793	58739	96483	60503	96157	15
46	56917	96788	58769	96477	60532	96151	14
47	56949	96783	58799	96472	60561	96146	13
48	56980	96778	58829	96467	60589	96140	12
49	57012	96772	58859	96461	60618	96135	11
50	9.57044	9.96767	9.58889	9.96456	9.60646	9.96129	10
51	57075	96762	58919	96451	60675	96123	9
52	57107	96757	58919	96445	60704	96118	8
53	57138	96752	58979	96440	60732	96112	7
54	57169	96747	59009	96435	60761	96107	6
55	57201	96742	59039	96429	60789	96101	5
56	57232	96737	59069	96424	60818	96095	4
57	57264	96732	59098	96419	60846	96090	3
58	57295	96727	59128	96413	60875	96084	2
59	57326	96722	59158	96408	60903	96079	1
60	57358	96717	59188	96403	60931	96073	0
′	Cosine 68°	Sine 68°	Cosine 67°	Sine 67°	Cosine 66°	Sine 66°	′

′	24°		25°		26°		′
	Sine	Cosine	Sine	Cosine	Sine	Cosine	
0	9.60931	9.96073	9.62595	9.95728	9.64184	9.95366	60
1	60960	96067	62622	95722	64210	95360	59
2	60988	96062	62649	95716	64236	95354	58
3	61016	96056	62676	95710	64262	95348	57
4	61045	96050	62703	95704	64288	95341	56
5	61073	96045	62730	95698	64313	95335	55
6	61101	96039	62757	95692	64339	95329	54
7	61129	96034	62784	95686	64365	95323	53
8	61158	96028	62811	95680	64391	95317	52
9	61186	96022	62838	95674	64417	95310	51
10	9.61214	9.96017	9.62865	9.95668	9.64442	9.95304	50
11	61242	96011	62892	95663	64468	95298	49
12	61270	96005	62918	95657	64494	95292	48
13	61298	96000	62945	95651	64519	95286	47
14	61326	95994	62972	95645	64545	95279	46
15	61354	95988	62999	95639	64571	95273	45
16	61382	95982	63026	95633	64596	95267	44
17	61411	95977	63052	95627	64622	95261	43
18	61438	95971	63079	95621	64647	95254	42
19	61466	95965	63106	95615	64673	95248	41
20	9.61494	9.95960	9.63133	9.95609	9.64698	9.95242	40
21	61522	95954	63159	95603	64724	95236	39
22	61550	95948	63186	95597	64749	95229	38
23	61578	95942	63213	95591	64775	95223	37
24	61606	95937	63239	95585	64800	95217	36
25	61634	95931	63266	95579	64826	95211	35
26	61662	95925	63292	95573	64851	95204	34
27	61689	95920	63319	95567	64877	95198	33
28	61717	95914	63345	95561	64902	95192	32
29	61745	95908	63372	95555	64927	95185	31
30	9.61773	9.95902	9.63398	9.95549	9.64953	9.95179	30
31	61800	95897	63425	95543	64978	95173	29
32	61828	95891	63451	95537	65003	95167	28
33	61856	95885	63478	95531	65029	95160	27
34	61883	95879	63504	95525	65054	95154	26
35	61911	95873	63531	95519	65079	95148	25
36	61939	95868	63557	95513	65104	95141	24
37	61966	95862	63583	95507	65130	95135	23
38	61994	95856	63610	95500	65155	95129	22
39	62021	95850	63636	95494	65180	95122	21
40	9.62049	9.95844	9.63662	9.95488	9.65205	9.95116	20
41	62076	95839	63689	95482	65230	95110	19
42	62104	95833	63715	95476	65255	95103	18
43	62131	95827	63741	95470	65281	95097	17
44	62159	95821	63767	95464	65306	95090	16
45	62186	95815	63794	95458	65331	95084	15
46	62214	95810	63820	95452	65356	95078	14
47	62241	95804	63846	95446	65381	95071	13
48	62268	95798	63872	95440	65406	95065	12
49	62296	95792	63898	95434	65431	95059	11
50	9.62323	9.95786	9.63924	9.95427	9.65456	9.95052	10
51	62350	95780	63950	95421	65481	95046	9
52	62377	95775	63976	95415	65506	95039	8
53	62405	95769	64002	95409	65531	95033	7
54	62432	95763	64028	95403	65556	95027	6
55	62459	95757	64054	95397	65580	95020	5
56	62486	95751	64080	95391	65605	95014	4
57	62513	95745	64106	95384	65630	95007	3
58	62541	95739	64132	95378	65655	95001	2
59	62568	95733	64158	95372	65680	94995	1
60	62595	95728	64184	95366	65705	94988	0
′	Cosine	Sine	Cosine	Sine	Cosine	Sine	′
	65°		64°		63°		

′	27° Sine	27° Cosine	28° Sine	28° Cosine	29° Sine	29° Cosine	′
0	9.65705	9.94988	9.67161	9.94593	9.68557	9.94182	60
1	65729	94982	67185	94587	68580	94175	59
2	65754	94975	67208	94580	68603	94168	58
3	65779	94969	67232	94573	68625	94161	57
4	65804	94962	67256	94567	68648	94154	56
5	65828	94956	67280	94560	68671	94147	55
6	65853	94949	67303	94553	68694	94140	54
7	65878	94943	67327	94546	68716	94133	53
8	65902	94936	67350	94540	68739	94126	52
9	65927	94930	67374	94533	68762	94119	51
10	9.65952	9.94923	9.67398	9.94526	9.68784	9.94112	50
11	65976	94917	67421	94519	68807	94105	49
12	66001	94911	67445	94513	68829	94098	48
13	66025	94904	67468	94506	68852	94090	47
14	66050	94898	67492	94499	68875	94083	46
15	66075	94891	67515	94492	68897	94076	45
16	66099	94885	67539	94485	68920	94069	44
17	66124	94878	67562	94479	68942	94062	43
18	66148	94871	67586	94472	68965	94055	42
19	66173	94865	67609	94465	68987	94048	41
20	9.66197	9.94858	9.67633	9.94458	9.69010	9.94041	40
21	66221	94852	67656	94451	69032	94034	39
22	66246	94845	67680	94445	69055	94027	38
23	66270	94839	67703	94438	69077	94020	37
24	66295	94832	67726	94431	69100	94012	36
25	66319	94826	67750	94424	69122	94005	35
26	66343	94819	67773	94417	69144	93998	34
27	66368	94813	67796	94410	69167	93991	33
28	66392	94806	67820	94404	69189	93984	32
29	66416	94799	67843	94397	69212	93977	31
30	9.66441	9.94793	9.67866	9.94390	9.69234	9.93970	30
31	66465	94786	67890	94383	69256	93963	29
32	66489	94780	67913	94376	69279	93955	28
33	66513	94773	67936	94369	69301	93948	27
34	66537	94767	67959	94362	69323	93941	26
35	66562	94760	67982	94355	69345	93934	25
36	66586	94753	68006	94349	69368	93927	24
37	66610	94747	68029	94342	69390	93920	23
38	66634	94740	68052	94335	69412	93912	22
39	66658	94734	68075	94328	69434	93905	21
40	9.66682	9.94727	9.68098	9.94321	9.69456	9.93898	20
41	66706	94720	68121	94314	69479	93891	19
42	66731	94714	68144	94307	69501	93884	18
43	66755	94707	68167	94300	69523	93876	17
44	66779	94700	68190	94293	69545	93869	16
45	66803	94694	68213	94286	69567	93862	15
46	66827	94687	68237	94279	69589	93855	14
47	66851	94680	68260	94273	69611	93847	13
48	66875	94674	68283	94266	69633	93840	12
49	66899	94667	68305	94259	69655	93833	11
50	9.66922	9.94660	9.68328	9.94252	9.69677	9.93826	10
51	66946	94654	68351	94245	69699	93819	9
52	66970	94647	68374	94238	69721	93811	8
53	66994	94640	68397	94231	69743	93804	7
54	67018	94634	68420	94224	69765	93797	6
55	67042	94627	68443	94217	69787	93789	5
56	67066	94620	68466	94210	69809	93782	4
57	67090	94614	68489	94203	69831	93775	3
58	67113	94607	68512	94196	69853	93768	2
59	67137	94600	68534	94189	69875	93760	1
60	67161	94593	68557	94182	69897	93753	0
′	Cosine	Sine	Cosine	Sine	Cosine	Sine	′
	62°		61°		60°		

′	30°		31°		32°		′
	Sine	Cosine	Sine	Cosine	Sine	Cosine	
0	9.69897	9.93753	9.71184	9.93307	9.72421	9.92842	60
1	69919	93746	71205	93299	72441	92834	59
2	69941	93738	71226	93291	72461	92826	58
3	69963	93731	71247	93284	72482	92818	57
4	69984	93724	71268	93276	72502	92810	56
5	70006	93717	71289	93269	72522	92803	55
6	70028	93709	71310	93261	72542	92795	54
7	70050	93702	71331	93253	72562	92787	53
8	70072	93695	71352	93246	72582	92779	52
9	70093	93687	71373	93238	72602	92771	51
10	9.70115	9.93680	9.71393	9.93230	9.72622	9.92763	50
11	70137	93673	71414	93228	72648	92755	49
12	70159	93665	71435	93215	72663	92747	48
13	70180	93658	71456	93207	72683	92739	47
14	70202	93650	71477	93200	72703	92731	46
15	70224	93643	71498	93192	72723	92723	45
16	70245	93636	71519	93184	72743	92715	44
17	70267	93628	71539	93177	72763	92707	43
18	70288	93621	71560	93169	72783	92699	42
19	70310	93614	71581	93161	72803	92691	41
20	9.70332	9.93606	9.71602	9.93154	9.72823	9.92683	40
21	70353	93599	71622	93146	72843	92675	39
22	70375	93591	71643	93138	72863	92667	38
23	70396	93584	71664	93131	72883	92659	37
24	70418	93577	71685	93123	72902	92651	36
25	70439	93569	71705	93115	72922	92643	35
26	70461	93562	71726	93108	72942	92635	34
27	70482	93554	71747	93100	72962	92627	33
28	70504	93547	71767	93092	72982	92619	32
29	70525	93539	71788	93084	73002	92611	31
30	9.70547	9.93532	9.71809	9.93077	9.73022	9.92603	30
31	70568	93525	71829	93069	73041	92595	29
32	70590	93517	71850	93061	73061	92587	28
33	70611	93510	71870	93053	73081	92579	27
34	70633	93502	71891	93046	73101	92571	26
35	70654	93495	71911	93038	73121	92563	25
36	70675	93487	71932	93030	73140	92555	24
37	70697	93480	71952	93022	73160	92546	23
38	70718	93472	71973	93014	73180	92538	22
39	70739	93465	71994	93007	73200	92530	21
40	9.70761	9.93457	9.72014	9.92999	9.73219	9.92522	20
41	70782	93450	72034	92991	73239	92514	19
42	70803	93442	72055	92983	73259	92506	18
43	70824	93435	72075	92976	73278	92498	17
44	70846	93427	72096	92968	73298	92490	16
45	70867	93420	72116	92960	73318	92482	15
46	70888	93412	72137	92952	73337	92473	14
47	70909	93405	72157	92944	73357	92465	13
48	70931	93397	72177	92936	73377	92457	12
49	70952	93390	72198	92929	73396	92449	11
50	9.70973	9.93382	9.72218	9.92921	9.73416	9.92441	10
51	70994	93375	72238	92913	73435	92433	9
52	71015	93367	72259	92905	73455	92425	8
53	71036	93360	72279	92897	73474	92416	7
54	71058	93352	72299	92889	73494	92408	6
55	71079	93344	72320	92881	73513	92400	5
56	71100	93337	72340	92874	73533	92392	4
57	71121	93329	72360	92866	73552	92384	3
58	71142	93322	72381	92858	73572	92376	2
59	71163	93314	72401	92850	73591	92367	1
60	71184	93307	72421	92842	73611	92359	0
′	Cosine	Sine	Cosine	Sine	Cosine	Sine	′
	59°		58°		57°		

′	33°		34°		35°		′
	Sine	Cosine	Sine	Cosine	Sine	Cosine	
0	9.73611	9.92359	9.74756	9.91857	9.75859	9.91336	60
1	73630	92351	74775	91849	75877	91328	59
2	73650	92343	74794	91840	75895	91319	58
3	73669	92335	74812	91832	75913	91310	57
4	73689	92326	74831	91823	75931	91301	56
5	73708	92318	74850	91815	75949	91292	55
6	73727	92310	74868	91806	75967	91283	54
7	73747	92302	74887	91798	75985	91274	53
8	73766	92293	74906	91789	76003	91266	52
9	73785	92285	74924	91781	76021	91257	51
10	9.73805	9.92277	9.74943	9.91772	9.76039	9.91248	50
11	73824	92269	74961	91763	76057	91239	49
12	73843	92260	74980	91755	76075	91230	48
13	73863	92252	74999	91746	76093	91221	47
14	73882	92244	75017	91738	76111	91212	46
15	73901	92235	75036	91729	76129	91203	45
16	73921	92227	75054	91720	76146	91194	44
17	73940	92219	75073	91712	76164	91185	43
18	73959	92211	75091	91703	76182	91176	42
19	73978	92202	75110	91695	76200	91167	41
20	9.73997	9.92194	9.75128	9.91686	9.76218	9.91158	40
21	74017	92186	75147	91677	76236	91149	39
22	74036	92177	75165	91669	76253	91141	38
23	74055	92169	75184	91660	76271	91132	37
24	74074	92161	75202	91651	76289	91123	36
25	74093	92152	75221	91643	76307	91114	35
26	74113	92144	75239	91634	76324	91105	34
27	74132	92136	75258	91625	76342	91096	33
28	74151	92127	75276	91617	76360	91087	32
29	74170	92119	75294	91608	76378	91078	31
30	9.74189	9.92111	9.75313	9.91599	9.76395	9.91069	30
31	74208	92102	75331	91591	76413	91060	29
32	74227	92094	75350	91582	76431	91051	28
33	74246	92086	75368	91573	76448	91042	27
34	74265	92077	75386	91565	76466	91033	26
35	74284	92069	75405	91556	76484	91023	25
36	74303	92060	75423	91547	76501	91014	24
37	74322	92052	75441	91538	76519	91005	23
38	74341	92044	75459	91530	76537	90996	22
39	74360	92035	75478	91521	76554	90987	21
40	9.74379	9.92027	9.75496	9.91512	9.76572	9.90978	20
41	74398	92018	75514	91504	76590	90969	19
42	74417	92010	75533	91495	76607	90960	18
43	74436	92002	75551	91486	76625	90951	17
44	74455	91993	75569	91477	76642	90942	16
45	74474	91985	75587	91469	76660	90933	15
46	74493	91976	75605	91460	76677	90924	14
47	74512	91968	75624	91451	76695	90915	13
48	74531	91959	75642	91442	76712	90906	12
49	74549	91951	75660	91433	76730	90896	11
50	9.74568	9.91942	9.75678	9.91425	9.76747	9.90887	10
51	74587	91934	75696	91416	76765	90878	9
52	74606	91925	75714	91407	76782	90869	8
53	74625	91917	75733	91398	76800	90860	7
54	74644	91908	75751	91389	76817	90851	6
55	74662	91900	75769	91381	76835	90842	5
56	74681	91891	75787	91372	76852	90832	4
57	74700	91883	75805	91363	76870	90823	3
58	74719	91874	75823	91354	76887	90814	2
59	74737	91866	75841	91345	76904	90805	1
60	74756	91857	75859	91336	76922	90796	0
′	Cosine	Sine	Cosine	Sine	Cosine	Sine	′
	56°		55°		54°		

′	36° Sine	36° Cosine	37° Sine	37° Cosine	38° Sine	38° Cosine	′
0	9.76922	9.90796	9.77946	9.90235	9.78934	9.89653	60
1	76939	90787	77963	90225	78950	89643	59
2	76957	90777	77980	90216	78967	89633	58
3	76974	90768	77997	90206	78983	89624	57
4	76991	90759	78013	90197	78999	89614	56
5	77009	90750	78030	90187	79015	89604	55
6	77026	90741	78047	90178	79031	89594	54
7	77043	90731	78063	90168	79047	89584	53
8	77061	90722	78080	90159	79063	89574	52
9	77078	90713	78097	90149	79079	89564	51
10	9.77095	9.90704	9.78113	9.90139	9.79095	9.89554	50
11	77112	90694	78130	90130	79111	89544	49
12	77130	90685	78147	90120	79128	89534	48
13	77147	90676	78163	90111	79144	89524	47
14	77164	90667	78180	90101	79160	89514	46
15	77181	90657	78197	90091	79176	89504	45
16	77199	90648	78213	90082	79192	89495	44
17	77216	90639	78230	90072	79208	89485	43
18	77233	90630	78246	90063	79224	89475	42
19	77250	90620	78263	90053	79240	89465	41
20	9.77268	9.90611	9.78280	9.90043	9.79256	9.89455	40
21	77285	90602	78296	90034	79272	89445	39
22	77302	90592	78313	90024	79288	89435	38
23	77319	90583	78329	90014	79304	89425	37
24	77336	90574	78346	90005	79319	89415	36
25	77353	90565	78362	89995	79335	89405	35
26	77370	90555	78379	89985	79351	89395	34
27	77387	90546	78395	89976	79367	89385	33
28	77405	90537	78412	89966	79383	89375	32
29	77422	90527	78428	89956	79399	89364	31
30	9.77439	9.90518	9.78445	9.89947	9.79415	9.89354	30
31	77456	90509	78461	89937	79431	89344	29
32	77473	90499	78478	89927	79447	89334	28
33	77490	90490	78494	89918	79463	89324	27
34	77507	90480	78510	89908	79478	89314	26
35	77524	90471	78527	89898	79494	89304	25
36	77541	90462	78543	89888	79510	89294	24
37	77558	90452	78560	89879	79526	89284	23
38	77575	90443	78576	89869	79542	89274	22
39	77592	90434	78592	89859	79558	89264	21
40	9.77609	9.90424	9.78609	9.89849	9.79573	9.89254	20
41	77626	90415	78625	89840	79589	89244	19
42	77643	90405	78642	89830	79605	89233	18
43	77660	90396	78658	89820	79621	89223	17
44	77677	90386	78674	89810	79636	89213	16
45	77694	90377	78691	89801	79652	89203	15
46	77711	90368	78707	89791	79668	89193	14
47	77728	90358	78723	89781	79684	89183	13
48	77744	90349	78739	89771	79699	89173	12
49	77761	90339	78756	89761	79715	89162	11
50	9.77778	9.90330	9.78772	9.89752	9.79731	9.89152	10
51	77795	90320	78788	89742	79746	89142	9
52	77812	90311	78805	89732	79762	89132	8
53	77829	90301	78821	89722	79778	89122	7
54	77846	90292	78837	89712	79793	89112	6
55	77862	90282	78853	89702	79809	89101	5
56	77879	90273	78869	89693	79825	89091	4
57	77896	90263	78886	89683	79840	89081	3
58	77913	90254	78902	89673	79856	89071	2
59	77930	90244	78918	89663	79872	89060	1
60	77946	90235	78934	89653	79887	89050	0
′	Cosine	Sine	Cosine	Sine	Cosine	Sine	′
	53°		52°		51°		

′	39° Sine	39° Cosine	40° Sine	40° Cosine	41° Sine	41° Cosine	′
0	9.79887	9.89050	9.80807	9.88425	9.81694	9.87778	60
1	79903	89040	80822	88415	81709	87767	59
2	79918	89030	80837	88404	81723	87756	58
3	79934	89020	80852	88394	81738	87745	57
4	79950	89009	80867	88383	81752	87734	56
5	79965	88999	80882	88372	81767	87723	55
6	79981	88989	80897	88362	81781	87712	54
7	79996	88978	80912	88351	81796	87701	53
8	80012	88968	80927	88340	81810	87690	52
9	80027	88958	80942	88330	81825	87679	51
10	9.80043	9.88948	9.80957	9.88319	9.81839	9.87668	50
11	80058	88937	80972	88308	81854	87657	49
12	80074	88927	80987	88298	81868	87646	48
13	80089	88917	81002	88287	81882	87635	47
14	80105	88906	81017	88276	81897	87624	46
15	80120	88896	81032	88266	81911	87613	45
16	80136	88886	81047	88255	81926	87601	44
17	80151	88875	81061	88244	81940	87590	43
18	80166	88865	81076	88234	81955	87579	42
19	80182	88855	81091	88223	81969	87568	41
20	9.80197	9.88844	9.81106	9.88212	9.81983	9.87557	40
21	80213	88834	81121	88201	81998	87546	39
22	80228	88824	81136	88191	82012	87535	38
23	80244	88813	81151	88180	82026	87524	37
24	80259	88803	81166	88169	82041	87513	36
25	80274	88793	81180	88158	82055	87501	35
26	80290	88782	81195	88148	82069	87490	34
27	80305	88772	81210	88137	82084	87479	33
28	80320	88761	81225	88126	82098	87468	32
29	80336	88751	81240	88115	82112	87457	31
30	9.80351	9.88741	9.81254	9.88105	9.82126	9.87446	30
31	80366	88730	81269	88094	82141	87434	29
32	80382	88720	81284	88083	82155	87423	28
33	80397	88709	81299	88072	82169	87412	27
34	80412	88699	81314	88061	82184	87401	26
35	80428	88688	81328	88051	82198	87390	25
36	80443	88678	81343	88040	82212	87378	24
37	80458	88668	81358	88029	82226	87367	23
38	80473	88657	81372	88018	82240	87356	22
39	80489	88647	81387	88007	82255	87345	21
40	9.80504	9.88636	9.81402	9.87996	9.82269	9.87334	20
41	80519	88626	81417	87985	82283	87322	19
42	80534	88615	81431	87975	82297	87311	18
43	80550	88605	81446	87964	82311	87300	17
44	80565	88594	81461	87953	82326	87288	16
45	80580	88584	81475	87942	82340	87277	15
46	80595	88573	81490	87931	82354	87266	14
47	80610	88563	81505	87920	82368	87255	13
48	80625	88552	81519	87909	82382	87243	12
49	80641	88542	81534	87898	82396	87232	11
50	9.80656	9.88531	9.81549	9.87887	9.82410	9.87221	10
51	80671	88521	81563	87877	82424	87209	9
52	80686	88510	81578	87866	82439	87198	8
53	80701	88499	81592	87855	82453	87187	7
54	80716	88489	81607	87844	82467	87175	6
55	80731	88478	81622	87833	82481	87164	5
56	80746	88468	81636	87822	82495	87153	4
57	80762	88457	81651	87811	82509	87141	3
58	80777	88447	81665	87800	82523	87130	2
59	80792	88436	81680	87789	82537	87119	1
60	80807	88425	81694	87778	82551	87107	0
′	50° Cosine	50° Sine	49° Cosine	49° Sine	48° Cosine	48° Sine	′

′	42° Sine	42° Cosine	43° Sine	43° Cosine	44° Sine	44° Cosine	′
0	9.82551	9.87107	9.83378	9.86413	9.84177	9.85693	60
1	82565	87096	83392	86401	84190	85681	59
2	82579	87085	83405	86389	84203	85669	58
3	82593	87073	83419	86377	84216	85657	57
4	82607	87062	83432	86366	84229	85645	56
5	82621	87050	83446	86354	84242	85632	55
6	82635	87039	83459	86342	84255	85620	54
7	82649	87028	83473	86330	84269	85608	53
8	82663	87016	83486	86318	84282	85596	52
9	82677	87005	83500	86306	84295	85583	51
10	9.82691	9.86993	9.83513	9.86295	9.84308	9.85571	50
11	82705	86982	83527	86283	84321	85559	49
12	82719	86970	83540	86271	84334	85547	48
13	82733	86959	83554	86259	84347	85534	47
14	82747	86947	83567	86247	84360	85522	46
15	82761	86936	83581	86235	84373	85510	45
16	82775	86924	83594	86223	84385	85497	44
17	82788	86913	83608	86211	84398	85485	43
18	82802	86902	83621	86200	84411	85473	42
19	82816	86890	83634	86188	84424	85460	41
20	9.82830	9.86879	9.83648	9.86176	9.84437	9.85448	40
21	82844	86867	83661	86164	84450	85436	39
22	82858	86855	83674	86152	84463	85423	38
23	82872	86844	83688	86140	84476	85411	37
24	82885	86832	83701	86128	84489	85399	36
25	82899	86821	83715	86116	84502	85386	35
26	82913	86809	83728	86104	84515	85374	34
27	82927	86798	83741	86092	84528	85361	33
28	82941	86786	83755	86080	84540	85349	32
29	82955	86775	83768	86068	84553	85337	31
30	9.82968	9.86763	9.83781	9.86056	9.84566	9.85324	30
31	82982	86752	83795	86044	84579	85312	29
32	82996	86740	83808	86032	84592	85299	28
33	83010	86728	83821	86020	84605	85287	27
34	83023	86717	83834	86008	84618	85274	26
35	83037	86705	83848	85996	84630	85262	25
36	83051	86694	83861	85984	84643	85250	24
37	83065	86682	83874	85972	84656	85237	23
38	83078	86670	83887	85960	84669	85225	22
39	83092	86659	83901	85948	84682	85212	21
40	9.83106	9.86647	9.83914	9.85936	9.84694	9.85200	20
41	83120	86635	83927	85924	84707	85187	19
42	83133	86624	83940	85912	84720	85175	18
43	83147	86612	83954	85900	84733	85162	17
44	83161	86600	83967	85888	84745	85150	16
45	83174	86589	83980	85876	84758	85137	15
46	83188	86577	83993	85864	84771	85125	14
47	83202	86565	84006	85851	84784	85112	13
48	83215	86554	84020	85839	84796	85100	12
49	83229	86542	84033	85827	84809	85087	11
50	9.83242	9.86530	9.84046	9.85815	9.84822	9.85074	10
51	83256	86518	84059	85803	84835	85062	9
52	83270	86507	84072	85791	84847	85049	8
53	83283	86495	84085	85779	84860	85037	7
54	83297	86483	84098	85766	84873	85024	6
55	83310	86472	84112	85754	84885	85012	5
56	83324	86460	84125	85742	84898	84999	4
57	83338	86448	84138	85730	84911	84986	3
58	83351	86436	84151	85718	84923	84974	2
59	83365	86425	84164	85706	84936	84961	1
60	83378	86413	84177	85693	84949	84949	0
′	47° Cosine	47° Sine	46° Cosine	46° Sine	45° Cosine	45° Sine	′

′	0° Tan	0° Cotan	1° Tan	1° Cotan	2° Tan	2° Cotan	′
0	−∞	∞	8.24192	11.75808	8.54308	11.45692	60
1	6.46373	13.53627	24910	75090	54669	45331	59
2	76476	23524	25616	74384	55027	44973	58
3	94085	05915	26312	73688	55382	44618	57
4	7.06579	12.93421	26996	73004	55734	44266	56
5	16270	83730	27669	72331	56083	43917	55
6	24188	75812	28332	71668	56429	43571	54
7	30882	69118	28986	71014	56773	43227	53
8	36682	63318	29629	70371	57114	42886	52
9	41797	58203	30263	69737	57452	42548	51
10	7.46373	12.53627	8.30888	11.69112	8.57788	11.42212	50
11	50512	49488	31505	68495	58121	41879	49
12	54291	45709	32112	67888	58451	41549	48
13	57767	42233	32711	67289	58779	41221	47
14	60986	39014	33302	66698	59105	40895	46
15	63982	36018	33886	66114	59428	40572	45
16	66785	33215	34461	65539	59749	40251	44
17	69418	30582	35029	64971	60068	39932	43
18	71900	28100	35590	64410	60384	39616	42
19	74248	25752	36143	63857	60698	39302	41
20	7.76476	12.23524	8.36689	11.63311	8.61009	11.38991	40
21	78595	21405	37229	62771	61319	38681	39
22	80615	19385	37762	62238	61626	38374	38
23	82546	17454	38289	61711	61931	38069	37
24	84394	15606	38809	61191	62234	37766	36
25	86167	13833	39323	60677	62535	37465	35
26	87871	12129	39832	60168	62834	37166	34
27	89510	10490	40334	59666	63131	36869	33
28	91089	08911	40830	59170	63426	36574	32
29	92613	07387	41321	58679	63718	36282	31
30	7.94086	12.05914	8.41807	11.58193	8.64009	11.35991	30
31	95510	04490	42287	57713	64298	35702	29
32	96889	03111	42762	57238	64585	35415	28
33	98225	01775	43232	56768	64870	35130	27
34	99522	00478	43696	56304	65154	34846	26
35	8.00781	11.99219	44156	55844	65435	34565	25
36	02004	97996	44611	55389	65715	34285	24
37	03194	96806	45061	54939	65993	34007	23
38	04353	95647	45507	54493	66269	33731	22
39	05481	94519	45948	54052	66543	33457	21
40	8.06581	11.93419	8.46385	11.53615	8.66816	11.33184	20
41	07653	92347	46817	53183	67087	32913	19
42	08700	91300	47245	52755	67356	32644	18
43	09722	90278	47669	52331	67624	32376	17
44	10720	89280	48089	51911	67890	32110	16
45	11696	88304	48505	51495	68154	31846	15
46	12651	87349	48917	51083	68417	31583	14
47	13585	86415	49325	50675	68678	31322	13
48	14500	85500	49729	50271	68938	31062	12
49	15395	84605	50130	49870	69196	30804	11
50	8.16273	11.83727	8.50527	11.49473	8.69453	11.30547	10
51	17133	82867	50920	49080	69708	30292	9
52	17976	82024	51310	48690	69962	30038	8
53	18804	81196	51696	48304	70214	29786	7
54	19616	80384	52079	47921	70465	29535	6
55	20413	79587	52459	47541	70714	29286	5
56	21195	78805	52835	47165	70962	29038	4
57	21964	78036	53208	46792	71208	28792	3
58	22720	77280	53578	46422	71453	28547	2
59	23462	76538	53945	46055	71697	28303	1
60	24192	75808	54308	45692	71940	28060	0
′	89° Cotan	89° Tan	88° Cotan	88° Tan	87° Cotan	87° Tan	′

′	3°		4°		5°		′
	Tan	Cotan	Tan	Cotan	Tan	Cotan	
0	8.71940	11.28060	8.84464	11.15536	8.94195	11.05805	60
1	72181	27819	84646	15354	94340	05660	59
2	72420	27580	84826	15174	94485	05515	58
3	72659	27341	85006	14994	94630	05370	57
4	72896	27104	85185	14815	94773	05227	56
5	73132	26868	85363	14637	94917	05083	55
6	73366	26634	85540	14460	95060	04940	54
7	73600	26400	85717	14283	95202	04798	53
8	73832	26168	85893	14107	95344	04656	52
9	74063	25937	86069	13931	95486	04514	51
10	8.74292	11.25708	8.86243	11.13757	8.95627	11.04373	50
11	74521	25479	86417	13583	95767	04233	49
12	74748	25252	86591	13409	95908	04092	48
13	74974	25026	86763	13237	96047	03953	47
14	75199	24801	86935	13065	96187	03813	46
15	75423	24577	87106	12894	96325	03675	45
16	75645	24355	87277	12723	96464	03536	44
17	75867	24133	87447	12553	96602	03398	43
18	76087	23913	87616	12384	96739	03261	42
19	76306	23694	87785	12215	96877	03123	41
20	8.76525	11.23475	8.87953	11.12047	8.97013	11.02987	40
21	76742	23258	88120	11880	97150	02850	39
22	76958	23042	88287	11713	97285	02715	38
23	77173	22827	88453	11547	97421	02579	37
24	77387	22613	88618	11382	97556	02444	36
25	77600	22400	88783	11217	97691	02309	35
26	77811	22189	88948	11052	97825	02175	34
27	78022	21978	89111	10889	97959	02041	33
28	78232	21768	89274	10726	98092	01908	32
29	78441	21559	89437	10563	98225	01775	31
30	8.78649	11.21351	8.89598	11.10402	8.98358	11.01642	30
31	78855	21145	89760	10240	98490	01510	29
32	79061	20939	89920	10080	98622	01378	28
33	79266	20734	90080	09920	98753	01247	27
34	79470	20530	90240	09760	98884	01116	26
35	79673	20327	90399	09601	99015	00985	25
36	79875	20125	90557	09443	99145	00855	24
37	80076	19924	90715	09285	99275	00725	23
38	80277	19723	90872	09128	99405	00595	22
39	80476	19524	91029	08971	99534	00466	21
40	8.80674	11.19326	8.91185	11.08815	8.99662	11.00338	20
41	80872	19128	91340	08660	99791	00209	19
42	81068	18932	91495	08505	99919	00081	18
43	81264	18736	91650	08350	9.00046	10.99954	17
44	81459	18541	91803	08197	00174	99826	16
45	81653	18347	91957	08043	00301	99699	15
46	81846	18154	92110	07890	00427	99573	14
47	82038	17962	92262	07738	00553	99447	13
48	82230	17770	92414	07586	00679	99321	12
49	82420	17580	92565	07435	00805	99195	11
50	8.82610	11.17390	8.92716	11.07284	9.00930	10.99070	10
51	82799	17201	92866	07134	01055	98945	9
52	82987	17013	93016	06984	01179	98821	8
53	83175	16825	93165	06835	01303	98697	7
54	83361	16639	93313	06687	01427	98573	6
55	83547	16453	93462	06538	01550	98450	5
56	83732	16268	93609	06391	01673	98327	4
57	83916	16084	93756	06244	01796	98204	3
58	84100	15900	93903	06097	01918	98082	2
59	84282	15718	94049	05951	02040	97960	1
60	84464	15536	94195	05805	02162	97838	0
′	Cotan	Tan	Cotan	Tan	Cotan	Tan	′
	86°		85°		84°		

′	6° Tan	6° Cotan	7° Tan	7° Cotan	8° Tan	8° Cotan	′
0	9.02162	10.97838	9.08914	10.91086	9.14780	10.85220	60
1	02283	97717	09019	90981	14872	85128	59
2	02404	97596	09123	90877	14963	85037	58
3	02525	97475	09227	90773	15054	84946	57
4	02645	97355	09330	90670	15145	84855	56
5	02766	97234	09434	90566	15236	84764	55
6	02885	97115	09537	90463	15327	84673	54
7	03005	96995	09640	90360	15417	84583	53
8	03124	96876	09742	90258	15508	84492	52
9	03242	96758	09845	90155	15598	84402	51
10	9.03361	10.96639	9.09947	10.90053	9.15688	10.84312	50
11	03479	96521	10049	89951	15777	84223	49
12	03597	96403	10150	89850	15867	84133	48
13	03714	96286	10252	89748	15956	84044	47
14	03832	96168	10353	89647	16046	83954	46
15	03948	96052	10454	89546	16135	83865	45
16	04065	95935	10555	89445	16224	83776	44
17	04181	95819	10656	89344	16312	83688	43
18	04297	95703	10756	89244	16401	83599	42
19	04413	95587	10856	89144	16489	83511	41
20	9.04528	10.95472	9.10956	10.89044	9.16577	10.83423	40
21	04643	95357	11056	88944	16665	83335	39
22	04758	95242	11155	88845	16753	83247	38
23	04873	95127	11254	88746	16841	83159	37
24	04987	95013	11353	88647	16928	83072	36
25	05101	94899	11452	88548	17016	82984	35
26	05214	94786	11551	88449	17103	82897	34
27	05328	94672	11649	88351	17190	82810	33
28	05441	94559	11747	88253	17277	82723	32
29	05553	94447	11845	88155	17363	82637	31
30	9.05666	10.94334	9.11943	10.88057	9.17450	10.82550	30
31	05778	94222	12040	87960	17536	82464	29
32	05890	94110	12138	87862	17622	82378	28
33	06002	93998	12235	87765	17708	82292	27
34	06113	93887	12332	87668	17794	82206	26
35	06224	93776	12428	87572	17880	82120	25
36	06335	93665	12525	87475	17965	82035	24
37	06445	93555	12621	87379	18051	81949	23
38	06556	93444	12717	87283	18136	81864	22
39	06666	93334	12813	87187	18221	81779	21
40	9.06775	10.93225	9.12909	10.87091	9.18306	10.81694	20
41	06885	93115	13004	86996	18391	81609	19
42	06994	93006	13099	86901	18475	81525	18
43	07103	92897	13194	86806	18560	81440	17
44	07211	92789	13289	86711	18644	81356	16
45	07320	92680	13384	86616	18728	81272	15
46	07428	92572	13478	86522	18812	81188	14
47	07536	92464	13573	86427	18896	81104	13
48	07643	92357	13667	86333	18979	81021	12
49	07751	92249	13761	86239	19063	80937	11
50	9.07858	10.92142	9.13854	10.86146	9.19146	10.80854	10
51	07964	92036	13948	86052	19229	80771	9
52	08071	91929	14041	85959	19312	80688	8
53	08177	91823	14134	85866	19395	80605	7
54	08283	91717	14227	85773	19478	80522	6
55	08389	91611	14320	85680	19561	80439	5
56	08495	91505	14412	85588	19643	80357	4
57	08600	91400	14504	85496	19725	80275	3
58	08705	91295	14597	85403	19807	80193	2
59	08810	91190	14688	85312	19889	80111	1
60	08914	91086	14780	85220	19971	80029	0
′	Cotan	Tan	Cotan	Tan	Cotan	Tan	′
	83°		82°		81°		

′	9°		10°		11°		′
	Tan	Cotan	Tan	Cotan	Tan	Cotan	
0	9.19971	10.80029	9.24632	10.75368	9.28865	10.71135	60
1	20053	79947	24706	75294	28933	71067	59
2	20134	79866	21770	75221	29000	71000	58
3	20216	79784	24853	75147	29067	70933	57
4	20297	79703	24926	75074	29134	70866	56
5	20378	79622	25000	75000	29201	70799	55
6	20459	79541	25073	74927	29268	70732	54
7	20540	79460	25146	74854	29335	70665	53
8	20621	79379	25219	74781	29402	70598	52
9	20701	79299	25292	74708	29468	70532	51
10	9.20782	10.79218	9.25365	10.74635	9.29535	10.70465	50
11	20862	79138	25437	74563	29601	70399	49
12	20942	79058	25510	74490	29668	70332	48
13	21022	78978	25582	74418	29734	70266	47
14	21102	78898	25655	74345	29800	70200	46
15	21182	78818	25727	74273	29866	70134	45
16	21261	78739	25799	74201	29932	70068	44
17	21341	78659	25871	74129	29998	70002	43
18	21420	78580	25943	74057	30064	69936	42
19	21499	78501	26015	73985	30130	69870	41
20	9.21578	10.78422	9.26086	10.73914	9.30195	10.69805	40
21	21657	78343	26158	73842	30261	69739	39
22	21736	78264	26229	73771	30326	69674	38
23	21814	78186	26301	73699	30391	69609	37
24	21893	78107	26372	73628	30457	69543	36
25	21971	78029	26443	73557	30522	69478	35
26	22049	77951	26514	73486	30587	69413	34
27	22127	77873	26585	73415	30652	69348	33
28	22205	77795	26655	73345	30717	69283	32
29	22283	77717	26726	73274	30782	69218	31
30	9.22361	10.77639	9.26797	10.73203	9.30846	10.69154	30
31	22438	77562	26867	73133	30911	69089	29
32	22516	77484	26937	73063	30975	69025	28
33	22593	77407	27008	72992	31040	68960	27
34	22670	77330	27078	72922	31104	68896	26
35	22747	77253	27148	72852	31168	68832	25
36	22824	77176	27218	72782	31233	68767	24
37	22901	77099	27288	72712	31297	68703	23
38	22977	77023	27357	72643	31361	68639	22
39	23054	76946	27427	72573	31425	68575	21
40	9.23130	10.76870	9.27496	10.72504	9.31489	10.68511	20
41	23206	76794	27566	72434	31552	68448	19
42	23283	76717	27635	72365	31616	68384	18
43	23359	76641	27704	72296	31679	68321	17
44	23435	76565	27773	72227	31743	68257	16
45	23510	76490	27842	72158	31806	68194	15
46	23586	76414	27911	72089	31870	68130	14
47	23661	76339	27980	72020	31933	68067	13
48	23737	76263	28049	71951	31996	68004	12
49	23812	76188	28117	71883	32059	67941	11
50	9.23887	10.76113	9.28186	10.71814	9.32122	10.67878	10
51	23962	76038	28254	71746	32185	67815	9
52	24037	75963	28323	71677	32248	67752	8
53	24112	75888	28391	71609	32311	67689	7
54	21186	75814	28459	71541	32373	67627	6
55	24261	75739	28527	71473	32436	67564	5
56	24335	75665	28595	71405	32498	67502	4
57	24410	75590	28662	71338	32561	67439	3
58	24484	75516	28730	71270	32623	67377	2
59	21558	75442	28798	71202	32685	67315	1
60	24632	75368	28865	71135	32747	67253	0
′	Cotan	Tan	Cotan	Tan	Cotan	Tan	
	80°		79°		78°		

′	12°		13°		14°		′
	Tan	Cotan	Tan	Cotan	Tan	Cotan	
0	9.32747	10.67253	9.36336	10.63664	9.39677	10.60323	60
1	32810	67190	36394	63606	39731	60269	59
2	32872	67128	36452	63548	39785	60215	58
3	32933	67067	36509	63491	39838	60162	57
4	32995	67005	36566	63434	39892	60108	56
5	33057	66943	36624	63376	39945	60055	55
6	33119	66881	36681	63319	39999	60001	54
7	33180	66820	36738	63262	40052	59948	53
8	33242	66758	36795	63205	40106	59894	52
9	33303	66697	36852	63148	40159	59841	51
10	9.33365	10.66635	9.36909	10.63091	9.40212	10.59788	50
11	33426	66574	36966	63034	40266	59734	49
12	33487	66513	37023	62977	40319	59681	48
13	33548	66452	37080	62920	40372	59628	47
14	33609	66391	37137	62863	40425	59575	46
15	33670	66330	37193	62807	40478	59522	45
16	33731	66269	37250	62750	40531	59469	44
17	33792	66208	37306	62694	40584	59416	43
18	33853	66147	37363	62637	40636	59364	42
19	33913	66087	37419	62581	40689	59311	41
20	9.33974	10.66026	9.37476	10.62524	9.40742	10.59258	40
21	34034	65966	37532	62468	40795	59205	39
22	34095	65905	37588	62412	40847	59153	38
23	34155	65845	37644	62356	40900	59100	37
24	34215	65785	37700	62300	40952	59048	36
25	34276	65724	37756	62244	41005	58995	35
26	34336	65664	37812	62188	41057	58943	34
27	34396	65604	37868	62132	41109	58891	33
28	34456	65544	37924	62076	41161	58839	32
29	34516	65484	37980	62020	41214	58786	31
30	9.34576	10.65424	9.38035	10.61965	9.41266	10.58734	30
31	34635	65365	38091	61909	41318	58682	29
32	34695	65305	38147	61853	41370	58630	28
33	34755	65245	38202	61798	41422	58578	27
34	34814	65186	38257	61743	41474	58526	26
35	34874	65126	38313	61687	41526	58474	25
36	34933	65067	38368	61632	41578	58422	24
37	34992	65008	38423	61577	41629	58371	23
38	35051	64949	38479	61521	41681	58319	22
39	35111	64889	38534	61466	41733	58267	21
40	9.35170	10.64830	9.38589	10.61411	9.41784	10.58216	20
41	35229	64771	38644	61356	41836	58164	19
42	35288	64712	38699	61301	41887	58113	18
43	35347	64653	38754	61246	41939	58061	17
44	35405	64595	38808	61192	41990	58010	16
45	35464	64536	38863	61137	42041	57959	15
46	35523	64477	38918	61082	42093	57907	14
47	35581	64419	38972	61028	42144	57856	13
48	35640	64360	39027	60973	42195	57805	12
49	35698	64302	39082	60918	42246	57754	11
50	9.35757	10.64243	9.39136	10.60864	9.42297	10.57703	10
51	35815	64185	39190	60810	42348	57652	9
52	35873	64127	39245	60755	42399	57601	8
53	35931	64069	39299	60701	42450	57550	7
54	35989	64011	39353	60647	42501	57499	6
55	36047	63953	39407	60593	42552	57448	5
56	36105	63895	39461	60539	42603	57397	4
57	36163	63837	39515	60485	42653	57347	3
58	36221	63779	39569	60431	42704	57296	2
59	36279	63721	39623	60377	42755	57245	1
60	36336	63664	39677	60323	42805	57195	0
′	Cotan	Tan	Cotan	Tan	Cotan	Tan	′
	77°		76°		75°		

′	15° Tan	15° Cotan	16° Tan	16° Cotan	17° Tan	17° Cotan	′
0	9.42805	10.57195	9.45750	10.54250	9.48534	10.51466	60
1	42856	57144	45797	54203	48579	51421	59
2	42906	57094	45845	54155	48624	51376	58
3	42957	57043	45892	54108	48669	51331	57
4	43007	56993	45940	54060	48714	51286	56
5	43057	56943	45987	54013	48759	51241	55
6	43108	56892	46035	53965	48804	51196	54
7	43158	56842	46082	53918	48849	51151	53
8	43208	56792	46130	53870	48894	51106	52
9	43258	56742	46177	53823	48939	51061	51
10	9.43308	10.56692	9.46224	10.53776	9.48984	10.51016	50
11	43358	56642	46271	53729	49029	50971	49
12	43408	56592	46319	53681	49073	50927	48
13	43458	56542	46366	53634	49118	50882	47
14	43508	56492	46413	53587	49163	50837	46
15	43558	56442	46460	53540	49207	50793	45
16	43607	56393	46507	53493	49252	50748	44
17	43657	56343	46554	53446	49296	50704	43
18	43707	56293	46601	53399	49341	50659	42
19	43756	56244	46648	53352	49385	50615	41
20	9.43806	10.56194	9.46694	10.53306	9.49430	10.50570	40
21	43855	56145	46741	53259	49474	50526	39
22	43905	56095	46788	53212	49519	50481	38
23	43954	56046	46835	53165	49563	50437	37
24	44004	55996	46881	53119	49607	50393	36
25	44053	55947	46928	53072	49652	50348	35
26	44102	55898	46975	53025	49696	50304	34
27	44151	55849	47021	52979	49740	50260	33
28	44201	55799	47068	52932	49784	50216	32
29	44250	55750	47114	52886	49828	50172	31
30	9.44299	10.55701	9.47160	10.52840	9.49872	10.50128	30
31	44348	55652	47207	52793	49916	50084	29
32	44397	55603	47253	52747	49960	50040	28
33	44446	55554	47299	52701	50004	49996	27
34	44495	55505	47346	52654	50048	49952	26
35	44544	55456	47392	52608	50092	49908	25
36	44592	55408	47438	52562	50136	49864	24
37	44641	55359	47484	52516	50180	49820	23
38	44690	55310	47530	52470	50223	49777	22
39	44738	55262	47576	52424	50267	49733	21
40	9.44787	10.55213	9.47622	10.52378	9.50311	10.49689	20
41	44836	55164	47668	52332	50355	49645	19
42	44884	55116	47714	52286	50398	49602	18
43	44933	55067	47760	52240	50442	49558	17
44	44981	55019	47806	52194	50485	49515	16
45	45029	54971	47852	52148	50529	49471	15
46	45078	54922	47897	52103	50572	49428	14
47	45126	54874	47943	52057	50616	49384	13
48	45174	54826	47989	52011	50659	49341	12
49	45222	54778	48035	51965	50703	49297	11
50	9.45271	10.54729	9.48080	10.51920	9.50746	10.49254	10
51	45319	54681	48126	51874	50789	49211	9
52	45367	54633	48171	51829	50833	49167	8
53	45415	54585	48217	51783	50876	49124	7
54	45463	54537	48262	51738	50919	49081	6
55	45511	54489	48307	51693	50962	49038	5
56	45559	54441	48353	51647	51005	48995	4
57	45606	54394	48398	51602	51048	48952	3
58	45654	54346	48443	51557	51092	48908	2
59	45702	54298	48489	51511	51135	48865	1
60	45750	54250	48534	51466	51178	48822	0
′	Cotan 74°	Tan 74°	Cotan 73°	Tan 73°	Cotan 72°	Tan 72°	′

′	18°		19°		20°		′
	Tan	Cotan	Tan	Cotan	Tan	Cotan	
0	9.51178	10.48822	9.53697	10.46303	9.56107	10.43893	60
1	51221	48779	53738	46262	56146	43854	59
2	51264	48736	53779	46221	56185	43815	58
3	51306	48694	53820	46180	56224	43776	57
4	51349	48651	53861	46139	56264	43736	56
5	51392	48608	53902	46098	56303	43697	55
6	51435	48565	53943	46057	56342	43658	54
7	51478	48522	53984	46016	56381	43619	53
8	51520	48480	54025	45975	56420	43580	52
9	51563	48437	54065	45935	56459	43541	51
10	9.51606	10.48394	9.54106	10.45894	9.56498	10.43502	50
11	51648	48352	54147	45853	56537	43463	49
12	51691	48309	54187	45813	56576	43424	48
13	51734	48266	54228	45772	56615	43385	47
14	51776	48224	54269	45731	56654	43346	46
15	51819	48181	54309	45691	56693	43307	45
16	51861	48139	54350	45650	56732	43268	44
17	51903	48097	54390	45610	56771	43229	43
18	51946	48054	54431	45569	56810	43190	42
19	51988	48012	54471	45529	56849	43151	41
20	9.52031	10.47969	9.54512	10.45488	9.56887	10.43113	40
21	52073	47927	54552	45448	56926	43074	39
22	52115	47885	54593	45407	56965	43035	38
23	52157	47843	54633	45367	57004	42996	37
24	52200	47800	54673	45327	57042	42958	36
25	52242	47758	54714	45286	57081	42919	35
26	52284	47716	54754	45246	57120	42880	34
27	52326	47674	54794	45206	57158	42842	33
28	52368	47632	54835	45165	57197	42803	32
29	52410	47590	54875	45125	57235	42765	31
30	9.52452	10.47548	9.54915	10.45085	9.57274	10.42726	30
31	52494	47506	54955	45045	57312	42688	29
32	52536	47464	54995	45005	57351	42649	28
33	52578	47422	55035	44965	57389	42611	27
34	52620	47380	55075	44925	57428	42572	26
35	52661	47339	55115	44885	57466	42534	25
36	52703	47297	55155	44845	57504	42496	24
37	52745	47255	55195	44805	57543	42457	23
38	52787	47213	55235	44765	57581	42419	22
39	52829	47171	55275	44725	57619	42381	21
40	9.52870	10.47130	9.55315	10.44685	9.57658	10.42342	20
41	52912	47088	55355	44645	57696	42304	19
42	52953	47047	55395	44605	57734	42266	18
43	52995	47005	55434	44566	57772	42228	17
44	53037	46963	55474	44526	57810	42190	16
45	53078	46922	55514	44486	57849	42151	15
46	53120	46880	55554	44446	57887	42113	14
47	53161	46839	55593	44407	57925	42075	13
48	53202	46798	55633	44367	57963	42037	12
49	53244	46756	55673	44327	58001	41999	11
50	9.53285	10.46715	9.55712	10.44288	9.58039	10.41961	10
51	53327	46673	55752	44248	58077	41923	9
52	53368	46632	55791	44209	58115	41885	8
53	53409	46591	55831	44169	58153	41847	7
54	53450	46550	55870	44130	58191	41809	6
55	53492	46508	55910	44090	58229	41771	5
56	53533	46467	55949	44051	58267	41733	4
57	53574	46426	55989	44011	58304	41696	3
58	53615	46385	56028	43972	58342	41658	2
59	53656	46344	56067	43933	58380	41620	1
60	53697	46303	56107	43893	58418	41582	0
′	Cotan	Tan	Cotan	Tan	Cotan	Tan	′
	71°		70°		69°		

′	21° Tan	21° Cotan	22° Tan	22° Cotan	23° Tan	23° Cotan	′
0	9.58418	10.41582	9.60641	10.39359	9.62785	10.37215	60
1	58455	41545	60677	39323	62820	37180	59
2	58493	41507	60714	39286	62855	37145	58
3	58531	41469	60750	39250	62890	37110	57
4	58569	41431	60786	39214	62926	37074	56
5	58606	41394	60823	39177	62961	37039	55
6	58644	41356	60859	39141	62996	37004	54
7	58681	41319	60895	39105	63031	36969	53
8	58719	41281	60931	39069	63066	36934	52
9	58757	41243	60967	39033	63101	36899	51
10	9.58794	10.41206	9.61004	10.38996	9.63135	10.36865	50
11	58832	41168	61040	38960	63170	36830	49
12	58869	41131	61076	38924	63205	36795	48
13	58907	41093	61112	38888	63240	36760	47
14	58944	41056	61148	38852	63275	36725	46
15	58981	41019	61184	38816	63310	36690	45
16	59019	40981	61220	38780	63345	36655	44
17	59056	40944	61256	38744	63379	36621	43
18	59094	40906	61292	38708	63414	36586	42
19	59131	40869	61328	38672	63449	36551	41
20	9.59168	10.40832	9.61364	10.38636	9.63484	10.36516	40
21	59205	40795	61400	38600	63519	36481	39
22	59243	40757	61436	38564	63553	36447	38
23	59280	40720	61472	38528	63588	36412	37
24	59317	40683	61508	38492	63623	36377	36
25	59354	40646	61544	38456	63657	36343	35
26	59391	40609	61579	38421	63692	36308	34
27	59429	40571	61615	38385	63726	36274	33
28	59466	40534	61651	38349	63761	36239	32
29	59503	40497	61687	38313	63796	36204	31
30	9.59540	10.40460	9.61722	10.38278	9.63830	10.36170	30
31	59577	40423	61758	38242	63865	36135	29
32	59614	40386	61794	38206	63899	36101	28
33	59651	40349	61830	38170	63934	36066	27
34	59688	40312	61865	38135	63968	36032	26
35	59725	40275	61901	38099	64003	35997	25
36	59762	40238	61936	38064	64037	35963	24
37	59799	40201	61972	38028	64072	35928	23
38	59835	40165	62008	37992	64106	35894	22
39	59872	40128	62043	37957	64140	35860	21
40	9.59909	10.40091	9.62079	10.37921	9.64175	10.35825	20
41	59946	40054	62114	37886	64209	35791	19
42	59983	40017	62150	37850	64243	35757	18
43	60019	39981	62185	37815	64278	35722	17
44	60056	39944	62221	37779	64312	35688	16
45	60093	39907	62256	37744	64346	35654	15
46	60130	39870	62292	37708	64381	35619	14
47	60166	39834	62327	37673	64415	35585	13
48	60203	39797	62362	37638	64449	35551	12
49	60240	39760	62398	37602	64483	35517	11
50	9.60276	10.39724	9.62433	10.37567	9.64517	10.35483	10
51	60313	39687	62468	37532	64552	35448	9
52	60349	39651	62504	37496	64586	35414	8
53	60386	39614	62539	37461	64620	35380	7
54	60422	39578	62574	37426	64654	35346	6
55	60459	39541	62609	37391	64688	35312	5
56	60495	39505	62645	37355	64722	35278	4
57	60532	39468	62680	37320	64756	35244	3
58	60568	39432	62715	37285	64790	35210	2
59	60605	39395	62750	37250	64824	35176	1
60	60641	39359	62785	37215	64858	35142	0
′	68° Cotan	68° Tan	67° Cotan	67° Tan	66° Cotan	66° Tan	′

′	24°		25°		26°		′
	Tan	Cotan	Tan	Cotan	Tan	Cotan	
0	9.64858	10.35142	9.66867	10.33133	9.68818	10.31182	60
1	64892	35108	66900	33100	68850	31150	59
2	64926	35074	66933	33067	68882	31118	58
3	64960	35040	66966	33034	68914	31086	57
4	64994	35006	66999	33001	68946	31054	56
5	65028	34972	67032	32968	68978	31022	55
6	65062	34938	67065	32935	69010	31990	54
7	65096	34904	67098	32902	69042	31958	53
8	65130	34870	67131	32869	69074	31926	52
9	65164	34836	67163	32837	69106	31894	51
10	9.65197	10.34803	9.67196	10.32804	9.69138	10.30862	50
11	65231	34769	67229	32771	69170	30830	49
12	65265	34735	67262	32738	69202	30798	48
13	65299	34701	67295	32705	69234	30766	47
14	65333	34667	67327	32673	69266	30734	46
15	65366	34634	67360	32640	69298	30702	45
16	65400	34600	67393	32607	69329	30671	44
17	65434	34566	67426	32574	69361	30639	43
18	65467	34533	67458	32542	69393	30607	42
19	65501	34499	67491	32509	69425	30575	41
20	9.65535	10.34465	9.67524	10.32476	9.69457	10.30543	40
21	65568	34432	67556	32444	69488	30512	39
22	65602	34398	67589	32411	69520	30480	38
23	65636	34364	67622	32378	69552	30448	37
24	65669	34331	67654	32346	69584	30416	36
25	65703	34297	67687	32313	69615	30385	35
26	65736	34264	67719	32281	69647	30353	34
27	65770	34230	67752	32248	69679	30321	33
28	65803	34197	67785	32215	69710	30290	32
29	65837	34163	67817	32183	69742	30258	31
30	9.65870	10.34130	9.67850	10.32150	9.69774	10.30226	30
31	65904	34096	67882	32118	69805	30195	29
32	65937	34063	67915	32085	69837	30163	28
33	65971	34029	67947	32053	69868	30132	27
34	66004	33996	67980	32020	69900	30100	26
35	66038	33962	68012	31988	69932	30068	25
36	66071	33929	68044	31956	69963	30037	24
37	66104	33896	68077	31923	69995	30005	23
38	66138	33862	68109	31891	70026	29974	22
39	66171	33829	68142	31858	70058	29942	21
40	9.66204	10.33796	9.68174	10.31826	9.70089	10.29911	20
41	66238	33762	68206	31794	70121	29879	19
42	66271	33729	68239	31761	70152	29848	18
43	66304	33696	68271	31729	70184	29816	17
44	66337	33663	68303	31697	70215	29785	16
45	66371	33629	68336	31664	70247	29753	15
46	66404	33596	68368	31632	70278	29722	14
47	66437	33563	68400	31600	70309	29691	13
48	66470	33530	68432	31568	70341	29659	12
49	66503	33497	68465	31535	70372	29628	11
50	9.66537	10.33463	9.68497	10.31503	9.70404	10.29596	10
51	66570	33430	68529	31471	70435	29565	9
52	66603	33397	68561	31439	70466	29534	8
53	66636	33364	68593	31407	70498	29502	7
54	66669	33331	68626	31374	70529	29471	6
55	66702	33298	68658	31342	70560	29440	5
56	66735	33265	68690	31310	70592	29408	4
57	66768	33232	68722	31278	70623	29377	3
58	66801	33199	68754	31246	70654	29346	2
59	66834	33166	68786	31214	70685	29315	1
60	66867	33133	68818	31182	70717	29283	0
′	Cotan	Tan	Cotan	Tan	Cotan	Tan	′
	65°		64°		63°		

′	27° Tan	27° Cotan	28° Tan	28° Cotan	29° Tan	29° Cotan	′
0	9.70717	10.29283	9.72567	10.27433	9.74375	10.25625	60
1	70748	29252	72598	27402	74405	25595	59
2	70779	29221	72628	27372	74435	25565	58
3	70810	29190	72659	27341	74465	25535	57
4	70841	29159	72689	27311	74494	25506	56
5	70873	29127	72720	27280	74524	25476	55
6	70904	29096	72750	27250	74554	25446	54
7	70935	29065	72780	27220	74583	25417	53
8	70966	29034	72811	27189	74613	25387	52
9	70997	29003	72841	27159	74643	25357	51
10	9.71028	10.28972	9.72872	10.27128	9.74673	10.25327	50
11	71059	28941	72902	27098	74702	25298	49
12	71090	28910	72932	27068	74732	25268	48
13	71121	28879	72963	27037	74762	25238	47
14	71153	28847	72993	27007	74791	25209	46
15	71184	28816	73023	26977	74821	25179	45
16	71215	28785	73054	26946	74851	25149	44
17	71246	28754	73084	26916	74880	25120	43
18	71277	28723	73114	26886	74910	25090	42
19	71308	28692	73144	26856	74939	25061	41
20	9.71339	10.28661	9.73175	10.26825	9.74969	10.25031	40
21	71370	28630	73205	26795	74998	25002	39
22	71401	28599	73235	26765	75028	24972	38
23	71431	28569	73265	26735	75058	24942	37
24	71462	28538	73295	26705	75087	24913	36
25	71493	28507	73326	26674	75117	24883	35
26	71524	28476	73356	26644	75146	24854	34
27	71555	28445	73386	26614	75176	24824	33
28	71586	28414	73416	26584	75205	24795	32
29	71617	28383	73446	26554	75235	24765	31
30	9.71648	10.28352	9.73476	10.26524	9.75264	10.24736	30
31	71679	28321	73507	26493	75294	24706	29
32	71709	28291	73537	26463	75323	24677	28
33	71740	28260	73567	26433	75353	24647	27
34	71771	28229	73597	26403	75382	24618	26
35	71802	28198	73627	26373	75411	24589	25
36	71833	28167	73657	26343	75441	24559	24
37	71863	28137	73687	26313	75470	24530	23
38	71894	28106	73717	26283	75500	24500	22
39	71925	28075	73747	26253	75529	24471	21
40	9.71955	10.28045	9.73777	10.26223	9.75558	10.24442	20
41	71986	28014	73807	26193	75588	24412	19
42	72017	27983	73837	26163	75617	24383	18
43	72048	27952	73867	26133	75647	24353	17
44	72078	27922	73897	26103	75676	24324	16
45	72109	27891	73927	26073	75705	24295	15
46	72140	27860	73957	26043	75735	24265	14
47	72170	27830	73987	26013	75764	24236	13
48	72201	27799	74017	25983	75793	24207	12
49	72231	27769	74047	25953	75822	24178	11
50	9.72262	10.27738	9.74077	10.25923	9.75852	10.24148	10
51	72293	27707	74107	25893	75881	24119	9
52	72323	27677	74137	25863	75910	24090	8
53	72354	27646	74166	25834	75939	24061	7
54	72384	27616	74196	25804	75969	24031	6
55	72415	27585	74226	25774	75998	24002	5
56	72445	27555	74256	25744	76027	23973	4
57	72476	27524	74286	25714	76056	23944	3
58	72506	27494	74316	25684	76086	23914	2
59	72537	27463	74345	25655	76115	23885	1
60	72567	27433	74375	25625	76144	23856	0
′	Cotan	Tan	Cotan	Tan	Cotan	Tan	′
	62°		61°		60°		

′	30°		31°		32°		′
	Tan	Cotan	Tan	Cotan	Tan	Cotan	
0	9.76144	10.23856	9.77877	10.22123	9.79579	10.20421	60
1	76173	23827	77906	22094	79607	20393	59
2	76202	23798	77935	22065	79635	20365	58
3	76231	23769	77963	22037	79663	20337	57
4	76261	23739	77992	22008	79691	20309	56
5	76290	23710	78020	21980	79719	20281	55
6	76319	23681	78049	21951	79747	20253	54
7	76348	23652	78077	21923	79776	20224	53
8	76377	23623	78106	21894	79804	20196	52
9	76406	23594	78135	21865	79832	20168	51
10	9.76435	10.23565	9.78163	10.21837	9.79860	10.20140	50
11	76464	23536	78192	21808	79888	20112	49
12	76493	23507	78220	21780	79916	20084	48
13	76522	23478	78249	21751	79944	20056	47
14	76551	23449	78277	21723	79972	20028	46
15	76580	23420	78306	21694	80000	20000	45
16	76609	23391	78334	21666	80028	19972	44
17	76639	23361	78363	21637	80056	19944	43
18	76668	23332	78391	21609	80084	19916	42
19	76697	23303	78419	21581	80112	19888	41
20	9.76725	10.23275	9.78448	10.21552	9.80140	10.19860	40
21	76754	23246	78476	21524	80168	19832	39
22	76783	23217	78505	21495	80195	19805	38
23	76812	23188	78533	21467	80223	19777	37
24	76841	23159	78562	21438	80251	19749	36
25	76870	23130	78590	21410	80279	19721	35
26	76899	23101	78618	21382	80307	19693	34
27	76928	23072	78647	21353	80335	19665	33
28	76957	23043	78675	21325	80363	19637	32
29	76986	23014	78704	21296	80391	19609	31
30	9.77015	10.22985	9.78732	10.21268	9.80419	10.19581	30
31	77044	22956	78760	21240	80447	19553	29
32	77073	22927	78789	21211	80474	19526	28
33	77101	22899	78817	21183	80502	19498	27
34	77130	22870	78845	21155	80530	19470	26
35	77159	22841	78874	21126	80558	19442	25
36	77188	22812	78902	21098	80586	19414	24
37	77217	22783	78930	21070	80614	19386	23
38	77246	22754	78959	21041	80642	19358	22
39	77274	22726	78987	21013	80669	19331	21
40	9.77303	10.22697	9.79015	10.20985	9.80697	10.19303	20
41	77332	22668	79043	20957	80725	19275	19
42	77361	22639	79072	20928	80753	19247	18
43	77390	22610	79100	20900	80781	19219	17
44	77418	22582	79128	20872	80808	19192	16
45	77447	22553	79156	20844	80836	19164	15
46	77476	22524	79185	20815	80864	19136	14
47	77505	22495	79213	20787	80892	19108	13
48	77533	22467	79241	20759	80919	19081	12
49	77562	22438	79269	20731	80947	19053	11
50	9.77591	10.22409	9.79297	10.20703	9.80975	10.19025	10
51	77619	22381	79326	20674	81003	18997	9
52	77648	22352	79354	20646	81030	18970	8
53	77677	22323	79382	20618	81058	18942	7
54	77706	22294	79410	20590	81086	18914	6
55	77734	22266	79438	20562	81113	18887	5
56	77763	22237	79466	20534	81141	18859	4
57	77791	22209	79495	20505	81169	18831	3
58	77820	22180	79523	20477	81196	18804	2
59	77849	22151	79551	20449	81224	18776	1
60	77877	22123	79579	20421	81252	18748	0
′	Cotan	Tan	Cotan	Tan	Cotan	Tan	′
	59°		58°		57°		

′	33° Tan	33° Cotan	34° Tan	34° Cotan	35° Tan	35° Cotan	′
0	9.81252	10.18748	9.82899	10.17101	9.84523	10.15477	60
1	81279	18721	82926	17074	84550	15450	59
2	81307	18693	82953	17047	84576	15424	58
3	81335	18665	82980	17020	84603	15397	57
4	81362	18638	83008	16992	84630	15370	56
5	81390	18610	83035	16965	84657	15343	55
6	81418	18582	83062	16938	84684	15316	54
7	81445	18555	83089	16911	84711	15289	53
8	81473	18527	83117	16883	84738	15262	52
9	81500	18500	83144	16856	84764	15236	51
10	9.81528	10.18472	9.83171	10.16829	9.84791	10.15209	50
11	81556	18444	83198	16802	84818	15182	49
12	81583	18417	83225	16775	84845	15155	48
13	81611	18389	83252	16748	84872	15128	47
14	81638	18362	83280	16720	84899	15101	46
15	81666	18334	83307	16693	84925	15075	45
16	81693	18307	83334	16666	84952	15048	44
17	81721	18279	83361	16639	84979	15021	43
18	81748	18252	83388	16612	85006	14994	42
19	81776	18224	83415	16585	85033	14967	41
20	9.81803	10.18197	9.83442	10.16558	9.85059	10.14941	40
21	81831	18169	83470	16530	85086	14914	39
22	81858	18142	83497	16503	85113	14887	38
23	81886	18114	83524	16476	85140	14860	37
24	81913	18087	83551	16449	85166	14834	36
25	81941	18059	83578	16422	85193	14807	35
26	81968	18032	83605	16395	85220	14780	34
27	81996	18004	83632	16368	85247	14753	33
28	82023	17977	83659	16341	85273	14727	32
29	82051	17949	83686	16314	85300	14700	31
30	9.82078	10.17922	9.83713	10.16287	9.85327	10.14673	30
31	82106	17894	83740	16260	85354	14646	29
32	82133	17867	83768	16232	85380	14620	28
33	82161	17839	83795	16205	85407	14593	27
34	82188	17812	83822	16178	85434	14566	26
35	82215	17785	83849	16151	85460	14540	25
36	82243	17757	83876	16124	85487	14513	24
37	82270	17730	83903	16097	85514	14486	23
38	82298	17702	83930	16070	85540	14460	22
39	82325	17675	83957	16043	85567	14433	21
40	9.82352	10.17648	9.83984	10.16016	9.85594	10.14406	20
41	82380	17620	84011	15989	85620	14380	19
42	82407	17593	84038	15962	85647	14353	18
43	82435	17565	84065	15935	85674	14326	17
44	82462	17538	84092	15908	85700	14300	16
45	82489	17511	84119	15881	85727	14273	15
46	82517	17483	84146	15854	85754	14246	14
47	82544	17456	84173	15827	85780	14220	13
48	82571	17429	84200	15800	85807	14193	12
49	82599	17401	84227	15773	85834	14166	11
50	9.82626	10.17374	9.84254	10.15746	9.85860	10.14140	10
51	82653	17347	84280	15720	85887	14113	9
52	82681	17319	84307	15693	85913	14087	8
53	82708	17292	84334	15666	85940	14060	7
54	82735	17265	84361	15639	85967	14033	6
55	82762	17238	84388	15612	85993	14007	5
56	82790	17210	84415	15585	86020	13980	4
57	82817	17183	84442	15558	86046	13954	3
58	82844	17156	84469	15531	86073	13927	2
59	82871	17129	84496	15504	86100	13900	1
60	82899	17101	84523	15477	86126	13874	0
′	Cotan 56°	Tan 56°	Cotan 55°	Tan 55°	Cotan 54°	Tan 54°	′

′	36° Tan	36° Cotan	37° Tan	37° Cotan	38° Tan	38° Cotan	′
0	9.86126	10.13874	9.87711	10.12289	9.89281	10.10719	60
1	86153	13847	87738	12262	89307	10693	59
2	86179	13821	87764	12236	89333	10667	58
3	86206	13794	87790	12210	89359	10641	57
4	86232	13768	87817	12183	89385	10615	56
5	86259	13741	87843	12157	89411	10589	55
6	86285	13715	87869	12131	89437	10563	54
7	86312	13688	87895	12105	89463	10537	53
8	86338	13662	87922	12078	89489	10511	52
9	86365	13635	87948	12052	89515	10485	51
10	9.86392	10.13608	9.87974	10.12026	9.89541	10.10459	50
11	86418	13582	88000	12000	89567	10433	49
12	86445	13555	88027	11973	89593	10407	48
13	86471	13529	88053	11947	89619	10381	47
14	86498	13502	88079	11921	89645	10355	46
15	86524	13476	88105	11895	89671	10329	45
16	86551	13449	88131	11869	89697	10303	44
17	86577	13423	88158	11842	89723	10277	43
18	86603	13397	88184	11816	89749	10251	42
19	86630	13370	88210	11790	89775	10225	41
20	9.86656	10.13344	9.88236	10.11764	9.89801	10.10199	40
21	86683	13317	88262	11738	89827	10173	39
22	86709	13291	88289	11711	89853	10147	38
23	86736	13264	88315	11685	89879	10121	37
24	86762	13238	88341	11659	89905	10095	36
25	86789	13211	88367	11633	89931	10069	35
26	86815	13185	88393	11607	89957	10043	34
27	86842	13158	88420	11580	89983	10017	33
28	86868	13132	88446	11554	90009	09991	32
29	86894	13106	88472	11528	90035	09965	31
30	9.86921	10.13079	9.88498	10.11502	9.90061	10.09939	30
31	86947	13053	88524	11476	90086	09914	29
32	86974	13026	88550	11450	90112	09888	28
33	87000	13000	88577	11423	90138	09862	27
34	87027	12973	88603	11397	90164	09836	26
35	87053	12947	88629	11371	90190	09810	25
36	87079	12921	88655	11345	90216	09784	24
37	87106	12894	88681	11319	90242	09758	23
38	87132	12868	88707	11293	90268	09732	22
39	87158	12842	88733	11267	90294	09706	21
40	9.87185	10.12815	9.88759	10.11241	9.90320	10.09680	20
41	87211	12789	88786	11214	90346	09654	19
42	87238	12762	88812	11188	90371	09629	18
43	87264	12736	88838	11162	90397	09603	17
44	87290	12710	88864	11136	90423	09577	16
45	87317	12683	88890	11110	90449	09551	15
46	87343	12657	88916	11084	90475	09525	14
47	87369	12631	88942	11058	90501	09499	13
48	87396	12604	88968	11032	90527	09473	12
49	87422	12578	88994	11006	90553	09447	11
50	9.87448	10.12552	9.89020	10.10980	9.90578	10.09422	10
51	87475	12525	89046	10954	90604	09396	9
52	87501	12499	89073	10927	90630	09370	8
53	87527	12473	89099	10901	90656	09344	7
54	87554	12446	89125	10875	90682	09318	6
55	87580	12420	89151	10849	90708	09292	5
56	87606	12394	89177	10823	90734	09266	4
57	87633	12367	89203	10797	90759	09241	3
58	87659	12341	89229	10771	90785	09215	2
59	87685	12315	89255	10745	90811	09189	1
60	87711	12289	89281	10719	90837	09163	0
′	Cotan 53°	Tan 53°	Cotan 52°	Tan 52°	Cotan 51°	Tan 51°	′

′	39° Tan	39° Cotan	40° Tan	40° Cotan	41° Tan	41° Cotan	′
0	9.90837	10.09163	9.92381	10.07619	9.93916	10.06084	60
1	90863	09137	92407	07593	93942	06058	59
2	90889	09111	92433	07567	93967	06033	58
3	90914	09086	92458	07542	93993	06007	57
4	90940	09060	92484	07516	94018	05982	56
5	90966	09034	92510	07490	94044	05956	55
6	90992	09008	92535	07465	94069	05931	54
7	91018	08982	92561	07439	94095	05905	53
8	91043	08957	92587	07413	94120	05880	52
9	91069	08931	92612	07388	94146	05854	51
10	9.91095	10.08905	9.92638	10.07362	9.94171	10.05829	50
11	91121	08879	92663	07337	94197	05803	49
12	91147	08853	92689	07311	94222	05778	48
13	91172	08828	92715	07285	94248	05752	47
14	91198	08802	92740	07260	94273	05727	46
15	91224	08776	92766	07234	94299	05701	45
16	91250	08750	92792	07208	94324	05676	44
17	91276	08724	92817	07183	94350	05650	43
18	91301	08699	92843	07157	94375	05625	42
19	91327	08673	92868	07132	94401	05599	41
20	9.91353	10.08647	9.92894	10.07106	9.94426	10.05574	40
21	91379	08621	92920	07080	94452	05548	39
22	91404	08596	92945	07055	94477	05523	38
23	91430	08570	92971	07029	94503	05497	37
24	91456	08544	92996	07004	94528	05472	36
25	91482	08518	93022	06978	94554	05446	35
26	91507	08493	93048	06952	94579	05421	34
27	91533	08467	93073	06927	94604	05396	33
28	91559	08441	93099	06901	94630	05370	32
29	91585	08415	93124	06876	94655	05345	31
30	9.91610	10.08390	9.93150	10.06850	9.94681	10.05319	30
31	91636	08364	93175	06825	94706	05294	29
32	91662	08338	93201	06799	94732	05268	28
33	91688	08312	93227	06773	94757	05243	27
34	91713	08287	93252	06748	94783	05217	26
35	91739	08261	93278	06722	94808	05192	25
36	91765	08235	93303	06697	94834	05166	24
37	91791	08209	93329	06671	94859	05141	23
38	91816	08184	93354	06646	94884	05116	22
39	91842	08158	93380	06620	94910	05090	21
40	9.91868	10.08132	9.93406	10.06594	9.94935	10.05065	20
41	91893	08107	93431	06569	94961	05039	19
42	91919	08081	93457	06543	94986	05014	18
43	91945	08055	93482	06518	95012	04988	17
44	91971	08029	93508	06492	95037	04963	16
45	91996	08004	93533	06467	95062	04938	15
46	92022	07978	93559	06441	95088	04912	14
47	92048	07952	93584	06416	95113	04887	13
48	92073	07927	93610	06390	95139	04861	12
49	92099	07901	93636	06364	95164	04836	11
50	9.92125	10.07875	9.93661	10.06339	9.95190	10.04810	10
51	92150	07850	93687	06313	95215	04785	9
52	92176	07824	93712	06288	95240	04760	8
53	92202	07798	93738	06262	95266	04734	7
54	92227	07773	93763	06237	95291	04709	6
55	92253	07747	93789	06211	95317	04683	5
56	92279	07721	93814	06186	95342	04658	4
57	92304	07696	93840	06160	95368	04632	3
58	92330	07670	93865	06135	95393	04607	2
59	92356	07644	93891	06109	95418	04582	1
60	92381	07619	93916	06084	95444	04556	0
′	Cotan	Tan	Cotan	Tan	Cotan	Tan	′
	50°		49°		48°		

′	42°		43°		44°		′
	Tan	Cotan	Tan	Cotan	Tan	Cotan	
0	9.95444	10.04556	9.96966	10.03034	9.98484	10.01516	60
1	95469	04531	96991	03009	98509	01491	59
2	95495	04505	97016	02984	98534	01466	58
3	95520	04480	97042	02958	98560	01440	57
4	95545	04455	97067	02933	98585	01415	56
5	95571	04429	97092	02908	98610	01390	55
6	95596	04404	97118	02882	98635	01365	54
7	95622	04378	97143	02857	98661	01339	53
8	95647	04353	97168	02832	98686	01314	52
9	95672	04328	97193	02807	98711	01289	51
10	9.95698	10.04302	9.97219	10.02781	9.98737	10.01263	50
11	95723	04277	97244	02756	98762	01238	49
12	95748	04252	97269	02731	98787	01213	48
13	95774	04226	97295	02705	98812	01188	47
14	95799	04201	97320	02680	98838	01162	46
15	95825	04175	97345	02655	98863	01137	45
16	95850	04150	97371	02629	98888	01112	44
17	95875	04125	97396	02604	98913	01087	43
18	95901	04099	97421	02579	98939	01061	42
19	95926	04074	97447	02553	98964	01036	41
20	9.95952	10.04048	9.97472	10.02528	9.98989	10.01011	40
21	95977	04023	97497	02503	99015	00985	39
22	96002	03998	97523	02477	99040	00960	38
23	96028	03972	97548	02452	99065	00935	37
24	96053	03947	97573	02427	99090	00910	36
25	96078	03922	97598	02402	99116	00884	35
26	96104	03896	97624	02376	99141	00859	34
27	96129	03871	97649	02351	99166	00834	33
28	96155	03845	97674	02326	99191	00809	32
29	96180	03820	97700	02300	99217	00783	31
30	9.96205	10.03795	9.97725	10.02275	9.99242	10.00758	30
31	96231	03769	97750	02250	99267	00733	29
32	96256	03744	97776	02224	99293	00707	28
33	96281	03719	97801	02199	99318	00682	27
34	96307	03693	97826	02174	99343	00657	26
35	96332	03668	97851	02149	99368	00632	25
36	96357	03643	97877	02123	99394	00606	24
37	96383	03617	97902	02098	99419	00581	23
38	96408	03592	97927	02073	99444	00556	22
39	96433	03567	97953	02047	99469	00531	21
40	9.96459	10.03541	9.97978	10.02022	9.99495	10.00505	20
41	96484	03516	98003	01997	99520	00480	19
42	96510	03490	98029	01971	99545	00455	18
43	96535	03465	98054	01946	99570	00430	17
44	96560	03440	98079	01921	99596	00404	16
45	96586	03414	98104	01896	99621	00379	15
46	96611	03389	98130	01870	99646	00354	14
47	96636	03364	98155	01845	99672	00328	13
48	96662	03338	98180	01820	99697	00303	12
49	96687	03313	98206	01794	99722	00278	11
50	9.96712	10.03288	9.98231	10.01769	9.99747	10.00253	10
51	96738	03262	98256	01744	99773	00227	9
52	96763	03237	98281	01719	99798	00202	8
53	96788	03212	98307	01693	99823	00177	7
54	96814	03186	98332	01668	99848	00152	6
55	96839	03161	98357	01643	99874	00126	5
56	96864	03136	98383	01617	99899	00101	4
57	96890	03110	98408	01592	99924	00076	3
58	96915	03085	98433	01567	99949	00051	2
59	96940	03060	98458	01542	99975	00025	1
60	96966	03034	98484	01516	10.00000	00000	0
′	Cotan	Tan	Cotan	Tan	Cotan	Tan	′
	47°		46°		45°		

TABLE IX.—FUNCTIONS OF A ONE-DEGREE CURVE.

The Long Chords, Mid-Ordinates, Externals, and Tangent Distances of this table are for a curve of 5730 feet radius. To find the corresponding functions of any other curve divide the tabular values by the degree of curve.

For metric curves having 20-metre chords, multiply the degree by 5 and enter the table with the result as a value of *D*, the tabular values being taken as metres instead of feet

Thus for a 1° 30′ metric curve having $I = 45$ the tangent distance is $T = \frac{2373.4}{1.5 \times 5} = 316.45$ metres. Again, suppose $I = 38°$ and the long chord = 373.1 m. known and *D* required. The tabular *L. C.* is 3731 m.; therefore $D = \frac{3731.0}{373.1 \times 5} = 2° \; 0'$.

	0°				1°				′
	L. C.	M.	E.	T.	L. C.	M.	E.	T.	
0	0.00	0.000	0.000	0.00	100.00	0.218	0.218	50.00	0
2	3.33	0.000	0.000	1.67	103.33	0.233	0.233	51.67	2
4	6.67	0.001	0.001	3.33	106.66	0.248	0.248	53.33	4
6	10.00	0.002	0.002	5.00	110.00	0.264	0.264	55.00	6
8	13.33	0.004	0.004	6.67	113.33	0.280	0.280	56.67	8
10	16.67	0.006	0.006	8.33	116.66	0.297	0.297	58.33	10
12	20.00	0.009	0.009	10.00	120.00	0.314	0.314	60.00	12
14	23.33	0.012	0.012	11.67	123.33	0.332	0.332	61.67	14
16	26.67	0.015	0.015	13.33	126.66	0.350	0.350	63.33	16
18	30.00	0.019	0.019	15.00	130.00	0.368	0.368	65.00	18
20	33.33	0.024	0.024	16.67	133.33	0.388	0.388	66.67	20
22	36.67	0.029	0.029	18.33	136.66	0.407	0.407	68.33	22
24	40.00	0.035	0.035	20.00	140.00	0.427	0.427	70.00	24
26	43.33	0.041	0.041	21.67	143.33	0.448	0.448	71.67	26
28	46.67	0.048	0.048	23.33	146.66	0.469	0.469	73.33	28
30	50.00	0.054	0.054	25.00	150.00	0.491	0.491	75.00	30
32	53.33	0.062	0.062	26.67	153.33	0.513	0.513	76.67	32
34	56.67	0.070	0.070	28.33	156.66	0.536	0.536	78.33	34
36	60.00	0.079	0.079	30.00	160.00	0.559	0.559	80.00	36
38	63.33	0.088	0.088	31.67	163.33	0.582	0.582	81.67	38
40	66.67	0.097	0.097	33.33	166.66	0.606	0.606	83.33	40
42	70.00	0.107	0.107	35.00	170.00	0.630	0.630	85.00	42
44	73.33	0.117	0.117	36.67	173.33	0.655	0.655	86.67	44
46	76.67	0.128	0.128	38.33	176.66	0.681	0.681	88.33	46
48	80.00	0.140	0.140	40.00	180.00	0.706	0.706	90.00	48
50	83.33	0.151	0.151	41.67	183.33	0.733	0.733	91.67	50
52	86.67	0.164	0.164	43.33	186.66	0.760	0.760	93.33	52
54	90.00	0.176	0.176	45.00	190.00	0.788	0.788	95.00	54
56	93.33	0.190	0.190	46.67	193.33	0.815	0.815	96.67	56
58	96.67	0.204	0.204	48.33	196.66	0.844	0.844	98.33	58
60	100.00	0.218	0.218	50.00	199.98	0.873	0.873	100.00	60

′	2° L. C.	2° M.	2° E.	2° T.	3° L. C.	3° M.	3° E.	3° T.	′
0	199.98	0.873	0.873	100.00	299.96	1.964	1.964	150.07	0
2	203.31	0.902	0.902	101.67	303.29	2.008	2.009	151.74	2
4	206.64	0.932	0.932	103.34	306.62	2.053	2.054	153.41	4
6	209.97	0.962	0.962	105.01	309.95	2.098	2.099	155.08	6
8	213.31	0.993	0.993	106.68	313.29	2.143	2.144	156.75	8
10	216.64	1.024	1.024	108.35	316.62	2.188	2.189	158.42	10
12	219.97	1.056	1.056	110.02	319.95	2.235	2.236	160.09	12
14	223.30	1.088	1.088	111.69	323.28	2.282	2.283	161.76	14
16	226.64	1.121	1.121	113.36	326.62	2.329	2.330	163.43	16
18	229.97	1.154	1.154	115.02	329.95	2.376	2.377	165.09	18
20	233.30	1.188	1.188	116.69	333.28	2.424	2.425	166.76	20
22	236.63	1.222	1.222	118.36	336.61	2.473	2.474	168.43	22
24	239.97	1.256	1.256	120.03	339.95	2.523	2.523	170.10	24
26	243.30	1.292	1.292	121.70	343.28	2.572	2.573	171.77	26
28	246.63	1.328	1.328	123.37	346.61	2.622	2.623	173.44	28
30	249.96	1.364	1.364	125.03	349.94	2.672	2.673	175.10	30
32	253.29	1.399	1.399	126.70	353.27	2.724	2.725	176.72	32
34	256.62	1.437	1.437	128.37	356.60	2.776	2.777	178.39	34
36	259.96	1.475	1.475	130.04	359.94	2.828	2.829	180.06	36
38	263.29	1.513	1.513	131.71	363.27	2.880	2.881	181.73	38
40	266.62	1.552	1.552	133.38	366.60	2.933	2.934	183.40	40
42	269.96	1.592	1.592	135.05	369.94	2.987	2.988	185.07	42
44	273.29	1.632	1.632	136.72	373.27	3.042	3.043	186.74	44
46	276.62	1.672	1.672	138.38	376.60	3.096	3.097	188.40	46
48	279.96	1.712	1.712	140.05	379.94	3.151	3.152	190.07	48
50	283.29	1.752	1.752	141.72	383.27	3.206	3.207	191.74	50
52	286.62	1.794	1.794	143.39	386.60	3.263	3.264	193.41	52
54	289.96	1.836	1.836	145.06	389.94	3.320	3.321	195.08	54
56	293.29	1.878	1.878	146.73	393.27	3.377	3.378	196.75	56
58	296.62	1.921	1.921	148.40	396.60	3.434	3.435	198.42	58
60	299.96	1.964	1.964	150.07	399.94	3.491	3.492	200.09	60

′	4° L. C.	4° M.	4° E.	4° T.	5° L. C.	5° M.	5° E.	5° T.	′
0	399.94	3.491	3.492	200.09	499.88	5.454	5.459	250.17	0
2	403.27	3.550	3.551	201.76	503.21	5.527	5.533	251.84	2
4	406.60	3.609	3.610	203.43	506.54	5.601	5.607	253.51	4
6	409.93	3.668	3.670	205.10	509.87	5.675	5.681	255.18	6
8	413.26	3.727	3.730	206.77	513.20	5.749	5.755	256.85	8
10	416.59	3.787	3.790	208.44	516.53	5.823	5.829	258.52	10
12	419.92	3.848	3.851	210.11	519.86	5.899	5.905	260.20	12
14	423.26	3.910	3.913	211.77	523.19	5.975	5.981	261.86	14
16	426.59	3.972	3.975	213.45	526.52	6.052	6.058	263.54	16
18	429.92	4.034	4.037	215.11	529.85	6.129	6.135	265.20	18
20	433.25	4.096	4.099	216.78	533.18	6.206	6.212	266.87	20
22	436.58	4.160	4.163	218.45	536.51	6.284	6.290	268.54	22
24	439.91	4.224	4.227	220.12	539.84	6.362	6.369	270.21	24
26	443.24	4.288	4.291	221.79	543.17	6.441	6.448	271.88	26
28	446.58	4.353	4.356	223.46	546.50	6.520	6.527	273.54	28
30	449.91	4.418	4.421	225.13	549.83	6.599	6.606	275.21	30
32	453.24	4.484	4.487	226.80	553.17	6.680	6.687	276.88	32
34	456.57	4.550	4.554	228.47	556.50	6.761	6.768	278.55	34
36	459.90	4.617	4.621	230.14	559.83	6.842	6.849	280.23	36
38	463.23	4.684	4.688	231.81	563.16	6.923	6.931	281.90	38
40	466.56	4.751	4.755	233.48	566.49	7.005	7.013	283.57	40
42	469.89	4.820	4.824	235.15	569.82	7.088	7.096	285.24	42
44	473.23	4.889	4.893	236.82	573.15	7.171	7.180	286.91	44
46	476.56	4.958	4.962	238.48	576.48	7.255	7.264	288.59	46
48	479.89	5.027	5.031	240.15	579.81	7.339	7.348	290.26	48
50	483.22	5.096	5.100	241.82	583.14	7.423	7.432	291.93	50
52	486.55	5.167	5.171	243.49	586.47	7.508	7.517	293.60	52
54	489.88	5.238	5.243	245.16	589.80	7.593	7.603	295.27	54
56	493.21	5.310	5.315	246.83	593.13	7.678	7.689	296.95	56
58	496.54	5.382	5.387	248.50	596.46	7.764	7.775	298.62	58
60	499.88	5.454	5.459	250.17	599.80	7.850	7.861	300.30	60

′	6° L. C.	6° M.	6° E.	6° T.	7° L. C.	7° M.	7° E.	7° T.	′
0	599.80	7.850	7.861	300.30	699.60	10.69	10.71	350.44	0
2	603.13	7.940	7.951	301.97	702.93	10.79	10.81	352.11	2
4	606.46	8.030	8.041	303.64	706.26	10.90	10.92	353.79	4
6	609.78	8.120	8.131	305.31	709.58	11.00	11.02	355.46	6
8	613.11	8.210	8.221	306.98	712.91	11.11	11.13	357.13	8
10	616.44	8.300	8.311	308.65	716.24	11.21	11.23	358.81	10
12	619.76	8.390	8.401	310.32	719.56	11.31	11.33	360.48	12
14	623.09	8.480	8.491	311.99	722.89	11.42	11.44	362.15	14
16	626.42	8.570	8.581	313.66	726.21	11.52	11.54	363.83	16
18	629.74	8.660	8.671	315.33	729.53	11.63	11.65	365.50	18
20	633.07	8.750	8.761	317.00	732.86	11.73	11.75	367.17	20
22	636.40	8.844	8.856	318.67	736.19	11.84	11.86	368.85	22
24	639.72	8.939	8.951	320.34	739.51	11.95	11.97	370.52	24
26	643.05	9.033	9.046	322.01	742.84	12.06	12.08	372.19	26
28	646.38	9.128	9.141	323.68	746.17	12.17	12.19	373.86	28
30	649.70	9.222	9.236	325.35	749.49	12.27	12.30	375.54	30
32	653.03	9.317	9.331	327.02	752.82	12.38	12.41	377.22	32
34	656.36	9.411	9.426	328.69	756.15	12.49	12.52	378.89	34
36	659.69	9.506	9.521	330.37	759.47	12.60	12.63	380.57	36
38	663.02	9.600	9.616	332.04	762.80	12.71	12.74	382.24	38
40	666.34	9.695	9.712	333.71	766.13	12.82	12.85	383.92	40
42	669.67	9.794	9.812	335.38	769.45	12.93	12.96	385.60	42
44	673.00	9.894	9.912	337.05	772.78	13.04	13.08	387.27	44
46	676.32	9.993	10.01	338.73	776.11	13.15	13.19	388.95	46
48	679.65	10.09	10.11	340.40	779.43	13.26	13.31	390.62	48
50	682.98	10.19	10.21	342.07	782.76	13.37	13.42	392.30	50
52	686.30	10.29	10.31	343.74	786.09	13.48	13.53	393.98	52
54	689.63	10.39	10.41	345.41	789.41	13.59	13.65	395.65	54
56	692.96	10.49	10.51	347.08	792.74	13.70	13.76	397.33	56
58	696.28	10.59	10.61	348.76	796.07	13.81	13.88	399.01	58
60	699.60	10.69	10.71	350.44	799.40	13.96	13.99	400.70	60

′	8° L. C.	8° M.	8° E.	8° T.	9° L. C.	9° M.	9° E.	9° T.	′
0	799.40	13.96	13.99	400.70	899.10	17.66	17.71	450.95	0
2	802.72	14.07	14.10	402.37	902.42	17.79	17.84	452.63	2
4	806.04	14.19	14.22	404.05	905.74	17.92	17.98	454.31	4
6	809.37	14.31	14.34	405.72	909.07	18.06	18.11	455.98	6
8	812.69	14.43	14.46	407.39	912.39	18.19	18.25	457.66	8
10	816.01	14.55	14.58	409.06	915.71	18.32	18.38	459.34	10
12	819.34	14.66	14.70	410.74	919.04	18.46	18.52	461.02	12
14	822.66	14.78	14.82	412.41	922.36	18.59	18.65	462.70	14
16	825.98	14.90	14.94	414.08	925.68	18.72	18.79	464.37	16
18	829.31	15.02	15.06	415.75	929.01	18.86	18.92	466.05	18
20	832.63	15.14	15.18	417.43	932.33	18.99	19.06	467.73	20
22	835.95	15.26	15.30	419.10	935.65	19.12	19.19	469.41	22
24	839.28	15.38	15.43	420.77	938.98	19.26	19.33	471.08	24
26	842.60	15.51	15.55	422.45	942.30	19.40	19.47	472.76	26
28	845.92	15.63	15.68	424.12	945.62	19.54	19.61	474.43	28
30	849.25	15.75	15.80	425.79	948.95	19.68	19.75	476.10	30
32	852.57	15.88	15.93	427.47	952.27	19.82	19.89	477.78	32
34	855.89	16.00	16.05	429.15	955.59	19.96	20.03	479.46	34
36	859.22	16.12	16.18	430.82	958.92	20.10	20.17	481.14	36
38	862.54	16.25	16.30	432.50	962.24	20.24	20.31	482.83	38
40	865.86	16.38	16.43	434.18	965.56	20.38	20.45	484.51	40
42	869.19	16.50	16.55	435.86	968.89	20.52	20.59	486.19	42
44	872.51	16.63	16.68	437.54	972.21	20.66	20.74	487.87	44
46	875.83	16.76	16.81	439.21	975.53	20.80	20.88	489.56	46
48	879.16	16.89	16.94	440.89	978.86	20.94	21.03	491.24	48
50	882.48	17.02	17.07	442.57	982.18	21.09	21.17	492.92	50
52	885.80	17.14	17.19	444.25	985.50	21.23	21.31	494.60	52
54	889.13	17.27	17.32	445.93	988.83	21.37	21.46	496.28	54
56	892.45	17.40	17.45	447.60	992.15	21.51	21.60	497.96	56
58	895.77	17.53	17.58	449.28	995.47	21.65	21.75	499.65	58
60	899.10	17.66	17.71	450.95	998.80	21.80	21.89	501.32	60

′	10° L. C.	10° M.	10° E.	10° T.	11° L. C.	11° M.	11° E.	11° T.	′
0	998.8	21.80	21.89	501.32	1098.4	26.38	26.50	551.74	0
2	1002.1	21.94	22.03	503.00	1101.7	26.54	26.66	553.42	2
4	1005.4	22.09	22.18	504.68	1105.0	26.70	26.83	555.10	4
6	1008.8	22.24	22.33	506.36	1108.3	26.86	26.99	556.78	6
8	1012.1	22.39	22.48	508.04	1111.7	27.02	27.16	558.46	8
10	1015.4	22.54	22.63	509.72	1115.0	27.19	27.32	560.14	10
12	1018.7	22.68	22.78	511.40	1118.3	27.35	27.48	561.82	12
14	1022.0	22.83	22.93	513.08	1121.6	27.51	27.65	563.50	14
16	1025.4	22.98	23.08	514.76	1124.9	27.67	27.81	565.18	16
18	1028.7	23.13	23.23	516.44	1128.2	27.83	27.98	566.86	18
20	1032.0	23.28	23.38	518.12	1131.6	28.00	28.14	568.54	20
22	1035.3	23.43	23.53	519.80	1134.9	28.17	28.30	570.22	22
24	1038.6	23.58	23.68	521.48	1138.2	28.34	28.47	571.90	24
26	1042.0	23.73	23.84	523.16	1141.5	28.50	28.64	573.58	26
28	1045.3	23.88	23.99	524.85	1144.8	28.67	28.81	575.27	28
30	1048.6	24.04	24.14	526.53	1148.1	28.84	28.98	576.95	30
32	1051.9	24.19	24.30	528.21	1151.5	29.00	29.14	578.63	32
34	1055.2	24.34	24.45	529.89	1154.8	29.17	29.31	580.32	34
36	1058.6	24.49	24.60	531.57	1158.1	29.34	29.48	582.00	36
38	1061.9	24.64	24.76	533.25	1161.4	29.50	29.65	583.69	38
40	1065.2	24.80	24.91	534.93	1164.7	29.67	29.82	585.37	40
42	1068.5	24.95	25.06	536.61	1168.0	29.84	29.99	587.05	42
44	1071.8	25.11	25.22	538.29	1171.4	30.01	30.17	588.74	44
46	1075.2	25.27	25.38	539.97	1174.7	30.18	30.34	590.42	46
48	1078.5	25.43	25.54	541.65	1178.0	30.35	30.52	592.11	48
50	1081.8	25.59	25.70	543.33	1181.3	30.53	30.69	593.79	50
52	1085.1	25.74	25.86	545.01	1184.6	30.70	30.86	595.47	52
54	1088.4	25.90	26.02	546.69	1187.9	30.87	31.04	597.16	54
56	1091.8	26.06	26.18	548.37	1191.3	31.04	31.21	598.84	56
58	1095.1	26.22	26.34	550.06	1194.6	31.21	31.39	600.53	58
60	1098.4	26.38	26.50	551.74	1197.9	31.39	31.56	602.22	60

′	12° L. C.	12° M.	12° E.	12° T.	13° L. C.	13° M.	13° E	13° T.	′
0	1197.9	31.39	31.56	602.22	1297.3	36.83	37.07	652.87	0
2	1201.2	31.57	31.73	603.91	1300.6	37.02	37.26	654.56	2
4	1204.5	31.74	31.91	605.60	1303.9	37.21	37.46	656.25	4
6	1207.8	31.92	32.09	607.28	1307.2	37.40	37.65	657.93	6
8	1211.1	32.09	32.27	608.97	1310.5	37.59	37.85	659.62	8
10	1214.5	32.27	32.45	610.66	1313.8	37.79	38.04	661.31	10
12	1217.8	32.45	32.63	612.35	1317.2	37.98	38.23	663.00	12
14	1221.1	32.62	32.81	614.04	1320.5	38.17	38.43	664.69	14
16	1224.4	32.80	32.99	615.72	1323.8	38.36	38.62	666.37	16
18	1227.7	32.97	33.17	617.41	1327.1	38.55	38.82	668.06	18
20	1231.0	33.15	33.35	619.10	1330.4	38.75	39.01	669.75	20
22	1234.3	33.33	33.53	620.79	1333.7	38.95	39.20	671.44	22
24	1237.7	33.51	33.72	622.48	1337.0	39.15	39.40	673.13	24
26	1241.0	33.69	33.90	624.16	1340.3	39.35	39.60	674.81	26
28	1244.3	33.87	34.09	625.85	1343.6	39.54	39.80	676.51	28
30	1247.6	34.06	34.27	627.55	1346.9	39.74	40.00	678.20	30
32	1250.9	34.24	34.45	629.24	1350.3	39.94	40.19	679.89	32
34	1254.2	34.42	34.64	630.93	1353.6	40.13	40.39	681.58	34
36	1257.5	34.60	34.82	632.61	1356.9	40.33	40.59	683.26	36
38	1260.8	34.78	35.01	634.30	1360.2	40.52	40.79	684.95	38
40	1264.2	34.97	35.19	635.99	1363.5	40.71	40.99	686.64	40
42	1267.5	35.16	35.37	637.68	1366.8	40.91	41.19	688.33	42
44	1270.8	35.34	35.56	639.37	1370.1	41.11	41.40	690.02	44
46	1274.1	35.53	35.75	641.05	1373.4	41.31	41.60	691.70	46
48	1277.4	35.71	35.94	642.74	1376.7	41.51	41.81	693.39	48
50	1280.7	35.90	36.13	644.43	1380.0	41.71	42.01	695.08	50
52	1284.0	36.09	36.31	646.12	1383.4	41.91	42.21	696.77	52
54	1287.4	36.27	36.50	647.81	1386.7	42.11	42.42	698.46	54
56	1290.7	36.46	36.69	649.49	1390.0	42.31	42.62	700.14	56
58	1294.0	36.64	36.88	651.18	1393.3	42.51	42.83	701.83	58
60	1297.3	36.83	37.07	652.87	1396.6	42.71	43.03	703.53	60

′	14° L. C.	14° M.	14° E.	14° T.	15° L. C.	15° M.	15° E.	15° T.	′
0	1396.6	42.71	43.03	703.53	1495.9	49.02	49.44	754.35	0
2	1399.9	42.92	43.23	705.23	1499.2	49.24	49.66	756.05	2
4	1403.2	43.12	43.44	706.92	1502.5	49.46	49.89	757.74	4
6	1406.5	43.33	43.65	708.62	1505.8	49.68	50.11	759.44	6
8	1409.8	43.53	43.86	710.31	1509.1	49.90	50.34	761.13	8
10	1413.1	43.74	44.07	712.01	1512.4	50.12	50.56	762.83	10
12	1416.5	43.94	44.28	713.71	1515.7	50.34	50.78	764.53	12
14	1419.8	44.15	44.49	715.40	1519.0	50.56	51.01	766.22	14
16	1423.1	44.35	44.70	717.10	1522.3	50.78	51.23	767.92	16
18	1426.4	44.56	44.91	718.79	1525.6	51.00	51.46	769.61	18
20	1429.7	44.77	45.12	720.49	1528.9	51.22	51.68	771.31	20
22	1433.0	44.98	45.33	722.20	1532.2	51.44	51.90	773.01	22
24	1436.3	45.19	45.54	723.89	1535.5	51.67	52.13	774.70	24
26	1439.6	45.40	45.76	725.59	1538.8	51.89	52.36	776.40	26
28	1442.9	45.61	45.97	727.28	1542.1	52.12	52.59	778.09	28
30	1446.2	45.82	46.18	728.97	1545.4	52.34	52.82	779.79	30
32	1449.6	46.03	46.40	730.66	1548.7	52.57	53.05	781.49	32
34	1452.9	46.24	46.61	732.35	1552.0	52.79	53.28	783.19	34
36	1456.2	46.45	46.82	734.05	1555.3	53.02	53.51	784.89	36
38	1459.5	46.66	47.04	735.74	1558.6	53.24	53.74	786.59	38
40	1462.8	46.87	47.25	737.43	1561.9	53.47	53.97	788.29	40
42	1466.1	47.08	47.46	739.12	1565.2	53.69	54.20	789.99	42
44	1469.4	47.30	47.68	740.81	1568.5	53.92	54.44	791.69	44
46	1472.7	47.51	47.90	742.51	1571.8	54.15	54.67	793.39	46
48	1476.0	47.73	48.12	744.20	1575.1	54.38	54.91	795.09	48
50	1479.3	47.94	48.34	745.89	1578.4	54.61	55.14	796.79	50
52	1482.7	48.16	48.56	747.58	1581.7	54.84	55.37	798.49	52
54	1486.0	48.37	48.78	749.27	1585.0	55.07	55.61	800.19	54
56	1489.3	48.59	49.00	750.97	1588.3	55.30	55.84	801.89	56
58	1492.6	48.80	49.22	752.66	1591.6	55.53	56.08	803.59	58
60	1495.9	49.02	49.44	754.35	1594.9	55.76	56.31	805.29	60

′	16° L. C.	16° M.	16° E.	16° T.	17° L. C.	17° M.	17° E.	17° T.	′
0	1594.9	55.76	56.31	805.29	1693.9	62.94	63.64	856.35	0
2	1598.2	55.99	56.54	806.99	1697.2	63.18	63.89	858.05	2
4	1601.5	56.23	56.78	808.64	1700.5	63.43	64.15	859.76	4
6	1604.8	56.46	57.02	810.39	1703.8	63.68	64.40	861.46	6
8	1608.1	56.70	57.26	812.09	1707.1	63.93	64.66	863.16	8
10	1611.4	56.93	57.50	813.79	1710.4	64.18	64.91	864.87	10
12	1614.7	57.17	57.74	815.49	1713.7	64.42	65.16	866.57	12
14	1618.0	57.40	57.98	817.19	1716.9	64.67	65.42	868.27	14
16	1621.3	57.64	58.22	818.89	1720.2	64.92	65.67	869.98	16
18	1624.6	57.87	58.46	820.59	1723.5	65.17	65.93	871.68	18
20	1627.9	58.11	58.70	822.29	1726.8	65.42	66.18	873.38	20
22	1631.2	58.34	58.94	823.99	1730.1	65.67	66.43	875.09	22
24	1634.5	58.58	59.19	825.69	1733.4	65.93	66.69	876.79	24
26	1637.8	58.82	59.43	827.39	1736.7	66.18	66.95	878.49	26
28	1641.1	59.06	59.68	829.09	1740.0	66.44	67.21	880.20	28
30	1644.4	59.30	59.92	830.79	1743.3	66.69	67.47	881.90	30
32	1647.7	59.54	60.16	832.49	1746.6	66.94	67.72	883.61	32
34	1651.0	59.78	60.41	834.20	1749.9	67.20	67.98	885.32	34
36	1654.3	60.02	60.65	835.90	1753.2	67.45	68.24	887.02	36
38	1657.6	60.26	60.90	837.61	1756.5	67.71	68.50	888.73	38
40	1660.9	60.50	61.14	839.31	1759.8	67.93	68.76	890.44	40
42	1664.2	60.74	61.39	841.01	1763.1	68.21	69.03	892.15	42
44	1667.5	60.99	61.64	842.72	1766.3	68.47	69.29	893.86	44
46	1670.8	61.23	61.89	844.42	1769.6	68.73	69.56	895.56	46
48	1674.1	61.48	62.14	846.13	1772.9	68.99	69.82	897.27	48
50	1677.4	61.72	62.39	847.83	1776.2	69.25	70.09	898.98	50
52	1680.7	61.96	62.64	849.53	1779.5	69.50	70.36	900.69	52
54	1684.0	62.21	62.89	851.24	1782.8	69.76	70.62	902.40	54
56	1687.3	62.45	63.14	852.94	1786.1	70.02	70.89	904.10	56
58	1690.6	62.70	63.39	854.65	1789.4	70.28	51.15	905.81	58
60	1693.9	62.91	63.64	856.35	1792.7	70.54	71.42	907.52	60

′	18° L. C.	18° M.	18° E.	18° T.	19° L. C.	19° M.	19° E.	19° T.	′
0	1792.7	70.54	71.42	907.52	1891.5	78.58	79.65	958.86	0
2	1796.0	70.80	71.69	909.23	1894.8	78.86	79.94	960.57	2
4	1799.3	71.06	71.96	910.94	1898.1	79.13	80.22	962.30	4
6	1802.6	71.33	72.23	912.65	1901.3	79.41	80.51	964.00	6
8	1805.9	71.59	72.50	914.36	1904.6	79.68	80.79	965.72	8
10	1809.2	71.85	72.77	916.07	1907.9	79.96	81.08	967.43	10
12	1812.5	72.12	73.04	917.78	1911.2	80.24	81.37	969.15	12
14	1815.7	72.38	73.31	919.49	1914.5	80.51	81.65	970.86	14
16	1819.0	72.64	73.58	921.20	1917.8	80.79	81.94	972.58	16
18	1822.3	72.91	73.85	922.91	1921.0	81.07	82.22	974.29	18
20	1825.6	73.17	74.12	924.63	1924.3	81.35	82.51	976.01	20
22	1828.9	73.43	74.39	926.34	1927.6	81.63	82.80	977.72	22
24	1832.2	73.70	74.67	928.05	1930.9	81.91	83.09	979.44	24
26	1835.5	73.97	74.94	929.76	1934.2	82.20	83.38	981.15	26
28	1838.8	74.24	75.22	931.47	1937.5	82.48	83.67	982.86	28
30	1842.1	74.51	75.49	933.18	1940.7	82.76	83.97	984.58	30
32	1845.4	74.77	75.77	934.89	1944.0	83.05	84.26	986.30	32
34	1848.7	75.04	76.04	936.60	1947.3	83.33	84.55	988.02	34
36	1852.0	75.31	76.32	938.32	1950.6	83.61	84.84	989.74	36
38	1855.3	75.58	76.59	940.03	1953.9	83.90	85.13	991.46	38
40	1858.6	75.85	76.87	941.74	1957.2	84.18	85.43	993.18	40
42	1861.9	76.12	77.14	943.45	1960.4	84.47	85.73	994.90	42
44	1865.1	76.39	77.42	945.16	1963.7	84.75	86.02	996.62	44
46	1868.4	76.67	77.70	946.88	1967.0	85.04	86.32	998.34	46
48	1871.7	76.94	77.98	948.59	1970.3	85.32	86.61	1000.0	48
50	1875.0	77.21	78.26	950.30	1973.6	85.61	86.91	1001.8	50
52	1878.3	77.49	78.53	952.01	1976.9	85.90	87.21	1003.5	52
54	1881.6	77.76	78.81	953.72	1980.1	86.19	87.50	1005.2	54
56	1884.9	78.03	79.09	955.44	1983.4	86.47	87.80	1006.9	56
58	1888.2	78.31	79.37	957.15	1986.7	86.76	88.09	1008.6	58
60	1891.5	78.58	79.65	958.86	1990.0	87.05	88.39	1010.4	60

′	20° L. C.	20° M.	20° E.	20° T.	21° L. C.	21° M.	21° E.	21° T.	′
0	1990.0	87.05	88.39	1010.4	2088.5	95.95	97.58	1062.0	0
2	1993.3	87.34	88.69	1012.1	2091.8	96.26	97.90	1063.7	2
4	1996.6	87.63	88.99	1013.8	2095.0	96.56	98.21	1065.4	4
6	1999.8	87.92	89.29	1015.5	2098.3	96.87	98.53	1067.2	6
8	2003.1	88.21	89.59	1017.2	2101.6	97.17	98.84	1068.9	8
10	2006.4	88.50	89.89	1019.0	2104.9	97.48	99.16	1070.6	10
12	2009.7	88.79	90.19	1020.7	2108.1	97.79	99.48	1072.4	12
14	2013.0	89.08	90.49	1022.4	2111.4	98.09	99.79	1074.1	14
16	2016.3	89.37	90.79	1024.1	2114.7	98.40	100.1	1075.8	16
18	2019.5	89.66	91.09	1025.8	2118.0	98.70	100.4	1077.5	18
20	2022.8	89.96	91.40	1027.6	2121.2	99.00	100.7	1079.3	20
22	2026.1	90.25	91.71	1029.3	2124.5	99.30	101.1	1081.0	22
24	2029.4	90.55	92.01	1031.0	2127.8	99.60	101.4	1082.7	24
26	2032.7	90.85	92.32	1032.7	2131.0	99.90	101.7	1084.4	26
28	2036.0	91.15	92.62	1034.4	2134.3	100.2	102.0	1086.2	28
30	2039.2	91.45	92.93	1036.1	2137.6	100.5	102.3	1087.9	30
32	2042.5	91.74	93.24	1037.9	2140.9	100.8	102.7	1089.6	32
34	2045.8	92.04	93.54	1039.6	2144.1	101.1	103.0	1091.3	34
36	2049.1	92.34	93.85	1041.3	2147.4	101.4	103.3	1093.1	36
38	2052.4	92.64	94.15	1043.0	2150.7	101.7	103.6	1094.8	38
40	2055.7	92.94	94.46	1044.8	2154.0	102.1	104.0	1096.5	40
42	2058.9	93.24	94.78	1046.5	2157.2	102.4	104.3	1098.3	42
44	2062.2	93.54	95.09	1048.2	2160.5	102.7	104.6	1100.0	44
46	2065.5	93.84	95.40	1049.9	2163.8	103.0	104.9	1101.7	46
48	2068.8	94.14	95.71	1051.7	2167.1	103.3	105.3	1103.4	48
50	2072.1	94.44	96.03	1053.4	2170.3	103.6	105.6	1105.2	50
52	2075.4	94.74	96.34	1055.1	2173.6	103.9	105.9	1106.9	52
54	2078.6	95.04	96.65	1056.8	2176.9	104.2	106.3	1108.6	54
56	2081.9	95.34	96.96	1058.6	2180.1	104.5	106.6	1110.3	56
58	2085.2	95.64	97.27	1060.3	2183.4	104.8	106.9	1112.1	58
60	2088.5	95.95	97.58	1062.0	2186.7	105.2	107.2	1113.8	60

′	22°				23°				′
	L. C.	M.	E.	T.	L. C.	M.	E.	T.	
0	2186.7	105.2	107.2	1113.8	2284.8	115.0	117.4	1165.8	0
2	2190.0	105.6	107.6	1115.5	2288.1	115.3	117.7	1167.5	2
4	2193.2	105.9	107.9	1117.3	2291.3	115.7	118.1	1169.2	4
6	2196.5	106.2	108.2	1119.0	2294.6	116.0	118.4	1171.0	6
8	2199.8	106.5	108.6	1120.7	2297.8	116.4	118.8	1172.7	8
10	2203.0	106.8	108.9	1122.4	2301.1	116.7	119.1	1174.4	10
12	2206.3	107.1	109.2	1124.2	2304.4	117.0	119.5	1176.2	12
14	2209.6	107.4	109.6	1125.9	2307.6	117.4	119.8	1177.9	14
16	2212.9	107.7	109.9	1127.6	2310.9	117.7	120.2	1179.7	16
18	2216.1	108.0	110.2	1129.4	2314.1	118.1	120.5	1181.4	18
20	2219.4	108.4	110.6	1131.1	2317.4	118.4	120.9	1183.1	20
22	2222.7	108.7	110.9	1132.8	2320.7	118.7	121.2	1184.9	22
24	2225.9	109.0	111.2	1134.6	2323.9	119.1	121.6	1186.6	24
26	2229.2	109.4	111.6	1136.3	2327.2	119.4	121.9	1188.4	26
28	2232.5	109.7	111.9	1138.0	2330.4	119.8	122.3	1190.1	28
30	2235.7	110.0	112.3	1139.7	2333.7	120.1	122.6	1191.8	30
32	2239.0	110.4	112.6	1141.5	2337.0	120.4	123.0	1193.6	32
34	2242.3	110.7	112.9	1143.2	2340.2	120.8	123.3	1195.3	34
36	2245.6	111.0	113.3	1144.9	2343.5	121.1	123.7	1197.1	36
38	2248.8	111.4	113.6	1146.7	2346.7	121.5	124.1	1198.8	38
40	2252.1	111.7	113.9	1148.4	2350.0	121.8	124.4	1200.5	40
42	2255.4	112.0	114.3	1150.1	2353.3	122.1	124.8	1202.3	42
44	2258.6	112.3	114.6	1151.9	2356.5	122.5	125.1	1204.0	44
46	2261.9	112.7	115.0	1153.6	2359.8	122.8	125.5	1205.8	46
48	2265.2	113.0	115.3	1155.4	2363.0	123.2	125.8	1207.5	48
50	2268.4	113.3	115.7	1157.1	2366.3	123.5	126.2	1209.2	50
52	2271.7	113.7	116.0	1158.8	2369.6	123.8	126.6	1211.0	52
54	2275.0	114.0	116.3	1160.6	2372.8	124.2	126.9	1212.7	54
56	2278.3	114.3	116.7	1162.3	2376.1	124.5	127.3	1214.5	56
58	2281.5	114.7	117.0	1164.0	2379.3	124.9	127.6	1216.2	58
60	2284.8	115.0	117.4	1165.8	2382.6	125.2	128.0	1218.0	60

′	24°				25°				′
	L. C.	M.	E.	T.	L. C.	M.	E.	T.	
0	2382.6	125.2	128.0	1218.0	2480.4	135.8	139.1	1270.3	0
2	2385.9	125.5	128.4	1219.7	2483.6	136.2	139.5	1272.0	2
4	2389.1	125.9	128.7	1221.4	2486.9	136.5	139.9	1273.8	4
6	2392.4	126.2	129.1	1223.2	2490.1	136.9	140.3	1275.5	6
8	2395.6	126.6	129.5	1224.9	2493.4	137.2	140.6	1277.3	8
10	2398.9	126.9	129.8	1226.7	2496.6	137.6	141.0	1279.0	10
12	2402.2	127.3	130.2	1228.4	2499.9	138.0	141.4	1280.8	12
14	2405.4	127.6	130.6	1230.2	2503.1	138.3	141.8	1282.5	14
16	2408.7	128.0	130.9	1231.9	2506.4	138.7	142.2	1284.3	16
18	2411.9	128.3	131.3	1233.6	2509.6	139.0	142.5	1286.1	18
20	2415.2	128.7	131.7	1235.4	2512.9	139.4	142.9	1287.8	20
22	2418.5	129.0	132.0	1237.1	2516.1	139.8	143.3	1289.6	22
24	2421.7	129.4	132.4	1238.9	2519.4	140.1	143.7	1291.3	24
26	2425.0	129.7	132.8	1240.6	2522.6	140.5	144.1	1293.1	26
28	2428.2	130.1	133.1	1242.4	2525.9	140.8	144.5	1294.8	28
30	2431.5	130.4	133.5	1244.1	2529.1	141.2	144.9	1296.6	30
32	2434.8	130.8	133.9	1245.8	2532.4	141.6	145.3	1298.3	32
34	2438.0	131.1	134.2	1247.6	2535.6	142.0	145.6	1300.1	34
36	2441.3	131.5	134.6	1249.3	2538.9	142.3	146.0	1301.8	36
38	2444.5	131.8	135.0	1251.1	2542.1	142.7	146.4	1303.6	38
40	2447.8	132.2	135.4	1252.8	2545.4	143.1	146.8	1305.3	40
42	2451.1	132.6	135.7	1254.6	2548.6	143.5	147.2	1307.1	42
44	2454.3	132.9	136.1	1256.3	2551.9	143.8	147.6	1308.8	44
46	2457.6	133.3	136.5	1258.1	2555.1	144.2	148.0	1310.6	46
48	2460.8	133.6	136.9	1259.8	2558.4	144.5	148.4	1312.4	48
50	2464.1	134.0	137.2	1261.5	2561.6	144.9	148.8	1314.1	50
52	2467.4	134.4	137.6	1263.3	2564.9	145.3	149.2	1315.9	52
54	2470.6	134.7	138.0	1265.0	2568.1	145.7	149.5	1317.6	54
56	2473.9	135.1	138.4	1266.8	2571.4	146.0	149.9	1319.4	56
58	2477.1	135.4	138.7	1268.5	2574.6	146.4	150.3	1321.1	58
60	2480.4	135.8	139.1	1270.3	2577.9	146.8	150.7	1322.9	60

′	26° L. C.	26° M.	26° E.	26° T.	27° L. C.	27° M.	27° E.	27° T.	′
0	2577.9	146.8	150.7	1322.9	2675.3	158.3	162.8	1375.6	0
2	2581.1	147.1	151.1	1324.6	2678.5	158.6	163.2	1377.4	2
4	2584.4	147.5	151.5	1326.4	2681.8	159.0	163.7	1379.2	4
6	2587.6	147.9	151.9	1328.1	2685.0	159.4	164.1	1380.9	6
8	2590.9	148.3	152.3	1329.9	2688.2	159.8	164.5	1382.7	8
10	2594.1	148.7	152.7	1331.6	2691.5	160.2	164.9	1384.5	10
12	2597.4	149.1	153.1	1333.4	2694.7	160.6	165.3	1386.2	12
14	2600.6	149.4	153.5	1335.2	2698.0	161.0	165.7	1388.0	14
16	2603.9	149.8	153.9	1336.9	2701.2	161.4	166.1	1389.8	16
18	2607.1	150.2	154.3	1338.7	2704.4	161.8	166.5	1391.5	18
20	2610.4	150.6	154.7	1340.4	2707.7	162.2	167.0	1393.3	20
22	2613.6	151.0	155.1	1342.2	2710.9	162.6	167.4	1395.0	22
24	2616.9	151.4	155.5	1343.9	2714.1	163.0	167.8	1396.8	24
26	2620.1	151.7	155.9	1345.7	2717.4	163.4	168.2	1398.6	26
28	2623.4	152.1	156.3	1347.4	2720.6	163.8	168.6	1400.3	28
30	2626.6	152.5	156.7	1349.2	2723.8	164.2	169.1	1402.1	30
32	2629.8	152.9	157.1	1351.0	2727.1	164.6	169.5	1403.9	32
34	2633.1	153.3	157.5	1352.7	2730.3	165.0	169.9	1405.6	34
36	2636.3	153.7	157.9	1354.5	2733.6	165.4	170.3	1407.4	36
38	2639.6	154.0	158.3	1356.2	2736.8	165.8	170.8	1409.2	38
40	2642.8	154.4	158.7	1358.0	2740.0	166.2	171.2	1410.9	40
42	2646.1	154.8	159.1	1359.8	2743.3	166.6	171.6	1412.7	42
44	2649.3	155.2	159.5	1361.5	2746.5	167.0	172.0	1414.5	44
46	2652.6	155.6	160.0	1363.3	2749.7	167.4	172.5	1416.3	46
48	2655.8	156.0	160.4	1365.1	2753.0	167.8	172.9	1418.0	48
50	2659.1	156.3	160.8	1366.8	2756.2	168.2	173.3	1419.8	50
52	2662.3	156.7	161.2	1368.6	2759.5	168.6	173.7	1421.6	52
54	2665.6	157.1	161.6	1370.4	2762.7	169.0	174.1	1423.3	54
56	2668.8	157.5	162.0	1372.1	2765.9	169.4	174.6	1425.1	56
58	2672.1	157.9	162.4	1373.9	2769.2	169.8	175.0	1426.9	58
60	2675.3	158.3	162.8	1375.6	2772.4	170.2	175.4	1428.6	60

′	28° L. C.	28° M.	28° E.	28° T.	29° L. C.	29° M.	29° E.	29° T.	′
0	2772.4	170.2	175.4	1428.6	2869.4	182.5	188.5	1481.9	0
2	2775.6	170.6	175.8	1430.4	2872.6	182.9	189.0	1483.7	2
4	2778.9	171.0	176.3	1432.2	2875.8	183.3	189.4	1485.4	4
6	2782.1	171.4	176.7	1434.0	2879.1	183.7	189.9	1487.2	6
8	2785.3	171.8	177.1	1435.7	2882.3	184.2	190.3	1489.0	8
10	2788.6	172.2	177.6	1437.5	2885.5	184.6	190.8	1490.8	10
12	2791.8	172.6	178.0	1439.3	2888.7	185.0	191.2	1492.6	12
14	2795.0	173.0	178.4	1441.1	2892.0	185.4	191.7	1494.3	14
16	2798.3	173.4	178.9	1442.8	2895.2	185.8	192.1	1496.1	16
18	2801.5	173.8	179.3	1444.6	2898.4	186.3	192.5	1497.9	18
20	2804.7	174.3	179.7	1446.4	2901.6	186.7	193.0	1499.7	20
22	2808.0	174.7	180.2	1448.2	2904.8	187.1	193.5	1501.5	22
24	2811.2	175.1	180.6	1449.9	2908.1	187.5	193.9	1503.2	24
26	2814.4	175.5	181.0	1451.7	2911.3	188.0	194.4	1505.0	26
28	2817.7	175.9	181.5	1453.5	2914.5	188.4	194.8	1506.8	28
30	2820.9	176.3	181.9	1455.2	2917.7	188.8	195.3	1508.6	30
32	2824.1	176.7	182.3	1457.0	2921.0	189.2	195.7	1510.4	32
34	2827.4	177.1	182.8	1458.8	2924.2	189.7	196.2	1512.1	34
36	2830.6	177.5	183.2	1460.6	2927.4	190.1	196.7	1513.9	36
38	2833.8	177.9	183.6	1462.3	2930.6	190.5	197.1	1515.7	38
40	2837.1	178.4	184.1	1464.1	2933.9	190.9	197.6	1517.5	40
42	2840.3	178.8	184.5	1465.9	2937.1	191.4	198.0	1519.3	42
44	2843.5	179.2	185.0	1467.7	2940.3	191.9	198.5	1521.0	44
46	2846.8	179.6	185.4	1469.5	2943.5	192.4	198.9	1522.8	46
48	2850.0	180.0	185.9	1471.2	2946.8	192.8	199.4	1524.6	48
50	2853.2	180.4	186.3	1473.0	2950.0	193.2	199.8	1526.4	50
52	2856.5	180.8	186.8	1474.8	2953.2	193.6	200.3	1528.2	52
54	2859.7	181.2	187.2	1476.6	2956.4	194.0	200.8	1530.0	54
56	2862.9	181.6	187.6	1478.3	2959.6	194.4	201.2	1531.7	56
58	2866.2	182.0	188.1	1480.1	2962.9	194.8	201.7	1533.5	58
60	2869.4	182.5	188.5	1481.9	2966.1	195.2	202.1	1535.3	60

′	30°				31°				′
	L. C.	M.	E.	T.	L. C.	M.	E.	T.	
0	2966.1	195.2	202.1	1535.3	3062.6	208.4	216.3	1589.0	0
2	2969.3	195.6	202.6	1537.1	3065.8	208.8	216.8	1590.8	2
4	2972.5	196.1	203.1	1538.9	3069.0	209.3	217.2	1592.6	4
6	2975.7	196.5	203.5	1540.7	3072.2	209.7	217.7	1594.4	6
8	2979.0	197.0	204.0	1542.5	3075.4	210.2	218.2	1506.2	8
10	2982.2	197.4	204.5	1544.3	3078.6	210.6	218.7	1598.0	10
12	2985.4	197.8	204.9	1546.0	3081.8	211.1	219.2	1599.8	12
14	2988.6	198.2	205.4	1547.8	3085.0	211.5	219.6	1601.6	14
16	2991.8	198.6	205.9	1549.6	3088.3	212.0	220.1	1603.4	16
18	2995.0	199.1	206.3	1551.4	3091.5	212.4	220.6	1605.2	18
20	2998.3	199.5	206.8	1553.2	3094.7	212.9	221.1	1607.0	20
22	3001.5	199.9	207.3	1555.0	3097.9	213.3	221.6	1608.8	22
24	3004.7	200.4	207.7	1556.8	3101.1	213.8	222.1	1610.6	24
26	3007.9	200.8	208.2	1558.6	3104.3	214.2	222.6	1612.4	26
28	3011.1	201.3	208.7	1560.4	3107.5	214.7	223.0	1614.2	28
30	3014.3	201.7	209.1	1562.2	3110.7	215.1	223.5	1616.0	30
32	3017.6	202.1	209.6	1564.0	3113.9	215.6	224.0	1617.8	32
34	3020.8	202.6	210.1	1565.7	3117.1	216.0	224.5	1619.6	34
36	3024.0	203.0	210.5	1567.5	3120.3	216.5	225.0	1621.4	36
38	3027.2	203.5	211.0	1569.3	3123.5	216.9	225.5	1623.2	38
40	3030.4	203.9	211.5	1571.1	3126.7	217.4	226.0	1625.0	40
42	3033.6	204.3	212.0	1572.9	3129.9	217.8	226.5	1626.8	42
44	3036.9	204.8	212.4	1574.7	3133.1	218.3	227.0	1628.6	44
46	3040.1	205.2	212.9	1576.5	3136.4	218.7	227.5	1630.5	46
48	3043.3	205.7	213.4	1578.3	3139.6	219.2	228.0	1632.3	48
50	3046.5	206.1	213.9	1580.1	3142.8	219.6	228.4	1634.1	50
52	3049.7	206.5	214.4	1581.9	3146.0	220.1	228.9	1635.9	52
54	3052.9	207.0	214.8	1583.7	3149.2	220.5	229.4	1637.7	54
56	3056.2	207.4	215.3	1585.5	3152.4	221.0	229.9	1639.5	56
58	3059.4	207.9	215.8	1587.2	3155.6	221.5	230.4	1641.3	58
60	3062.6	208.4	216.3	1589.0	3158.8	222.0	230.9	1643.1	60

′	32°				33°				′
	L. C.	M.	E.	T.	L. C.	M.	E.	T.	
0	3158.8	222.0	230.9	1643.1	3254.9	236.0	246.1	1697.3	0
2	3162.0	222.5	231.4	1644.9	3258.1	236.4	246.6	1699.1	2
4	3165.2	222.9	231.9	1646.7	3261.3	236.9	247.1	1700.9	4
6	3168.4	223.4	232.4	1648.5	3264.5	237.4	247.7	1702.7	6
8	3171.6	223.8	232.9	1650.3	3267.7	237.9	248.2	1704.5	8
10	3174.8	224.3	233.4	1652.1	3270.8	238.4	248.7	1706.4	10
12	3178.0	224.8	233.9	1653.9	3274.0	238.9	249.2	1708.2	12
14	3181.2	225.2	234.4	1655.7	3277.2	239.3	249.7	1710.0	14
16	3184.4	225.7	234.9	1657.5	3280.4	239.8	250.2	1711.8	16
18	3187.6	226.1	235.4	1659.3	3283.6	240.3	250.8	1713.6	18
20	3190.8	226.6	235.9	1661.1	3286.8	240.8	251.3	1715.5	20
22	3194.0	227.1	236.4	1662.9	3290.0	241.2	251.8	1717.3	22
24	3197.2	227.5	236.9	1664.7	3293.2	241.7	252.3	1719.1	24
26	3200.4	228.0	237.4	1666.5	3296.4	242.2	252.9	1720.9	26
28	3203.6	228.4	237.9	1668.3	3299.6	242.7	253.4	1722.7	28
30	3206.8	228.9	238.4	1670.1	3302.7	243.2	253.9	1724.6	30
32	3210.0	229.4	239.0	1671.9	3305.9	243.6	254.4	1726.4	32
34	3213.2	229.8	239.5	1673.7	3309.1	244.1	255.0	1728.2	34
36	3216.5	230.3	240.0	1675.5	3312.3	244.6	255.5	1730.0	36
38	3219.7	230.7	240.5	1677.4	3315.5	245.1	256.0	1731.8	38
40	3222.9	231.2	241.0	1679.2	3318.7	245.6	256.5	1733.6	40
42	3226.1	231.7	241.5	1681.0	3321.9	246.0	257.1	1735.5	42
44	3629.3	232.2	242.0	1682.8	3325.1	246.5	257.6	1737.3	44
46	3232.5	232.6	242.5	1684.6	3328.3	247.0	258.1	1739.1	46
48	3235.7	233.1	243.0	1686.4	3331.5	247.5	258.6	1740.9	48
50	3238.9	233.5	243.5	1688.2	3334.6	248.0	259.2	1742.7	50
52	3242.1	234.0	244.1	1690.0	3337.8	248.4	259.7	1744.6	52
54	3245.3	234.5	244.6	1691.8	3341.0	218.9	260.2	1746.4	54
56	3248.5	235.0	245.1	1693.7	3344.2	219.4	260.8	1748.2	56
58	3251.7	235.5	245.6	1695.5	3347.4	249.9	261.3	1750.0	58
60	3254.9	236.0	246.1	1697.3	3350.6	250.4	261.8	1751.8	60

′	34° L. C.	34° M.	34° E.	34° T.	35° L. C.	35° M.	35° E.	35° T.	′
0	3350.6	250.4	261.8	1751.8	3446.1	265.2	278.1	1806.7	0
2	3353.8	250.8	262.3	1753.7	3449.3	265.7	278.6	1808.5	2
4	3357.0	251.2	262.9	1755.5	3452.5	266.2	279.2	1810.3	4
6	3360.1	251.7	263.4	1757.3	3455.6	266.7	279.7	1812.2	6
8	3363.3	252.2	264.0	1759.1	3458.8	267.2	280.3	1814.0	8
10	3366.5	252.7	264.5	1761.0	3462.0	267.7	280.8	1815.8	10
12	3369.7	253.2	265.0	1762.8	3465.2	268.2	281.4	1817.7	12
14	3372.9	253.7	265.6	1764.6	3468.3	268.7	281.9	1819.5	14
16	3376.1	254.2	266.1	1766.4	3471.5	269.2	282.5	1821.3	16
18	3379.2	254.7	266.7	1768.3	3474.7	269.7	283.0	1823.2	18
20	3382.4	255.2	267.2	1770.1	3477.9	270.2	283.6	1825.0	20
22	3385.6	255.7	267.7	1771.9	3481.0	270.7	284.2	1826.8	22
24	3388.8	256.2	268.3	1773.7	3484.2	271.2	284.7	1828.7	24
26	3392.0	256.7	268.8	1775.6	3487.4	271.7	285.3	1830.5	26
28	3395.2	257.2	269.3	1777.4	3490.6	272.2	285.9	1832.3	28
30	3398.3	257.7	269.9	1779.2	3493.7	272.7	286.4	1834.2	30
32	3401.5	258.2	270.4	1781.0	3496.9	273.2	287.0	1836.0	32
34	3404.7	258.7	271.0	1782.9	3500.1	273.7	287.5	1837.8	34
36	3407.9	259.2	271.5	1784.7	3503.3	274.2	288.1	1839.7	36
38	3411.1	259.7	272.0	1786.5	3506.5	274.7	288.7	1841.5	38
40	3414.3	260.2	272.6	1788.4	3509.6	275.2	289.2	1843.4	40
42	3417.4	260.7	273.1	1790.2	3512.8	275.7	289.8	1845.2	42
44	3420.6	261.2	273.7	1792.0	3516.0	276.2	290.4	1847.1	44
46	3423.8	261.7	274.2	1793.9	3519.2	276.7	290.9	1848.9	46
48	3427.0	262.2	274.8	1795.7	3522.3	277.2	291.5	1850.7	48
50	3430.2	262.7	275.3	1797.5	3525.5	277.7	292.0	1852.6	50
52	3433.4	263.2	275.9	1799.3	3528.7	278.2	292.6	1854.4	52
54	3436.5	263.7	276.4	1801.2	3531.9	278.7	293.2	1856.3	54
56	3439.7	264.2	277.0	1803.0	3535.0	279.2	293.7	1858.1	56
58	3442.9	264.7	277.5	1804.8	3538.2	279.8	294.3	1859.9	58
60	3446.1	265.2	278.1	1806.7	3541.4	280.4	294.9	1861.8	60

′	36° L. C.	36° M.	36° E.	36° T.	37° L. C.	37° M.	37° E.	37° T.	′
0	3541.4	280.4	294.9	1861.8	3636.3	296.1	312.3	1917.3	0
2	3544.6	280.9	295.4	1863.6	3639.5	296.6	312.8	1919.1	2
4	3547.7	281.4	296.0	1865.5	3642.6	297.1	313.4	1921.0	4
6	3550.9	281.9	296.6	1867.3	3645.8	297.7	314.0	1922.8	6
8	3554.0	282.5	297.2	1869.2	3648.9	298.2	314.6	1924.7	8
10	3557.2	283.0	297.7	1871.0	3652.1	298.7	315.2	1926.5	10
12	3560.4	283.5	298.3	1872.9	3655.2	299.3	315.8	1928.4	12
14	3563.5	284.0	298.9	1874.7	3658.4	299.8	316.4	1930.2	14
16	3566.7	284.6	299.5	1876.5	3661.6	300.3	317.0	1932.1	16
18	3569.9	285.1	300.0	1878.4	3664.7	300.9	317.5	1933.9	18
20	3573.0	285.6	300.6	1880.2	3667.9	301.4	318.1	1935.8	20
22	3576.2	286.1	301.2	1882.1	3671.0	301.9	318.7	1937.6	22
24	3579.4	286.7	301.8	1883.9	3674.2	302.5	319.3	1939.5	24
26	3582.5	287.2	302.3	1885.8	3677.3	303.0	319.9	1941.3	26
28	3585.7	287.7	302.9	1887.6	3680.5	303.5	320.5	1943.2	28
30	3588.8	288.2	303.5	1889.5	3683.6	304.1	321.1	1945.0	30
32	3592.0	288.8	304.1	1891.3	3686.8	304.6	321.7	1946.9	32
34	3595.2	289.3	304.6	1893.2	3690.0	305.1	322.3	1948.8	34
36	3598.3	289.8	305.2	1895.0	3693.1	305.7	322.9	1950.6	36
38	3601.5	290.3	305.8	1896.9	3696.3	306.2	323.5	1952.5	38
40	3604.7	290.9	306.4	1898.7	3699.4	306.7	324.2	1954.4	40
42	3607.8	291.4	307.0	1900.6	3702.6	307.3	324.8	1956.2	42
44	3611.0	291.9	307.5	1902.4	3705.7	307.8	325.4	1958.1	44
46	3614.1	292.4	308.1	1904.3	3708.9	308.3	326.0	1960.0	46
48	3617.3	293.0	308.7	1906.1	3712.1	308.9	326.6	1961.8	48
50	3620.5	293.5	309.3	1908.0	3715.2	309.4	327.2	1963.7	50
52	3623.6	294.0	309.9	1909.8	3718.4	309.9	327.8	1965.5	52
54	3626.8	294.5	310.5	1911.7	3721.5	310.5	328.4	1967.4	54
56	3630.0	295.1	311.1	1913.5	3724.7	311.0	329.0	1969.3	56
58	3633.1	295.6	311.7	1915.4	3727.8	311.6	329.6	1971.1	58
60	3636.3	296.1	312.3	1917.3	3731.0	312.2	330.2	1973.0	60

′	38°				39°				′
	L. C.	M.	E.	T.	L. C.	M.	E.	T.	
0	3731.0	312.2	330.2	1973.0	3825.5	328.7	348.7	2029.1	0
2	3734.1	312.7	330.8	1974.9	3828.6	329.2	349.3	2031.0	2
4	3737.3	313.3	331.4	1976.7	3831.8	329.8	349.9	2032.9	4
6	3740.4	313.8	332.0	1978.6	3834.9	330.3	350.6	2034.7	6
8	3743.6	314.4	332.6	1980.5	3838.0	330.9	351.2	2036.6	8
10	3746.7	314.9	333.2	1982.3	3841.2	331.5	351.8	2038.5	10
12	3749.9	315.5	333.8	1984.2	3844.3	332.0	352.4	2040.4	12
14	3753.0	316.0	334.5	1986.1	3847.4	332.6	353.1	2042.3	14
16	3756.2	316.6	335.1	1987.9	3850.6	333.2	353.7	2044.1	16
18	3759.3	317.1	335.7	1989.8	3853.7	333.7	354.3	2046.0	18
20	3762.5	317.7	336.3	1991.7	3856.8	334.3	354.9	2047.9	20
22	3765.6	318.2	336.9	1993.6	3860.0	334.9	355.6	2049.8	22
24	3768.8	318.8	337.5	1995.4	3863.1	335.4	356.2	2051.7	24
26	3771.9	319.3	338.1	1997.3	3866.2	336.0	356.9	2053.5	26
28	3775.1	319.9	338.7	1999.2	3869.4	336.6	357.5	2055.4	28
30	3778.2	320.4	339.4	2001.0	3872.5	337.1	358.1	2057.3	30
32	3781.4	321.0	340.0	2002.9	3875.6	337.7	358.8	2059.2	32
34	3784.5	321.5	340.6	2004.8	3878.8	338.3	359.4	2061.1	34
36	3787.7	322.1	341.2	2006.6	3881.9	338.8	360.1	2063.0	36
38	3790.8	322.6	341.8	2008.5	3885.0	339.4	360.7	2064.8	38
40	3794.0	323.2	342.4	2010.4	3888.2	340.0	361.3	2066.7	40
42	3797.1	323.7	343.1	2012.3	3891.3	340.5	362.0	2068.6	42
44	3800.3	324.3	343.7	2014.1	3894.4	341.1	362.6	2070.5	44
46	3803.4	324.8	344.3	2016.0	3897.6	341.7	363.3	2072.4	46
48	3806.6	325.4	344.9	2017.9	3900.7	342.2	363.9	2074.2	48
50	3809.7	325.9	345.6	2019.7	3903.8	342.8	364.5	2076.1	50
52	3812.9	326.5	346.2	2021.6	3907.0	343.4	365.2	2078.0	52
54	3816.0	327.0	346.8	2023.5	3910.1	343.9	365.8	2079.9	54
56	3819.2	327.6	347.4	2025.4	3913.2	344.5	366.5	2081.8	56
68	3822.3	328.1	348.1	2027.2	3916.4	345.1	367.1	2083.7	58
60	3825.5	328.7	348.7	2029.1	3919.5	345.6	367.7	2085.5	60

′	40°				41°				′
	L. C.	M.	E.	T.	L. C.	M.	E.	T.	
0	3919.5	345.6	367.7	2085.5	4013.4	362.9	387.4	2142.3	0
2	3922.6	346.1	368.4	2087.4	4016.5	363.4	388.1	2144.2	2
4	3925.8	346.7	369.0	2089.3	4019.6	364.0	388.8	2146.1	4
6	3928.9	347.2	369.7	2091.2	4022.7	364.5	389.4	2148.0	6
8	3932.0	347.8	370.3	2093.1	4025.9	365.1	390.1	2149.9	8
10	3935.1	348.4	371.0	2095.0	4029.0	365.6	390.7	2151.9	10
12	3938.3	348.9	371.6	2096.9	4032.1	366.2	391.4	2153.8	12
14	3941.4	349.5	372.3	2098.8	4035.2	366.8	392.1	2155.7	14
16	3944.5	350.1	372.9	2100.7	4038.3	367.4	392.7	2157.6	16
18	3947.7	350.7	373.6	2102.6	4041.4	368.0	393.4	2159.5	18
20	3950.8	351.3	374.3	2104.5	4044.6	368.6	394.1	2161.4	20
22	3953.9	351.8	374.9	2106.3	4047.7	369.2	394.7	2163.3	22
24	3957.1	352.4	375.6	2108.2	4050.8	369.8	395.4	2165.2	24
26	3960.2	353.0	376.2	2110.1	4053.9	370.4	396.1	2167.1	26
28	3963.3	353.6	376.9	2112.0	4057.0	371.0	396.8	2169.0	28
30	3966.4	354.2	377.5	2113.9	4060.1	371.6	397.5	2170.9	30
32	3969.6	354.7	378.2	2115.8	4063.3	372.2	398.1	2172.8	32
34	3972.7	355.3	378.8	2117.7	4066.4	372.8	398.8	2174.7	34
36	3975.8	355.9	379.5	2119.6	4069.5	373.4	399.5	2176.6	36
38	3979.0	356.5	380.1	2121.5	4072.6	374.0	400.2	2178.5	38
40	3982.1	357.1	380.8	2123.4	4075.7	374.6	400.9	2180.4	40
42	3985.2	357.6	381.4	2125.3	4078.8	375.2	401.5	2182.4	42
44	3988.4	358.2	382.1	2127.2	4082.0	375.8	402.2	2184.3	44
46	3991.5	358.8	382.8	2129.1	4085.1	376.4	402.9	2186.2	46
48	3994.6	359.4	383.4	2131.0	4088.2	377.0	403.6	2188.1	48
50	3997.7	360.0	384.1	2132.9	4091.3	377.6	404.3	2190.0	50
52	4000.9	360.5	384.8	2134.7	4094.4	378.2	404.9	2191.9	52
54	4004.0	361.1	385.4	2136.6	4097.5	378.8	405.6	2193.8	54
56	4007.1	361.7	386.1	2138.5	4100.7	379.4	406.3	2195.7	56
58	4010.3	362.3	386.8	2140.4	4103.8	380.0	407.0	2197.6	58
60	4013.4	362.9	387.4	2142.3	4106.9	380.6	407.7	2199.5	60

′	42°				43°				′
	L. C.	M.	E.	T.	L. C.	M.	E.	T.	
0	4106.9	380.6	407.7	2199.5	4200.1	398.7	428.6	2257.1	0
2	4110.0	381.2	408.3	2201.4	4203.2	399.3	429.3	2259.0	2
4	4113.1	381.8	409.0	2203.3	4206.3	399.9	430.0	2261.0	4
6	4116.2	382.4	409.7	2205.3	4209.4	400.5	430.7	2262.9	6
8	4119.3	383.0	410.4	2207.2	4212.5	401.1	431.4	2264.8	8
10	4122.4	383.6	411.1	2209.1	4215.6	401.7	432.1	2266.7	10
12	4125.5	384.2	411.8	2211.0	4218.7	402.4	432.8	2268.7	12
14	4128.6	384.8	412.5	2212.9	4221.8	403.0	433.5	2270.6	14
16	4131.8	385.4	413.2	2214.9	4224.9	403.6	434.2	2272.5	16
18	4134.9	386.0	413.9	2216.8	4228.0	404.2	434.9	2274.5	18
20	4138.0	386.6	414.6	2218.7	4231.1	404.8	435.6	2276.4	20
22	4141.1	387.2	415.3	2220.6	4234.2	405.4	436.3	2278.3	22
24	4144.2	387.8	416.0	2222.5	4237.3	406.1	437.0	2280.2	24
26	4147.3	388.4	416.6	2224.4	4240.4	406.7	437.8	2282.2	26
28	4150.4	389.0	417.3	2226.4	4243.5	407.3	438.5	2284.1	28
30	4153.5	389.6	418.0	2228.3	4246.5	407.9	439.2	2286.0	30
32	4156.6	390.2	418.7	2230.2	4249.6	408.5	439.9	2288.0	32
34	4159.7	390.8	419.4	2232.1	4252.7	409.1	440.6	2289.9	34
36	4162.8	391.4	420.1	2234.0	4255.8	409.8	441.4	2291.8	36
38	4165.9	392.0	420.8	2236.0	4258.9	410.4	442.1	2293.8	38
40	4169.0	392.6	421.5	2237.9	4262.0	411.0	442.8	2295.7	40
42	4172.1	393.2	422.2	2239.8	4265.1	411.6	443.5	2297.7	42
44	4175.2	393.8	422.9	2241.7	4268.2	412.2	444.2	2299.6	44
46	4178.4	394.4	423.6	2243.6	4271.3	412.8	445.0	2301.5	46
48	4181.5	395.0	424.3	2245.6	4274.4	413.5	445.7	2303.5	48
50	4184.6	395.6	425.0	2247.5	4277.5	414.1	446.4	2305.4	50
52	4187.7	396.2	425.7	2249.4	4280.6	414.7	447.1	2307.3	52
54	4190.8	396.8	426.4	2251.3	4283.7	415.3	447.8	2309.3	54
56	4193.9	397.4	427.1	2253.3	4286.8	415.9	448.6	2311.2	56
58	4197.0	398.0	427.8	2255.2	4289.9	416.5	449.3	2313.1	58
60	4200.1	398.7	428.6	2257.1	4293.0	417.2	450.0	2315.1	60

′	44°				45°				′
	L. C.	M.	E.	T.	L. C.	M.	E.	T.	
0	4293.0	417.2	450.0	2315.1	4385.5	436.2	472.1	2373.4	0
2	4296.1	417.8	450.7	2317.0	4388.6	436.8	472.9	2375.4	2
4	4299.2	418.4	451.5	2319.0	4391.7	437.5	473.6	2377.3	4
6	4302.2	419.1	452.2	2320.9	4394.7	438.1	474.4	2379.3	6
8	4305.3	419.7	452.9	2322.8	4397.8	438.8	475.1	2381.2	8
10	4308.4	420.3	453.7	2324.8	4400.9	439.4	475.9	2383.2	10
12	4311.5	421.0	454.4	2326.7	4404.0	440.0	476.6	2385.2	12
14	4314.6	421.6	455.1	2328.7	4407.0	440.7	477.4	2387.1	14
16	4317.7	422.2	455.9	2330.6	4410.1	441.3	478.1	2389.1	16
18	4320.7	422.9	456.6	2332.6	4413.2	442.0	478.9	2391.0	18
20	4323.8	423.5	457.3	2334.5	4416.3	442.6	479.6	2393.0	20
22	4326.9	424.1	458.1	2336.4	4419.3	443.2	480.4	2394.9	22
24	4330.0	424.8	458.8	2338.4	4422.4	443.9	481.1	2396.9	24
26	4333.1	425.4	459.5	2340.3	4425.5	444.5	481.9	2398.8	26
28	4336.2	426.0	460.3	2342.3	4428.6	445.2	482.6	2400.8	28
30	4339.2	426.7	461.0	2344.2	4431.6	445.8	483.4	2402.8	30
32	4342.3	427.3	461.7	2346.1	4434.7	446.4	484.2	2404.7	32
34	4345.4	427.9	462.5	2348.1	4437.8	447.1	484.9	2406.7	34
36	4348.5	428.6	463.2	2350.0	4440.9	447.7	485.7	2408.6	36
38	4351.6	429.2	463.9	2352.0	4444.0	448.3	486.5	2410.6	38
40	4354.7	429.8	464.7	2353.9	4447.0	448.9	487.2	2412.6	40
42	4357.7	430.5	465.4	2355.9	4450.1	449.5	488.0	2414.5	42
44	4360.8	431.1	466.2	2357.8	4453.2	450.2	488.7	2416.5	44
46	4363.9	431.7	466.9	2359.8	4456.3	450.8	489.5	2418.5	46
48	4367.0	432.4	467.7	2361.7	4459.3	451.5	490.3	2420.4	48
50	4370.1	433.0	468.4	2363.7	4462.4	452.1	491.0	2422.4	50
52	4373.2	433.6	469.1	2365.6	4465.5	452.7	491.8	2424.4	52
54	4376.2	434.3	469.9	2367.6	4468.6	453.4	492.5	2426.3	54
56	4379.3	434.9	470.6	2369.5	4471.6	454.1	493.3	2428.3	56
58	4382.4	435.6	471.4	2371.5	4474.7	454.8	494.1	2430.2	58
60	4385.5	436.2	472.1	2373.4	4477.8	455.5	494.8	2432.2	60

′	46° L. C.	46° M.	46° E.	46° T.	47° L. C.	47° M.	47° E.	47° T.	′
0	4477.8	455.5	494.8	2432.2	4569.7	475.2	518.3	2491.5	0
2	4480.9	456.1	495.6	2434.2	4572.7	475.9	519.0	2493.4	2
4	4483.9	456.8	496.5	2436.1	4575.8	476.5	519.8	2495.4	4
6	4487.0	457.4	497.2	2438.1	4578.8	477.2	520.6	2497.4	6
8	4490.0	458.1	497.9	2440.1	4581.9	477.8	521.4	2499.4	8
10	4493.1	458.7	498.7	2442.1	4584.9	478.5	522.2	2501.4	10
12	4496.2	459.4	499.5	2444.0	4588.0	479.2	523.0	2503.4	12
14	4499.2	460.0	500.3	2446.0	4591.0	479.8	523.8	2505.4	14
16	4502.3	460.7	501.0	2448.0	4594.1	480.5	524.6	2507.3	16
18	4505.4	461.3	501.8	2449.9	4597.1	481.1	525.4	2509.3	18
20	4508.4	462.0	502.6	2451.9	4600.2	481.7	526.2	2511.3	20
22	4511.5	462.7	503.4	2453.9	4603.2	482.3	527.0	2513.3	22
24	4514.6	463.3	504.1	2455.9	4606.3	483.0	527.8	2515.3	24
26	4517.6	464.0	504.9	2457.8	4609.3	483.7	528.6	2517.3	26
28	4520.7	464.6	505.7	2459.8	4612.4	484.3	529.4	2519.3	28
30	4523.7	465.3	506.5	2461.8	4615.4	485.0	530.2	2521.2	30
32	4526.8	466.0	507.3	2463.8	4618.5	485.7	531.0	2523.2	32
34	4529.9	466.6	508.0	2465.7	4621.5	486.3	531.8	2525.2	34
36	4532.9	467.3	508.8	2467.7	4624.6	487.0	532.6	2527.2	36
38	4536.0	467.9	509.6	2469.7	4627.6	487.7	533.4	2529.2	38
40	4539.1	468.6	510.4	2471.7	4630.7	488.4	534.2	2531.2	40
42	4542.1	469.3	511.1	2473.6	4633.7	489.1	535.0	2533.2	42
44	4545.2	469.9	511.9	2475.6	4636.8	489.8	535.8	2535.2	44
46	4548.2	470.6	512.7	2477.6	4639.8	490.5	536.6	2537.2	46
48	4551.3	471.2	513.5	2479.6	4642.9	491.2	537.4	2539.2	48
50	4554.4	471.9	514.3	2481.6	4645.9	491.9	538.2	2541.2	50
52	4557.4	472.6	515.1	2483.5	4649.0	492.6	539.0	2543.1	52
54	4560.5	473.2	515.9	2485.5	4652.0	493.3	539.8	2545.1	54
56	4563.6	473.9	516.7	2487.5	4655.1	494.0	540.6	2547.1	56
58	4566.6	474.5	517.5	2489.5	4658.1	494.7	541.4	2549.1	58
60	4569.7	475.2	518.3	2491.5	4661.2	495.4	542.3	2551.1	60

′	48° L. C.	48° M.	48° E.	48° T.	49° L. C.	49° M.	49° E.	49° T.	′
0	4661.2	495.4	542.3	2551.1	4752.3	515.9	567.0	2611.3	0
2	4664.2	496.0	543.1	2553.1	4755.3	516.5	567.8	2613.3	2
4	4667.3	496.7	543.9	2555.1	4758.4	517.2	568.7	2615.3	4
6	4670.3	497.4	544.7	2557.1	4761.4	517.9	569.5	2617.3	6
8	4673.3	498.1	545.5	2559.1	4764.4	518.6	570.3	2619.3	8
10	4676.4	498.8	546.4	2561.1	4767.4	519.3	571.2	2621.4	10
12	4679.4	499.4	547.2	2563.1	4770.5	520.0	572.0	2623.4	12
14	4682.5	500.1	548.0	2565.1	4773.5	520.7	572.8	2625.4	14
16	4685.5	500.8	548.8	2567.1	4776.5	521.4	573.7	2627.4	16
18	4688.5	501.5	549.6	2569.1	4779.6	522.1	574.5	2629.4	18
20	4691.6	502.2	550.5	2571.1	4782.6	522.8	575.3	2631.4	20
22	4694.6	502.8	551.3	2573.1	4785.6	523.5	576.2	2633.5	22
24	4697.6	503.5	552.1	2575.1	4788.7	524.2	577.0	2635.5	24
26	4700.7	504.2	552.9	2577.1	4791.7	524.9	577.9	2637.5	26
28	4703.7	504.9	553.7	2579.1	4794.7	525.6	578.7	2639.5	28
30	4706.7	505.6	554.6	2581.1	4797.7	526.3	579.6	2641.5	30
32	4709.8	506.2	555.4	2583.1	4800.8	527.0	580.4	2643.5	32
34	4712.8	506.9	556.2	2585.1	4803.8	527.7	581.3	2645.6	34
36	4715.9	507.6	557.0	2587.2	4806.8	528.4	582.1	2647.6	36
38	4718.9	508.3	557.8	2589.2	4809.9	529.1	583.0	2649.6	38
40	4721.9	509.0	558.7	2591.2	4812.9	529.8	583.8	2651.6	40
42	4725.0	509.6	559.5	2593.2	4815.9	530.5	584.7	2653.7	42
44	4728.0	510.3	560.3	2595.2	4819.0	531.2	585.5	2655.7	44
46	4731.0	511.0	561.2	2597.2	4822.0	531.9	586.4	2657.7	46
48	4734.1	511.7	562.0	2599.2	4825.0	532.6	587.2	2659.7	48
50	4737.1	512.4	562.8	2601.2	4828.0	533.3	588.1	2661.8	50
52	4740.2	513.1	563.7	2603.2	4831.1	534.0	588.9	2663.8	52
54	4743.2	513.8	564.5	2605.2	4834.1	534.7	589.8	2665.8	54
56	4746.2	514.5	565.3	2607.2	4837.1	535.4	590.6	2667.8	56
58	4749.3	515.2	566.2	2609.3	4840.2	536.1	591.5	2669.9	58
60	4752.3	515.9	567.0	2611.3	4843.2	536.8	592.4	2671.9	60

′	50°				51°				′
	L. C.	M.	E.	T.	L. C.	M.	E.	T.	
0	4843.2	536.8	592.4	2671.9	4933.6	558.2	618.5	2733.0	0
2	4846.2	537.5	593.2	2673.9	4936.6	558.9	619.3	2735.1	2
4	4849.2	538.2	594.1	2676.0	4939.6	559.7	620.2	2737.1	4
6	4852.2	538.9	594.9	2678.0	4942.6	560.4	621.1	2739.2	6
8	4855.2	539.6	595.8	2680.0	4945.6	561.1	622.0	2741.2	8
10	4858.3	540.3	596.7	2682.1	4948.6	561.8	622.9	2743.3	10
12	4861.3	541.0	597.5	2684.1	4951.6	562.5	623.7	2745.3	12
14	4864.3	541.7	598.4	2686.1	4954.6	563.3	624.6	2747.4	14
16	4867.3	542.4	599.3	2688.2	4957.6	564.0	625.5	2749.4	16
18	4870.3	543.1	600.1	2690.2	4960.6	564.7	626.4	2751.5	18
20	4873.3	543.9	601.0	2692.3	4963.6	565.4	627.3	2753.5	20
22	4876.3	544.6	601.9	2694.3	4966.6	566.2	628.2	2755.6	22
24	4879.4	545.3	602.7	2696.3	4969.6	566.9	629.9	2757.7	24
26	4882.4	546.0	603.6	2698.4	4972.6	567.6	630.0	2759.7	26
28	4885.4	546.7	604.5	2700.4	4975.6	568.3	630.9	2761.8	28
30	4888.4	547.4	605.3	2702.4	4978.6	569.1	631.8	2763.8	30
32	4891.4	548.1	606.2	2704.5	4981.6	569.8	632.7	2765.9	32
34	4894.4	548.8	607.0	2706.5	4984.6	570.5	633.6	2767.9	34
36	4897.4	549.5	607.9	2708.6	4987.7	571.2	634.5	2770.0	36
38	4900.4	550.2	608.8	2710.6	4990.7	572.0	635.3	2772.0	38
40	4903.5	551.0	609.7	2712.6	4993.7	572.7	636.2	2774.1	40
42	4906.5	551.7	610.5	2714.7	4996.7	573.4	637.1	2776.2	42
44	4909.5	552.4	611.4	2716.7	4999.7	574.1	638.0	2778.2	44
46	4912.5	553.1	612.3	2718.8	5002.7	574.9	638.9	2780.3	46
48	4915.5	553.8	613.2	2720.8	5005.7	575.6	639.8	2782.3	48
50	4918.5	554.5	614.1	2722.8	5008.7	576.3	640.7	2784.4	50
52	4921.5	555.2	614.9	2724.9	5011.7	577.0	641.6	2786.4	52
54	4924.6	555.9	615.8	2726.9	5014.7	577.8	642.5	2788.5	54
56	4927.6	556.6	616.7	2729.0	5017.7	578.5	643.4	2790.6	56
58	4930.6	557.4	617.6	2731.0	5020.7	579.2	644.3	2792.6	58
60	4933.6	558.2	618.5	2733.0	5023.7	579.9	645.2	2794.7	60

′	52°				53°				′
	L. C.	M.	E.	T.	L. C.	M.	E.	T.	
0	5023.7	579.9	645.2	2794.7	5113.5	602.0	672.7	2856.9	0
2	5026.7	580.6	646.1	2796.8	5116.5	602.8	673.7	2858.9	2
4	5029.7	581.3	647.0	2798.8	5119.4	603.5	674.6	2861.0	4
6	5032.7	582.1	647.9	2800.9	5122.4	604.3	675.5	2863.1	6
8	5035.7	582.8	648.9	2803.0	5125.4	605.0	676.4	2865.2	8
10	5038.7	583.5	649.8	2805.0	5128.4	605.8	677.4	2867.3	10
12	5041.7	584.3	650.7	2807.1	5131.3	606.5	678.3	2869.4	12
14	5044.7	585.0	651.6	2809.2	5134.3	607.3	679.2	2871.5	14
16	5047.7	585.7	652.5	2811.2	5137.3	608.0	680.2	2873.5	16
18	5050.7	586.5	653.4	2813.3	5140.3	608.8	681.1	2875.6	18
20	5053.6	587.2	654.3	2815.4	5143.2	609.5	682.0	2877.7	20
22	5056.6	587.9	655.2	2817.4	5146.2	610.3	683.0	2879.8	22
24	5059.6	588.7	656.2	2819.5	5149.2	611.0	683.9	2881.9	24
26	5062.6	589.4	657.1	2821.6	5152.1	611.8	684.9	2884.0	26
28	5065.6	590.1	658.0	2823.6	5155.1	612.5	685.8	2886.1	28
30	5068.6	590.9	658.9	2825.7	5158.1	613.3	686.7	2888.1	30
32	5071.6	591.6	659.8	2827.8	5161.1	614.0	687.7	2890.2	32
34	5074.6	592.3	660.7	2829.8	5164.0	614.8	688.6	2892.3	34
36	5077.6	593.1	661.6	2831.9	5167.0	615.5	689.6	2894.4	36
38	5080.6	593.8	662.5	2834.0	5170.0	616.3	690.5	2896.5	38
40	5083.6	594.5	663.5	2836.1	5173.0	617.0	691.5	2898.6	40
42	5086.6	595.3	664.4	2838.2	5175.9	617.8	692.4	2900.7	42
44	5089.6	596.0	665.3	2840.2	5178.9	618.5	693.4	2902.8	44
46	5092.6	596.7	666.2	2842.3	5181.9	619.3	694.3	2904.9	46
48	5095.6	597.5	667.2	2844.4	5184.9	620.1	695.3	2907.0	48
50	5098.6	598.2	668.1	2846.5	5187.8	620.8	696.2	2909.1	50
52	5101.6	598.9	669.0	2848.5	5190.8	621.5	697.1	2911.2	52
54	5104.6	599.7	669.9	2850.6	5193.8	622.3	698.1	2913.3	54
56	5107.6	600.4	670.9	2852.7	5196.7	623.0	699.0	2915.4	56
58	5110.6	601.2	671.8	2854.8	5199.7	623.8	700.0	2917.5	58
60	5113.5	602.0	672.7	2856.9	5202.7	624.6	700.9	2919.5	60

′	54°				55°				′
	L. C.	M.	E.	T.	L. C.	M.	E.	T.	
0	5202.7	621.6	700.9	2919.5	5291.7	647.4	729.9	2982.8	0
2	5205.7	625.4	701.9	2921.6	5294.6	648.1	730.9	2984.9	2
4	5208.6	626.1	702.8	2923.8	5297.6	648.9	731.9	2987.1	4
6	5211.6	626.9	703.8	2925.9	5300.5	649.6	732.9	2989.2	6
8	5214.6	627.6	704.8	2928.0	5303.5	650.4	733.8	2991.3	8
10	5217.5	628.4	705.7	2930.1	5306.4	651.2	734.8	2993.4	10
12	5220.5	629.2	706.7	2932.2	5309.4	652.0	735.8	2995.5	12
14	5223.5	620.9	707.7	2934.3	5312.3	652.7	736.8	2997.7	14
16	5226.4	630.7	708.6	2936.4	5315.3	653.5	737.8	2999.8	16
18	5229.4	631.4	709.6	2938.5	5318.2	654.3	738.7	3001.9	18
20	5232.4	632.2	710.5	2940.6	5321.2	655.1	739.7	3004.0	20
22	5235.3	633.0	711.5	2942.7	5324.1	655.8	740.7	3006.2	22
24	5238.3	633.7	712.5	2944.8	5327.1	656.6	741.7	3008.3	24
26	5241.3	634.5	713.4	2946.9	5330.0	657.4	742.7	3010.4	26
28	5244.2	635.2	714.4	2949.0	5333.0	658.2	743.7	3012.5	28
30	5247.2	636.0	715.3	2951.1	5335.9	658.9	744.7	3014.7	30
32	5250.2	636.8	716.3	2953.2	5338.8	659.7	745.7	3016.8	32
34	5253.1	637.5	717.3	2955.3	5341.8	660.5	746.7	3018.9	34
36	5256.1	638.3	718.2	2957.5	5344.7	661.3	747.7	3021.1	36
38	5259.1	639.0	719.2	2959.6	5347.7	662.0	748.7	3023.2	38
40	5262.0	639.8	720.2	2961.7	5350.6	662.8	749.7	3025.3	40
42	5265.0	640.6	721.1	2963.8	5353.6	663.6	750.7	3027.5	42
44	5268.0	641.3	722.1	2965.9	5356.5	664.4	751.7	3029.6	44
46	5270.9	642.1	723.1	2968.0	5359.5	665.1	752.6	3031.7	46
48	5273.9	642.8	724.1	2970.1	5362.4	665.9	753.6	3033.8	48
50	5276.9	643.6	725.0	2972.2	5365.4	666.7	754.6	3036.0	50
52	5279.8	644.4	726.0	2974.4	5368.3	667.5	755.6	3038.1	52
54	5282.8	645.1	727.0	2976.5	5371.3	668.3	756.6	3040.2	54
56	5285.8	645.9	728.0	2978.6	5374.2	669.1	757.6	3042.4	56
58	5288.7	646.6	729.0	2980.7	5377.2	669.9	758.6	3044.5	58
60	5291.7	647.4	729.9	2982.8	5380.1	670.7	759.6	3046.6	60

′	56°				57°				′
	L. C.	M.	E.	T.	L. C.	M.	E.	T.	
0	5380.1	670.7	759.6	3046.6	5468.2	694.4	790.2	3111.1	0
2	5383.0	671.4	760.6	3048.8	5471.1	695.2	791.2	3113.3	2
4	5386.0	672.2	761.6	3050.9	5474.0	696.0	792.2	3115.4	4
6	5388.9	672.9	762.7	3053.1	5477.0	696.8	793.3	3117.6	6
8	5391.8	673.7	763.7	3055.2	5479.9	697.6	794.3	3119.7	8
10	5394.8	674.4	764.7	3057.4	5482.8	698.4	795.3	3121.9	10
12	5397.7	675.2	765.7	3059.5	5485.7	699.2	796.3	3124.1	12
14	5400.7	676.0	766.7	3061.6	5488.7	700.0	797.4	3126.2	14
16	5403.6	676.8	767.7	3063.8	5491.6	700.8	798.4	3128.4	16
18	5406.5	677.6	768.7	3065.9	5494.5	701.6	799.4	3130.6	18
20	5409.5	678.4	769.7	3068.1	5497.4	702.4	800.5	3132.7	20
22	5412.4	679.2	770.8	3070.2	5500.3	703.2	801.5	3134.9	22
24	5415.3	680.0	771.8	3072.4	5503.3	704.0	802.6	3137.0	24
26	5418.3	680.8	772.8	3074.5	5506.2	704.8	803.6	3139.2	26
28	5421.2	681.6	773.8	3076.6	5509.1	705.6	804.7	3141.4	28
30	5424.1	682.4	774.8	3078.8	5512.0	706.4	805.7	3143.5	30
32	5427.1	683.2	775.8	3080.9	5515.0	707.2	806.8	3145.7	32
34	5430.0	684.0	776.8	3083.1	5517.9	708.0	807.8	3147.9	34
36	5433.0	684.8	777.8	3085.2	5520.8	708.8	808.8	3150.0	36
38	5435.9	685.6	778.9	3087.4	5523.7	709.6	809.9	3152.2	38
40	5438.8	686.4	779.9	3089.6	5526.7	710.4	810.9	3154.4	40
42	5441.8	687.2	780.9	3091.7	5529.6	711.2	812.0	3156.6	42
44	5444.7	688.0	781.9	3093.9	5532.5	712.0	813.0	3158.7	44
46	5447.6	688.8	783.0	3096.0	5535.4	712.8	814.1	3160.9	46
48	5450.6	689.6	784.0	3098.2	5538.4	713.6	815.1	3163.1	48
50	5453.5	690.4	785.0	3100.3	5541.3	714.4	816.2	3165.3	50
52	5456.5	691.2	786.0	3102.5	5544.2	715.2	817.2	3167.4	52
54	5459.4	692.0	787.1	3104.6	5547.1	716.0	818.3	3169.6	54
56	5462.3	692.8	788.1	3106.8	5550.0	716.8	819.3	3171.8	56
58	5465.3	693.6	789.1	3108.9	5553.0	717.6	820.4	3174.0	58
60	5468.2	694.1	790.2	3111.1	5555.9	718.4	821.4	3176.1	60

′	58°				59°				′
	L. C.	M.	E.	T.	L. C.	M.	E.	T.	
0	5555.9	718.4	821.4	3176.1	5643.1	742.8	853.5	3241.9	0
2	5558.8	719.2	822.5	3178.3	5646.0	743.6	854.6	3244.1	2
4	5561.7	720.0	823.5	3180.5	5648.9	744.4	855.7	3246.3	4
6	5564.6	720.8	824.6	3182.7	5651.8	745.3	856.8	3248.5	6
8	5567.5	721.6	825.7	3184.9	5654.7	746.1	857.9	3250.7	8
10	5570.4	722.4	826.7	3187.1	5657.6	746.9	859.0	3252.9	10
12	5573.3	723.2	827.8	3189.2	5660.5	747.7	860.0	3255.1	12
14	5576.2	724.0	828.9	3191.4	5663.4	748.6	861.1	3257.3	14
16	5579.2	724.8	829.9	3193.6	5666.3	749.4	862.2	3259.5	16
18	5582.1	725.6	831.0	3195.8	5669.2	750.2	863.3	3261.7	18
20	5585.0	726.5	832.1	3198.0	5672.1	751.1	864.4	3263.9	20
22	5587.9	727.3	833.1	3200.2	5675.0	751.9	865.5	3266.1	22
24	5590.8	728.1	834.2	3202.4	5677.9	752.7	866.6	3268.3	24
26	5593.7	728.9	835.3	3204.5	5680.8	753.5	867.7	3270.5	26
28	5596.6	729.7	836.3	3206.7	5683.7	754.4	868.8	3272.7	28
30	5599.5	730.5	837.4	3208.9	5686.5	755.2	869.9	3274.9	30
32	5602.4	731.3	838.4	3211.1	5689.4	756.0	871.0	3277.1	32
34	5605.3	732.1	839.5	3213.3	5692.3	756.9	872.1	3279.4	34
36	5608.2	732.9	840.6	3215.5	5695.2	757.7	873.2	3281.6	36
38	5611.1	733.7	841.6	3217.7	5698.1	758.5	874.3	3283.8	38
40	5614.0	734.6	842.7	3219.9	5701.0	759.4	875.4	3286.0	40
42	5616.9	735.4	843.8	3222.1	5703.9	760.2	876.5	3288.2	42
44	5619.8	736.2	844.9	3224.3	5706.8	761.0	877.6	3290.5	44
46	5622.8	737.0	846.0	3226.5	5709.7	761.9	878.7	3292.7	46
48	5625.7	737.8	847.0	3228.7	5712.6	762.7	879.8	3294.9	48
50	5628.6	738.6	848.1	3230.9	5715.5	763.5	880.9	3297.1	50
52	5631.5	739.4	849.2	3233.1	5718.4	764.4	882.0	3299.3	52
54	5634.4	740.2	850.3	3235.3	5721.3	765.2	883.1	3301.5	54
56	5637.3	741.0	851.4	3237.5	5724.2	766.0	884.2	3303.8	56
58	5640.2	741.9	852.5	3239.7	5727.1	766.8	885.3	3306.0	58
60	5643.1	742.8	853.5	3241.9	5730.0	767.7	886.4	3308.2	60

′	60°				61°				′
	L. C.	M.	E.	T.	L. C.	M.	E.	T.	
0	5730.0	767.7	886.4	3308.2	5816.4	792.9	920.2	3375.2	0
2	5732.9	768.5	887.5	3310.4	5819.3	793.7	921.4	3377.4	2
4	5735.8	769.4	888.7	3312.7	5822.1	794.6	922.5	3379.7	4
6	5738.6	770.2	889.8	3314.9	5825.0	795.4	923.6	3381.9	6
8	5741.5	771.1	890.9	3317.1	5827.9	796.3	924.8	3384.2	8
10	5744.4	771.9	892.0	3319.3	5830.7	797.1	925.9	3386.4	10
12	5747.3	772.7	893.1	3321.6	5833.6	798.0	927.1	3388.7	12
14	5750.2	773.6	894.3	3323.8	5836.5	798.8	928.2	3390.9	14
16	5753.0	774.4	895.4	3326.0	5839.3	799.7	929.3	3393.2	16
18	5755.9	775.3	896.5	3328.3	5842.2	800.5	930.5	3395.4	18
20	5758.8	776.1	897.6	3330.5	5845.1	801.4	931.6	3397.7	20
22	5761.7	776.9	898.8	3332.7	5847.9	802.2	932.8	3399.9	22
24	5764.6	777.8	899.9	3334.9	5850.8	803.1	933.9	3402.2	24
26	5767.4	778.6	901.0	3337.2	5853.7	803.9	935.1	3404.4	26
28	5770.3	779.5	902.1	3339.4	5856.5	804.8	936.3	3406.7	28
30	5773.2	780.3	903.2	3341.6	5859.4	805.6	937.4	3408.9	30
32	5776.1	781.1	904.4	3343.9	5862.3	806.5	938.6	3411.2	32
34	5779.0	782.0	905.5	3346.1	5865.1	807.3	939.7	3413.5	34
36	5781.8	782.8	906.6	3348.3	5868.0	808.2	940.9	3415.7	36
38	5784.7	783.7	907.7	3350.6	5870.9	809.0	942.1	3418.0	38
40	5787.6	784.5	908.8	3352.8	5873.7	809.9	943.2	3420.3	40
42	5790.5	785.3	910.0	3355.0	5876.6	810.7	944.4	3422.5	42
44	5793.4	786.2	911.1	3357.3	5879.5	811.6	945.5	3424.8	44
46	5796.2	787.0	912.3	3359.5	5882.3	812.4	946.7	3427.1	46
48	5799.1	787.9	913.4	3361.8	5885.2	813.3	947.8	3429.3	48
50	5802.0	788.7	914.5	3364.0	5888.1	814.1	949.0	3431.6	50
52	5804.9	789.5	915.7	3366.2	5890.9	815.0	950.2	3433.9	52
54	5807.8	790.4	916.8	3368.5	5893.8	815.8	951.3	3436.1	54
56	5810.6	791.2	918.0	3370.7	5896.7	816.7	952.5	3438.4	56
58	5813.5	792.1	919.1	3373.0	5899.5	817.5	953.6	3440.7	58
60	5816.4	792.9	920.2	3375.2	5902.4	818.4	954.8	3442.9	60

′	62°				63°				′
	L. C.	M.	E.	T.	L. C.	M.	E.	T.	
0	5902.4	818.4	954.8	3442.9	5987.8	844.4	990.3	3511.3	0
2	5905.2	819.3	956.0	3445.2	5990.6	845.3	991.5	3513.6	2
4	5908.1	820.1	957.2	3447.5	5993.5	846.2	992.7	3515.9	4
6	5910.9	821.0	958.3	3449.7	5996.3	847.1	993.9	3518.2	6
8	5913.8	821.8	959.5	3452.0	5999.1	847.9	995.1	3520.5	8
10	5916.6	822.7	960.7	3454.3	6002.0	848.8	996.3	3522.8	10
12	5919.5	823.6	961.9	3456.6	6004.8	849.7	997.5	3525.1	12
14	5922.3	824.4	963.0	3458.8	6007.7	850.6	998.7	3527.4	14
16	5925.2	825.3	964.2	3461.1	6010.5	851.4	999.9	3529.7	16
18	5928.0	826.1	965.4	3463.4	6013.3	852.3	1001.1	3532.0	18
20	5930.9	827.0	966.6	3465.7	6016.2	853.2	1002.3	3534.3	20
22	5933.7	827.9	967.8	3467.9	6019.0	854.1	1003.5	3536.6	22
24	5936.6	828.7	968.9	3470.2	6021.8	854.9	1004.7	3538.9	24
26	5939.4	829.6	970.1	3472.5	6024.7	855.8	1005.9	3541.2	26
28	5942.3	830.4	971.3	3474.7	6027.5	856.7	1007.1	3543.5	28
30	5945.1	831.3	972.5	3477.0	6030.3	857.6	1008.4	3545.8	30
32	5947.9	832.2	973.6	3479.3	6033.2	858.4	1009.6	3548.1	32
34	5950.8	833.0	974.8	3481.6	6036.0	859.3	1010.8	3550.4	34
36	5953.6	833.9	976.0	3483.9	6038.9	860.2	1012.0	3552.7	36
38	5956.5	834.7	977.2	3486.2	6041.7	861.1	1013.2	3555.0	38
40	5959.3	835.6	978.4	3488.5	6044.5	861.9	1014.5	3557.3	40
42	5962.2	836.5	979.6	3490.7	6047.4	862.8	1015.7	3559.6	42
44	5965.0	837.4	980.8	3493.0	6050.2	863.7	1016.9	3562.0	44
46	5967.9	838.3	982.0	3495.3	6053.0	864.6	1018.1	3564.3	46
48	5970.7	839.1	983.2	3497.6	6055.9	865.4	1019.3	3566.6	48
50	5973.6	840.0	984.4	3499.9	6058.7	866.3	1020.6	3568.9	50
52	5976.4	840.9	985.5	3502.2	6061.6	867.2	1021.8	3571.2	52
54	5979.3	841.7	986.7	3504.5	6064.4	868.1	1023.0	3573.5	54
56	5982.1	842.6	987.9	3506.8	6067.2	868.9	1024.2	3575.8	56
58	5985.0	843.5	989.1	3509.0	6070.1	869.8	1025.4	3578.1	58
60	5987.8	844.4	990.3	3511.3	6072.9	870.7	1026.7	3580.4	60

′	64°				65°				′
	L. C.	M.	E.	T.	L. C.	M.	E.	T.	
0	6072.9	870.7	1026.7	3580.4	6157.5	897.3	1064.0	3650.4	0
2	6075.7	871.5	1027.9	3582.8	6160.3	898.2	1065.2	3652.8	2
4	6078.5	872.4	1029.2	3585.1	6163.1	899.1	1066.5	3655.1	4
6	6081.4	873.3	1030.4	3587.4	6165.9	900.0	1067.7	3657.5	6
8	6084.2	874.2	1031.7	3589.7	6168.7	900.9	1069.0	3659.8	8
10	6087.0	875.1	1032.9	3592.1	6171.5	901.8	1070.2	3662.2	10
12	6089.8	875.9	1034.1	3594.4	6174.3	902.7	1071.5	3664.5	12
14	6092.6	876.8	1035.4	3596.7	6177.1	903.6	1072.7	3666.9	14
16	6095.5	877.7	1036.6	3599.1	6179.9	904.5	1074.0	3669.2	16
18	6098.3	878.6	1037.9	3601.4	6182.7	905.4	1075.2	3671.6	18
20	6101.1	879.5	1039.1	3603.7	6185.5	906.3	1076.6	3673.9	20
22	6103.9	880.3	1040.3	3606.0	6188.3	907.2	1077.8	3676.2	22
24	6106.7	881.2	1041.6	3608.4	6191.1	908.1	1079.1	3678.6	24
26	6109.6	882.1	1042.8	3610.7	6193.9	909.0	1080.4	3680.9	26
28	6112.4	883.0	1044.1	3613.0	6196.7	909.9	1081.7	3683.3	28
30	6115.2	883.9	1045.3	3615.3	6199.5	910.8	1083.0	3685.6	30
32	6118.0	884.7	1046.5	3617.7	6202.3	911.7	1084.2	3688.0	32
34	6120.8	885.6	1047.8	3620.0	6205.1	912.6	1085.5	3690.4	34
36	6123.7	886.5	1049.0	3622.3	6208.0	913.5	1086.8	3692.7	36
38	6126.5	887.4	1050.3	3624.7	6210.8	914.4	1088.1	3695.1	38
40	6129.3	888.3	1051.5	3627.0	6213.6	915.3	1089.4	3697.4	40
42	6132.1	889.2	1052.7	3629.4	6216.4	916.2	1090.6	3699.8	42
44	6134.9	890.1	1054.0	3631.7	6219.2	917.1	1091.9	3702.2	44
46	6137.8	891.0	1055.2	3634.0	6222.0	918.0	1093.2	3704.5	46
48	6140.6	891.9	1056.5	3636.4	6224.8	918.9	1094.5	3706.9	48
50	6143.4	892.8	1057.7	3638.7	6227.6	919.8	1095.8	3709.3	50
52	6146.2	893.7	1059.0	3641.1	6230.4	920.7	1097.0	3711.6	52
54	6149.0	894.6	1060.2	3643.4	6233.2	921.6	1098.3	3714.0	54
56	6151.9	895.5	1061.5	3645.7	6236.0	922.5	1099.6	3716.3	56
58	6154.7	896.4	1062.7	3648.1	6238.8	923.4	1100.9	3718.7	58
60	6157.5	897.3	1064.0	3650.4	6241.6	924.3	1102.2	3721.1	60

′	66°				67°				′
	L. C.	M.	E.	T.	L. C.	M.	E.	T.	
0	6241.6	924.3	1102.2	3721.1	6325.2	951.8	1141.5	3792.6	0
2	6244.4	925.2	1103.5	3723.4	6328.0	952.7	1142.8	3795.0	2
4	6247.2	926.1	1104.8	3725.8	6330.7	953.6	1144.1	3797.4	4
6	6250.0	927.0	1106.1	3728.2	6333.5	954.5	1145.4	3799.8	6
8	6252.7	927.9	1107.4	3730.6	6336.3	955.5	1146.7	3802.2	8
10	6255.5	928.8	1108.7	3732.9	6339.0	956.4	1148.1	3804.6	10
12	6258.3	929.8	1110.0	3735.3	6341.8	957.3	1149.4	3807.0	12
14	6261.1	930.7	1111.3	3737.7	6344.6	958.2	1150.7	3809.4	14
16	6263.9	931.6	1112.6	3740.1	6347.4	959.2	1152.0	3811.8	16
18	6266.7	932.5	1113.9	3742.4	6350.1	960.1	1153.3	3814.2	18
20	6269.5	933.4	1115.2	3744.8	6352.9	961.0	1154.7	3816.6	20
22	6272.3	934.3	1116.5	3747.2	6355.7	961.9	1156.0	3819.0	22
24	6275.0	935.3	1117.8	3749.6	6358.4	962.9	1157.4	3821.4	24
26	6277.8	936.2	1119.1	3751.9	6361.2	963.8	1158.7	3823.8	26
28	6280.6	937.1	1120.4	3754.3	6364.0	964.7	1160.1	3826.2	28
30	6283.4	938.0	1121.7	3756.7	6366.7	965.6	1161.4	3828.6	30
32	6286.2	938.9	1123.0	3759.1	6369.5	966.6	1162.8	3831.0	32
34	6289.0	939.8	1124.3	3761.5	6372.3	967.5	1164.1	3833.4	34
36	6291.8	940.8	1125.6	3763.9	6375.1	968.4	1165.5	3835.9	36
38	6294.5	941.7	1126.9	3766.3	6377.8	969.3	1166.8	3838.3	38
40	6297.3	942.6	1128.3	3768.7	6380.6	970.3	1168.2	3840.7	40
42	6300.1	943.5	1129.6	3771.0	6383.4	971.2	1169.5	3843.1	42
44	6302.9	944.4	1130.9	3773.4	6386.1	972.1	1170.9	3845.5	44
46	6305.7	945.3	1132.2	3775.8	6388.9	973.0	1172.2	3847.9	46
48	6308.5	946.3	1133.5	3778.2	6391.7	974.0	1173.6	3850.4	48
50	6311.3	947.2	1134.9	3780.6	6394.4	974.9	1174.9	3852.8	50
52	6314.1	948.1	1136.2	3783.0	6397.2	975.8	1176.3	3855.2	52
54	6316.8	949.0	1137.5	3785.4	6400.0	976.8	1177.6	3857.6	54
56	6319.6	949.9	1138.8	3787.8	6402.8	977.7	1179.0	3860.0	56
58	6322.4	950.8	1140.1	3790.2	6405.5	978.6	1180.3	3862.5	58
60	6325.2	951.8	1141.5	3792.6	6408.3	979.6	1181.6	3864.9	60

′	68°				69°				′
	L. C.	M.	E.	T.	L. C.	M.	E.	T.	
0	6408.3	979.6	1181.6	3864.9	6491.1	1007.7	1222.9	3938.1	0
2	6411.1	980.5	1183.0	3867.3	6493.8	1008.7	1224.3	3940.6	2
4	6413.8	981.4	1184.4	3869.7	6496.6	1009.6	1225.7	3943.0	4
6	6416.6	982.4	1185.7	3872.2	6499.3	1010.6	1227.1	3945.5	6
8	6419.3	983.3	1187.1	3874.6	6502.1	1011.5	1228.5	3947.9	8
10	6422.1	984.2	1188.5	3877.0	6504.8	1012.5	1229.9	3950.4	10
12	6424.9	985.2	1189.8	3879.5	6507.5	1013.4	1231.3	3952.9	12
14	6427.6	986.1	1191.2	3881.9	6510.3	1014.4	1232.7	3955.3	14
16	6430.4	987.0	1192.6	3884.3	6513.0	1015.3	1234.1	3957.8	16
18	6433.1	988.0	1193.9	3886.8	6515.8	1016.3	1235.5	3960.2	18
20	6435.9	988.9	1195.3	3889.2	6518.5	1017.2	1236.9	3962.7	20
22	6438.7	989.8	1196.7	3891.6	6521.2	1018.2	1238.3	3965.2	22
24	6441.4	990.8	1198.0	3894.1	6524.0	1019.1	1239.7	3967.6	24
26	6444.2	991.7	1199.4	3896.5	6526.7	1020.1	1241.1	3970.1	26
28	6446.9	992.6	1200.8	3898.9	6529.5	1021.0	1242.5	3972.5	28
30	6449.7	993.6	1202.1	3901.4	6532.2	1022.0	1243.9	3975.0	30
32	6452.5	994.5	1203.5	3903.8	6534.9	1022.9	1245.3	3977.5	32
34	6455.2	995.4	1204.9	3906.3	6537.7	1023.9	1246.7	3980.0	34
36	6458.0	996.4	1206.2	3908.7	6540.4	1024.8	1248.1	3982.4	36
38	6460.7	997.3	1207.6	3911.2	6543.2	1025.8	1249.5	3984.9	38
40	6463.5	998.2	1209.0	3913.6	6545.9	1026.7	1250.9	3987.4	40
42	6466.3	999.2	1210.3	3916.1	6548.6	1027.7	1252.3	3989.9	42
44	6469.0	1000.1	1211.7	3918.5	6551.4	1028.6	1253.7	3992.3	44
46	6471.8	1001.0	1213.1	3921.0	6554.1	1029.6	1255.1	3994.8	46
48	6474.5	1002.0	1214.5	3923.4	6556.9	1030.5	1256.5	3997.3	48
50	6477.3	1002.9	1215.9	3925.9	6559.6	1031.5	1257.9	3999.8	50
52	6480.1	1003.8	1217.3	3928.3	6562.3	1032.4	1259.3	4002.2	52
54	6482.8	1004.8	1218.7	3930.8	6565.1	1033.4	1260.7	4004.7	54
56	6485.6	1005.7	1220.1	3933.2	6567.8	1034.3	1262.1	4007.2	56
58	6488.3	1006.7	1221.5	3935.7	6570.6	1035.3	1263.5	4009.7	58
60	6491.1	1007.7	1222.9	3938.1	6573.3	1036.3	1265.0	4012.1	60

′	70°				71°				′
	L. C.	M.	E.	T.	L. C.	M.	E.	T.	
0	6573.3	1036.3	1265.0	4012.1	6654.9	1065.1	1308.4	4087.1	0
2	6576.0	1037.3	1266.4	4014.6	6657.6	1066.1	1309.9	4089.7	2
4	6578.7	1038.2	1267.9	4017.1	6660.3	1067.0	1311.3	4092.2	4
6	6581.5	1039.2	1269.3	4019.6	6663.0	1068.0	1312.8	4094.7	6
8	6584.2	1040.1	1270.8	4022.1	6665.7	1068.9	1314.2	4097.2	8
10	6586.9	1041.1	1272.2	4024.6	6668.4	1069.9	1315.7	4099.8	10
12	6589.6	1042.1	1273.6	4027.1	6671.1	1070.9	1317.2	4102.3	12
14	6592.3	1043.0	1275.1	4029.6	6673.8	1071.9	1318.6	4104.8	14
16	6595.1	1044.0	1276.5	4032.1	6676.6	1072.9	1320.1	4107.3	16
18	6597.8	1044.9	1278.0	4034.6	6679.3	1073.8	1321.5	4109.8	18
20	6600.5	1045.9	1279.4	4037.1	6682.0	1074.8	1323.0	4112.4	20
22	6603.2	1046.9	1280.8	4039.6	6684.7	1075.8	1324.4	4114.9	22
24	6605.9	1047.8	1282.3	4042.1	6687.4	1076.8	1325.9	4117.4	24
26	6608.7	1048.8	1283.7	4044.6	6690.1	1077.7	1327.4	4119.9	26
28	6611.4	1049.7	1285.2	4047.1	6692.8	1078.7	1328.9	4122.4	28
30	6614.1	1050.7	1286.6	4049.6	6695.5	1079.7	1330.4	4125.0	30
32	6616.8	1051.7	1288.0	4052.1	6698.2	1080.7	1331.8	4127.5	32
34	6619.5	1052.6	1289.5	4054.6	6700.9	1081.6	1333.3	4130.4	34
36	6622.3	1053.6	1290.9	4057.1	6703.6	1082.6	1334.8	4132.6	36
38	6625.0	1054.5	1292.4	4059.6	6706.3	1083.6	1336.3	4135.1	38
40	6627.7	1055.5	1293.8	4062.1	6709.0	1084.5	1337.8	4137.7	40
42	6630.4	1056.5	1295.3	4064.6	6711.7	1085.5	1339.2	4140.2	42
44	6633.1	1057.4	1296.7	4067.1	6714.4	1086.5	1340.7	4142.7	44
46	6635.9	1058.4	1298.2	4069.6	6717.2	1087.5	1342.2	4145.3	46
48	6638.6	1059.3	1299.6	4072.1	6719.9	1088.4	1343.7	4147.8	48
50	6641.3	1060.3	1301.1	4074.6	6722.6	1089.4	1345.2	4150.4	50
52	6644.0	1061.3	1302.6	4077.1	6725.3	1090.4	1346.7	4152.9	52
54	6646.7	1062.2	1304.0	4079.6	6728.0	1091.3	1348.2	4155.4	54
56	6649.5	1063.2	1305.5	4082.1	6730.7	1092.3	1349.7	4158.0	56
58	6652.2	1064.1	1306.9	4084.6	6733.4	1093.3	1351.2	4160.5	58
60	6654.9	1065.1	1308.4	4087.1	6736.1	1094.3	1352.7	4163.1	60

′	72°				73°				′
	L. C.	M.	E.	T.	L. C.	M.	E.	T.	
0	6736.1	1094.3	1352.7	4163.1	6816.6	1123.9	1398.1	4240.0	0
2	6738.8	1095.2	1354.2	4165.6	6819.3	1124.8	1399.6	4242.6	2
4	6741.5	1096.2	1355.7	4168.2	6821.9	1125.8	1401.2	4245.1	4
6	6744.1	1097.2	1357.2	4170.7	6824.6	1126.8	1402.7	4247.7	6
8	6746.8	1098.2	1358.7	4173.3	6827.3	1127.8	1404.2	4250.3	8
10	6749.5	1099.2	1360.2	4175.8	6830.0	1128.8	1405.8	4252.9	10
12	6752.2	1100.1	1361.7	4178.4	6832.6	1129.8	1407.3	4255.5	12
14	6754.9	1101.1	1363.2	4181.0	6835.3	1130.8	1408.8	4258.1	14
16	6757.6	1102.1	1364.7	4183.5	6838.0	1131.8	1410.4	4260.7	16
18	6760.2	1103.1	1366.2	4186.1	6840.7	1132.8	1411.9	4263.2	18
20	6762.9	1104.1	1367.7	4188.6	6843.3	1133.8	1413.5	4265.8	20
22	6765.6	1105.1	1369.2	4191.2	6846.0	1134.8	1415.1	4268.4	22
24	6768.3	1106.0	1370.7	4193.7	6848.7	1135.8	1416.6	4271.0	24
26	6771.0	1107.0	1372.2	4196.3	6851.3	1136.8	1418.2	4273.6	26
28	6773.7	1108.0	1373.7	4198.8	6854.0	1137.8	1419.7	4276.2	28
30	6776.3	1109.0	1375.2	4201.4	6856.7	1138.8	1421.3	4278.8	30
32	6779.0	1109.9	1376.7	4204.0	6859.4	1139.8	1422.9	4281.4	32
34	6781.7	1110.9	1378.2	4206.5	6862.0	1140.8	1424.4	4284.0	34
36	6784.4	1111.9	1379.7	4209.1	6864.7	1141.8	1426.0	4286.6	36
38	6787.1	1112.9	1381.2	4211.7	6867.4	1142.8	1427.5	4289.2	38
40	6789.8	1113.9	1382.8	4214.3	6870.1	1143.8	1429.1	4291.8	40
42	6792.4	1114.9	1384.3	4216.8	6872.7	1144.8	1430.7	4204.4	42
44	6795.1	1115.9	1385.8	4219.4	6875.4	1145.8	1432.2	4297.0	44
46	6797.8	1116.9	1387.4	4222.0	6878.1	1146.8	1433.8	4299.6	46
48	6800.5	1117.9	1388.9	4224.5	6880.8	1147.8	1435.3	4302.2	48
50	6803.2	1118.9	1390.4	4227.1	6883.4	1148.8	1436.9	4304.8	50
52	6805.9	1119.9	1392.0	4229.7	6886.1	1149.8	1438.5	4307.4	52
54	6808.5	1120.9	1393.5	4232.3	6888.8	1150.8	1440.0	4310.0	54
56	6811.2	1121.9	1395.0	4234.8	6891.4	1151.8	1441.6	4312.6	56
58	6813.9	1122.9	1396.6	4237.4	6894.1	1152.8	1443.1	4315.2	58
60	6816.6	1123.9	1398.1	4240.0	6896.8	1153.8	1444.7	4317.8	60

	74°				75°				
′	L. C.	M.	E.	T.	L. C.	M	E.	T.	′
0	6896.8	1153.8	1444.7	4317.8	6976.4	1184.1	1492.5	4396.7	0
2	6899.4	1154.8	1446.2	4320.5	6979.0	1185.1	1494.1	4399.4	2
4	6902.1	1155.8	1447.8	4323.1	6981.7	1186.1	1495.7	4402.1	4
6	6904.8	1156.8	1449.4	4325.7	6984.3	1187.1	1497.3	4404.7	6
8	6907.4	1157.8	1451.0	4328.3	6986.9	1188.1	1499.0	4407.4	8
10	6910.1	1158.8	1452.6	4330.9	6989.6	1189.2	1500.6	4410.0	10
12	6912.7	1159.8	1454.1	4333.6	6992.2	1190.2	1502.2	4412.7	12
14	6915.4	1160.8	1455.7	4336.2	6994.9	1191.2	1503.8	4415.3	14
16	6918.0	1161.8	1457.3	4338.8	6997.5	1192.2	1505.4	4418.0	16
18	6920.7	1162.8	1458.9	4341.4	7000.1	1193.2	1507.0	4420.7	18
20	6923.3	1163.9	1460.5	4344.0	7002.8	1194.3	1508.7	4423.3	20
22	6926.0	1164.9	1462.0	4346.7	7005.4	1195.3	1510.3	4426.0	22
24	6928.6	1165.9	1463.6	4349.3	7008.0	1196.3	1512.0	4428.6	24
26	6931.3	1166.9	1465.2	4351.9	7010.7	1197.3	1513.6	4431.3	26
28	6933.9	1167.9	1466.8	4354.5	7013.3	1198.3	1515.3	4434.0	28
30	6936.6	1168.9	1468.4	4357.1	7015.9	1199.4	1516.9	4436.6	30
32	6939.2	1169.9	1469.9	4359.8	7018.6	1200.4	1518.5	4439.3	32
34	6941.9	1170.9	1471.5	4362.4	7021.2	1201.4	1520.2	4442.0	34
36	6944.6	1171.9	1473.1	4365.1	7023.9	1202.4	1521.8	4444.6	36
38	6947.2	1172.9	1474.7	4367.7	7026.5	1203.4	1523.5	4447.3	38
40	6949.9	1174.0	1476.4	4370.3	7029.1	1204.5	1525.1	4450.0	40
42	6952.5	1175.0	1478.0	4373.0	7031.8	1205.5	1526.7	4452.7	42
44	6955.2	1176.0	1479.6	4375.6	7034.4	1206.5	1528.4	4455.3	44
46	6957.8	1177.0	1481.2	4378.3	7037.0	1207.5	1530.0	4458.0	46
48	6960.5	1178.0	1482.8	4380.9	7039.7	1208.5	1531.7	4460.7	48
50	6963.1	1179.0	1484.4	4383.5	7042.3	1209.6	1533.3	4463.4	50
52	6965.8	1180.0	1486.0	4386.2	7045.0	1210.6	1534.9	4466.0	52
54	6968.4	1181.0	1487.7	4388.8	7047.6	1211.6	1536.6	4468.7	54
56	6971.1	1182.0	1489.3	4391.5	7050.2	1212.6	1538.2	4471.4	56
58	6973.7	1183.0	1490.9	4394.1	7052.9	1213.6	1539.9	4474.1	58
60	6976.4	1184.1	1492.5	4396.7	7055.5	1214.7	1541.5	4476.7	60

′	76°				77°				′
	L. C.	M.	E.	T.	L. C.	M	E.	T.	
0	7055.5	1214.7	1541.5	4476.7	7134.0	1245.6	1591.7	4557.8	0
2	7058.1	1215.7	1543.2	4479.4	7136.6	1246.6	1593.4	4560.5	2
4	7060.7	1216.7	1544.9	4482.1	7139.2	1247.7	1595.1	4563.3	4
6	7063.3	1217.8	1546.5	4484.8	7141.8	1248.7	1596.8	4566.0	6
8	7066.0	1218.8	1548.2	4487.5	7144.4	1249.8	1598.5	4568.7	8
10	7068.6	1219.8	1549.9	4490.2	7147.0	1250.8	1600.2	4571.5	10
12	7071.2	1220.9	1551.5	4492.9	7149.6	1251.8	1601.9	4574.2	12
14	7073.8	1221.9	1553.2	4495.6	7152.2	1252.9	1603.6	4576.9	14
16	7076.4	1222.9	1554.9	4498.3	7154.8	1253.9	1605.3	4579.7	16
18	7079.0	1224.0	1556.5	4501.0	7157.4	1255.0	1607.0	4582.4	18
20	7081.7	1225.0	1558.2	4503.7	7160.0	1256.0	1608.7	4585.1	20
22	7084.3	1226.0	1559.9	4506.3	7162.6	1257.0	1610.4	4587.9	22
24	7086.9	1227.1	1561.5	4509.0	7165.2	1258.1	1612.1	4590.6	24
26	7089.5	1228.1	1563.2	4511.7	7167.8	1259.1	1613.8	4593.3	26
28	7092.1	1229.1	1564.9	4514.4	7170.4	1260.2	1615.5	4596.0	28
30	7094.7	1230.2	1566.5	4517.1	7173.0	1261.2	1617.3	4598.8	30
32	7097.4	1231.2	1568.2	4519.8	7175.6	1262.2	1619.0	4601.5	32
34	7100.0	1232.2	1569.9	4522.5	7178.2	1263.3	1620.7	4604.3	34
36	7102.6	1233.3	1571.5	4525.3	7180.8	1264.3	1622.4	4607.0	36
38	7105.2	1234.3	1573.2	4528.0	7183.4	1265.4	1624.1	4609.8	38
40	7107.8	1235.3	1574.8	4530.7	7186.0	1266.4	1625.9	4612.5	40
42	7110.4	1236.4	1576.4	4533.4	7188.6	1267.4	1627.6	4615.3	42
44	7113.1	1237.4	1578.1	4536.1	7191.2	1268.5	1629.3	4618.0	44
46	7115.7	1238.4	1579.8	4538.8	7193.8	1269.5	1631.0	4620.8	46
48	7118.3	1239.5	1581.5	4541.5	7196.4	1270.6	1632.7	4623.5	48
50	7120.9	1240.5	1583.2	4544.2	7199.0	1271.6	1634.5	4626.3	50
52	7123.5	1241.5	1584.9	4547.0	7201.6	1272.7	1636.2	4629.0	52
54	7126.1	1242.6	1536.6	4549.7	7204.2	1273.7	1637.9	4631.8	54
56	7128.8	1243.6	1588.3	4552.4	7206.8	1274.8	1639.6	4634.5	56
58	7131.4	1244.6	1590.0	4555.1	7209.4	1275.8	1641.3	4637.3	58
60	7134.0	1245.6	1591.7	4557.8	7212.0	1276.9	1643.1	4640.0	60

′	78°				79°				′
	L. C.	M.	E.	T.	L. C.	M.	E.	T.	
0	7212.0	1276.9	1643.1	4640.0	7289.5	1308.5	1696.0	4723.4	0
2	7214.6	1278.0	1644.8	4642.8	7292.1	1309.5	1697.7	4726.2	2
4	7217.2	1279.0	1646.6	4645.6	7294.6	1310.6	1699.5	4729.0	4
6	7219.7	1280.1	1648.3	4648.3	7297.2	1311.7	1701.3	4731.8	6
8	7222.3	1281.1	1650.1	4651.1	7299.7	1312.7	1703.1	4734.7	8
10	7224.9	1282.2	1651.8	4653.9	7302.3	1313.8	1704.9	4737.5	10
12	7227.5	1283.2	1653.6	4656.7	7304.9	1314.9	1706.6	4740.3	12
14	7230.1	1284.3	1655.3	4659.4	7307.4	1315.9	1708.4	4743.1	14
16	7232.7	1285.3	1657.1	4662.2	7310.0	1317.0	1710.2	4745.9	16
18	7235.2	1286.4	1658.8	4665.0	7312.6	1318.1	1712.0	4748.7	18
20	7237.8	1287.4	1660.6	4667.7	7315.1	1319.1	1713.8	4751.5	20
22	7240.4	1288.5	1662.3	4670.5	7317.7	1320.2	1715.6	4754.3	22
24	7243.0	1289.5	1664.1	4673.3	7320.3	1321.3	1717.4	4757.1	24
26	7245.6	1290.6	1665.8	4676.0	7322.8	1322.3	1719.2	4760.0	26
28	7248.2	1291.6	1667.6	4678.8	7325.4	1323.4	1721.0	4762.8	28
30	7250.7	1292.7	1669.3	4681.6	7327.9	1324.5	1722.8	4765.6	30
32	7253.3	1293.7	1671.1	4684.4	7330.5	1325.5	1724.6	4768.4	32
34	7255.9	1294.8	1672.8	4687.2	7333.1	1326.6	1726.	4771.2	34
36	7258.5	1295.8	1674.6	4689.9	7335.6	1327.7	1728.2	4774.1	36
38	7261.1	1296.9	1676.3	4692.7	7338.2	1328.7	1730.0	4776.9	38
40	7263.7	1297.9	1678.2	4695.5	7340.8	1329.8	1731.9	4779.7	40
42	7266.2	1299.0	1679.9	4698.3	7343.3	1330.8	1733.7	4782.6	42
44	7268.8	1300.0	1681.7	4701.1	7345.9	1331.9	1735.5	4785.4	44
46	7271.4	1301.1	1683.5	4703.9	7348.4	1333.0	1737.3	4788.2	46
48	7274.0	1302.1	1685.3	4706.7	7351.0	1334.1	1739.1	4791.0	48
50	7276.6	1303.2	1687.1	4709.5	7353.6	1335.2	1740.9	4793.9	50
52	7279.2	1304.2	1688.8	4712.2	7356.1	1336.2	1742.7	4796.7	52
54	7281.7	1305.3	1690.6	4715.0	7358.7	1337.3	1744.5	4799.5	54
56	7284.3	1306.3	1692.4	4717.8	7361.3	1338.4	1746.3	4802.4	56
58	7286.9	1307.4	1694.2	4720.6	7363.8	1339.5	1748.1	4805.2	58
60	7289.5	1308.5	1696.0	4723.4	7366.4	1340.6	1750.0	4808.0	60

′	80°				81°				′
	L. C.	M.	E.	T.	L. C.	M.	E.	T.	
0	7366.4	1340.6	1750.0	4808.0	7442.7	1372.8	1805.5	4893.9	0
2	7368.9	1341.7	1751.8	4810.9	7445.2	1373.9	1807.3	4896.8	2
4	7371.5	1342.7	1753.7	4813.7	7447.7	1375.0	1809.2	4899.7	4
6	7374.0	1343.8	1755.5	4816.6	7450.3	1376.1	1811.1	4902.6	6
8	7376.6	1344.9	1757.4	4819.4	7452.8	1377.1	1813.0	4905.4	8
10	7379.1	1346.0	1759.2	4822.3	7455.3	1378.2	1814.9	4908.3	10
12	7381.7	1347.0	1761.0	4825.1	7457.8	1379.3	1816.8	4911.2	12
14	7384.2	1348.1	1762.9	4828.0	7460.4	1380.4	1818.6	4914.1	14
16	7386.7	1349.2	1764.7	4830.8	7462.9	1381.4	1820.5	4917.0	16
18	7389.3	1350.3	1766.6	4833.7	7465.4	1382.5	1822.4	4919.9	18
20	7391.8	1351.3	1768.4	4836.5	7467.9	1383.6	1824.2	4922.8	20
22	7394.4	1352.4	1770.2	4839.4	7470.4	1384.7	1826.1	4925.7	22
24	7396.9	1353.5	1772.1	4842.2	7473.0	1385.7	1828.0	4928.6	24
26	7399.5	1354.6	1773.9	4845.1	7475.5	1386.8	1829.9	4931.5	26
28	7402.0	1355.6	1775.8	4847.9	7478.0	1387.9	1831.8	4934.4	28
30	7404.5	1356.7	1777.6	4850.8	7480.5	1389.0	1833.7	4937.2	30
32	7407.1	1357.8	1779.4	4853.7	7483.1	1390.1	1835.6	4940.2	32
34	7409.6	1358.9	1781.3	4856.5	7485.6	1391.2	1837.5	4943.1	34
36	7412.2	1359.9	1783.1	4859.4	7488.1	1392.3	1839.4	4946.0	36
38	7414.7	1361.0	1785.0	4862.3	7490.6	1393.4	1841.3	4948.9	38
40	7417.3	1362.1	1786.8	4865.1	7493.2	1394.5	1843.2	4951.8	40
42	7419.8	1363.2	1788.6	4868.0	7495.7	1395.6	1845.1	4954.7	42
44	7422.3	1364.2	1790.5	4870.9	7498.2	1396.7	1847.0	4957.6	44
46	7424.9	1365.3	1792.4	4873.8	7500.7	1397.8	1848.9	4960.6	46
48	7427.4	1366.4	1794.3	4876.6	7503.3	1398.9	1850.8	4963.5	48
50	7430.0	1367.5	1796.2	4879.5	7505.8	1400.0	1852.7	4966.4	50
52	7432.5	1368.5	1798.0	4882.4	7508.3	1401.1	1854.6	4969.3	52
54	7435.1	1369.6	1799.9	4885.3	7510.8	1402.2	1856.5	4972.2	54
56	7437.6	1370.7	1801.8	4888.1	7513.3	1403.3	1858.4	4975.1	56
58	7440.1	1371.8	1803.7	4891.0	7515.9	1404.4	1860.3	4978.0	58
60	7442.7	1372.8	1805.5	4893.9	7518.4	1405.5	1862.3	4981.0	60

′	82° L. C.	82° M.	82° E.	82° T.	83° L. C.	83° M.	83° E.	83° T.	′
0	7518.4	1405.5	1862.3	4981.0	7593.6	1438.5	1920.6	5069.4	0
2	7520.9	1406.6	1864.2	4983.9	7596.1	1439.6	1922.6	5072.4	2
4	7523.4	1407.7	1866.1	4986.8	7598.6	1440.7	1924.6	5075.4	4
6	7525.9	1408.8	1868.1	4989.8	7601.1	1441.8	1926.5	5078.4	6
8	7528.4	1409.9	1870.0	4992.7	7603.6	1442.9	1928.5	5081.4	8
10	7530.9	1411.0	1871.9	4995.7	7606.0	1444.0	1930.5	5084.4	10
12	7533.4	1412.1	1873.9	4998.6	7608.5	1445.1	1932.4	5087.3	12
14	7535.9	1413.2	1875.8	5001.5	7611.0	1446.2	1934.4	5090.3	14
16	7538.5	1414.3	1877.7	5004.5	7613.5	1447.3	1936.4	5093.3	16
18	7541.0	1415.4	1879.7	5007.4	7616.0	1448.4	1938.4	5096.3	18
20	7543.5	1416.5	1881.6	5010.3	7618.5	1449.6	1940.4	5099.3	20
22	7546.0	1417.6	1883.5	5013.3	7621.0	1450.7	1942.4	5102.3	22
24	7548.5	1418.7	1885.5	5016.2	7623.5	1451.8	1944.4	5105.2	24
26	7551.0	1419.8	1887.4	5019.2	7626.0	1452.9	1946.4	5108.2	26
28	7553.5	1420.9	1889.3	5022.1	7628.5	1454.0	1948.4	5111.2	28
30	7556.0	1422.0	1891.3	5025.0	7630.9	1455.1	1950.4	5114.2	30
32	7558.5	1423.1	1893.2	5028.0	7633.4	1456.2	1952.4	5117.2	32
34	7561.0	1424.2	1895.1	5031.0	7635.9	1457.3	1954.4	5120.2	34
36	7563.5	1425.3	1897.1	5033.9	7638.4	1458.4	1956.4	5123.2	36
38	7566.0	1426.4	1899.0	5036.9	7640.9	1459.5	1958.4	5126.2	38
40	7568.5	1427.5	1901.0	5039.8	7643.4	1460.7	1960.4	5129.2	40
42	7571.0	1428.6	1902.9	5042.8	7645.9	1461.8	1962.4	5132.2	42
44	7573.5	1429.7	1904.9	5045.8	7648.4	1462.9	1964.4	5135.2	44
46	7576.1	1430.8	1906.9	5048.7	7650.9	1464.0	1966.4	5138.2	46
48	7578.6	1431.9	1908.8	5051.7	7653.4	1465.1	1968.4	5141.2	48
50	7581.1	1433.0	1910.8	5054.6	7655.8	1466.2	1970.4	5144.3	50
52	7583.6	1434.1	1912.8	5057.6	7658.3	1467.3	1972.4	5147.3	52
54	7586.1	1435.2	1914.7	5060.6	7660.8	1468.4	1974.4	5150.3	54
56	7588.6	1436.3	1916.7	5063.5	7663.3	1469.5	1976.4	5153.3	56
58	7591.1	1437.4	1918.7	5066.5	7665.8	1470.6	1978.4	5156.3	58
60	7593.6	1438.5	1920.6	5069.4	7668.3	1471.8	1980.5	5159.3	60

′	84° L. C.	84° M.	84° E.	84° T.	85° L. C.	85° M.	85° E.	85° T.	′
0	7668.3	1471.8	1980.5	5159.3	7742.4	1505.4	2041.8	5250.6	0
2	7670.8	1472.9	1982.5	5162.3	7744.8	1506.5	2043.9	5253.6	2
4	7673.2	1474.0	1984.5	5165.3	7747.3	1507.6	2046.0	5256.7	4
6	7675.7	1475.1	1986.6	5168.4	7749.7	1508.8	2048.0	5259.8	6
8	7678.2	1476.2	1988.6	5171.4	7752.2	1509.9	2050.1	5262.9	8
10	7680.6	1477.4	1990.6	5174.4	7754.6	1511.0	2052.2	5266.0	10
12	7683.1	1478.5	1992.7	5177.5	7757.1	1512.2	2054.2	5269.0	12
14	7685.6	1479.6	1994.7	5180.5	7759.5	1513.3	2056.3	5272.1	14
16	7688.1	1480.7	1996.7	5183.5	7762.0	1514.4	2058.4	5275.2	16
18	7690.5	1481.8	1998.8	5186.6	7764.4	1515.6	2060.5	5278.3	18
20	7693.0	1483.0	2000.8	5189.6	7766.9	1516.7	2062.6	5281.4	20
22	7695.5	1484.1	2002.8	5192.6	7769.3	1517.8	2064.7	5284.4	22
24	7697.9	1485.2	2004.9	5195.6	7771.8	1519.0	2066.8	5287.5	24
26	7700.4	1486.3	2006.9	5198.7	7774.2	1520.1	2068.9	5290.6	26
28	7702.9	1487.4	2008.9	5201.7	7776.7	1521.2	2071.0	5293.7	28
30	7705.3	1488.6	2011.0	5204.7	7779.1	1522.4	2073.1	5296.7	30
32	7707.8	1489.7	2013.0	5207.8	7781.5	1523.5	2075.2	5299.8	32
34	7710.3	1490.8	2015.0	5210.8	7784.0	1524.6	2077.3	5302.9	34
36	7712.8	1491.9	2017.0	5213.9	7786.4	1525.8	2079.4	5306.1	36
38	7715.2	1493.0	2019.1	5216.9	7788.9	1526.9	2081.5	5309.2	38
40	7717.7	1494.2	2021.2	5220.0	7791.3	1528.0	2083.7	5312.3	40
42	7720.2	1495.3	2023.2	5223.1	7793.8	1529.2	2085.8	5315.4	42
44	7722.6	1496.4	2025.3	5226.1	7796.2	1530.3	2087.9	5318.5	44
46	7725.1	1497.5	2027.4	5229.2	7798.7	1531.4	2090.0	5321.6	46
48	7727.6	1498.6	2029.4	5232.2	7801.1	1532.6	2092.1	5324.7	48
50	7730.0	1499.8	2031.5	5235.3	7803.6	1533.7	2094.2	5327.8	50
52	7732.5	1500.9	2033.6	5238.3	7806.0	1534.8	2096.3	5330.9	52
54	7735.0	1502.0	2035.6	5241.4	7808.5	1536.0	2098.4	5334.0	54
56	7737.5	1503.1	2037.7	5244.5	7810.9	1537.1	2100.6	5337.1	56
58	7739.9	1504.2	2039.8	5247.5	7813.4	1538.2	2102.7	5340.2	58
60	7742.4	1505.4	2041.8	5250.6	7815.8	1539.3	2104.8	5343.3	60

′	86° L. C.	86° M.	86° E.	86° T.	87° L. C.	87° M.	87° E.	87° T.	′
0	7815.8	1539.3	2104.8	5343.3	7888.5	1573.6	2169.5	5437.5	0
2	7818.2	1540.4	2106.9	5346.4	7890.9	1574.8	2171.6	5440.7	2
4	7820.6	1541.6	2109.1	5349.5	7893.3	1575.9	2173.8	5443.9	4
6	7823.1	1542.7	2111.2	5352.7	7895.7	1577.1	2176.0	5447.1	6
8	7825.5	1543.9	2113.4	5355.8	7898.1	1578.2	2178.2	5450.3	8
10	7827.9	1545.0	2115.5	5358.9	7900.5	1579.4	2180.4	5453.4	10
12	7830.3	1546.1	2117.6	5362.0	7903.0	1580.5	2182.5	5456.6	12
14	7832.8	1547.3	2119.8	5365.2	7905.4	1581.7	2184.7	5459.8	14
16	7835.2	1548.4	2121.9	5368.3	7907.8	1582.9	2186.9	5463.0	16
18	7837.6	1549.6	2124.1	5371.4	7910.2	1584.0	2189.1	5466.2	18
20	7840.0	1550.7	2126.2	5374.6	7912.6	1585.1	2191.3	5469.4	20
22	7842.4	1551.8	2128.3	5377.7	7915.0	1586.3	2193.5	5472.5	22
24	7844.9	1553.0	2130.5	5 80.8	7917.4	1587.4	2195.7	5475.7	24
26	7847.3	1554.1	2132.6	5383.9	7919.8	1588.6	2197.9	5478.9	26
28	7849.7	1555.3	2134.8	5387.1	7922.2	1589.7	2200.1	5482.1	28
30	7852.1	1556.4	2136.9	5390.2	7924.6	1590.9	2202.3	5485.3	30
32	7854.6	1557.5	2139.0	5393.4	7927.1	1592.0	2204.5	5488.5	32
34	7857.0	1558.7	2141.2	5396.5	7929.5	1593.2	2206.8	5491.7	34
36	7859.4	1559.8	2143.3	5399.7	7931.9	1594.3	2209.0	5494.9	36
38	7861.8	1561.0	2145.5	5402.8	7934.3	1595.5	2211.2	5498.1	38
40	7864.3	1562.1	2147.7	5406.0	7936.7	1596.6	2213.4	5501.3	40
42	7866.7	1563.2	2149.8	5409.1	7939.1	1597.8	2215.6	5504.5	42
44	7869.1	1564.4	2152.0	5412.3	7941.5	1598.9	2217.8	5507.7	44
46	7871.5	1565.5	2154.2	5415.4	7943.9	1600.1	2220.0	5510.9	46
48	7874.0	1566.7	2156.4	5418.6	7946.3	1601.2	2222.3	5514.1	48
50	7876.4	1567.8	2158.6	5421.8	7948.7	1602.4	2224.5	5517.3	50
52	7878.8	1568.9	2160.7	5424.9	7951.2	1603.5	2226.7	5520.5	52
54	7881.2	1570.1	2162.9	5428.1	7953.6	1604.7	2228.9	5523.7	54
56	7883.6	1571.2	2165.1	5431.2	7956.0	1605.8	2231.1	5526.9	56
58	7886.1	1572.4	2167.3	5434.4	7958.4	1607.0	2233.3	5530.1	58
60	7888.5	1573.6	2169.5	5437.5	7960.8	1608.2	2235.6	5533.3	60

′	88° L. C.	88° M.	88° E.	88° T.	89° L. C.	89° M.	89° E.	89° T.	′
0	7960.8	1608.2	2235.6	5533.3	8032.4	1643.0	2303.6	5630.8	0
2	7963.2	1609.4	2237.8	5536.6	8034.8	1644.1	2305.9	5634.1	2
4	7965.6	1610.5	2240.1	5539.8	8037.1	1645.3	2308.2	5637.4	4
6	7968.0	1611.7	2242.3	5543.1	8039.5	1646.5	2310.5	5640.7	6
8	7970.3	1612.8	2244.6	5546.3	8041.9	1647.7	2312.8	5644.0	8
10	7972.7	1614.0	2246.8	5549.5	8044.2	1648.9	2315.1	5647.3	10
12	7975.1	1615.2	2249.1	5552.8	8046.6	1650.0	2317.4	5650.6	12
14	7977.5	1616.3	2251.3	5556.0	8049.0	1651.2	2319.7	5653.9	14
16	7979.9	1617.5	2253.6	5559.2	8051.4	1652.4	2322.0	5657.1	16
18	7982.3	1618.6	2255.8	5562.5	8053.7	1653.6	2324.3	5660.4	18
20	7984.7	1619.8	2258.1	5565.7	8056.1	1654.8	2326.7	5663.7	20
22	7987.1	1621.0	2260.4	5568.9	8058.5	1655.9	2329.0	5667.0	22
24	7989.4	1622.1	2262.7	5572.2	8060.8	1657.1	2331.3	5670.3	24
26	7991.8	1623.3	2264.9	5575.4	8063.2	1658.3	2333.7	5673.6	26
28	7994.2	1624.4	2267.2	5578.6	8065.6	1659.5	2336.0	5676.9	28
30	7996.6	1625.6	2269.5	5581.9	8067.9	1660.7	2338.3	5680.2	30
32	7999.0	1626.8	2271.7	5585.1	8070.3	1661.8	2340.7	5683.5	32
34	8001.4	1627.9	2273.9	5588.4	8073.7	1663.0	2343.0	5686.8	34
36	8003.8	1629.1	2276.2	5591.7	8075.1	1664.2	2345.3	5690.2	36
38	8006.1	1630.2	2278.5	5594.9	8077.4	1665.4	2347.7	5693.5	38
40	8008.5	1631.4	2280.8	5598.2	8079.8	1666.6	2350.0	5696.8	40
42	8010.9	1632.6	2283.0	5601.4	8082.2	1667.7	2352.3	5700.1	42
44	8013.3	1633.7	2285.3	5604.7	8084.5	1668.8	2354.7	5703.4	44
46	8015.7	1634.9	2287.6	5608.0	8086.9	1670.0	2357.0	5706.8	46
48	8018.1	1636.0	2289.9	5611.2	8089.3	1671.2	2359.3	5710.1	48
50	8020.5	1637.2	2292.2	5614.5	8091.6	1672.4	2361.7	5713.4	50
52	8022.9	1638.4	2294.4	5617.8	8094.0	1673.5	2364.0	5716.7	52
54	8025.2	1639.5	2296.7	5621.0	8096.4	1674.7	2366.3	5720.0	54
56	8027.6	1640.7	2299.0	5624.3	8098.8	1675.9	2368.7	5723.4	56
58	8030.0	1641.8	2301.3	5627.5	8101.1	1677.1	2371.0	5726.7	58
60	8032.4	1643.0	2303.6	5630.8	8103.5	1678.3	2373.4	5730.0	60

′	90° L. C.	90° M.	90° E.	90° T.	91° L. C.	91° M.	91° E.	91° T.	′
0	8103.5	1678.3	2373.4	5730.0	8173.9	1713.8	2445.1	5830.9	0
2	8105.8	1679.5	2375.8	5733.3	8176.2	1715.0	2447.5	5834.3	2
4	8108.2	1680.6	2378.2	5736.7	8178.5	1716.2	2450.0	5837.7	4
6	8110.5	1681.8	2380.5	5740.0	8180.9	1717.4	2452.4	5841.1	6
8	8112.9	1683.0	2382.9	5743.4	8183.2	1718.6	2454.8	5844.5	8
10	8115.2	1684.2	2385.3	5746.7	8185.5	1719.7	2457.2	5847.9	10
12	8117.6	1685.4	2387.6	5750.0	8187.9	1720.9	2459.7	5851.3	12
14	8119.9	1686.5	2390.0	5753.4	8190.2	1722.1	2462.1	5854.7	14
16	8122.3	1687.7	2392.4	5756.7	8192.5	1723.3	2464.5	5858.1	16
18	8124.6	1688.9	2394.7	5760.1	8194.8	1724.5	2467.0	5861.5	18
20	8127.0	1690.1	2397.1	5763.4	8197.2	1725.7	2469.4	5864.9	20
22	8129.3	1691.3	2399.5	5766.8	8199.5	1726.9	2471.9	5868.3	22
24	8131.7	1692.5	2401.9	5770.1	8201.8	1728.1	2474.3	5871.8	24
26	8134.0	1693.6	2404.3	5773.5	8204.2	1729.3	2476.7	5875.2	26
28	8136.4	1694.8	2406.6	5776.9	8206.5	1730.5	2479.2	5878.6	28
30	8138.7	1696.0	2409.0	5780.2	8208.8	1731.7	2481.6	5882.0	30
32	8141.1	1697.2	2411.4	5783.6	8211.1	1732.9	2484.1	5885.4	32
34	8143.4	1698.4	2413.8	5787.0	8213.5	1734.1	2486.5	5888.9	34
36	8145.8	1699.6	2416.2	5790.3	8215.8	1735.3	2489.0	5892.3	36
38	8148.1	1700.7	2418.6	5793.7	8218.1	1736.4	2491.5	5895.7	38
40	8150.4	1701.9	2421.0	5797.1	8220.4	1737.6	2493.9	5899.2	40
42	8152.8	1703.1	2423.4	5800.4	8222.8	1738.8	2496.4	5902.6	42
44	8155.1	1704.3	2425.8	5803.8	8225.1	1740.0	2498.9	5906.0	44
46	8157.5	1705.5	2428.2	5807.2	8227.4	1741.2	2501.3	5909.4	46
48	8159.8	1706.7	2430.6	5810.6	8229.7	1742.4	2503.8	5912.9	48
50	8162.2	1707.9	2433.0	5814.0	8232.0	1743.6	2506.3	5916.3	50
52	8164.5	1709.0	2435.4	5817.3	8234.3	1744.8	2508.7	5919.8	52
54	8166.8	1710.2	2437.9	5820.7	8236.7	1746.0	2511.2	5923.2	54
56	8169.2	1711.4	2440.3	5824.1	8239.0	1747.2	2513.7	5926.7	56
58	8171.5	1712.6	2442.7	5827.5	8241.3	1748.4	2516.2	5930.1	58
60	8173.9	1713.8	2445.1	5830.9	8243.6	1749.6	2518.7	5933.6	60

′	92° L. C	92° M.	92° E.	92° T.	93° L. C.	93° M.	93° E.	93° T.	′
0	8243.6	1749.6	2518.7	5933.6	8312.8	1785.7	2594.2	6038.2	0
2	8245.9	1750.8	2521.2	5937.0	8315.1	1786.9	2596.8	6041.7	2
4	8248.2	1752.0	2523.6	5940.5	8317.4	1788.2	2599.3	6045.2	4
6	8250.6	1753.2	2526.1	5944.0	8319.7	1789.4	2601.9	6048.7	6
	8252.9	1754.4	2528.6	5947.4	8322.0	1790.6	2604.4	6052.2	8
10	8255.2	1755.6	2531.1	5950.9	8324.3	1791.8	2607.0	6055.8	10
12	8257.5	1756.8	2533.6	5954.4	8326.6	1793.0	2609.6	6059.3	12
14	8259.8	1758.0	2536.1	5957.8	8328.8	1794.2	2612.1	6062.8	14
16	8262.2	1759.2	2538.6	5961.3	8331.1	1795.4	2614.7	6066.4	16
18	8264.5	1760.4	2541.1	5964.8	8333.3	1796.6	2617.3	6069.9	18
20	8266.8	1761.6	2543.6	5968.2	8335.6	1797.8	2619.8	6073.4	20
22	8269.1	1762.8	2546.1	5971.7	8337.9	1799.1	2622.4	6077.0	22
24	8271.4	1764.0	2548.6	5975.2	8340.2	1800.3	2625.0	6080.5	24
26	8273.7	1765.2	2551.2	5978.7	8342.5	1801.5	2627.6	6084.1	26
28	8276.0	1766.4	2553.7	5982.2	8341.8	1802.7	2630.2	6087.6	28
30	8278.3	1767.6	2556.2	5985.6	8347.1	1803.9	2632.7	6091.2	30
32	8280.6	1768.8	2558.7	5989.1	8349.4	1805.1	2635.3	6094.7	32
34	8282.9	1770.0	2561.2	5992.6	8351.7	1806.3	2637.9	6098.3	34
36	8285.2	1771.2	2563.8	5996.1	8354.0	1807.6	2640.5	6101.8	36
38	8287.5	1772.5	2566.3	5999.6	8356.3	1808.8	2643.1	6105.4	38
40	8289.8	1773.7	2568.8	6003.1	8358.5	1810.0	2645.7	6109.0	40
42	8292.1	1774.9	2571.3	6006.6	8360.8	1811.2	2648.3	6112.5	42
44	8291.4	1776.1	2573.9	6010.1	8363.1	1812.4	2650.9	6116.1	44
46	8296.7	1777.3	2576.4	6013.6	8365.4	1813.6	2653.5	6119.7	46
48	8290.0	1778.5	2578.9	6017.1	8367.7	1814.9	2656.1	6123.2	48
50	8301.3	1779.7	2581.5	6020.6	8369.9	1816.1	2658.7	6126.8	50
52	8303.6	1780.9	2584.0	6024.1	8372.2	1817.3	2661.3	6130.4	52
54	8305.9	1782.1	2586.6	6027.6	8374.5	1818.5	2663.9	6133.9	54
56	8308.2	1783.3	2589.1	6031.1	8376.8	1819.7	2666.6	6137.5	56
58	8310.5	1781.5	2591.7	6034.6	8379.1	1820.9	2669.2	6141.1	58
60	8312.8	1785.7	2594.2	6038.2	8381.3	1822.2	2671.8	6144.7	60

′	94°				95°				′
	L. C.	M.	E.	T.	L. C.	M.	E.	T.	
0	8381.3	1822.2	2671.8	6144.7	8449.2	1858.9	2751.5	6253.2	0
2	8383.6	1823.4	2674.4	6148.3	8451.5	1860.1	2754.2	6256.9	2
4	8385.9	1824.6	2677.0	6151.9	8453.7	1861.3	2756.9	6260.5	4
6	8388.1	1825.8	2679.7	6155.4	8456.0	1862.6	2759.6	6264.2	6
8	8390.4	1827.0	2682.3	6159.0	8458.2	1863.8	2762.3	6267.8	8
10	8392.7	1828.3	2684.9	6162.6	8460.4	1865.0	2765.0	6271.5	10
12	8395.0	1829.5	2687.6	6166.2	8462.7	1866.3	2767.7	6275.2	12
14	8397.2	1830.7	2690.2	6169.8	8464.9	1867.5	2770.4	6278.8	14
16	8399.5	1831.9	2692.8	6173.4	8467.2	1868.7	2773.1	6282.5	16
18	8401.7	1833.1	2695.6	6177.0	8469.4	1869.9	2775.8	6286.2	18
20	8404.0	1834.4	2698.1	6180.6	8471.7	1871.2	2778.5	6289.8	20
22	8406.3	1835.6	2700.8	6184.2	8473.9	1872.4	2781.2	6293.5	22
24	8408.5	1836.8	2703.4	6187.8	8476.2	1873.6	2784.0	6297.2	24
26	8410.8	1838.0	2706.1	6191.5	8478.4	1874.9	2786.7	6300.9	26
28	8413.1	1839.3	2708.7	6195.1	8480.7	1876.1	2789.4	6304.6	28
30	8415.3	1840.5	2711.4	6198.7	8482.9	1877.3	2792.1	6308.2	30
32	8417.6	1841.7	2714.0	6202.3	8485.1	1878.6	2794.9	6311.9	32
34	8419.9	1842.9	2716.7	6205.9	8487.4	1879.8	2797.6	6315.6	34
36	8422.1	1844.2	2719.3	6209.5	8489.6	1881.0	2800.3	6319.3	36
38	8424.4	1845.4	2722.0	6213.2	8491.9	1882.3	2803.1	6323.0	38
40	8426.6	1846.6	2724.7	6216.8	8494.1	1883.5	2805.8	6326.7	40
42	8428.9	1847.8	2727.3	6220.4	8496.3	1884.8	2808.6	6330.4	42
44	8431.2	1849.1	2730.0	6224.1	8498.6	1886.0	2811.3	6334.1	44
46	8433.4	1850.3	2732.7	6227.7	8500.8	1887.2	2814.1	6337.8	46
48	8435.7	1851.5	2735.4	6231.3	8503.0	1888.5	2816.8	6341.5	48
50	8437.9	1852.7	2738.0	6235.0	8505.3	1889.7	2819.6	6345.2	50
52	8440.2	1854.0	2740.7	6238.6	8507.5	1890.9	2822.3	6349.0	52
54	8442.4	1855.2	2743.4	6242.3	8509.8	1892.2	2825.1	6352.7	54
56	8444.7	1856.4	2746.1	6245.0	8512.0	1893.4	2827.8	6356.4	56
58	8447.0	1857.6	2748.8	6249.6	8514.2	1894.6	2830.6	6360.1	58
60	8449.2	1858.9	2751.5	6253.2	8516.4	1895.9	2833.4	6363.8	60

′	96°				97°				′
	L. C.	M.	E.	T.	L. C.	M.	E.	T.	
0	8516.4	1895.9	2833.4	6363.8	8583.0	1933.2	2917.5	6476.6	0
2	8518.7	1897.1	2836.1	6367.5	8585.2	1934.4	2920.3	6480.4	2
4	8520.9	1898.4	2838.9	6371.3	8587.5	1935.7	2923.2	6484.2	4
6	8523.1	1899.6	2841.7	6375.0	8589.7	1936.9	2926.0	6488.0	6
8	8525.4	1900.8	2844.5	6378.7	8591.9	1938.2	2928.9	6491.8	8
10	8527.6	1902.1	2847.2	6382.5	8594.1	1939.4	2931.7	6495.6	10
12	8529.8	1903.3	2850.0	6386.2	8596.3	1940.7	2934.6	6499.4	12
14	8532.0	1904.6	2852.8	6389.9	8598.5	1941.9	2937.5	6503.2	14
16	8534.3	1905.8	2855.6	6393.7	8600.7	1943.2	2940.3	6507.1	16
18	8536.5	1907.0	2858.4	6397.4	8602.9	1944.4	2943.2	6510.9	18
20	8538.7	1908.3	2861.2	6401.2	8605.1	1945.7	2946.1	6514.7	20
22	8540.9	1909.5	2864.0	6404.9	8607.3	1946.9	2948.9	6518.5	22
24	8543.2	1910.8	2866.7	6408.7	8609.5	1948.2	2951.8	6522.3	24
26	8545.4	1912.0	2869.5	6412.4	8611.7	1949.4	2954.7	6526.2	26
28	8547.6	1913.3	2872.3	6416.2	8613.9	1950.7	2957.6	6530.0	28
30	8549.8	1914.5	2875.1	6419.9	8616.1	1952.0	2960.4	6533.8	30
32	8552.0	1915.7	2877.9	6423.7	8618.3	1953.2	2963.3	6537.7	32
34	8554.3	1917.0	2880.8	6427.5	8620.5	1954.5	2966.2	6541.5	34
36	8556.5	1918.2	2883.6	6431.2	8622.7	1955.7	2969.1	6545.3	36
38	8558.7	1919.5	2886.4	6435.0	8624.9	1957.0	2972.0	6549.2	38
40	8560.9	1920.7	2889.2	6438.8	8627.1	1958.2	2974.9	6553.0	40
42	8563.1	1922.0	2892.0	6442.5	8629.3	1959.5	2977.8	6556.9	42
44	8565.3	1923.2	2894.8	6446.3	8631.5	1960.7	2980.7	6560.7	44
46	8567.6	1924.5	2897.7	6450.1	8633.7	1962.0	2983.6	6564.6	46
48	8569.8	1925.7	2900.5	6453.9	8635.8	1963.2	2986.5	6568.4	48
50	8572.0	1927.0	2903.3	6457.6	8638.0	1964.5	2989.4	6572.3	50
52	8574.2	1928.2	2906.1	6461.4	8640.2	1965.8	2992.3	6576.2	52
54	8576.4	1929.4	2909.0	6465.2	8642.4	1967.0	2995.2	6580.0	54
56	8578.6	1930.7	2911.8	6469.0	8644.6	1968.3	2998.1	6583.9	56
58	8580.8	1931.9	2914.7	6472.8	8646.8	1969.5	3001.1	6587.7	58
60	8583.0	1933.2	2917.5	6476.6	8649.0	1970.8	3004.0	6591.6	60

′	98° L. C.	98° M.	98° E.	98° T.	99° L. C.	99° M.	99° E.	99° T.	′
0	8649.0	1970.8	3001.0	6591.6	8714.3	2008.7	3092.9	6709.0	0
2	8651.2	1972.0	3006.9	6595.5	8716.4	2009.9	3095.9	6712.9	2
4	8653.3	1973.3	3009.8	6599.4	8718.6	2011.2	3098.9	6716.9	4
6	8655.5	1974.6	3012.8	6603.2	8720.7	2012.5	3101.9	6720.8	6
8	8657.7	1975.8	3015.7	6607.1	8722.9	2013.7	3104.9	6724.8	8
10	8659.9	1977.1	3018.6	6611.0	8725.1	2015.0	3107.9	6728.8	10
12	8662.1	1978.3	3021.6	6614.9	8727.2	2016.3	3111.0	6732.7	12
14	8664.3	1979.6	3024.5	6618.8	8729.4	2017.5	3114.0	6736.7	14
16	8666.4	1980.9	3027.5	6622.7	8731.5	2018.8	3117.0	6740.7	16
18	8668.6	1982.1	3030.4	6626.6	8733.7	2020.1	3120.0	6744.6	18
20	8670.8	1983.4	3033.3	6630.5	8735.9	2021.4	3123.1	6748.6	20
22	8673.0	1984.6	3036.3	6634.4	8738.0	2022.6	3126.1	6752.6	22
24	8675.2	1985.9	3039.3	6638.3	8740.2	2023.9	3129.1	6756.6	24
26	8677.3	1987.2	3042.2	6642.2	8742.3	2025.2	3132.2	6760.6	26
28	8679.5	1988.4	3045.2	6646.1	8744.5	2026.4	3135.2	6764.6	28
30	8681.7	1989.7	3048.1	6650.0	8746.6	2027.7	3138.3	6768.6	30
32	8683.9	1991.0	3051.1	6653.9	8748.8	2029.0	3141.3	6772.6	32
34	8686.0	1992.2	3054.1	6657.8	8750.9	2030.3	3144.4	6776.6	34
36	8688.2	1993.5	3057.0	6661.7	8753.1	2031.5	3147.4	6780.6	36
38	8690.4	1994.7	3060.0	6665.7	8755.3	2032.8	3150.5	6781.6	38
40	8692.6	1996.0	3063.0	6669.6	8757.4	2034.1	3153.5	6788.6	40
42	8694.7	1997.3	3066.0	6673.5	8759.5	2035.4	3156.6	6792.6	42
44	8696.9	1998.5	3068.7	6677.4	8761.7	2036.6	3159.7	6796.6	44
46	8699.1	1999.8	3071.9	6681.4	8763.8	2037.9	3162.7	6800.6	46
48	8701.2	2001.1	3074.9	6685.3	8766.0	2039.2	3165.8	6804.6	48
50	8703.4	2002.3	3077.9	6689.2	8768.1	2040.5	3168.9	6808.6	50
52	8705.6	2003.6	3080.9	6693.2	8770.3	2041.7	3172.0	6812.6	52
54	8707.8	2004.9	3083.9	6697.1	8772.4	2043.0	3175.1	6816.7	54
56	8709.9	2006.1	3086.9	6701.1	8774.6	2044.3	3178.1	6820.7	56
58	8712.1	2007.4	3089.9	6705.2	8776.7	2045.6	3181.2	6824.7	58
60	8714.3	2008.7	3092.9	6709.0	8778.9	2046.8	3184.3	6828.8	60

′	100° L. C.	100° M.	100° E.	100° T.	101° L. C.	101° M.	101° E.	101° T.	′
0	8778.9	2046.8	3184.3	6828.8	8842.8	2085.3	3278.3	6951.0	0
2	8781.0	2048.1	3187.4	6832.8	8844.9	2086.6	3281.5	6955.2	2
4	8783.1	2049.4	3190.5	6836.8	8847.0	2087.8	3284.7	6959.3	4
6	8785.3	2050.7	3193.6	6840.9	8849.2	2089.1	3287.9	6963.4	6
8	8787.4	2051.9	3196.7	6844.9	8851.3	2090.4	3291.1	6967.6	8
10	8789.6	2053.2	3199.8	6849.0	8853.4	2091.7	3294.3	6971.7	10
12	8791.7	2054.5	3202.9	6853.0	8855.5	2093.0	3297.5	6975.8	12
14	8793.9	2055.8	3206.0	6857.1	8857.6	2094.3	3300.7	6980.0	14
16	8796.0	2057.1	3209.1	6861.1	8859.8	2095.6	3303.9	6984.1	16
18	8798.9	2058.3	3212.2	6865.2	8861.9	2096.9	3307.1	6988.2	18
20	8800.3	2059.6	3215.4	6869.2	8864.0	2098.2	3310.3	6992.4	20
22	8802.4	2060.9	3218.5	6873.3	8866.1	2099.4	3313.5	6996.6	22
24	8804.5	2062.2	3221.6	6877.4	8868.2	2100.7	3316.7	7000.7	24
26	8806.7	2063.5	3224.7	6881.4	8870.3	2102.0	3319.9	7004.9	26
28	8808.8	2064.7	3227.9	6885.5	8872.4	2103.3	3323.1	7009.0	28
30	8810.9	2066.0	3231.0	6889.6	8874.5	2104.6	3326.4	7013.2	30
32	8813.1	2067.3	3234.1	6893.7	8876.7	2105.9	3329.6	7017.3	32
34	8815.2	2068.6	3237.3	6897.8	8878.8	2107.2	3332.8	7021.5	34
36	8817.3	2069.9	3240.4	6901.8	8880.9	2108.5	3336.0	7025.7	36
38	8819.5	2071.1	3243.5	6905.9	8883.0	2109.8	3339.3	7029.9	38
40	8821.6	2072.4	3246.7	6910.0	8885.1	2111.1	3342.5	7034.0	40
42	8823.7	2073.7	3249.8	6914.1	8887.2	2112.4	3345.8	7038.2	42
44	8825.8	2075.0	3253.0	6918.2	8889.3	2113.6	3349.0	7042.4	44
46	8828.0	2076.3	3256.2	6922.3	8891.4	2114.9	3352.3	7046.6	46
48	8830.1	2077.6	3259.3	6926.4	8893.5	2116.2	3355.5	7050.8	48
50	8832.2	2078.9	3262.5	6930.5	8895.6	2117.5	3358.8	7055.0	50
52	8834.3	2080.1	3265.7	6934.6	8897.7	2118.8	3362.0	7059.2	52
54	8836.4	2081.4	3268.8	6938.7	8899.8	2120.1	3365.5	7063.4	54
56	8838.6	2082.7	3272.0	6942.8	8901.9	2121.4	3368.7	7067.6	56
58	8340.7	2084.0	3275.2	6946.9	8904.0	2122.7	3372.0	7071.8	58
60	8842.8	2085.3	3278.3	6951.0	8906.1	2124.0	3375.1	7076.0	60

′	102°				103°				′
	L. C.	M.	E.	T.	L. C.	M.	E.	T.	
0	8906.1	2124.0	3375.1	7076.0	8968.7	2163.0	3474.6	7203.6	0
2	8908.2	2125.3	3378.3	7080.2	8970.8	2164.3	3478.0	7207.9	2
4	8910.3	2126.6	3381.6	7084.4	8972.9	2165.6	3481.4	7212.2	4
6	8912.4	2127.9	3384.9	7088.6	8974.9	2166.9	3484.7	7216.5	6
8	8914.5	2129.2	3388.2	7092.8	8977.0	2168.2	3488.1	7220.8	8
10	8916.6	2130.5	3391.5	7097.1	8979.1	2169.5	3491.5	7225.1	10
12	8918.7	2131.8	3394.7	7101.3	8981.1	2170.8	3494.9	7229.5	12
14	8920.8	2133.1	3398.0	7105.5	8983.2	2172.1	3498.3	7233.8	14
16	8922.9	2134.4	3401.3	7109.7	8985.3	2173.4	3501.6	7238.1	16
18	8925.0	2135.7	3404.6	7114.0	8987.3	2174.7	3505.3	7242.4	18
20	8927.0	2137.0	3407.9	7118.2	8989.4	2176.1	3508.4	7246.8	20
22	8929.1	2138.3	3411.2	7122.4	8991.5	2177.4	3511.8	7251.1	22
24	8931.2	2139.6	3414.5	7126.7	8993.5	2178.7	3515.2	7255.4	24
26	8933.3	2140.9	3417.9	7130.9	8995.6	2180.0	3518.7	7259.8	26
28	8935.4	2142.2	3421.2	7135.2	8997.7	2181.3	3522.1	7264.1	28
30	8937.5	2143.5	3424.5	7139.4	8999.7	2182.6	3525.5	7268.5	30
32	8939.6	2144.8	3427.8	7143.7	9001.8	2183.9	3528.9	7272.8	32
34	8941.6	2146.1	3431.1	7148.0	9003.9	2185.2	3532.3	7277.2	34
36	8943.7	2147.4	3434.5	7152.2	9005.9	2186.5	3535.7	7281.5	36
38	8945.8	2148.7	3437.8	7156.5	9008.0	2187.8	3539.2	7285.9	38
40	8947.9	2150.0	3441.1	7160.7	9010.0	2189.1	3542.6	7290.3	40
42	8950.0	2151.3	3444.4	7165.0	9012.1	2190.5	3546.0	7294.6	42
44	8952.1	2152.6	3447.8	7169.3	9014.2	2191.8	3549.5	7299.0	44
46	8954.1	2153.9	3451.1	7173.6	9016.2	2193.1	3552.9	7303.4	46
48	8956.2	2155.2	3454.5	7177.9	9018.3	2194.4	3556.3	7307.7	48
50	8958.3	2156.5	3457.8	7182.1	9020.3	2195.7	3559.8	7312.1	50
52	8960.4	2157.8	3461.2	7186.4	9022.4	2197.0	3563.2	7316.5	52
54	8962.5	2159.1	3464.5	7190.7	9024.5	2198.3	3566.7	7320.9	54
56	8964.5	2160.4	3467.9	7195.0	9026.5	2199.6	3570.2	7325.3	56
58	8966.6	2161.7	3471.2	7199.3	9028.6	2200.9	3573.6	7329.7	58
60	8968.7	2163.0	3474.6	7203.6	9030.6	2202.3	3577.1	7334.1	60

′	104°				105°				′
	L. C.	M.	E.	T.	L. C.	M.	E.	T.	
0	9030.6	2202.3	3577.1	7334.1	9091.8	2241.8	3682.6	7467.5	0
2	9032.7	2203.6	3580.5	7338.5	9093.9	2243.1	3686.1	7472.0	2
4	9034.7	2204.9	3584.0	7342.9	9095.9	2244.4	3689.7	7476.5	4
6	9036.8	2206.2	3587.5	7347.3	9097.9	2245.8	3693.3	7481.0	6
8	9038.8	2207.5	3591.0	7351.7	9099.9	2247.1	3696.9	7485.5	8
10	9040.9	2208.8	3594.4	7356.1	9102.0	2248.4	3700.4	7490.0	10
12	9042.9	2210.2	3597.9	7360.5	9104.0	2249.7	3704.0	7494.5	12
14	9045.0	2211.5	3601.4	7364.9	9106.0	2251.1	3707.6	7499.1	14
16	9047.0	2212.8	3604.9	7369.4	9108.0	2252.4	3711.2	7503.6	16
18	9049.1	2214.1	3608.4	7373.8	9110.1	2253.7	3714.8	7508.1	18
20	9051.1	2215.4	3611.9	7378.2	9112.1	2255.0	3718.4	7512.6	20
22	9053.1	2216.7	3615.4	7382.6	9114.1	2256.4	3722.0	7517.2	22
24	9055.2	2218.0	3618.9	7387.1	9116.1	2257.7	3725.6	7521.7	24
26	9057.2	2219.4	3622.4	7391.5	9118.1	2259.0	3729.3	7526.3	26
28	9059.3	2220.7	3625.9	7396.0	9120.2	2260.3	3732.9	7530.8	28
30	9061.3	2222.0	3629.4	7400.4	9122.2	2261.7	3736.5	7535.3	30
32	9063.3	2223.3	3633.0	7404.8	9124.2	2263.0	3740.1	7539.9	32
34	9065.4	2224.6	3636.5	7409.3	9126.2	2264.3	3743.7	7544.4	34
36	9067.4	2226.0	3640.0	7413.8	9128.2	2265.7	3747.4	7549.0	36
38	9069.5	2227.3	3643.5	7418.2	9130.2	2267.0	3751.0	7553.6	38
40	9071.5	2228.6	3647.1	7422.7	9132.3	2268.3	3754.6	7558.1	40
42	9073.5	2229.9	3650.6	7427.1	9134.3	2269.6	3758.3	7562.7	42
44	9075.6	2231.2	3654.1	7431.6	9136.3	2271.0	3761.9	7567.3	44
46	9077.6	2232.6	3657.7	7436.1	9138.3	2272.3	3765.6	7571.8	46
48	9079.6	2233.9	3661.2	7440.6	9140.3	2273.6	3769.2	7576.4	48
50	9081.7	2235.2	3664.8	7445.0	9142.3	2275.0	3772.9	7581.0	50
52	9083.7	2236.5	3668.3	7449.5	9144.3	2276.3	3776.5	7585.6	52
54	9085.7	2237.8	3671.9	7454.0	9146.3	2277.6	3780.2	7590.2	54
56	9087.8	2239.2	3675.4	7458.5	9148.3	2278.9	3783.9	7594.8	56
58	9089.8	2240.5	3679.0	7463.0	9150.4	2280.3	3787.5	7599.4	58
60	9091.8	2241.8	3682.6	7467.5	9152.4	2281.6	3791.2	7604.0	60

′	106° L. C.	106° M.	106° E.	106° T.	107° L. C.	107° M.	107° E.	107° T.	′
0	9152.4	2281.6	3791.2	7604.0	9212.2	2321.7	3903.1	7743.7	0
2	9154.4	2282.9	3794.9	7608.6	9214.2	2323.0	3906.9	7748.4	2
4	9156.4	2284.3	3798.6	7613.2	9216.2	2324.4	3910.7	7753.1	4
6	9158.4	2285.6	3802.3	7617.8	9218.1	2325.7	3914.5	7757.8	6
8	9160.4	2286.9	3805.9	7622.4	9220.1	2327.0	3918.3	7762.5	8
10	9162.4	2288.3	3809.6	7627.0	9222.1	2328.4	3922.1	7767.3	10
12	9164.4	2289.6	3813.3	7631.7	9224.1	2329.7	3925.9	7772.0	12
14	9166.4	2290.9	3817.0	7636.3	9226.1	2331.1	3929.7	7776.7	14
16	9168.4	2292.3	3820.7	7640.9	9228.1	2332.4	3933.6	7781.5	16
18	9170.4	2293.6	3824.4	7645.5	9230.0	2333.7	3937.4	7786.2	18
20	9172.4	2294.9	3828.1	7650.2	9232.0	2335.1	3941.2	7791.0	20
22	9174.4	2296.3	3831.9	7654.8	9234.0	2336.4	3945.0	7795.7	22
24	9176.4	2297.6	3835.6	7659.5	9235.9	2337.8	3948.9	7800.5	24
26	9178.4	2298.9	3839.3	7664.1	9237.9	2339.1	3952.7	7805.2	26
28	9180.4	2300.3	3843.0	7668.8	9239.9	2340.5	3956.5	7810.0	28
30	9182.4	2301.6	3846.7	7673.4	9241.9	2341.8	3960.4	7814.7	30
32	9184.4	2302.9	3850.5	7678.1	9243.8	2343.1	3964.2	7819.5	32
34	9186.4	2304.3	3854.2	7682.7	9245.8	2344.5	3968.1	7824.3	34
36	9188.4	2305.6	3858.0	7687.4	9247.8	2345.8	3971.9	7829.1	36
38	9190.4	2306.9	3861.7	7692.1	9249.7	2347.2	3975.8	7833.8	38
40	9192.4	2308.3	3865.4	7696.7	9251.7	2348.5	3979.6	7838.6	40
42	9194.4	2309.6	3869.2	7701.4	9253.7	2349.9	3983.5	7843.4	42
44	9196.3	2311.0	3873.0	7706.1	9255.6	2351.2	3987.4	7848.2	44
46	9198.3	2312.3	3876.7	7710.8	9257.6	2352.6	3991.3	7853.0	46
48	9200.3	2313.6	3880.5	7715.5	9259.6	2353.9	3995.1	7857.8	48
50	9202.3	2315.0	3884.2	7720.1	9261.5	2355.3	3999.0	7862.6	50
52	9204.3	2316.3	3888.0	7724.8	9263.5	2356.6	4002.9	7867.4	52
54	9206.3	2317.7	3891.8	7729.5	9265.4	2358.0	4006.8	7872.2	54
56	9208.2	2319.0	3895.6	7734.2	9267.4	2359.3	4010.7	7877.0	56
58	9210.2	2320.3	3899.3	7739.0	9269.4	2360.7	4014.6	7881.9	58
60	9212.2	2321.7	3903.1	7743.7	9271.3	2362.0	4018.5	7886.7	60

′	108° L. C.	108° M.	108° E.	108° T.	109° L. C.	109° M.	109° E.	109° T.	′
0	9271.3	2362.0	4018.5	7886.7	9329.8	2402.6	4137.4	8033.2	0
2	9273.3	2363.3	4022.4	7891.5	9331.7	2403.9	4141.4	8038.1	2
4	9275.3	2364.7	4026.3	7896.3	9333.6	2405.3	4145.4	8043.1	4
6	9277.2	2366.0	4030.2	7901.2	9335.6	2406.6	4149.5	8048.0	6
8	9279.2	2367.4	4034.1	7906.0	9337.5	2408.0	4153.5	8053.0	8
10	9281.1	2368.7	4038.0	7910.8	9339.4	2409.4	4157.5	8057.9	10
12	9283.1	2370.1	4042.0	7915.7	9341.4	2410.7	4161.6	8062.9	12
14	9285.0	2371.4	4045.9	7920.5	9343.3	2412.1	4165.6	8067.9	14
16	9287.0	2372.8	4049.8	7925.4	9345.2	2413.4	4169.7	8072.8	16
18	9288.9	2374.1	4053.8	7930.3	9347.2	2414.8	4173.8	8077.8	18
20	9290.9	2375.5	4057.7	7935.1	9349.1	2416.2	4177.8	8082.8	20
22	9292.8	2376.8	4061.6	7940.0	9351.0	2417.5	4181.9	8087.8	22
24	9294.8	2378.2	4065.6	7944.8	9352.9	2418.9	4186.0	8092.8	24
26	9296.7	2379.5	4069.5	7949.7	9354.9	2420.2	4190.0	8097.8	26
28	9298.7	2380.9	4073.5	7954.6	9356.8	2421.6	4193.1	8102.8	28
30	9300.6	2382.3	4077.5	7959.5	9358.7	2423.0	4198.2	8107.8	30
32	9302.6	2383.6	4081.4	7964.4	9360.6	2424.3	4202.3	8112.8	32
34	9304.5	2385.0	4085.4	7969.3	9362.6	2425.7	4206.4	8117.8	34
36	9306.5	2386.3	4089.4	7974.1	9364.5	2427.0	4210.5	8122.8	36
38	9308.4	2387.7	4093.4	7979.0	9366.4	2428.4	4214.6	8127.8	38
40	9310.4	2389.0	4097.3	7983.9	9368.3	2430.0	4218.7	8132.8	40
42	9312.3	2390.4	4101.3	7988.8	9370.2	2431.1	4222.8	8137.9	42
44	9314.2	2391.7	4105.3	7993.8	9372.2	2432.5	4226.9	8142.9	44
46	9316.2	2393.1	4109.3	7998.7	9374.1	2433.9	4231.0	8147.9	46
48	9318.1	2394.4	4113.3	8003.6	9376.0	2435.2	4235.1	8153.0	48
50	9320.1	2395.8	4117.3	8008.5	9377.9	2436.6	4239.3	8158.0	50
52	9322.0	2397.2	4121.3	8013.4	9379.8	2438.0	4243.4	8163.1	52
54	9323.9	2398.5	4125.3	8018.4	9381.7	2439.3	4247.5	8168.1	54
56	9325.9	2399.9	4129.3	8023.3	9383.7	2440.7	4251.7	8173.2	56
58	9327.8	2401.2	4133.4	8028.2	9385.6	2442.1	4255.8	8178.2	58
60	9329.8	2402.6	4137.4	8033.2	9387.5	2443.4	4260.0	8183.3	60

′	110° L. C.	110° M.	110° E.	110° T.	111° L. C.	111° M.	111° E.	111° T.	′
0	9387.5	2443.4	4260.0	8183.3	9444.5	2484.5	4386.4	8337.2	0
2	9389.4	2444.8	4264.1	8188.4	9446.4	2485.9	4390.7	8342.4	2
4	9391.3	2446.1	4268.3	8193.4	9448.3	2487.2	4395.0	8347.6	4
6	9393 2	2447.5	4272.4	8198.5	9450.1	2488.6	4399.3	8352.8	6
8	9395.1	2448.9	4276.6	8203.6	9452.0	2490.0	4403.6	8358.0	8
10	9397.0	2450.2	4280.8	8208.7	9453.9	2491.4	4407.9	8363.2	10
12	9398.9	2451 6	4284.9	8213.8	9455.8	2492.7	4412.2	8368.5	12
14	9400.8	2453.0	4289.1	8218.9	9457.7	2494.1	4416.5	8273.7	14
16	9402.7	2454.3	4293.3	8224.0	9459.6	2495.5	4420.8	8378 9	16
18	9404.7	2455.7	4297.5	8229.1	9461.4	2496.9	4425.1	8384 1	18
20	9406.6	2457.1	4301.7	8234.2	9463.3	2498.2	4429.5	8389.4	20
22	9408.5	2458.4	4305.9	8239.3	9465.2	2499.6	4433.8	8394 6	22
24	9410.4	2459 8	4310.1	8244.4	9467.1	2501.0	4438.1	8399.9	24
26	9412.3	2461.2	4314.3	8249.5	9469.0	2502.4	4442.5	8405.1	26
28	9414.2	2462.6	4318.5	8254.6	9470.8	2503.8	4446.8	8410.4	28
30	9416.1	2463.9	4322.7	8259.8	9472.7	2505.1	4451.2	8415.6	30
32	9418.0	2465.3	4326.9	8264.9	9474.6	2506.5	4455.5	8420.9	32
34	9419.9	2466.7	4331.1	8270.0	9476.5	2507.9	4459.9	8426.2	34
36	9421.8	2468.0	4335.4	8275.2	9478.3	2509.3	4464.2	8431.4	36
38	9423.7	2469.4	4339.6	8280.3	9480.2	2510.6	4468.6	8436.7	38
40	9425.6	2470.8	4343.8	8285.5	9482.1	2512.0	4473.0	8442.0	40
42	9427.5	2472.1	4348.1	8290.6	9484.0	2513.4	4477.3	8447.3	42
44	9429.3	2473.5	4352.3	8295.8	9485.8	2514.8	4481.7	8452.6	44
46	9431.2	2474.9	4356.6	8300.9	9487.7	2516.2	4486.1	8457.9	46
48	9433.1	2476.3	4360.8	8306.1	9489.6	2517.5	4490.5	8463.2	48
50	9435.0	2477.6	4365.1	8311.3	9491.4	2518.9	4494.9	8468.5	50
52	9436.9	2479.0	4369.3	8316.5	9493.3	2520.3	4499.3	8473.8	52
54	9438.8	2480.4	4373.6	8321.6	9495.2	2521.7	4503.7	8479.1	54
56	9440.7	2481.7	4377.9	8326.8	9497.0	2523.1	4508.1	8484.4	56
58	9442.6	2483.1	4382.2	8332.0	9498.9	2524.5	4512.5	8489.7	58
60	9444.5	2484.5	4386.4	8337 2	9500.8	2525.8	4516.9	8495.1	60

′	112° L. C.	112° M.	112° E.	112° T.	113° L. C.	113° M.	113° E.	113° T.	′
0	9500.8	2525.8	4516.9	8495.1	9556.3	2567.4	4651.6	8657.1	0
2	9502.6	2527.2	4521.4	8500.4	9558.2	2568.8	4656.2	8662.6	2
4	9504.5	2528.6	4525.8	8505.8	9560.0	2570.2	4660.8	8668.0	4
6	9506.4	2530.0	4530.2	8511.1	9561.8	2571.6	4665.4	8673.5	6
8	9508.2	2531.4	4534.6	8516.4	9563.7	2573.0	4669.9	8679.0	8
10	9510.1	2532.7	4539.1	8521.8	9565.5	2574.4	4674.5	8684.5	10
12	9511.9	2534.1	4543.5	8527.1	9567.4	2575.8	4679.1	8690.0	12
14	9513.8	2535.5	4548.0	8532.5	9569.2	2577.1	4683.7	8695.5	14
16	9515.7	2536.9	4552.4	8537.9	9571.0	2578.5	4688.3	8701.0	16
18	9517.5	2538.3	4556.9	8543.2	9572.9	2579.9	4692.9	8706.5	18
20	9519.4	2539.7	4561.3	8548.6	9574.7	2581.3	4697.5	8712.0	20
22	9521.2	2541.0	4565.8	8554.0	9576.5	2582.7	4702.1	8717.6	22
24	9523.1	2542.4	4570.3	8559 4	9578.4	2584.1	4706.8	8723.1	24
26	9524 9	2543.8	4574.8	8564.8	9580 2	2585.5	4711.4	8728.6	26
28	9526.8	2545.2	4579.3	8570 2	9582.0	2586.9	4716.0	8734.2	28
30	9528.6	2546.6	4583.7	8575.6	9583 8	2588.3	4720.6	8739.7	30
32	9530 5	2548.0	4588.2	8581.0	9585.7	2589.7	4725.3	8745.3	32
34	9532.3	2549.4	4592.7	8586.4	9587.5	2591.1	4729.9	8750.8	34
36	9534.2	2550.7	4597.2	8591.8	9589.3	2592.5	4734.6	8756.4	36
38	9536 0	2552.1	4601.7	8597.2	9591.1	2593.9	4739.2	8761.9	38
40	9537.9	2553.5	4606 2	8602.6	9593.0	2595.3	4743.9	8767.5	40
42	9539.7	2554.9	4610 8	8608.0	9594.8	2596.7	4748.5	8773.1	42
44	9541.6	2556.3	4615.3	8613.5	9596.6	2598.1	4753.2	8778.6	44
46	9543.4	2557.7	4619.8	8618.9	9598.4	2599.4	4757.9	8784.2	46
48	9545.3	2559.1	4624.3	8624.3	9600.3	2600.8	4762.6	8789.8	48
50	9547.1	2560.5	4628.9	8629.8	9602.1	2602.2	4767.2	8795.4	50
52	9549.0	2561.8	4633.4	8635.2	9603.9	2603.6	4771.9	8801.0	52
54	9550.8	2563 2	4638.0	8640.7	9605.7	2605 0	4776.6	8806.6	54
56	9552.6	2564.6	4642.5	8646.2	9607.5	2606.4	4781.3	8812.2	56
58	9554 5	2566.0	4647.1	8651.6	9609.4	2607.8	4786.0	8817.8	58
60	9556.3	2567.4	4651.6	8657.1	9611.2	2609.2	4790.7	8823.4	60

′	114°				115°				′
	L. C.	M.	E.	T.	L. C.	M.	E.	T.	
0	9611.2	2609.2	4790.7	8823.4	9665.3	2651.3	4934.4	8994.3	0
2	9613.0	2610.6	4795.5	8829.1	9667.1	2652.7	4939.3	9000.1	2
4	9614.8	2612.0	4800.2	8834.7	9668.8	2654.1	4944.2	9005.9	4
6	9616.6	2613.4	4804.9	8840.3	9670.6	2655.5	4949.1	9011.6	6
8	9618.4	2614.8	4809.6	8846.0	9672.4	2656.9	4954.0	9017.4	8
10	9620.2	2616.2	4814.4	8851.6	9674.2	2658.3	4958.9	9023.2	10
12	9622.0	2617.6	4819.1	8857.2	9676.0	2659.7	4963.8	9029.0	12
14	9623.8	2619.0	4823.9	8862.9	9677.8	2661.1	4968.7	9034.8	14
16	9625.7	2620.4	4828.6	8868.5	9679.6	2662.5	4973.6	9040.7	16
18	9627.5	2621.8	4833.4	8874.2	9681.4	2663.9	4978.5	9046.5	18
20	9629.3	2623.2	4838.1	8879.9	9683.1	2665.4	4983.4	9052.3	20
22	9631.1	2624.6	4812.9	8885.5	9684.9	2666.8	4968.3	9058.1	22
24	9632.9	2626.0	4847.7	8891.2	9686.7	2668.2	4993.3	9064.0	24
26	9634.7	2627.4	4852.4	8896.9	9688.5	2669.6	4998.2	9069.8	26
28	9636.5	2628.8	4857.2	8902.6	9690.3	2671.0	5003.2	9075.7	28
30	9638.3	2630.2	4862.0	8908.3	9692.0	2672.4	5008.1	9081.5	30
32	9640.1	2631.6	4866.8	8914.0	9693.8	2673.8	5013.1	9087.4	32
34	9641.9	2633.0	4871.6	8919.7	9695.6	2675.2	5018.0	9093.2	34
36	9643.7	2634.4	4876.4	8925.4	9697.4	2676.6	5023.0	9099.1	36
38	9645.5	2635.8	4881.2	8931.1	9699.1	2678.0	5028.0	9105.0	38
40	9647.3	2637.2	4885.0	8936.8	9700.9	2679.5	5032.9	9110.8	40
42	9649.1	2638.6	4890.9	8942.6	9702.7	2680.9	5037.9	9116.7	42
44	9650.9	2640.0	4895.7	8948.3	9704.5	2682.3	5042.9	9122.6	44
46	9652.7	2641.4	4900.5	8954.0	9706.2	2683.7	5047.9	9128.5	46
48	9654.5	2642.9	4905.3	8959.8	9708.0	2685.1	5052.9	9134.4	48
50	9656.3	2644.3	4910.2	8965.5	9709.8	2686.5	5057.9	9140.3	50
52	9658.1	2615.7	4915.0	8971.3	9711.6	2687.9	5062.9	9146.2	52
54	9659.9	2647.1	4919.9	8977.0	9713.3	2689.3	5067.9	9152.1	54
56	9661.7	2648.5	4924.7	8982.8	9715.1	2690.7	5072.9	9158.1	56
58	9663.5	2649.9	4929.6	8988.5	9716.9	2692.2	5078.0	9164.0	58
60	9665.3	2651.3	4934.4	8994.3	9718.6	2693.6	5083.0	9169.9	60

′	116°				117°				′
	L. C.	M.	E.	T.	L. C.	M.	E.	T.	
0	9718.6	2693.6	5083.0	9169.9	9771.3	2736.1	5236.6	9350.5	0
2	9720.4	2695.0	5088.0	9175.9	9773.0	2737.5	5241.8	9356.6	2
4	9722.2	2696.4	5093.1	9181.8	9774.7	2738.9	5247.0	9362.7	4
6	9723.9	2697.8	5098.1	9188.8	9776.5	2740.4	5252.2	9368.9	6
8	9725.7	2699.2	5103.2	9193.7	9778.2	2741.8	5257.4	9375.0	8
10	9727.4	2700.6	5108.2	9199.7	9779.9	2743.2	5262.6	9381.1	10
12	9729.2	2702.1	5113.3	9205.6	9781.7	2744.6	5267.9	9387.3	12
14	9731.0	2703.5	5118.4	9211.6	9783.4	2746.0	5273.1	9393.4	14
16	9732.7	2704.9	5123.4	9217.6	9785.2	2747.5	5278.4	9399.5	16
18	9734.5	2706.3	5128.5	9223.6	9786.9	2748.9	5283.6	9405.7	18
20	9736.3	2707.7	5133.6	9229.6	9788.6	2750.3	5288.9	9411.9	20
22	9738.0	2709.1	5138.7	9235.5	9790.4	2751.7	5294.2	9418.0	22
24	9739.8	2710.6	5143.8	9241.5	9792.1	2753.2	5299.5	9424.2	24
26	9741.5	2712.0	5148.9	9247.6	9793.8	2754.6	5304.7	9430.4	26
28	9743.3	2713.4	5154.0	9253.6	9795.6	2756.0	5310.0	9436.6	28
30	9745.0	2714.8	5159.1	9259.6	9797.3	2757.4	5315.3	9442.8	30
32	9746.8	2716.2	5164.2	9265.6	9790.0	2758.9	5320.6	9449.0	32
34	9748.5	2717.6	5169.4	9271.6	9800.7	2760.3	5325.9	9455.2	34
36	9750.3	2719.1	5174.5	9277.7	9802.5	2761.7	5331.2	9461.4	36
38	9752.0	2720.5	5179.7	9283.7	9804.2	2763.1	5336.5	9467.6	38
40	9753.8	2721.9	5184.8	9289.8	9805.9	2764.6	5341.8	9473.8	40
42	9755.6	2723.3	5190.0	9295.8	9807.7	2766.0	5347.2	9480.0	42
44	9757.3	2724.7	5195.1	9301.9	9809.4	2767.4	5352.5	9486.3	44
46	9759.0	2726.2	5200.3	9307.9	9811.1	2768.8	5357.9	9192.5	46
48	9760.8	2727.6	5205.4	9314.0	9812.8	2770.3	5363.2	9498.7	48
50	9762.5	2729.0	5210.6	9320.1	9814.5	2771.7	5368.5	9505.0	50
52	9764.3	2730.4	5215.8	9326.1	9816.3	2773.1	5373.9	9511.2	52
54	9766.0	2731.8	5221.0	9332.2	9818.0	2774.6	5379.3	9517.5	54
56	9767.8	2733.3	5226.2	9338.3	9819.7	2776.0	5384.7	9523.8	56
58	9769.5	2734.7	5231.4	9314.4	9821.4	2777.4	5390.0	9530.0	58
60	9771.3	2736.1	5236.6	9350.5	9823.1	2778.8	5395.4	9536.3	60

′	0°		1°		2°		3°		4°		
	Sine	Cosin	Sine	Cosin	Sine	Cosin	Sine	Cosin	Sine	Cosin	
0	.00000	One.	.01745	.99985	.03490	.99939	.05234	.99863	.06976	.99756	60
1	.00029	One.	.01774	.99984	.03519	.99938	.05263	.99861	.07005	.99754	59
2	.00058	One.	.01803	.99984	.03548	.99937	.05292	.99860	.07034	.99752	58
3	.00087	One.	.01832	.99983	.03577	.99936	.05321	.99858	.07063	.99750	57
4	.00116	One.	.01862	.99983	.03606	.99935	.05350	.99857	.07092	.99748	56
5	.00145	One.	.01891	.99982	.03635	.99934	.05379	.99855	.07121	.99746	55
6	.00175	One.	.01920	.99982	.03664	.99933	.05408	.99854	.07150	.99744	54
7	.00204	One.	.01949	.99981	.03693	.99932	.05437	.99852	.07179	.99742	53
8	.00233	One.	.01978	.99980	.03723	.99931	.05466	.99851	.07208	.99740	52
9	.00262	One.	.02007	.99980	.03752	.99930	.05495	.99849	.07237	.99738	51
10	.00291	One.	.02036	.99979	.03781	.99929	.05524	.99847	.07266	.99736	50
11	.00320	.99999	.02065	.99979	.03810	.99927	.05553	.99846	.07295	.99734	49
12	.00349	.99999	.02094	.99978	.03839	.99926	.05582	.99844	.07324	.99731	48
13	.00378	.99999	.02123	.99977	.03868	.99925	.05611	.99842	.07353	.99729	47
14	.00407	.99999	.02152	.99977	.03897	.99924	.05640	.99841	.07382	.99727	46
15	.00436	.99999	.02181	.99976	.03926	.99923	.05669	.99839	.07411	.99725	45
16	.00465	.99999	.02211	.99976	.03955	.99922	.05698	.99838	.07440	.99723	44
17	.00495	.99999	.02240	.99975	.03984	.99921	.05727	.99836	.07469	.99721	43
18	.00524	.99999	.02269	.99974	.04013	.99919	.05756	.99834	.07498	.99719	42
19	.00553	.99998	.02298	.99974	.04042	.99918	.05785	.99833	.07527	.99716	41
20	.00582	.99998	.02327	.99973	.04071	.99917	.05814	.99831	.07556	.99714	40
21	.00611	.99998	.02356	.99972	.04100	.99916	.05944	.99829	.07585	.99712	39
22	.00640	.99998	.02385	.99972	.04129	.99915	.05873	.99827	.07614	.99710	38
23	.00669	.99998	.02414	.99971	.04159	.99913	.05902	.99826	.07643	.99708	37
24	.00698	.99998	.02443	.99970	.04188	.99912	.05931	.99824	.07672	.99705	36
25	.00727	.99997	.02472	.99969	.04217	.99911	.05960	.99822	.07701	.99703	35
26	.00756	.99997	.02501	.99969	.04246	.99910	.05989	.99821	.07730	.99701	34
27	.00785	.99997	.02530	.99968	.04275	.99909	.06018	.99819	.07759	.99699	33
28	.00814	.99997	.02560	.99967	.04304	.99907	.06047	.99817	.07788	.99696	32
29	.00844	.99996	.02589	.99966	.04333	.99906	.06076	.99815	.07817	.99694	31
30	.00873	.99996	.02618	.99966	.04362	.99905	.06105	.99813	.07846	.99692	30
31	.00902	.99996	.02647	.99965	.04391	.99904	.06134	.99812	.07875	.99689	29
32	.00931	.99996	.02676	.99964	.04420	.99902	.06163	.99810	.07904	.99687	28
33	.00960	.99995	.02705	.99963	.04449	.99901	.06192	.99808	.07933	.99685	27
34	.00989	.99995	.02734	.99963	.04478	.99900	.06221	.99806	.07962	.99683	26
35	.01018	.99995	.02763	.99962	.04507	.99898	.06250	.99804	.07991	.99680	25
36	.01047	.99995	.02792	.99961	.04536	.99897	.06279	.99803	.08020	.99678	24
37	.01076	.99994	.02821	.99960	.04565	.99896	.06308	.99801	.08049	.99676	23
38	.01105	.99994	.02850	.99959	.04594	.99894	.06337	.99799	.08078	.99673	22
39	.01134	.99994	.02879	.99959	.04623	.99893	.06366	.99797	.08107	.99671	21
40	.01164	.99993	.02908	.99958	.04653	.99892	.06395	.99795	.08136	.99668	20
41	.01193	.99993	.02938	.99957	.04682	.99890	.06424	.99793	.08165	.99666	19
42	.01222	.99993	.02967	.99956	.04711	.99889	.06453	.99792	.08194	.99664	18
43	.01251	.99992	.02996	.99955	.04740	.99888	.06482	.99790	.08223	.99661	17
44	.01280	.99992	.03025	.99954	.04769	.99886	.06511	.99788	.08252	.99659	16
45	.01309	.99991	.03054	.99953	.04798	.99885	.06540	.99786	.08281	.99657	15
46	.01338	.99991	.03083	.99952	.04827	.99883	.06569	.99784	.08310	.99651	14
47	.01367	.99991	.03112	.99952	.04856	.99882	.06598	.99782	.08339	.99652	13
48	.01396	.99990	.03141	.99951	.04885	.99881	.06627	.99780	.08368	.99649	12
49	.01425	.99990	.03170	.99950	.04914	.99879	.06656	.99778	.08397	.99647	11
50	.01454	.99989	.03199	.99949	.04943	.99878	.06685	.99776	.08426	.99644	10
51	.01483	.99989	.03228	.99948	.04972	.99876	.06714	.99774	.08455	.99642	9
52	.01513	.99989	.03257	.99947	.05001	.99875	.06743	.99772	.08484	.99639	8
53	.01542	.99988	.03286	.99946	.05030	.99873	.06773	.99770	.08513	.99637	7
51	.01571	.99988	.03316	.99945	.05059	.99872	.06802	.99768	.08542	.99635	6
55	.01600	.99987	.03345	.99944	.05088	.99870	.06831	.99766	.08571	.99632	5
56	.01629	.99987	.03374	.99943	.05117	.99869	.06860	.99764	.08600	.99630	4
57	.01658	.99986	.03403	.99942	.05146	.99867	.06889	.99762	.08629	.99627	3
58	.01687	.99986	.03432	.99941	.05175	.99866	.06918	.99760	.08658	.99625	2
59	.01716	.99985	.03461	.99940	.05205	.99864	.06947	.99758	.08687	.99622	1
60	.01745	.99985	.03490	.99939	.05234	.99863	.06976	.99756	.08716	.99619	0
	Cosin	Sine	Cosin	Sine	Cosin	Sine	Cosin	Sine	Cosin	Sine	′
′	89°		88°		87°		86°		85°		

′	5° Sine	5° Cosin	6° Sine	6° Cosin	7° Sine	7° Cosin	8° Sine	8° Cosin	9° Sine	9° Cosin	′
0	.08716	.99619	.10453	.99452	.12187	.99255	.13917	.99027	.15643	.98769	60
1	.08745	.99617	.10482	.99449	.12216	.99251	.13946	.99023	.15672	.98764	59
2	.08774	.99614	.10511	.99446	.12245	.99248	.13975	.99019	.15701	.98760	58
3	.08803	.99612	.10540	.99443	.12274	.99244	.14004	.99015	.15730	.98755	57
4	.08831	.99609	.10569	.99440	.12302	.99240	.14033	.99011	.15758	.98751	56
5	.08860	.99607	.10597	.99437	.12331	.99237	.14061	.99006	.15787	.98746	55
6	.08889	.99604	.10626	.99434	.12360	.99233	.14090	.99002	.15816	.98741	54
7	.08918	.99602	.10655	.99431	.12389	.99230	.14119	.98998	.15845	.98737	53
8	.08947	.99599	.10684	.99428	.12418	.99226	.14148	.98994	.15873	.98732	52
9	.08976	.99596	.10713	.99424	.12447	.99222	.14177	.98990	.15902	.98728	51
10	.09005	.99594	.10742	.99421	.12476	.99219	.14205	.98986	.15931	.98723	50
11	.09034	.99591	.10771	.99418	.12504	.99215	.14234	.98982	.15959	.98718	49
12	.09063	.99588	.10800	.99415	.12533	.99211	.14263	.98978	.15988	.98714	48
13	.09092	.99586	.10829	.99412	.12562	.99208	.14292	.98973	.16017	.98709	47
14	.09121	.99583	.10858	.99409	.12591	.99204	.14320	.98969	.16046	.98704	46
15	.09150	.99580	.10887	.99406	.12620	.99200	.14349	.98965	.16074	.98700	45
16	.09179	.99578	.10916	.99402	.12649	.99197	.14378	.98961	.16103	.98695	44
17	.09208	.99575	.10945	.99399	.12678	.99193	.14407	.98957	.16132	.98690	43
18	.09237	.99572	.10973	.99396	.12706	.99189	.14436	.98953	.16160	.98686	42
19	.09266	.99570	.11002	.99393	.12735	.99186	.14464	.98948	.16189	.98681	41
20	.09295	.99567	.11031	.99390	.12764	.99182	.14493	.98944	.16218	.98676	40
21	.09324	.99564	.11060	.99386	.12793	.99178	.14522	.98940	.16246	.98671	39
22	.09353	.99562	.11089	.99383	.12822	.99175	.14551	.98936	.16275	.98667	38
23	.09382	.99559	.11118	.99380	.12851	.99171	.14580	.98931	.16304	.98662	37
24	.09411	.99556	.11147	.99377	.12880	.99167	.14608	.98927	.16333	.98657	36
25	.09440	.99553	.11176	.99374	.12908	.99163	.14637	.98923	.16361	.98652	35
26	.09469	.99551	.11205	.99370	.12937	.99160	.14666	.98919	.16390	.98648	34
27	.09498	.99548	.11234	.99367	.12966	.99156	.14695	.98914	.16419	.98643	33
28	.09527	.99545	.11263	.99364	.12995	.99152	.14723	.98910	.16447	.98638	32
29	.09556	.99542	.11291	.99360	.13024	.99148	.14752	.98906	.16476	.98633	31
30	.09585	.99540	.11320	.99357	.13053	.99144	.14781	.98902	.16505	.98629	30
31	.09614	.99537	.11349	.99354	.13081	.99141	.14810	.98897	.16533	.98624	29
32	.09642	.99534	.11378	.99351	.13110	.99137	.14838	.98893	.16562	.98619	28
33	.09671	.99531	.11407	.99347	.13139	.99133	.14867	.98889	.16591	.98614	27
34	.09700	.99528	.11436	.99344	.13168	.99129	.14896	.98884	.16620	.98609	26
35	.09729	.99526	.11465	.99341	.13197	.99125	.14925	.98880	.16648	.98604	25
36	.09758	.99523	.11494	.99337	.13226	.99122	.14954	.98876	.16677	.98600	24
37	.09787	.99520	.11523	.99334	.13254	.99118	.14982	.98871	.16706	.98595	23
38	.09816	.99517	.11552	.99331	.13283	.99114	.15011	.98867	.16734	.98590	22
39	.09845	.99514	.11580	.99327	.13312	.99110	.15040	.98863	.16763	.98585	21
40	.09874	.99511	.11609	.99324	.13341	.99106	.15069	.98858	.16792	.98580	20
41	.09903	.99508	.11638	.99320	.13370	.99102	.15097	.98854	.16820	.98575	19
42	.09932	.99506	.11667	.99317	.13399	.99098	.15126	.98849	.16849	.98570	18
43	.09961	.99503	.11696	.99314	.13427	.99094	.15155	.98845	.16878	.98565	17
44	.09990	.99500	.11725	.99310	.13456	.99091	.15184	.98841	.16906	.98561	16
45	.10019	.99497	.11754	.99307	.13485	.99087	.15212	.98836	.16935	.98556	15
46	.10048	.99494	.11783	.99303	.13514	.99083	.15241	.98832	.16964	.98551	14
47	.10077	.99491	.11812	.99300	.13543	.99079	.15270	.98827	.16992	.98546	13
48	.10106	.99488	.11840	.99297	.13572	.99075	.15299	.98823	.17021	.98541	12
49	.10135	.99485	.11869	.99293	.13600	.99071	.15327	.98818	.17050	.98536	11
50	.10164	.99482	.11898	.99290	.13629	.99067	.15356	.98814	.17078	.98531	10
51	.10192	.99479	.11927	.99286	.13658	.99063	.15385	.98809	.17107	.98526	9
52	.10221	.99476	.11956	.99283	.13687	.99059	.15414	.98805	.17136	.98521	8
53	.10250	.99473	.11985	.99279	.13716	.99055	.15442	.98800	.17164	.98516	7
54	.10279	.99470	.12014	.99276	.13744	.99051	.15471	.98796	.17193	.98511	6
55	.10308	.99467	.12043	.99272	.13773	.99047	.15500	.98791	.17222	.98506	5
56	.10337	.99464	.12071	.99269	.13802	.99043	.15529	.98787	.17250	.98501	4
57	.10366	.99461	.12100	.99265	.13831	.99039	.15557	.98782	.17279	.98496	3
58	.10395	.99458	.12129	.99262	.13860	.99035	.15586	.98778	.17308	.98491	2
59	.10424	.99455	.12158	.99258	.13889	.99031	.15615	.98773	.17336	.98486	1
60	.10453	.99452	.12187	.99255	.13917	.99027	.15643	.98769	.17365	.98481	0
′	Cosin	Sine	Cosin	Sine	Cosin	Sine	Cosin	Sine	Cosin	Sine	′
	84°		83°		82°		81°		80°		

′	10° Sine	10° Cosin	11° Sine	11° Cosin	12° Sine	12° Cosin	13° Sine	13° Cosin	14° Sine	14° Cosin	′
0	.17365	.98481	.19081	.98163	.20791	.97815	.22495	.97437	.24192	.97030	60
1	.17393	.98476	.19109	.98157	.20820	.97809	.22523	.97430	.24220	.97023	59
2	.17422	.98471	.19138	.98152	.20848	.97803	.22552	.97424	.24249	.97015	58
3	.17451	.98466	.19167	.98146	.20877	.97797	.22580	.97417	.24277	.97008	57
4	.17479	.98461	.19195	.98140	.20905	.97791	.22608	.97411	.24305	.97001	56
5	.17508	.98455	.19224	.98135	.20933	.97784	.22637	.97404	.24333	.96994	55
6	.17537	.98450	.19252	.98129	.20962	.97778	.22665	.97398	.24362	.96987	54
7	.17565	.98445	.19281	.98124	.20990	.97772	.22693	.97391	.24390	.96980	53
8	.17594	.98440	.19309	.98118	.21019	.97766	.22722	.97384	.24418	.96973	52
9	.17623	.98435	.19338	.98112	.21047	.97760	.22750	.97378	.24446	.96966	51
10	.17651	.98430	.19366	.98107	.21076	.97754	.22778	.97371	.24474	.96959	50
11	.17680	.98425	.19395	.98101	.21104	.97748	.22807	.97365	.24503	.96952	49
12	.17708	.98420	.19423	.98096	.21132	.97742	.22835	.97358	.24531	.96945	48
13	.17737	.98414	.19452	.98090	.21161	.97735	.22863	.97351	.24559	.96937	47
14	.17766	.98409	.19481	.98084	.21189	.97729	.22892	.97345	.24587	.96930	46
15	.17794	.98404	.19509	.98079	.21218	.97723	.22920	.97338	.24615	.96923	45
16	.17823	.98399	.19538	.98073	.21246	.97717	.22948	.97331	.24644	.96916	44
17	.17852	.98394	.19566	.98067	.21275	.97711	.22977	.97325	.24672	.96909	43
18	.17880	.98389	.19595	.98061	.21303	.97705	.23005	.97318	.24700	.96902	42
19	.17909	.98383	.19623	.98056	.21331	.97698	.23033	.97311	.24728	.96894	41
20	.17937	.98378	.19652	.98050	.21360	.97692	.23062	.97304	.24756	.96887	40
21	.17966	.98373	.19680	.98044	.21388	.97686	.23090	.97298	.24784	.96880	39
22	.17995	.98368	.19709	.98039	.21417	.97680	.23118	.97291	.24813	.96873	38
23	.18023	.98362	.19737	.98033	.21445	.97673	.23146	.97284	.24841	.96866	37
24	.18052	.98357	.19766	.98027	.21474	.97667	.23175	.97278	.24869	.96858	36
25	.18081	.98352	.19794	.98021	.21502	.97661	.23203	.97271	.24897	.96851	35
26	.18109	.98347	.19823	.98016	.21530	.97655	.23231	.97264	.24925	.96844	34
27	.18138	.98341	.19851	.98010	.21559	.97648	.23260	.97257	.24954	.96837	33
28	.18166	.98336	.19880	.98004	.21587	.97642	.23288	.97251	.24982	.96829	32
29	.18195	.98331	.19908	.97998	.21616	.97636	.23316	.97244	.25010	.96822	31
30	.18224	.98325	.19937	.97992	.21644	.97630	.23345	.97237	.25038	.96815	30
31	.18252	.98320	.19965	.97987	.21672	.97623	.23373	.97230	.25066	.96807	29
32	.18281	.98315	.19994	.97981	.21701	.97617	.23401	.97223	.25094	.96800	28
33	.18309	.98310	.20022	.97975	.21729	.97611	.23429	.97217	.25122	.96793	27
34	.18338	.98304	.20051	.97969	.21758	.97604	.23458	.97210	.25151	.96786	26
35	.18367	.98299	.20079	.97963	.21786	.97598	.23486	.97203	.25179	.96778	25
36	.18395	.98294	.20108	.97958	.21814	.97592	.23514	.97196	.25207	.96771	24
37	.18424	.98288	.20136	.97952	.21843	.97585	.23542	.97189	.25235	.96764	23
38	.18452	.98283	.20165	.97946	.21871	.97579	.23571	.97182	.25263	.96756	22
39	.18481	.98277	.20193	.97940	.21899	.97573	.23599	.97176	.25291	.96749	21
40	.18509	.98272	.20222	.97934	.21928	.97566	.23627	.97169	.25320	.96742	20
41	.18538	.98267	.20250	.97928	.21956	.97560	.23656	.97162	.25348	.96734	19
42	.18567	.98261	.20279	.97922	.21985	.97553	.23684	.97155	.25376	.96727	18
43	.18595	.98256	.20307	.97916	.22013	.97547	.23712	.97148	.25404	.96719	17
44	.18624	.98250	.20336	.97910	.22041	.97541	.23740	.97141	.25432	.96712	16
45	.18652	.98245	.20364	.97905	.22070	.97534	.23769	.97134	.25460	.96705	15
46	.18681	.98240	.20393	.97899	.22098	.97528	.23797	.97127	.25488	.96697	14
47	.18710	.98234	.20421	.97893	.22126	.97521	.23825	.97120	.25516	.96690	13
48	.18738	.98229	.20450	.97887	.22155	.97515	.23853	.97113	.25545	.96682	12
49	.18767	.98223	.20478	.97881	.22183	.97508	.23882	.97106	.25573	.96675	11
50	.18795	.98218	.20507	.97875	.22212	.97502	.23910	.97100	.25601	.96667	10
51	.18824	.98212	.20535	.97869	.22240	.97496	.23938	.97093	.25629	.96660	9
52	.18852	.98207	.20563	.97863	.22268	.97489	.23966	.97086	.25657	.96653	8
53	.18881	.98201	.20592	.97857	.22297	.97483	.23995	.97079	.25685	.96645	7
54	.18910	.98196	.20620	.97851	.22325	.97476	.24023	.97072	.25713	.96638	6
55	.18938	.98190	.20649	.97845	.22353	.97470	.24051	.97065	.25741	.96630	5
56	.18967	.98185	.20677	.97839	.22382	.97463	.24079	.97058	.25769	.96623	4
57	.18995	.98179	.20706	.97833	.22410	.97457	.24108	.97051	.25798	.96615	3
58	.19024	.98174	.20734	.97827	.22438	.97450	.24136	.97044	.25826	.96608	2
59	.19052	.98168	.20763	.97821	.22467	.97444	.24164	.97037	.25854	.96600	1
60	.19081	.98163	.20791	.97815	.22495	.97437	.24192	.97030	.25882	.96593	0
′	Cosin	Sine	Cosin	Sine	Cosin	Sine	Cosin	Sine	Cosin	Sine	′
	79°		78°		77°		76°		75°		

′	15° Sine	15° Cosin	16° Sine	16° Cosin	17° Sine	17° Cosin	18° Sine	18° Cosin	19° Sine	19° Cosin	′
0	.25882	.96593	.27564	.96126	.29237	.95630	.30902	.95106	.32557	.94552	60
1	.25910	.96585	.27592	.96118	.29265	.95622	.30929	.95097	.32584	.94542	59
2	.25938	.96578	.27620	.96110	.29293	.95613	.30957	.95088	.32612	.94533	58
3	.25966	.96570	.27648	.96102	.29321	.95605	.30985	.95079	.32639	.94523	57
4	.25994	.96562	.27676	.96094	.29348	.95596	.31012	.95070	.32667	.94514	56
5	.26022	.96555	.27704	.96086	.29376	.95588	.31040	.95061	.32694	.94504	55
6	.26050	.96547	.27731	.96078	.29404	.95579	.31068	.95052	.32722	.94495	54
7	.26079	.96540	.27759	.96070	.29432	.95571	.31095	.95043	.32749	.94485	53
8	.26107	.96532	.27787	.96062	.29460	.95562	.31123	.95033	.32777	.94476	52
9	.26135	.96524	.27815	.96054	.29487	.95554	.31151	.95024	.32804	.94466	51
10	.26163	.96517	.27843	.96046	.29515	.95545	.31178	.95015	.32832	.94457	50
11	.26191	.96509	.27871	.96037	.29543	.95536	.31206	.95006	.32859	.94447	49
12	.26219	.96502	.27899	.96029	.29571	.95528	.31233	.94997	.32887	.94438	48
13	.26247	.96494	.27927	.96021	.29599	.95519	.31261	.94988	.32914	.94428	47
14	.26275	.96486	.27955	.96013	.29626	.95511	.31289	.94979	.32942	.94418	46
15	.26303	.96479	.27983	.96005	.29654	.95502	.31316	.94970	.32969	.94409	45
16	.26331	.96471	.28011	.95997	.29682	.95493	.31344	.94961	.32997	.94399	44
17	.26359	.96463	.28039	.95989	.29710	.95485	.31372	.94952	.33024	.94390	43
18	.26387	.96456	.28067	.95981	.29737	.95476	.31399	.94943	.33051	.94380	42
19	.26415	.96448	.28095	.95972	.29765	.95467	.31427	.94933	.33079	.94370	41
20	.26443	.96440	.28123	.95964	.29793	.95459	.31454	.94924	.33106	.94361	40
21	.26471	.96433	.28150	.95956	.29821	.95450	.31482	.94915	.33134	.94351	39
22	.26500	.96425	.28178	.95948	.29849	.95441	.31510	.94906	.33161	.94342	38
23	.26528	.96417	.28206	.95940	.29876	.95433	.31537	.94897	.33189	.94332	37
24	.26556	.96410	.28234	.95931	.29904	.95424	.31565	.94888	.33216	.94322	36
25	.26584	.96402	.28262	.95923	.29932	.95415	.31593	.94878	.33244	.94313	35
26	.26612	.96394	.28290	.95915	.29960	.95407	.31620	.94869	.33271	.94303	34
27	.26640	.96386	.28318	.95907	.29987	.95398	.31648	.94860	.33298	.94293	33
28	.26668	.96379	.28346	.95898	.30015	.95389	.31675	.94851	.33326	.94284	32
29	.26696	.96371	.28374	.95890	.30043	.95380	.31703	.94842	.33353	.94274	31
30	.26724	.96363	.28402	.95882	.30071	.95372	.31730	.94832	.33381	.94264	30
31	.26752	.96355	.28429	.95874	.30098	.95363	.31758	.94823	.33408	.94254	29
32	.26780	.96347	.28457	.95865	.30126	.95354	.31786	.94814	.33436	.94245	28
33	.26808	.96340	.28485	.95857	.30154	.95345	.31813	.94805	.33463	.94235	27
34	.26836	.96332	.28513	.95849	.30182	.95337	.31841	.94795	.33490	.94225	26
35	.26864	.96324	.28541	.95841	.30209	.95328	.31868	.94786	.33518	.94215	25
36	.26892	.96316	.28569	.95832	.30237	.95319	.31896	.94777	.33545	.94206	24
37	.26920	.96308	.28597	.95824	.30265	.95310	.31923	.94768	.33573	.94196	23
38	.26948	.96301	.28625	.95816	.30292	.95301	.31951	.94758	.33600	.94186	22
39	.26976	.96293	.28652	.95807	.30320	.95293	.31979	.94749	.33627	.94176	21
40	.27004	.96285	.28680	.95799	.30348	.95284	.32006	.94740	.33655	.94167	20
41	.27032	.96277	.28708	.95791	.30376	.95275	.32034	.94730	.33682	.94157	19
42	.27060	.96269	.28736	.95782	.30403	.95266	.32061	.94721	.33710	.94147	18
43	.27088	.96261	.28764	.95774	.30431	.95257	.32089	.94712	.33737	.94137	17
44	.27116	.96253	.28792	.95766	.30459	.95248	.32116	.94702	.33764	.94127	16
45	.27144	.96246	.28820	.95757	.30486	.95240	.32144	.94693	.33792	.94118	15
46	.27172	.96238	.28847	.95749	.30514	.95231	.32171	.94684	.33819	.94108	14
47	.27200	.96230	.28875	.95740	.30542	.95222	.32199	.94674	.33846	.94098	13
48	.27228	.96222	.28903	.95732	.30570	.95213	.32227	.94665	.33874	.94088	12
49	.27256	.96214	.28931	.95724	.30597	.95204	.32254	.94656	.33901	.94078	11
50	.27284	.96206	.28959	.95715	.30625	.95195	.32282	.94646	.33929	.94068	10
51	.27312	.96198	.28987	.95707	.30653	.95186	.32309	.94637	.33956	.94058	9
52	.27340	.96190	.29015	.95698	.30680	.95177	.32337	.94627	.33983	.94049	8
53	.27368	.96182	.29042	.95690	.30708	.95168	.32364	.94618	.34011	.94039	7
54	.27396	.96174	.29070	.95681	.30736	.95159	.32392	.94609	.34038	.94029	6
55	.27424	.96166	.29098	.95673	.30763	.95150	.32419	.94599	.34065	.94019	5
56	.27452	.96158	.29126	.95664	.30791	.95142	.32447	.94590	.34093	.94009	4
57	.27480	.96150	.29154	.95656	.30819	.95133	.32474	.93580	.34120	.93999	3
58	.27508	.96142	.29182	.95647	.30846	.95124	.32502	.94571	.34147	.93989	2
59	.27536	.96134	.29209	.95639	.30874	.95115	.32529	.94561	.34175	.93979	1
60	.27564	.96126	.29237	.95630	.30902	.95106	.32557	.94552	.34202	.93969	0
′	Cosin	Sine	Cosin	Sine	Cosin	Sine	Cosin	Sine	Cosin	Sine	′
	74°		73°		72°		71°		70°		

′	20° Sine	20° Cosin	21° Sine	21° Cosin	22° Sine	22° Cosin	23° Sine	23° Cosin	24° Sine	24° Cosin	′
0	.34202	.93969	.35837	.93358	.37461	.92718	.39073	.92050	.40674	.91355	60
1	.34229	.93959	.35864	.93348	.37488	.92707	.39100	.92039	.40700	.91343	59
2	.34257	.93949	.35891	.93337	.37515	.92697	.39127	.92028	.40727	.91331	58
3	.34284	.93939	.35918	.93327	.37542	.92686	.39153	.92016	.40753	.91319	57
4	.34311	.93929	.35945	.93316	.37569	.92675	.39180	.92005	.40780	.91307	56
5	.34339	.93919	.35973	.93306	.37595	.92664	.39207	.91994	.40806	.91295	55
6	.34366	.93909	.36000	.93295	.37622	.92653	.39234	.91982	.40833	.91283	54
7	.34393	.93899	.36027	.93285	.37649	.92642	.39260	.91971	.40860	.91272	53
8	.34421	.93889	.36054	.93274	.37676	.92631	.39287	.91959	.40886	.91260	52
9	.34448	.93879	.36081	.93264	.37703	.92620	.39314	.91948	.40913	.91248	51
10	.34475	.93869	.36108	.93253	.37730	.92609	.39341	.91936	.40939	.91236	50
11	.34503	.93859	.36135	.93243	.37757	.92598	.39367	.91925	.40966	.91224	49
12	.34530	.93849	.36162	.93232	.37784	.92587	.39394	.91914	.40992	.91212	48
13	.34557	.93839	.36190	.93222	.37811	.92576	.39421	.91902	.41019	.91200	47
14	.34584	.93829	.36217	.93211	.37838	.92565	.39448	.91891	.41045	.91188	46
15	.34612	.93819	.36244	.93201	.37865	.92554	.39474	.91879	.41072	.91176	45
16	.34639	.93809	.36271	.93190	.37892	.92543	.39501	.91868	.41098	.91164	44
17	.34666	.93799	.36298	.93180	.37919	.92532	.39528	.91856	.41125	.91152	43
18	.34694	.93789	.36325	.93169	.37946	.92521	.39555	.91845	.41151	.91140	42
19	.34721	.93779	.36352	.93159	.37973	.92510	.39581	.91833	.41178	.91128	41
20	.34748	.93769	.36379	.93148	.37999	.92499	.39608	.91822	.41204	.91116	40
21	.34775	.93759	.36406	.93137	.38026	.92488	.39635	.91810	.41231	.91104	39
22	.34803	.93748	.36434	.93127	.38053	.92477	.39661	.91799	.41257	.91092	38
23	.34830	.93738	.36461	.93116	.38080	.92466	.39688	.91787	.41284	.91080	37
24	.34857	.93728	.36488	.93106	.38107	.92455	.39715	.91775	.41310	.91068	36
25	.34884	.93718	.36515	.93095	.38134	.92444	.39741	.91764	.41337	.91056	35
26	.34912	.93708	.36542	.93084	.38161	.92432	.39768	.91752	.41363	.91044	34
27	.34939	.93698	.36569	.93074	.38188	.92421	.39795	.91741	.41390	.91032	33
28	.34966	.93688	.36596	.93063	.38215	.92410	.39822	.91729	.41416	.91020	32
29	.34993	.93677	.36623	.93052	.38241	.92399	.39848	.91718	.41443	.91008	31
30	.35021	.93667	.36650	.93042	.38268	.92388	.39875	.91706	.41469	.90996	30
31	.35048	.93657	.36677	.93031	.38295	.92377	.39902	.91694	.41496	.90984	29
32	.35075	.93647	.36704	.93020	.38322	.92366	.39928	.91683	.41522	.90972	28
33	.35102	.93637	.36731	.93010	.38349	.92355	.39955	.91671	.41549	.90960	27
34	.35130	.93626	.36758	.92999	.38376	.92343	.39982	.91660	.41575	.90948	26
35	.35157	.93616	.36785	.92988	.38403	.92332	.40008	.91648	.41602	.90936	25
36	.35184	.93606	.36812	.92978	.38430	.92321	.40035	.91636	.41628	.90924	24
37	.35211	.93596	.36839	.92967	.38456	.92310	.40062	.91625	.41655	.90911	23
38	.35239	.93585	.36867	.92956	.38483	.92299	.40088	.91613	.41681	.90899	22
39	.35266	.93575	.36894	.92945	.38510	.92287	.40115	.91601	.41707	.90887	21
40	.35293	.93565	.36921	.92935	.38537	.92276	.40141	.91590	.41734	.90875	20
41	.35320	.93555	.36948	.92924	.38564	.92265	.40168	.91578	.41760	.90863	19
42	.35347	.93544	.36975	.92913	.38591	.92254	.40195	.91566	.41787	.90851	18
43	.35375	.93534	.37002	.92902	.38617	.92243	.40221	.91555	.41813	.90839	17
44	.35402	.93524	.37029	.92892	.38644	.92231	.40248	.91543	.41840	.90826	16
45	.35429	.93514	.37056	.92881	.38671	.92220	.40275	.91531	.41866	.90814	15
46	.35456	.93503	.37083	.92870	.38698	.92209	.40301	.91519	.41892	.90802	14
47	.35484	.93493	.37110	.92859	.38725	.92198	.40328	.91508	.41919	.90790	13
48	.35511	.93483	.37137	.92849	.38752	.92186	.40355	.91496	.41945	.90778	12
49	.35538	.93472	.37164	.92838	.38778	.92175	.40381	.91484	.41972	.90766	11
50	.35565	.93462	.37191	.92827	.38805	.92164	.40408	.91472	.41998	.90753	10
51	.35592	.93452	.37218	.92816	.38832	.92152	.40434	.91461	.42024	.90741	9
52	.35619	.93441	.37245	.92805	.38859	.92141	.40461	.91449	.42051	.90729	8
53	.35647	.93431	.37272	.92794	.38886	.92130	.40488	.91437	.42077	.90717	7
54	.35674	.93420	.37299	.92784	.38912	.92119	.40514	.91425	.42104	.90704	6
55	.35701	.93410	.37326	.92773	.38939	.92107	.40541	.91414	.42130	.90692	5
56	.35728	.93400	.37353	.92762	.38966	.92096	.40567	.91402	.42156	.90680	4
57	.35755	.93389	.37380	.92751	.38993	.92085	.40594	.91390	.42183	.90668	3
58	.35782	.93379	.37407	.92740	.39020	.92073	.40621	.91378	.42209	.90655	2
59	.35810	.93368	.37434	.92729	.39046	.92062	.40647	.91366	.42235	.90643	1
60	.35837	.93358	.37461	.92718	.39073	.92050	.40674	.91355	.42262	.90631	0
′	Cosin	Sine	Cosin	Sine	Cosin	Sine	Cosin	Sine	Cosin	Sine	′
	69°		68°		67°		66°		65°		

′	25° Sine	25° Cosin	26° Sine	26° Cosin	27° Sine	27° Cosin	28° Sine	28° Cosin	29° Sine	29° Cosin	′
0	.42262	.90631	.43837	.89879	.45399	.89101	.46947	.88295	.48481	.87462	60
1	.42288	.90618	.43863	.89867	.45425	.89087	.46973	.88281	.48506	.87448	59
2	.42315	.90606	.43889	.89854	.45451	.89074	.46999	.88267	.48532	.87434	58
3	.42341	.90594	.43916	.89841	.45477	.89061	.47024	.88254	.48557	.87420	57
4	.42367	.90582	.43942	.89828	.45503	.89048	.47050	.88240	.48583	.87406	56
5	.42394	.90569	.43968	.89816	.45529	.89035	.47076	.88226	.48608	.87391	55
6	.42420	.90557	.43994	.89803	.45554	.89021	.47101	.88213	.48634	.87377	54
7	.42446	.90545	.44020	.89790	.45580	.89008	.47127	.88199	.48659	.87363	53
8	.42473	.90532	.44046	.89777	.45606	.88995	.47153	.88185	.48684	.87349	52
9	.42499	.90520	.44072	.89764	.45632	.88981	.47178	.88172	.48710	.87335	51
10	.42525	.90507	.44098	.89752	.45658	.88968	.47204	.88158	.48735	.87321	50
11	.42552	.90495	.44124	.89739	.45684	.88955	.47229	.88144	.48761	.87306	49
12	.42578	.90483	.44151	.89726	.45710	.88942	.47255	.88130	.48786	.87292	48
13	.42604	.90470	.44177	.89713	.45736	.88928	.47281	.88117	.48811	.87278	47
14	.42631	.90458	.44203	.89700	.45762	.88915	.47306	.88103	.48837	.87264	46
15	.42657	.90446	.44229	.89687	.45787	.88902	.47332	.88089	.48862	.87250	45
16	.42683	.90433	.44255	.89674	.45813	.88888	.47358	.88075	.48888	.87235	44
17	.42709	.90421	.44281	.89662	.45839	.88875	.47383	.88062	.48913	.87221	43
18	.42736	.90408	.44307	.89649	.45865	.88862	.47409	.88048	.48938	.87207	42
19	.42762	.90396	.44333	.89636	.45891	.88848	.47434	.88034	.48964	.87193	41
20	.42788	.90383	.44359	.89623	.45917	.88835	.47460	.88020	.48989	.87178	40
21	.42815	.90371	.44385	.89610	.45942	.88822	.47486	.88006	.49014	.87164	39
22	.42841	.90358	.44411	.89597	.45968	.88808	.47511	.87993	.49040	.87150	38
23	.42867	.90346	.44437	.89584	.45994	.88795	.47537	.87979	.49065	.87136	37
24	.42894	.90334	.44464	.89571	.46020	.88782	.47562	.87965	.49090	.87121	36
25	.42920	.90321	.44490	.89558	.46046	.88768	.47588	.87951	.49116	.87107	35
26	.42946	.90309	.44516	.89545	.46072	.88755	.47614	.87937	.49141	.87093	34
27	.42972	.90296	.44542	.89532	.46097	.88741	.47639	.87923	.49166	.87079	33
28	.42999	.90284	.44568	.89519	.46123	.88728	.47665	.87909	.49192	.87064	32
29	.43025	.90271	.44594	.89506	.46149	.88715	.47690	.87896	.49217	.87050	31
30	.43051	.90259	.44620	.89493	.46175	.88701	.47716	.87882	.49242	.87036	30
31	.43077	.90246	.44646	.89480	.46201	.88688	.47741	.87868	.49268	.87021	29
32	.43104	.90233	.44672	.89467	.46226	.88674	.47767	.87854	.49293	.87007	28
33	.43130	.90221	.44698	.89454	.46252	.88661	.47793	.87840	.49318	.86993	27
34	.43156	.90208	.44724	.89441	.46278	.88647	.47818	.87826	.49344	.86978	26
35	.43182	.90196	.44750	.89428	.46304	.88634	.47844	.87812	.49369	.86964	25
36	.43209	.90183	.44776	.89415	.46330	.88620	.47869	.87798	.49394	.86949	24
37	.43235	.90171	.44802	.89402	.46355	.88607	.47895	.87784	.49419	.86935	23
38	.43261	.90158	.44828	.89389	.46381	.88593	.47920	.87770	.49445	.86921	22
39	.43287	.90146	.44854	.89376	.46407	.88580	.47946	.87756	.49470	.86906	21
40	.43313	.90133	.44880	.89363	.46433	.88566	.47971	.87743	.49495	.86892	20
41	.43340	.90120	.44906	.89350	.46458	.88553	.47997	.87729	.49521	.86878	19
42	.43366	.90108	.44932	.89337	.46484	.88539	.48022	.87715	.49546	.86863	18
43	.43392	.90095	.44958	.89324	.46510	.88526	.48048	.87701	.49571	.86849	17
44	.43418	.90082	.44984	.89311	.46536	.88512	.48073	.87687	.49596	.86834	16
45	.43445	.90070	.45010	.89298	.46561	.88499	.48099	.87673	.49622	.86820	15
46	.43471	.90057	.45036	.89285	.46587	.88485	.48124	.87659	.49647	.86805	14
47	.43497	.90045	.45062	.89272	.46613	.88472	.48150	.87645	.49672	.86791	13
48	.43523	.90032	.45088	.89259	.46639	.88458	.48175	.87631	.49697	.86777	12
49	.43549	.90019	.45114	.89245	.46664	.88445	.48201	.87617	.49723	.86762	11
50	.43575	.90007	.45140	.89232	.46690	.88431	.48226	.87603	.49748	.86748	10
51	.43602	.89994	.45166	.89219	.46716	.88417	.48252	.87589	.49773	.86733	9
52	.43628	.89981	.45192	.89206	.46742	.88404	.48277	.87575	.49798	.86719	8
53	.43654	.89968	.45218	.89193	.46767	.88390	.48303	.87561	.49824	.86704	7
54	.43680	.89956	.45243	.89180	.46793	.88377	.48328	.87546	.49849	.86690	6
55	.43706	.89943	.45269	.89167	.46819	.88363	.48354	.87532	.49874	.86675	5
56	.43733	.89930	.45295	.89153	.46844	.88349	.48379	.87518	.49899	.86661	4
57	.43759	.89918	.45321	.89140	.46870	.88336	.48405	.87504	.49924	.86646	3
58	.43785	.89905	.45347	.89127	.46896	.88322	.48430	.87490	.49950	.86632	2
59	.43811	.89892	.45373	.89114	.46921	.88308	.48456	.87476	.49975	.86617	1
60	.43837	.89879	.45399	.89101	.46947	.88295	.48481	.87462	.50000	.86603	0
′	Cosin	Sine	Cosin	Sine	Cosin	Sine	Cosin	Sine	Cosin	Sine	′
	64°		63°		62°		61°		60°		

′	30°		31°		32°		33°		34°		′
	Sine	Cosin	Sine	Cosin	Sine	Cosin	Sine	Cosin	Sine	Cosin	
0	.50000	.86603	.51504	.85717	.52992	.84805	.54464	.83867	.55919	.82904	60
1	.50025	.86588	.51529	.85702	.53017	.84789	.54488	.83851	.55943	.82887	59
2	.50050	.86573	.51554	.85687	.53041	.84774	.54513	.83835	.55968	.82871	58
3	.50076	.86559	.51579	.85672	.53066	.84759	.54537	.83819	.55992	.82855	57
4	.50101	.86544	.51604	.85657	.53091	.84743	.54561	.83804	.56016	.82839	56
5	.50126	.86530	.51628	.85642	.53115	.84728	.54586	.83788	.56040	.82822	55
6	.50151	.86515	.51653	.85627	.53140	.84712	.54610	.83772	.56064	.82806	54
7	.50176	.86501	.51678	.85612	.53164	.84697	.54635	.83756	.56088	.82790	53
8	.50201	.86486	.51703	.85597	.53189	.84681	.54659	.83740	.56112	.82773	52
9	.50227	.86471	.51728	.85582	.53214	.84666	.54683	.83724	.56136	.82757	51
10	.50252	.86457	.51753	.85567	.53238	.84650	.54708	.83708	.56160	.82741	50
11	.50277	.86442	.51778	.85551	.53263	.84635	.54732	.83692	.56184	.82724	49
12	.50302	.86427	.51803	.85536	.53288	.84619	.54756	.83676	.56208	.82708	48
13	.50327	.86413	.51828	.85521	.53312	.84604	.54781	.83660	.56232	.82692	47
14	.50352	.86398	.51852	.85506	.53337	.84588	.54805	.83645	.56256	.82675	46
15	.50377	.86384	.51877	.85491	.53361	.84573	.54829	.83629	.56280	.82659	45
16	.50403	.86369	.51902	.85476	.53386	.84557	.54854	.83613	.56305	.82643	44
17	.50428	.86354	.51927	.85461	.53411	.84542	.54878	.83597	.56329	.82626	43
18	.50453	.86340	.51952	.85446	.53435	.84526	.54902	.83581	.56353	.82610	42
19	.50478	.86325	.51977	.85431	.53460	.84511	.54927	.83565	.56377	.82593	41
20	.50503	.86310	.52002	.85416	.53484	.84495	.54951	.83549	.56401	.82577	40
21	.50528	.86295	.52026	.85401	.53509	.84480	.54975	.83533	.56425	.82561	39
22	.50553	.86281	.52051	.85385	.53534	.84464	.54999	.83517	.56449	.82544	38
23	.50578	.86266	.52076	.85370	.53558	.84448	.55024	.83501	.56473	.82528	37
24	.50603	.86251	.52101	.85355	.53583	.84433	.55048	.83485	.56497	.82511	36
25	.50628	.86237	.52126	.85340	.53607	.84417	.55072	.83469	.56521	.82495	35
26	.50654	.86222	.52151	.85325	.53632	.84402	.55097	.83453	.56545	.82478	34
27	.50679	.86207	.52175	.85310	.53656	.84386	.55121	.83437	.56569	.82462	33
28	.50704	.86192	.52200	.85294	.53681	.84370	.55145	.83421	.56593	.82446	32
29	.50729	.86178	.52225	.85279	.53705	.84355	.55169	.83405	.56617	.82429	31
30	.50754	.86163	.52250	.85264	.53730	.84339	.55194	.83389	.56641	.82413	30
31	.50779	.86148	.52275	.85249	.53754	.84324	.55218	.83373	.56665	.82396	29
32	.50804	.86133	.52299	.85234	.53779	.84308	.55242	.83356	.56689	.82380	28
33	.50829	.86119	.52324	.85218	.53804	.84292	.55266	.83340	.56713	.82363	27
34	.50854	.86104	.52349	.85203	.53828	.84277	.55291	.83324	.56736	.82347	26
35	.50879	.86089	.52374	.85188	.53853	.84261	.55315	.83308	.56760	.82330	25
36	.50904	.86074	.52399	.85173	.53877	.84245	.55339	.83292	.56784	.82314	24
37	.50929	.86059	.52423	.85157	.53902	.84230	.55363	.83276	.56808	.82297	23
38	.50954	.86045	.52448	.85142	.53926	.84214	.55388	.83260	.56832	.82281	22
39	.50979	.86030	.52473	.85127	.53951	.84198	.55412	.83244	.56856	.82264	21
40	.51004	.86015	.52498	.85112	.53975	.84182	.55436	.83228	.56880	.82248	20
41	.51029	.86000	.52522	.85096	.54000	.84167	.55460	.83212	.56904	.82231	19
42	.51054	.85985	.52547	.85081	.54024	.84151	.55484	.83195	.56928	.82214	18
43	.51079	.85970	.52572	.85066	.54049	.84135	.55509	.83179	.56952	.82198	17
44	.51104	.85956	.52597	.85051	.54073	.84120	.55533	.83163	.56976	.82181	16
45	.51129	.85941	.52621	.85035	.54097	.84104	.55557	.83147	.57000	.82165	15
46	.51154	.85926	.52646	.85020	.54122	.84088	.55581	.83131	.57024	.82148	14
47	.51179	.85911	.52671	.85005	.54146	.84072	.55605	.83115	.57047	.82132	13
48	.51204	.85896	.52696	.84989	.54171	.84057	.55630	.83098	.57071	.82115	12
49	.51229	.85881	.52720	.84974	.54195	.84041	.55654	.83082	.57095	.82098	11
50	.51254	.85866	.52745	.84959	.54220	.84025	.55678	.83066	.57119	.82082	10
51	.51279	.85851	.52770	.84943	.54244	.84009	.55702	.83050	.57143	.82065	9
52	.51304	.85836	.52794	.84928	.54269	.83994	.55726	.83034	.57167	.82048	8
53	.51329	.85821	.52819	.84913	.54293	.83978	.55750	.83017	.57191	.82032	7
54	.51354	.85806	.52844	.84897	.54317	.83962	.55775	.83001	.57215	.82015	6
55	.51379	.85792	.52869	.84882	.54342	.83946	.55799	.82985	.57238	.81999	5
56	.51404	.85777	.52893	.84866	.54366	.83930	.55823	.82969	.57262	.81982	4
57	.51429	.85762	.52918	.84851	.54391	.83915	.55847	.82953	.57286	.81965	3
58	.51454	.85747	.52943	.84836	.54415	.83899	.55871	.82936	.57310	.81949	2
59	.51479	.85732	.52967	.84820	.54440	.83883	.55895	.82920	.57334	.81932	1
60	.51504	.85717	.52992	.84805	.54464	.83867	.55919	.82904	.57358	.81915	0
	Cosin	Sine	Cosin	Sine	Cosin	Sine	Cosin	Sine	Cosin	Sine	
′	59°		58°		57°		56°		55°		′

′	35° Sine	35° Cosin	36° Sine	36° Cosin	37° Sine	37° Cosin	38° Sine	38° Cosin	39° Sine	39° Cosin	′
0	.57358	.81915	.58779	.80902	.60182	.79864	.61566	.78801	.62932	.77715	60
1	.57381	.81899	.58802	.80885	.60205	.79846	.61589	.78783	.62955	.77696	59
2	.57405	.81882	.58826	.80867	.60228	.79829	.61612	.78765	.62977	.77678	58
3	.57429	.81865	.58849	.80850	.60251	.79811	.61635	.78747	.63000	.77660	57
4	.57453	.81848	.58873	.80833	.60274	.79793	.61658	.78729	.63022	.77641	56
5	.57477	.81832	.58896	.80816	.60298	.79776	.61681	.78711	.63045	.77623	55
6	.57501	.81815	.58920	.80799	.60321	.79758	.61704	.78694	.63068	.77605	54
7	.57524	.81798	.58943	.80782	.60344	.79741	.61726	.78676	.63090	.77586	53
8	.57548	.81782	.58967	.80765	.60367	.79723	.61749	.78658	.63113	.77568	52
9	.57572	.81765	.58990	.80748	.60390	.79706	.61772	.78640	.63135	.77550	51
10	.57596	.81748	.59014	.80730	.60414	.79688	.61795	.78622	.63158	.77531	50
11	.57619	.81731	.59037	.80713	.60437	.79671	.61818	.78604	.63180	.77513	40
12	.57643	.81714	.59061	.80696	.60460	.79653	.61841	.78586	.63203	.77494	48
13	.57667	.81698	.59084	.80679	.60483	.79635	.61864	.78568	.63225	.77476	47
14	.57691	.81681	.59108	.80662	.60506	.79618	.61887	.78550	.63248	.77458	46
15	.57715	.81664	.59131	.80644	.60529	.79600	.61909	.78532	.63271	.77439	45
16	.57738	.81647	.59154	.80627	.60553	.79583	.61932	.78514	.63293	.77421	44
17	.57762	.81631	.59178	.80610	.60576	.79565	.61955	.78496	.63316	.77402	43
18	.57786	.81614	.59201	.80593	.60599	.79547	.61978	.78478	.63338	.77384	42
19	.57810	.81597	.59225	.80576	.60622	.79530	.62001	.78460	.63361	.77366	41
20	.57833	.81580	.59248	.80558	.60645	.79512	.62024	.78442	.63383	.77347	40
21	.57857	.81563	.59272	.80541	.60668	.79494	.62046	.78424	.63406	.77329	39
22	.57881	.81546	.59295	.80524	.60691	.79477	.62069	.78405	.63428	.77310	38
23	.57904	.81530	.59318	.80507	.60714	.79459	.62092	.78387	.63451	.77292	37
24	.57928	.81513	.59342	.80489	.60738	.79441	.62115	.78369	.63473	.77273	36
25	.57952	.81496	.59365	.80472	.60761	.79424	.62138	.78351	.63496	.77255	35
26	.57976	.81479	.59389	.80455	.60784	.79406	.62160	.78333	.63518	.77236	34
27	.57999	.81462	.59412	.80438	.60807	.79388	.62183	.78315	.63540	.77218	33
28	.58023	.81445	.59436	.80420	.60830	.79371	.62206	.78297	.63563	.77199	32
29	.58047	.81428	.59459	.80403	60853	.79353	.62229	.78279	.63585	.77181	31
30	.58070	.81412	.59482	.80386	.60876	.79335	.62251	.78261	.63608	.77162	30
31	.58094	.81395	.59506	.80368	.60899	.79318	.62274	.78243	.63630	.77144	29
32	.58118	.81378	.59529	.80351	.60922	.79300	.62297	.78225	.63653	.77125	28
33	.58141	.81361	.59552	.80334	.60945	.79282	.62320	.78206	.63675	.77107	27
34	.58165	.81344	.59576	.80316	.60968	.79264	.62342	.78188	.63698	.77088	26
35	.58189	.81327	.59599	.80299	.60991	.79247	.62365	.78170	.63720	.77070	25
36	.58212	.81310	.59622	.80282	.61015	.79229	.62388	.78152	.63742	.77051	24
37	.58236	.81293	.59646	.80264	.61038	.79211	.62411	.78134	.63765	.77033	23
38	.58260	.81276	.59669	.80247	.61061	.79193	.62433	.78116	.63787	.77014	22
39	.58283	.81259	.59693	.80230	.61084	.79176	.62456	.78098	.63810	.76996	21
40	.58307	.81242	.59716	.80212	.61107	.79158	.62479	.78079	.63832	.76977	20
41	.58330	.81225	.59739	.80195	.61130	.79140	.62502	.78061	.63854	.76959	19
42	.58354	.81208	.59763	.80178	.61153	.79122	.62524	.78043	.63877	.76940	18
43	.58378	.81191	.59786	.80160	.61176	.79105	.62547	.78025	.63899	.76921	17
44	.58401	.81174	.59809	.80143	.61199	.79087	.62570	.78007	.63922	.76903	16
45	.58425	.81157	.59832	.80125	.61222	.79069	.62592	.77988	.63944	.76884	15
46	.58449	.81140	.59856	.80108	.61245	.79051	.62615	.77970	.63966	.76866	14
47	.58472	.81123	.59879	.80091	.61268	.79033	.62638	.77952	.63989	.76847	13
48	.58496	.81106	.59902	.80073	.61291	.79016	.62660	.77934	.64011	.76828	12
49	.58519	.81089	.59926	.80056	.61314	.78998	.62683	.77916	.64033	.76810	11
50	.58543	.81072	.59949	.80038	.61337	.78980	.62706	.77897	.64056	.76791	10
51	.58567	.81055	.59972	.80021	.61360	.78962	.62728	.77879	.64078	.76772	9
52	.58590	.81038	.59995	80003	.61383	.78944	.62751	.77861	.64100	.76754	8
53	.58614	.81021	.60019	.79986	.61406	.78926	.62774	.77843	.64123	.76735	7
54	.58637	.81004	.60042	.79968	.61429	.78908	.62796	.77824	.64145	.76717	6
55	.58661	.80987	.60065	.79951	.61451	.78891	.62819	.77806	.64167	.76698	5
56	.58684	.80970	.60089	.79934	.61474	.78873	.62842	.77788	.64190	.76679	4
57	.58708	.80953	.60112	.79916	.61497	.78855	.62864	.77769	.64212	.76661	3
58	.58731	.80936	.60135	.79899	.61520	.78837	.62887	.77751	.64234	.76642	2
59	.58755	.80919	.60158	.79881	.61543	.78819	.62909	.77733	.64256	.76623	1
60	.58779	.80902	.60182	.79864	.61566	.78801	.62932	.77715	.64279	.76604	0
′	Cosin	Sine	Cosin	Sine	Cosin	Sine	Cosin	Sine	Cosin	Sine	′
	54°		53°		52°		51°		50°		

′	40° Sine	40° Cosin	41° Sine	41° Cosin	42° Sine	42° Cosin	43° Sine	43° Cosin	44° Sine	44° Cosin	′
0	.64279	.76604	.65606	.75471	.66913	.74314	.68200	.73135	.69466	.71934	60
1	.64301	.76586	.65628	.75452	.66935	.74295	.68221	.73116	.69487	.71914	59
2	.64323	.76567	.65650	.75433	.66956	.74276	.68242	.73096	.69508	.71894	58
3	.64346	.76548	.65672	.75414	.66978	.74256	.68264	.73076	.69529	.71873	57
4	.64368	.76530	.65694	.75395	.66999	.74237	.68285	.73056	.69549	.71853	56
5	.64390	.76511	.65716	.75375	.67021	.74217	.68306	.73036	.69570	.71833	55
6	.64412	.76492	.65738	.75356	.67043	.74198	.68327	.73016	.69591	.71813	54
7	.64435	.76473	.65759	.75337	.67064	.74178	.68349	.72996	.69612	.71792	53
8	.64457	.76455	.65781	.75318	.67086	.74159	.68370	.72976	.69633	.71772	52
9	.64479	.76436	.65803	.75299	.67107	.74139	.68391	.72957	.69654	.71752	51
10	.64501	.76417	.65825	.75280	.67129	.74120	.68412	.72937	.69675	.71732	50
11	.64524	.76398	.65847	.75261	.67151	.74100	.68434	.72917	.69696	.71711	49
12	.64546	.76380	.65869	.75241	.67172	.74080	.68455	.72897	.69717	.71691	48
13	.64568	.76361	.65891	.75222	.67194	.74061	.68476	.72877	.69737	.71671	47
14	.64590	.76342	.65913	.75203	.67215	.74041	.68497	.72857	.69758	.71650	46
15	.64612	.76323	.65935	.75184	.67237	.74022	.68518	.72837	.69779	.71630	45
16	.64635	.76304	.65956	.75165	.67258	.74002	.68539	.72817	.69800	.71610	44
17	.64657	.76286	.65978	.75146	.67280	.73983	.68561	.72797	.69821	.71590	43
18	.64679	.76267	.66000	.75126	.67301	.73963	.68582	.72777	.69842	.71569	42
19	.64701	.76248	.66022	.75107	.67323	.73944	.68603	.72757	.69862	.71549	41
20	.64723	.76229	.66044	.75088	.67344	.73924	.68624	.72737	.69883	.71529	40
21	.64746	.76210	.66066	.75069	.67366	.73904	.68645	.72717	.69904	.71508	39
22	.64768	.76192	.66088	.75050	.67387	.73885	.68666	.72697	.69925	.71488	38
23	.64790	.76173	.66109	.75030	.67409	.73865	.68688	.72677	.69946	.71468	37
24	.64812	.76154	.66131	.75011	.67430	.73846	.68709	.72657	.69966	.71447	36
25	.64834	.76135	.66153	.74992	.67452	.73826	.68730	.72637	.69987	.71427	35
26	.64856	.76116	.66175	.74973	.67473	.73806	.68751	.72617	.70008	.71407	34
27	.64878	.76097	.66197	.74953	.67495	.73787	.68772	.72597	.70029	.71386	33
28	.64901	.76078	.66218	.74934	.67516	.73767	.68793	.72577	.70049	.71366	32
29	.64923	.76059	.66240	.74915	.67538	.73747	.68814	.72557	.70070	.71345	31
30	.64945	.76041	.66262	.74896	.67559	.73728	.68835	.72537	.70091	.71325	30
31	.64967	.76022	.66284	.74876	.67580	.73708	.68857	.72517	.70112	.71305	29
32	.64989	.76003	.66306	.74857	.67602	.73688	.68878	.72497	.70132	.71284	28
33	.65011	.75984	.66327	.74838	.67623	.73669	.68899	.72477	.70153	.71264	27
34	.65033	.75965	.66349	.74818	.67645	.73649	.68920	.72457	.70174	.71242	26
35	.65055	.75946	.66371	.74799	.67666	.73629	.68941	.72437	.70195	.71223	25
36	.65077	.75927	.66393	.74780	.67688	.73610	.68962	.72417	.70215	.71203	24
37	.65100	.75908	.66414	.74760	.67709	.73590	.68983	.72397	.70236	.71182	23
38	.65122	.75889	.66436	.74741	.67730	.73570	.69004	.72377	.70257	.71162	22
39	.65144	.75870	.66458	.74722	.67752	.73551	.69025	.72357	.70277	.71141	21
40	.65166	.75851	.66480	.74703	.67773	.73531	.69046	.72337	.70298	.71121	20
41	.65188	.75832	.66501	.74683	.67795	.73511	.69067	.72317	.70319	.71100	19
42	.65210	.75813	.66523	.74664	.67816	.73491	.69088	.72297	.70339	.71080	18
43	.65232	.75794	.66545	.74644	.67837	.73472	.69109	.72277	.70360	.71059	17
44	.65254	.75775	.66566	.74625	.67859	.73452	.69130	.72257	.70381	.71039	16
45	.65276	.75756	.66588	.74606	.67880	.73432	.69151	.72236	.70401	.71019	15
46	.65298	.75738	.66610	.74586	.67901	.73413	.69172	.72216	.70422	.70998	14
47	.65320	.75719	.66632	.74567	.67923	.73393	.69193	.72196	.70443	.70978	13
48	.65342	.75700	.66653	.74548	.67944	.73373	.69214	.72176	.70463	.70957	12
49	.65364	.75680	.66675	.74528	.67965	.73353	.69235	.72156	.70484	.70937	11
50	.65386	.75661	.66697	.74509	.67987	.73333	.69256	.72136	.70505	.70916	10
51	.65408	.75642	.66718	.74489	.68008	.73314	.69277	.72116	.70525	.70896	9
52	.65430	.75623	.66740	.74470	.68029	.73294	.69298	.72095	.70546	.70875	8
53	.65452	.75604	.66762	.74451	.68051	.73274	.69319	.72075	.70567	.70855	7
54	.65474	.75585	.66783	.74431	.68072	.73254	.69340	.72055	.70587	.70834	6
55	.65496	.75566	.66805	.74412	.68093	.73234	.69361	.72035	.70608	.70813	5
56	.65518	.75547	.66827	.74392	.68115	.73215	.69382	.72015	.70628	.70793	4
57	.65540	.75528	.66848	.74373	.68136	.73195	.69403	.71995	.70649	.70772	3
58	.65562	.75509	.66870	.74352	.68157	.73175	.69424	.71974	.70670	.70752	2
59	.65584	.75490	.66891	.74334	.68179	.73155	.69445	.71954	.70690	.70731	1
60	.65606	.75471	.66913	.74314	.68200	.73135	.69466	.71934	.70711	.70711	0
′	Cosin	Sine	Cosin	Sine	Cosin	Sine	Cosin	Sine	Cosin	Sine	′
	49°		48°		47°		46°		45°		

′	SECANTS.							′
	0°	1°	2°	3°	4°	5°	6°	
0	1.00000	1.00015	1.00061	1.00137	1.00244	1.00382	1.00551	60
1	00000	00016	00062	00139	00246	00385	00554	59
2	00000	00016	00063	00140	00248	00387	00557	58
3	00000	00017	00064	00142	00250	00390	00560	57
4	00000	00017	00065	00143	00252	00392	00563	56
5	00000	00018	00066	00145	00254	00395	00566	55
6	00000	00018	00067	00147	00257	00397	00569	54
7	00000	00019	00068	00148	00259	00400	00573	53
8	00000	00020	00069	00150	00261	00403	00576	52
9	00000	00020	00070	00151	00263	00405	00579	51
10	00000	00021	00072	00153	00265	00408	00582	50
11	1.00001	1.00021	1.00073	1.00155	1.00267	1.00411	1.00585	49
12	00001	00022	00074	00156	00269	00413	00588	48
13	00001	00023	00075	00158	00271	00416	00592	47
14	00001	00023	00076	00159	00274	00419	00595	46
15	00001	00024	00077	00161	00276	00421	00598	45
16	00001	00024	00078	00163	00278	00424	00601	44
17	00001	00025	00079	00164	00280	00427	00604	43
18	00001	00026	00081	00166	00282	00429	00608	42
19	00002	00026	00082	00168	00284	00432	00611	41
20	00002	00027	00083	00169	00287	00435	00614	40
21	1.00002	1.00028	1.00084	1.00171	1.00289	1.00438	1.00617	39
22	00002	00028	00085	00173	00291	00440	00621	38
23	00002	00029	00087	00175	00293	00443	00624	37
24	00002	00030	00088	00176	00296	00446	00627	36
25	00003	00031	00089	00178	00298	00449	00630	35
26	00003	00031	00090	00180	00300	00451	00634	34
27	00003	00032	00091	00182	00302	00454	00637	33
28	00003	00033	00093	00183	00305	00457	00640	32
29	00004	00034	00094	00185	00307	00460	00644	31
30	00004	00034	00095	00187	00309	00463	00647	30
31	1.00004	1.00035	1.00097	1.00189	1.00312	1.00465	1.00650	29
32	00004	00036	00098	00190	00314	00468	00654	28
33	00005	00037	00099	00192	00316	00471	00657	27
34	00005	00037	00100	00194	00318	00474	00660	26
35	00005	00038	00102	00196	00321	00477	00664	25
36	00005	00039	00103	00198	00323	00480	00667	24
37	00006	00040	00104	00200	00326	00482	00671	23
38	00006	00041	00106	00201	00328	00485	00674	22
39	00006	00041	00107	00203	00330	00488	00677	21
40	00007	00042	00108	00205	00333	00491	00681	20
41	1.00007	1.00043	1.00110	1.00207	1.00335	1.00494	1.00684	19
42	00007	00044	00111	00209	00337	00497	00688	18
43	00008	00045	00113	00211	00340	00500	00691	17
44	00008	00046	00114	00213	00342	00503	00695	16
45	00009	00047	00115	00215	00345	00506	00698	15
46	00009	00048	00117	00216	00347	00509	00701	14
47	00009	00048	00118	00218	00350	00512	00705	13
48	00010	00049	00120	00220	00352	00515	00708	12
49	00010	00050	00121	00222	00354	00518	00712	11
50	00011	00051	00122	00224	00357	00521	00715	10
51	1.00011	1.00052	1.00124	1.00226	1.00359	1.00524	1.00719	9
52	00011	00053	00125	00228	00362	00527	00722	8
53	00012	00054	00127	00230	00364	00530	00726	7
54	00012	00055	00128	00232	00367	00533	00730	6
55	00013	00056	00130	00234	00369	00536	00733	5
56	00013	00057	00131	00236	00372	00539	00737	4
57	00014	00058	00133	00238	00374	00542	00740	3
58	00014	00059	00134	00240	00377	00545	00744	2
59	00015	00060	00136	00242	00379	00548	00747	1
60	00015	00061	00137	00244	00382	00551	00751	0
′	89°	88°	87°	86°	85°	84°	83°	′
	COSECANTS.							

′	SECANTS.							′
	7°	8°	9°	10°	11°	12°	13°	
0	1.00751	1.00983	1.01247	1.01543	1.01872	1.02234	1.02630	60
1	00755	00987	01251	01548	01877	02240	02637	59
2	00758	00991	01256	01553	01883	02247	02644	58
3	00762	00995	01261	01558	01889	02253	02651	57
4	00765	00999	01265	01564	01895	02259	02658	56
5	00769	01004	01270	01569	01901	02266	02665	55
6	00773	01008	01275	01574	01906	02272	02672	54
7	00776	01012	01279	01579	01912	02279	02679	53
8	00780	01016	01284	01585	01918	02285	02686	52
9	00784	01020	01289	01590	01924	02291	02693	51
10	00787	01024	01294	01595	01930	02298	02700	50
11	1.00791	1.01029	1.01298	1.01601	1.01936	1.02304	1.02707	49
12	00795	01033	01303	01606	01941	02311	02714	48
13	00799	01037	01308	01611	01947	02317	02721	47
14	00802	01041	01313	01616	01953	02323	02728	46
15	00806	01046	01318	01622	01959	02330	02735	45
16	00810	01050	01322	01627	01965	02336	02742	44
17	00813	01054	01327	01633	01971	02343	02749	43
18	00817	01059	01332	01638	01977	02349	02756	42
19	00821	01063	01337	01643	01983	02356	02763	41
20	00825	01067	01342	01649	01989	02362	02770	40
21	1.00828	1.01071	1.01346	1.01654	1.01995	1.02369	1.02777	39
22	00832	01076	01351	01659	02001	02375	02784	38
23	00836	01080	01356	01665	02007	02382	02791	37
24	00840	01084	01361	01670	02013	02388	02799	36
25	00844	01089	01366	01676	02019	02395	02806	35
26	00848	01093	01371	01681	02025	02402	02813	34
27	00851	01097	01376	01687	02031	02408	02820	33
28	00855	01102	01381	01692	02037	02415	02827	32
29	00859	01106	01386	01698	02043	02421	02834	31
30	00863	01111	01391	01703	02049	02428	02842	30
31	1.00867	1.01115	1.01395	1.01709	1.02055	1.02435	1.02849	29
32	00871	01119	01400	01714	02061	02441	02856	28
33	00875	01124	01405	01720	02067	02448	02863	27
34	00878	01128	01410	01725	02073	02454	02870	26
35	00882	01133	01415	01731	02079	02461	02878	25
36	00886	01137	01420	01736	02085	02468	02885	24
37	00890	01142	01425	01742	02091	02474	02892	23
38	00894	01146	01430	01747	02097	02481	02899	22
39	00898	01151	01435	01753	02103	02488	02907	21
40	00902	01155	01440	01758	02110	02494	02914	20
41	1.00906	1.01160	1.01445	1.01764	1.02116	1.02501	1.02921	19
42	00910	01164	01450	01769	02122	02508	02928	18
43	00914	01169	01455	01775	02128	02515	02936	17
44	00918	01173	01461	01781	02134	02521	02943	16
45	00922	01178	01466	01786	02140	02528	02950	15
46	00926	01182	01471	01792	02146	02535	02958	14
47	00930	01187	01476	01798	02153	02542	02965	13
48	00934	01191	01481	01803	02159	02548	02972	12
49	00938	01196	01486	01809	02165	02555	02980	11
50	00942	01200	01491	01815	02171	02562	02987	10
51	1.00946	1.01205	1.01496	1.01820	1.02178	1.02569	1.02994	9
52	00950	01209	01501	01826	02184	02576	03002	8
53	00954	01214	01506	01832	02190	02582	03009	7
54	00958	01219	01512	01837	02196	02589	03017	6
55	00962	01223	01517	01843	02203	02596	03024	5
56	00966	01228	01522	01849	02209	02603	03032	4
57	00970	01233	01527	01854	02215	02610	03039	3
58	00975	01237	01532	01860	02221	02617	03046	2
59	00979	01242	01537	01866	02228	02624	03054	1
60	00983	01247	01543	01872	02234	02630	03061	0
′	82°	81°	80°	79°	78°	77°	76°	′
	COSECANTS.							

′	SECANTS.							′
	14°	15°	16°	17°	18°	19°	20°	
0	1.03061	1.03528	1.04030	1.04569	1.05146	1.05762	1.06418	60
1	03069	03536	04039	04578	05156	05773	06429	59
2	03076	03544	04047	04588	05166	05783	06440	58
3	03084	03552	04056	04597	05176	05794	06452	57
4	03091	03560	04065	04606	05186	05805	06463	56
5	03099	03568	04073	04616	05196	05815	06474	55
6	03106	03576	04082	04625	05206	05826	06486	54
7	03114	03584	04091	04635	05216	05836	06497	53
8	03121	03592	04100	04644	05226	05847	06508	52
9	03129	03601	04108	04653	05236	05858	06520	51
10	03137	03609	04117	04663	05246	05869	06531	50
11	1.03144	1.03617	1.04126	1.04672	1.05256	1.05879	1.06542	49
12	03152	03625	04135	04682	05266	05890	06554	48
13	03159	03633	04144	04691	05276	05901	06565	47
14	03167	03642	04152	04700	05286	05911	06577	46
15	03175	03650	04161	04710	05297	05922	06588	45
16	03182	03658	04170	04719	05307	05933	06600	44
17	03190	03666	04179	04729	05317	05944	06611	43
18	03198	03674	04188	04738	05327	05955	06622	42
19	03205	03683	04197	04748	05337	05965	06634	41
20	03213	03691	04206	04757	05347	05976	06645	40
21	1.03221	1.03699	1.04214	1.04767	1.05357	1.05987	1.06657	39
22	03228	03708	04223	04776	05367	05998	06668	38
23	03236	03716	04232	04786	05378	06009	06680	37
24	03244	03724	04241	04795	05388	06020	06691	36
25	03251	03732	04250	04805	05398	06030	06703	35
26	03259	03741	04259	04815	05408	06041	06715	34
27	03267	03749	04268	04824	05418	06052	06726	33
28	03275	03758	04277	04834	05429	06063	06738	32
29	03282	03766	04286	04843	05439	06074	06749	31
30	03290	03774	04295	04853	05449	06085	06761	30
31	1.03298	1.03783	1.04304	1.04863	1.05460	1.06096	1.06773	29
32	03306	03791	04313	04872	05470	06107	06784	28
33	03313	03799	04322	04882	05480	06118	06796	27
34	03321	03808	04331	04891	05490	06129	06807	26
35	03329	03816	04340	04901	05501	06140	06819	25
36	03337	03825	04349	04911	05511	06151	06831	24
37	03345	03833	04358	04920	05521	06162	06843	23
38	03353	03842	04367	04930	05532	06173	06854	22
39	03360	03850	04376	04940	05542	06184	06866	21
40	03368	03858	04385	04950	05552	06195	06878	20
41	1.03376	1.03867	1.04394	1.04959	1.05563	1.06206	1.06889	19
42	03384	03875	04403	04969	05573	06217	06901	18
43	03392	03884	04413	04979	05584	06228	06913	17
44	03400	03892	04422	04989	05594	06239	06925	16
45	03408	03901	04431	04998	05604	06250	06936	15
46	03416	03909	04440	05008	05615	06261	06948	14
47	03424	03918	04449	05018	05625	06272	06960	13
48	03432	03927	04458	05028	05636	06283	06972	12
49	03439	03935	04468	05038	05646	06295	06984	11
50	03447	03944	04477	05047	05657	06306	06995	10
51	1.03455	1.03952	1.04486	1.05057	1.05667	1.06317	1.07007	9
52	03463	03961	04495	05067	05678	06328	07019	8
53	03471	03969	04504	05077	05688	06339	07031	7
54	03479	03978	04514	05087	05699	06350	07043	6
55	03487	03987	04523	05097	05709	06362	07055	5
56	03495	03995	04532	05107	05720	06373	07067	4
57	03503	04004	04541	05116	05730	06384	07079	3
58	03512	04013	04551	05126	05741	06395	07091	2
59	03520	04021	04560	05136	05751	06407	07103	1
60	03528	04030	04569	05146	05762	06418	07115	0
′	75°	74°	73°	72°	71°	70°	69°	′
	COSECANTS.							

′	SECANTS.							′
	21°	22°	23°	24°	25°	26°	27°	
0	1.07115	1.07853	1.08636	1.09464	1.10338	1.11260	1.12233	60
1	07126	07866	08649	09478	10353	11276	12249	59
2	07138	07879	08663	09492	10368	11292	12266	58
3	07150	07892	08676	09506	10383	11308	12283	57
4	07162	07904	08690	09520	10398	11323	12299	56
5	07174	07917	08703	09535	10413	11339	12316	55
6	07186	07930	08717	09549	10428	11355	12333	54
7	07199	07943	08730	09563	10443	11371	12349	53
8	07211	07955	08744	09577	10458	11387	12366	52
9	07223	07968	08757	09592	10473	11403	12383	51
10	07235	07981	08771	09606	10488	11419	12400	50
11	1.07247	1.07994	1.08784	1.09620	1.10503	1.11435	1.12416	49
12	07259	08006	08798	09635	10518	11451	12433	48
13	07271	08019	08811	09649	10533	11467	12450	47
14	07283	08032	08825	09663	10549	11483	12467	46
15	07295	08045	08839	09678	10564	11499	12484	45
16	07307	08058	08852	09692	10579	11515	12501	44
17	07320	08071	08866	09707	10594	11531	12518	43
18	07332	08084	08880	09721	10609	11547	12534	42
19	07344	08097	08893	09735	10625	11563	12551	41
20	07356	08109	08907	09750	10640	11579	12568	40
21	1.07368	1.08122	1.08921	1.09764	1.10655	1.11595	1.12585	39
22	07380	08135	08934	09779	10670	11611	12602	38
23	07393	08148	08948	09793	10686	11627	12619	37
24	07405	08161	08962	09808	10701	11643	12636	36
25	07417	08174	08975	09822	10716	11659	12653	35
26	07429	08187	08989	09837	10731	11675	12670	34
27	07442	08200	09003	09851	10747	11691	12687	33
28	07454	08213	09017	09866	10762	11708	12704	32
29	07466	08226	09030	09880	10777	11724	12721	31
30	07479	08239	09044	09895	10793	11740	12738	30
31	1.07491	1.08252	1.09058	1.09909	1.10808	1.11756	1.12755	29
32	07503	08265	09072	09924	10824	11772	12772	28
33	07516	08278	09086	09939	10839	11789	12789	27
34	07528	08291	09099	09953	10854	11805	12807	26
35	07540	08305	09113	09968	10870	11821	12824	25
36	07553	08318	09127	09982	10885	11838	12841	24
37	07565	08331	09141	09997	10901	11854	12858	23
38	07578	08344	09155	10012	10916	11870	12875	22
39	07590	08357	09169	10026	10932	11886	12892	21
40	07602	08370	09183	10041	10947	11903	12910	20
41	1.07615	1.08383	1 09197	1.10055	1.10963	1.11919	1.12927	19
42	07627	08397	09211	10071	10978	11936	12944	18
43	07640	08410	09224	10085	10994	11952	12961	17
44	07652	08423	09238	10100	11009	11968	12979	16
45	07665	08436	09252	10115	11025	11985	12996	15
46	07677	08449	09266	10130	11041	12001	13013	14
47	07690	08463	09280	10144	11056	12018	13031	13
48	07702	08476	09294	10159	11072	12034	13048	12
49	07715	08489	09308	10174	11087	12051	13065	11
50	07727	08503	09323	10189	11103	12067	13083	10
51	1.07740	1.08516	1.09337	1.10204	1.11119	1.12084	1.13100	9
52	07752	08529	09351	10218	11134	12100	13117	8
53	07765	08542	09365	10233	11150	12117	13135	7
54	07778	08556	09379	10248	11166	12133	13152	6
55	07790	08569	09393	10263	11181	12150	13170	5
56	07803	08582	09407	10278	11197	12166	13187	4
57	07816	08596	09421	10293	11213	12183	13205	3
58	07828	08609	09435	10308	11229	12199	13222	2
59	07841	08623	09449	10323	11244	12216	13240	1
60	07853	08636	09464	10338	11260	12233	13257	0
′	68°	67°	66°	65°	64°	63°	62°	′
	COSECANTS.							

′	SECANTS.							′
	28°	29°	30°	31°	32°	33°	34°	
0	1.13257	1.14335	1.15470	1.16663	1.17918	1.19236	1.20622	60
1	13275	14354	15489	16684	17939	19259	20645	59
2	13292	14372	15509	16704	17961	19281	20669	58
3	13310	14391	15528	16725	17982	19304	20693	57
4	13327	14409	15548	16745	18004	19327	20717	56
5	13345	14428	15567	16766	18025	19349	20740	55
6	13362	14446	15587	16786	18047	19372	20764	54
7	13380	14465	15606	16806	18068	19394	20788	53
8	13398	14483	15626	16827	18090	19417	20812	52
9	13415	14502	15645	16848	18111	19440	20836	51
10	13433	14521	15665	16868	18133	19463	20859	50
11	1.13451	1.14539	1.15684	1.16889	1.18155	1.19485	1.20883	49
12	13468	14558	15704	16909	18176	19508	20907	48
13	13486	14576	15724	16930	18198	19531	20931	47
14	13504	14595	15743	16950	18220	19554	20955	46
15	13521	14614	15763	16971	18241	19576	20979	45
16	13539	14632	15782	16992	18263	19599	21003	44
17	13557	14651	15802	17012	18285	19622	21027	43
18	13575	14670	15822	17033	18307	19645	21051	42
19	13593	14689	15841	17054	18328	19668	21075	41
20	13610	14707	15861	17075	18350	19691	21099	40
21	1.13628	1.14726	1.15881	1.17095	1.18372	1.19713	1.21123	39
22	13646	14745	15901	17116	18394	19736	21147	38
23	13664	14764	15920	17137	18416	19759	21171	37
24	13682	14782	15940	17158	18437	19782	21195	36
25	13700	14801	15960	17178	18459	19805	21220	35
26	13718	14820	15980	17199	18481	19828	21244	34
27	13735	14839	16000	17220	18503	19851	21268	33
28	13753	14858	16019	17241	18525	19874	21292	32
29	13771	14877	16039	17262	18547	19897	21316	31
30	13789	14896	16059	17283	18569	19920	21341	30
31	1.13807	1.14914	1.16079	1.17304	1.18591	1.19944	1.21365	29
32	13825	14933	16099	17325	18613	19967	21389	28
33	13843	14952	16119	17346	18635	19990	21414	27
34	13861	14971	16139	17367	18657	20013	21438	26
35	13879	14990	16159	17388	18679	20036	21462	25
36	13897	15009	16179	17409	18701	20059	21487	24
37	13916	15028	16199	17430	18723	20083	21511	23
38	13934	15047	16219	17451	18745	20106	21535	22
39	13952	15066	16239	17472	18767	20129	21560	21
40	13970	15085	16259	17493	18790	20152	21584	20
41	1.13988	1.15105	1.16279	1.17514	1.18812	1.20176	1.21609	19
42	14006	15124	16299	17535	18834	20199	21633	18
43	14024	15143	16319	17556	18856	20222	21658	17
44	14042	15162	16339	17577	18878	20246	21682	16
45	14061	15181	16359	17598	18901	20269	21707	15
46	14079	15200	16380	17620	18923	20292	21731	14
47	14097	15219	16400	17641	18945	20316	21756	13
48	14115	15239	16420	17662	18967	20339	21781	12
49	14134	15258	16440	17683	18990	20363	21805	11
50	14152	15277	16460	17704	19012	20386	21830	10
51	1.14170	1.15296	1.16481	1.17726	1.19034	1.20410	1.21855	9
52	14188	15315	16501	17747	19057	20433	21879	8
53	14207	15335	16521	17768	19079	20457	21904	7
54	14225	15354	16541	17790	19102	20480	21929	6
55	14243	15373	16562	17811	19124	20504	21953	5
56	14262	15393	16582	17832	19146	20527	21978	4
57	14280	15412	16602	17854	19169	20551	22003	3
58	14299	15431	16623	17875	19191	20575	22028	2
59	14317	15451	16643	17896	19214	20598	22053	1
60	14335	15470	16663	17918	19236	20622	22077	0
	61°	60°	59°	58°	57°	56°	55°	
′	COSECANTS.							′

′	SECANTS.							′
	35°	36°	37°	38°	39°	40°	41°	
0	1.22077	1.23607	1.25214	1.26902	1.28676	1.30541	1.32501	60
1	22102	23633	25241	26931	28706	30573	32535	59
2	22127	23659	25269	26960	28737	30605	32568	58
3	22152	23685	25296	26988	28767	30636	32602	57
4	22177	23711	25324	27017	28797	30668	32636	56
5	22202	23738	25351	27046	28828	30700	32669	55
6	22227	23764	25379	27075	28858	30732	32703	54
7	22252	23790	25406	27104	28889	30764	32737	53
8	22277	23816	25434	27133	28919	30796	32770	52
9	22302	23843	25462	27162	28950	30829	32804	51
10	22327	23869	25489	27191	28980	30861	32838	50
11	1.22352	1.23895	1.25517	1.27221	1.29011	1.30893	1.32872	49
12	22377	23922	25545	27250	29042	30925	32905	48
13	22402	23948	25572	27279	29072	30957	32939	47
14	22428	23975	25600	27308	29103	30989	32973	46
15	22453	24001	25628	27337	29133	31022	33007	45
16	22478	24028	25656	27366	29164	31054	33041	44
17	22503	24054	25683	27396	29195	31086	33075	43
18	22528	24081	25711	27425	29226	31119	33109	42
19	22554	24107	25739	27454	29256	31151	33143	41
20	22579	24134	25767	27483	29287	31183	33177	40
21	1.22604	1.24160	1.25795	1.27513	1.29318	1.31216	1.33211	39
22	22629	24187	25823	27542	29349	31248	33245	38
23	22655	24213	25851	27572	29380	31281	33279	37
24	22680	24240	25879	27601	29411	31313	33314	36
25	22706	24267	25907	27630	29442	31346	33348	35
26	22731	24293	25935	27660	29473	31378	33382	34
27	22756	24320	25963	27689	29504	31411	33416	33
28	22782	24347	25991	27719	29535	31443	33451	32
29	22807	24373	26019	27748	29566	31476	33485	31
30	22833	24400	26047	27778	29597	31509	33519	30
31	1.22858	1.24427	1.26075	1.27807	1.29628	1.31541	1.33554	29
32	22884	24454	26104	27837	29659	31574	33588	28
33	22909	24481	26132	27867	29690	31607	33622	27
34	22935	24508	26160	27896	29721	31640	33657	26
35	22960	24534	26188	27926	29752	31672	33691	25
36	22986	24561	26216	27956	29784	31705	33726	24
37	23012	24588	26245	27985	29815	31738	33760	23
38	23037	24615	26273	28015	29846	31771	33795	22
39	23063	24642	26301	28045	29877	31804	33830	21
40	23089	24669	26330	28075	29909	31837	33864	20
41	1.23114	1.24696	1.26358	1.28105	1.29940	1.31870	1.33899	19
42	23140	24723	26387	28134	29971	31903	33934	18
43	23166	24750	26415	28164	30003	31936	33968	17
44	23192	24777	26443	28194	30034	31969	34003	16
45	23217	24804	26472	28224	30066	32002	34038	15
46	23243	24832	26500	28254	30097	32035	34073	14
47	23269	24859	26529	28284	30129	32068	34108	13
48	23295	24886	26557	28314	30160	32101	34142	12
49	23321	24913	26586	28344	30192	32134	34177	11
50	23347	24940	26615	28374	30223	32168	34212	10
51	1.23373	1.24967	1.26643	1.28404	1.30255	1.32201	1.34247	9
52	23399	24995	26672	28434	30287	32234	34282	8
53	23424	25022	26701	28464	30318	32267	34317	7
54	23450	25049	26729	28495	30350	32301	34352	6
55	23476	25077	26758	28525	30382	32334	34387	5
56	23502	25104	26787	28555	30413	32368	34423	4
57	23529	25131	26815	28585	30445	32401	34458	3
58	23555	25159	26844	28615	30477	32434	34493	2
59	23581	25186	26873	28646	30509	32468	34528	1
60	23607	25214	26902	28676	30541	32501	34563	0
′	54°	53°	52°	51°	50°	49°	48°	′
	COSECANTS.							

′	SECANTS.							′
	42°	43°	44°	45°	46°	47°	48°	
0	1.34563	1.36733	1.39016	1.41421	1.43956	1.46628	1.49448	60
1	34599	36770	39055	41463	43999	46674	49496	59
2	34634	36807	39095	41504	44042	46719	49544	58
3	34669	36844	39134	41545	44086	46765	49593	57
4	34704	36881	39173	41586	44129	46811	49641	56
5	34740	36919	39212	41627	44173	46857	49690	55
6	34775	36956	39251	41669	44217	46903	49738	54
7	34811	36993	39291	41710	44260	46949	49787	53
8	34846	37030	39330	41752	44304	46995	49835	52
9	34882	37068	39369	41793	44347	47041	49884	51
10	34917	37105	39409	41835	44391	47087	49933	50
11	1.34953	1.37143	1.39448	1.41876	1.44435	1.47134	1.49981	49
12	34988	37180	39487	41918	44479	47180	50030	48
13	35024	37218	39527	41959	44523	47226	50079	47
14	35060	37255	39566	42001	44567	47272	50128	46
15	35095	37293	39606	42042	44610	47319	50177	45
16	35131	37330	39646	42084	44654	47365	50226	44
17	35167	37368	39685	42126	44698	47411	50275	43
18	35203	37406	39725	42168	44742	47458	50324	42
19	35238	37443	39764	42210	44787	47504	50373	41
20	35274	37481	39804	42251	44831	47551	50422	40
21	1.35310	1.37519	1.39844	1.42293	1.44875	1.47598	1.50471	39
22	35346	37556	39884	42335	44919	47644	50521	38
23	35382	37594	39924	42377	44963	47691	50570	37
24	35418	37632	39963	42419	45007	47738	50619	36
25	35454	37670	40003	42461	45052	47784	50669	35
26	35490	37708	40043	42503	45096	47831	50718	34
27	35526	37746	40083	42545	45141	47878	50767	33
28	35562	37784	40123	42587	45185	47925	50817	32
29	35598	37822	40163	42630	45229	47972	50866	31
30	35634	37860	40203	42672	45274	48019	50916	30
31	1.35670	1.37898	1.40243	1.42714	1.45319	1.48066	1.50966	29
32	35707	37936	40283	42756	45363	48113	51015	28
33	35743	37974	40324	42799	45408	48160	51065	27
34	35779	38012	40364	42841	45452	48207	51115	26
35	35815	38051	40404	42883	45497	48254	51165	25
36	35852	38089	40444	42926	45542	48301	51215	24
37	35888	38127	40485	42968	45587	48349	51265	23
38	35924	38165	40525	43011	45631	48396	51314	22
39	35961	38204	40565	43053	45676	48443	51364	21
40	35997	38242	40606	43096	45721	48491	51415	20
41	1.36034	1.38280	1.40646	1.43139	1.45766	1.48538	1.51465	19
42	36070	38319	40687	43181	45811	48586	51515	18
43	36107	38357	40727	43224	45856	48633	51565	17
44	36143	38396	40768	43267	45901	48681	51615	16
45	36180	38434	40808	43310	45946	48728	51665	15
46	36217	38473	40849	43352	45992	48776	51716	14
47	36253	38512	40890	43395	46037	48824	51766	13
48	36290	38550	40930	43438	46082	48871	51817	12
49	36327	38589	40971	43481	46127	48919	51867	11
50	36363	38628	41012	43524	46173	48967	51918	10
51	1.36400	1.38666	1.41053	1.43567	1.46218	1.49015	1.51968	9
52	36437	38705	41093	43610	46263	49063	52019	8
53	36474	38744	41134	43653	46309	49111	52069	7
54	36511	38783	41175	43696	46354	49159	52120	6
55	36548	38822	41216	43739	46400	49207	52171	5
56	36585	38860	41257	43783	46445	49255	52222	4
57	36622	38899	41298	43826	46491	49303	52273	3
58	36659	38938	41339	43869	46537	49351	52323	2
59	36696	38977	41380	43912	46582	49399	52374	1
60	36733	39016	41421	43956	46628	49448	52425	0
′	47°	46°	45°	44°	43°	42°	41°	′
	COSECANTS.							

′	SECANTS.							′
	49°	50°	51°	52°	53°	54°	55°	
0	1.52425	1.55572	1.58902	1.62427	1.66164	1.70130	1.74345	60
1	52476	55626	58959	62487	66228	70198	74417	59
2	52527	55680	59016	62548	66292	70267	74490	58
3	52579	55734	59073	62609	66357	70335	74562	57
4	52630	55789	59130	62669	66421	70403	74635	56
5	52681	55843	59188	62730	66486	70472	74708	55
6	52732	55897	59245	62791	66550	70540	74781	54
7	52784	55951	59302	62852	66615	70609	74851	53
8	52835	56005	59360	62913	66679	70677	74927	52
9	52886	56060	59418	62974	66744	70746	75000	51
10	52938	56114	59475	63035	66809	70815	75073	50
11	1.52989	1.56169	1.59533	1.63096	1.66873	1.70884	1.75146	49
12	53041	56223	59590	63157	66938	70953	75219	48
13	53092	56278	59648	63218	67003	71022	75293	47
14	53144	56332	59706	63279	67068	71091	75366	46
15	53196	56387	59764	63341	67133	71160	75440	45
16	53247	56442	59822	63402	67199	71229	75513	44
17	53299	56497	59880	63464	67264	71298	75587	43
18	53351	56551	59938	63525	67329	71368	75661	42
19	53403	56606	59996	63587	67394	71437	75734	41
20	53455	56661	60054	63648	67460	71506	75808	40
21	1.53507	1.56716	1.60112	1.63710	1.67525	1.71576	1.75882	39
22	53559	56771	60171	63772	67591	71646	75956	38
23	53611	56826	60229	63834	67656	71715	76031	37
24	53663	56881	60287	63895	67722	71785	76105	36
25	53715	56937	60346	63957	67788	71855	76179	35
26	53768	56992	60404	64019	67853	71925	76253	34
27	53820	57047	60463	64081	67919	71995	76328	33
28	53872	57103	60521	64144	67985	72065	76402	32
29	53924	57158	60580	64206	68051	72135	76477	31
30	53977	57213	60639	64268	68117	72205	76552	30
31	1.54029	1.57269	1.60698	1.64330	1.68183	1.72275	1.76626	29
32	54082	57324	60756	64393	68250	72346	76701	28
33	54134	57380	60815	64455	68316	72416	76776	27
34	54187	57436	60874	64518	68382	72487	76851	26
35	54240	57491	60933	64580	68449	72557	76926	25
36	54292	57547	60992	64643	68515	72628	77001	24
37	54345	57603	61051	64705	68582	72698	77077	23
38	54398	57659	61111	64768	68648	72769	77152	22
39	54451	57715	61170	64831	68715	72840	77227	21
40	54504	57771	61229	64894	68782	72911	77303	20
41	1.54557	1.57827	1.61288	1.64957	1.68848	1.72982	1.77378	19
42	54610	57883	61348	65020	68915	73053	77454	18
43	54663	57939	61407	65083	68982	73124	77530	17
44	54716	57995	61467	65146	69049	73195	77606	16
45	54769	58051	61526	65209	69116	73267	77681	15
46	54822	58108	61586	65272	69183	73338	77757	14
47	54876	58164	61646	65336	69250	73409	77833	13
48	54929	58221	61705	65399	69318	73481	77910	12
49	54982	58277	61765	65462	69385	73552	77986	11
50	55036	58333	61825	65526	69452	73624	78062	10
51	1.55089	1.58390	1.61885	1.65589	1.69520	1.73696	1.78138	9
52	55143	58447	61945	65653	69587	73768	78215	8
53	55196	58503	62005	65717	69655	73840	78291	7
54	55250	58560	62065	65780	69723	73911	78368	6
55	55303	58617	62125	65844	69790	73983	78445	5
56	55357	58674	62185	65908	69858	74056	78521	4
57	55411	58731	62246	65972	69926	74128	78598	3
58	55465	58788	62306	66036	69994	74200	78675	2
59	55518	58845	62366	66100	70062	74272	78752	1
60	55572	58902	62427	66164	70130	74345	78829	0
′	40°	39°	38°	37°	36°	35°	34°	′
	COSECANTS.							

′	SECANTS.							′
	56°	57°	58°	59°	60°	61°	62°	
0	1.78829	1.83608	1.88708	1.94160	2.00000	2.06267	2.13005	60
1	78906	83690	88796	94254	00101	06375	13122	59
2	78984	83773	88884	94349	00202	06483	13239	58
3	79061	83855	88972	94443	00303	06592	13356	57
4	79138	83938	89060	94537	00404	06701	13473	56
5	79216	84020	89148	94632	00505	06809	13590	55
6	79293	84103	89237	94726	00607	06918	13707	54
7	79371	84186	89325	94821	00708	07027	13825	53
8	79449	84269	89414	94916	00810	07137	13942	52
9	79527	84352	89503	95011	00912	07246	14060	51
10	79604	84435	89591	95106	01014	07356	14178	50
11	1.79682	1.84518	1.89680	1.95201	2.01116	2.07465	2.14296	49
12	79761	84601	89769	95296	01218	07575	14414	48
13	79839	84685	89858	95392	01320	07685	14533	47
14	79917	84768	89948	95487	01422	07795	14651	46
15	79995	84852	90037	95583	01525	07905	14770	45
16	80074	84935	90126	95678	01628	08015	14889	44
17	80152	85019	90216	95774	01730	08126	15008	43
18	80231	85103	90305	95870	01833	08236	15127	42
19	80309	85187	90395	95966	01936	08347	15246	41
20	80388	85271	90485	96062	02039	08458	15366	40
21	1.80467	1.85355	1.90575	1.96158	2.02143	2.08569	2.15485	39
22	80546	85439	90665	96255	02246	08680	15605	38
23	80625	85523	90755	96351	02349	08791	15725	37
24	80704	85608	90845	96448	02453	08903	15845	36
25	80783	85692	90935	96544	02557	09014	15965	35
26	80862	85777	91026	96641	02661	09126	16085	34
27	80942	85861	91116	96738	02765	09238	16206	33
28	81021	85946	91207	96835	02869	09350	16326	32
29	81101	86031	91297	96932	02973	09462	16447	31
30	81180	86116	91388	97029	03077	09574	16568	30
31	1.81260	1.86201	1.91479	1.97127	2.03182	2.09686	2.16689	29
32	81340	86286	91570	97224	03286	09799	16810	28
33	81419	86371	91661	97322	03391	09911	16932	27
34	81499	86457	91752	97420	03496	10024	17053	26
35	81579	86542	91844	97517	03601	10137	17175	25
36	81659	86627	91935	97615	03706	10250	17297	24
37	81740	86713	92027	97713	03811	10363	17419	23
38	81820	86799	92118	97811	03916	10477	17541	22
39	81900	86885	92210	97910	04022	10590	17663	21
40	81981	86990	92302	98008	04128	10704	17786	20
41	1.82061	1.87056	1.92394	1.98107	2.04233	2.10817	2.17909	19
42	82142	87142	92486	98205	04339	10931	18031	18
43	82222	87229	92578	98304	04145	11045	18154	17
44	82303	87315	92670	98403	04551	11159	18277	16
45	82384	87401	92762	98502	04658	11274	18401	15
46	82465	87488	92855	98601	04764	11388	18524	14
47	82546	87574	92947	98700	04870	11503	18648	13
48	82627	87661	93040	98799	04977	11617	18772	12
49	82709	87748	93133	98899	05084	11732	18895	11
50	82790	87834	93226	98998	05191	11847	19019	10
51	1.82871	1.87921	1.93319	1.99098	2.05298	2.11963	2.19144	9
52	82953	88008	93412	99198	05405	12078	19268	8
53	83034	88095	93505	99298	05512	12193	19393	7
54	83116	88183	93598	99398	05619	12309	19517	6
55	83198	88270	93692	99498	05727	12425	19642	5
56	83280	88357	93785	99598	05835	12540	19767	4
57	83362	88445	93879	99698	05942	12657	19892	3
58	83444	88532	93973	99799	06050	12773	20018	2
59	83526	88620	94066	99899	06158	12889	20143	1
60	83608	88708	94160	2.00000	06267	13005	20269	0
′	33°	32°	31°	30°	29°	28°	27°	′
	COSECANTS.							

′	SECANTS.							′
	63°	64°	65°	66°	67°	68°	69°	
0	2.20269	2.28117	2.36620	2.45859	2.55930	2.66947	2.79043	60
1	20395	28253	36768	46020	56106	67139	79254	59
2	20521	28390	36916	46181	56282	67332	79466	58
3	20647	28526	37064	46342	56458	67525	79679	57
4	20773	28663	37212	46504	56634	67718	79891	56
5	20900	28800	37361	46665	56811	67911	80104	55
6	21026	28937	37509	46827	56988	68105	80318	54
7	21153	29074	37658	46989	57165	68299	80531	53
8	21280	29211	37808	47152	57342	68494	80746	52
9	21407	29349	37957	47314	57520	68689	80960	51
10	21535	29487	38107	47477	57698	68884	81175	50
11	2.21662	2.29625	2.38256	2.47640	2.57876	2.69079	2.81390	49
12	21790	29763	38406	47804	58054	69275	81605	48
13	21918	29901	38556	47967	58233	69471	81821	47
14	22045	30040	38707	48131	58412	69667	82037	46
15	22174	30179	38857	48295	58591	69864	82254	45
16	22302	30318	39008	48459	58771	70061	82471	44
17	22430	30457	39159	48624	58950	70258	82688	43
18	22559	30596	39311	48789	59130	70455	82906	42
19	22688	30735	39462	48954	59311	70653	83124	41
20	22817	30875	39614	49119	59491	70851	83342	40
21	2.22946	2.31015	2.39766	2.49284	2.59672	2.71050	2.83561	39
22	23075	31155	39918	49450	59853	71249	83780	38
23	23205	31295	40070	49616	60035	71448	83999	37
24	23334	31436	40222	49782	60217	71647	84219	36
25	23464	31576	40375	49948	60399	71847	84439	35
26	23594	31717	40528	50115	60581	72047	84659	34
27	23724	31858	40681	50282	60763	72247	84880	33
28	23855	31999	40835	50449	60946	72448	85102	32
29	23985	32140	40988	50617	61129	72649	85323	31
30	24116	32282	41142	50784	61313	72850	85545	30
31	2.24247	2.32424	2.41296	2.50952	2.61496	2.73052	2.85767	29
32	24378	32566	41450	51120	61680	73254	85990	28
33	24509	32708	41605	51289	61864	73456	86213	27
34	24640	32850	41760	51457	62049	73659	86437	26
35	24772	32993	41914	51626	62234	73862	86661	25
36	24903	33135	42070	51795	62419	74065	86885	24
37	25035	33278	42225	51965	62604	74269	87109	23
38	25167	33422	42380	52134	62790	74473	87334	22
39	25300	33565	42536	52304	62976	74677	87560	21
40	25432	33708	42692	52474	63162	74881	87785	20
41	2.25565	2.33852	2.42848	2.52645	2.63348	2.75086	2.88011	19
42	25697	33996	43005	52815	63535	75292	88238	18
43	25830	34140	43162	52986	63722	75497	88465	17
44	25963	34284	43318	53157	63909	75703	88692	16
45	26097	34429	43476	53329	64097	75909	88920	15
46	26230	34573	43633	53500	64285	76116	89148	14
47	26364	34718	43790	53672	64473	76323	89376	13
48	26498	34863	43948	53845	64662	76530	89605	12
49	26632	35009	44106	54017	64851	76737	89834	11
50	26766	35154	44264	54190	65040	76945	90063	10
51	2.26900	2.35300	2.44423	2.54363	2.65229	2.77154	2.90293	9
52	27035	35446	44582	54536	65419	77362	90524	8
53	27169	35592	44741	54709	65609	77571	90754	7
54	27304	35738	44900	54883	65799	77780	90986	6
55	27439	35885	45059	55057	65989	77990	91217	5
56	27574	36031	45219	55231	66180	78200	91449	4
57	27710	36178	45378	55405	66371	78410	91681	3
58	27845	36325	45539	55580	66563	78621	91914	2
59	27981	36473	45699	55755	66755	78832	92147	1
60	28117	36620	45859	55930	66947	79043	92380	0
	26°	25°	24°	23°	22°	21°	20°	
′	COSECANTS.							′

′	SECANTS.							′
	70°	71°	72°	73°	74°	75°	76°	
0	2.92380	3.07155	3.23607	3.42030	3.62796	3.86370	4.13357	60
1	92614	07415	23897	42356	63164	86790	13839	59
2	92849	07675	24187	42683	63533	87211	14323	58
3	93083	07936	24478	43010	63903	87633	14809	57
4	93318	08197	24770	43337	64274	88056	15295	56
5	93554	08459	25062	43666	64645	88479	15782	55
6	93790	08721	25355	43995	65018	88904	16271	54
7	94026	08983	25648	44324	65391	89330	16761	53
8	94263	09246	25942	44655	65765	89756	17252	52
9	94500	09510	26237	44986	66140	90184	17744	51
10	94737	09774	26531	45317	66515	90613	18238	50
11	2.94975	3.10038	3.26827	3.45650	3.66892	3.91042	4.18733	49
12	95213	10303	27123	45983	67269	91473	19228	48
13	95452	10568	27420	46316	67647	91904	19725	47
14	95691	10834	27717	46651	68025	92337	20224	46
15	95931	11101	28015	46986	68405	92770	20723	45
16	96171	11367	28313	47321	68785	93204	21224	44
17	96411	11635	28612	47658	69167	93640	21726	43
18	96652	11903	28912	47995	69549	94076	22229	42
19	96 93	12171	29212	48333	69931	94514	22734	41
20	97135	12440	29512	48671	70315	94952	23239	40
21	2.97377	3.12709	3.29814	3.49010	3.70700	3.95392	4.23746	39
22	97619	12979	30115	49350	71085	95832	24255	38
23	97862	13249	30418	49691	71471	96274	24764	37
24	98106	13520	30721	50032	71858	96716	25275	36
25	98349	13791	31024	50374	72246	97160	25787	35
26	98594	14063	31328	50716	72635	97604	26300	34
27	98838	14335	31633	51060	73024	98050	26814	33
28	99083	14608	31939	51404	73414	98497	27330	32
29	99329	14881	32244	51748	73806	98944	27847	31
30	99574	15155	32551	52094	74198	99393	28366	30
31	2.99821	3.15429	3.32858	3.52440	3.74591	3.99843	4.28865	29
32	3.00067	15704	33166	52787	74984	4.00293	29406	28
33	00315	15979	33474	53134	75379	00745	29929	27
34	00562	16255	33783	53482	75775	01198	30452	26
35	00810	16531	34092	53831	76171	01652	30977	25
36	01059	16808	34403	54181	76568	02107	31503	24
37	01308	17085	34713	54531	76966	02563	32031	23
38	01557	17363	35025	54883	77365	03020	32560	22
39	01807	17641	35336	55235	77765	03479	33090	21
40	02057	17920	35649	55587	78166	03938	33622	20
41	3.02308	3.18199	3.35962	3.55940	3.78568	4.04398	4.34154	19
42	02559	18479	36276	56294	78970	04860	34689	18
43	02810	18759	36590	56649	79374	05322	35224	17
44	03062	19040	36905	57005	79778	05786	35761	16
45	03315	19322	37221	57361	80183	06251	36299	15
46	03568	19604	37537	57718	80589	06717	36839	14
47	03821	19886	37854	58076	80996	07184	37380	13
48	04075	20169	38171	58434	81404	07652	37923	12
49	04329	20453	38489	58794	81813	08121	38466	11
50	04584	20737	38808	59154	82223	08591	39012	10
51	3.04839	3.21021	3.39128	3.59514	3.82633	4.09063	4.39558	9
52	05094	21306	39448	59876	83045	09535	40106	8
53	05350	21592	39768	60238	83457	10009	40656	7
54	05607	21878	40089	60601	83871	10484	41206	6
55	05864	22165	40411	60965	84285	10960	41759	5
56	06121	22452	40734	61330	84700	11437	42312	4
57	06379	22740	41057	61695	85116	11915	42867	3
58	06637	23028	41381	62061	85533	12394	43424	2
59	06896	23317	41705	62428	85951	12875	43982	1
60	07155	23607	42030	62796	86370	13357	44541	0
′	19°	18°	17°	16°	15°	14°	13°	′
	COSECANTS.							

′	SECANTS.							′
	77°	78°	79°	80°	81°	82°	83°	
0	4.44541	4.80973	5.24084	5.75877	6.39245	7.18530	8.20551	60
1	45102	81633	24870	76829	40122	20020	22500	59
2	45664	82294	25658	77784	41602	21517	24457	58
3	46228	82956	26448	78742	42787	23019	26425	57
4	46793	83621	27241	79703	43977	24529	28402	56
5	47360	84288	28036	80667	45171	26044	30388	55
6	47928	84956	28833	81635	46369	27566	32384	54
7	48498	85627	29634	82606	47572	29095	34390	53
8	49069	86299	30436	83581	48779	30630	36405	52
9	49642	86973	31241	84558	49991	32171	38431	51
10	50216	87649	32049	85539	51208	33719	40466	50
11	4.50791	4.88327	5.32859	5.86524	6.52429	7.35274	8.42511	49
12	51368	89007	33671	87511	53655	36835	44566	48
13	51947	89689	34486	88502	54886	38403	46632	47
14	52527	90373	35304	89497	56121	39978	48707	46
15	53109	91058	36124	90495	57361	41560	50793	45
16	53692	91746	36947	91496	58606	43148	52889	44
17	54277	92436	37772	92501	59855	44743	54996	43
18	54863	93128	38600	93509	61110	46346	57113	42
19	55451	93821	39430	94521	62369	47955	59241	41
20	56041	94517	40263	95536	63633	49571	61379	40
21	4.56632	4.95215	5.41099	5.96555	6.64902	7.51194	8.63528	39
22	57224	95914	41937	97577	66176	52825	65688	38
23	57819	96616	42778	98603	67454	54462	67859	37
24	58414	97320	43622	99633	68738	56107	70041	36
25	59012	98025	44468	6.00666	70027	57759	72234	35
26	59611	98733	45317	01703	71321	59418	74438	34
27	60211	99443	46169	02743	72620	61085	76653	33
28	60813	5.00155	47023	03787	73924	62759	78880	32
29	61417	00869	47881	04834	75233	64441	81118	31
30	62023	01585	48740	05886	76547	66130	83367	30
31	4.62630	5.02303	5.49603	6.06941	6.77866	7.67826	8.85628	29
32	63238	03024	50468	08000	79191	69530	87901	28
33	63849	03746	51337	09062	80521	71242	90186	27
34	64461	04471	52208	10129	81856	72962	92482	26
35	65074	05197	53081	11199	83196	74689	94791	25
36	65690	05926	53958	12273	84542	76424	97111	24
37	66307	06657	54837	13350	85893	78167	99414	23
38	66925	07390	55720	14432	87250	79918	9.01788	22
39	67545	08125	56605	15517	88612	81677	04146	21
40	68167	08863	57493	16607	89979	83443	06515	20
41	4.68791	5.09602	5.58383	6.17700	6.91352	7.85218	9.08897	19
42	69417	10344	59277	18797	92731	87001	11292	18
43	70044	11088	60174	19898	94115	88792	13699	17
44	70673	11835	61073	21004	95505	90592	16120	16
45	71303	12588	61976	22113	96900	92400	18553	15
46	71935	13334	62881	23226	98301	94216	20999	14
47	72569	14087	63790	24343	99708	96040	23459	13
48	73205	14842	64701	25464	7.01120	97873	25931	12
49	73843	15599	65616	26590	02538	99714	28417	11
50	74482	16359	66533	27719	03962	8.01565	30917	10
51	4.75123	5.17121	5.67454	6.28853	7.05392	8.03423	9.33430	9
52	75766	17886	68377	29991	06828	05291	35957	8
53	76411	18652	69304	31133	08269	07167	38497	7
54	77057	19421	70234	32279	09717	09052	41052	6
55	77705	20193	71166	33429	11171	10946	43620	5
56	78355	20966	72102	34584	12630	12849	46203	4
57	79007	21742	73041	35743	14096	14760	48800	3
58	79661	22521	73983	36906	15568	16681	51411	2
59	80316	23301	74929	38073	17046	18612	54037	1
60	80973	24084	75877	39245	18530	20551	56677	0
′	12°	11°	10°	9°	8°	7°	6°	′
	COSECANTS.							

′	SECANTS.						′
	84°	85°	86°	87°	88°	89°	
0	9.56677	11.47371	14.33559	19.10732	28.65371	57.29869	60
1	59332	51199	39547	21397	89440	58.26976	59
2	62002	55052	45586	32182	29.13917	59.27431	58
3	64687	58932	51676	43088	38812	60.31411	57
4	67387	62837	57817	54119	64137	61.39105	56
5	70103	66769	64011	65275	89903	62.50715	55
6	72833	70728	70258	76560	30.16120	63.66460	54
7	75579	74714	76558	87976	42802	64.86572	53
8	78341	78727	82913	99524	69960	66.11304	52
9	81119	82768	89323	20.11208	97607	67.40927	51
10	83912	86837	95788	23028	31.25758	68.75736	50
11	9.86722	11.90934	15.02310	20.34989	31.54425	70.16047	49
12	89547	95060	08890	47093	83623	71.62285	48
13	92389	99214	15527	59341	32.13366	73.14583	47
14	95248	12.03397	22223	71737	43671	74.73586	46
15	98123	07610	28979	84283	74554	76.39655	45
16	10.01015	11852	35795	96982	33.06030	78.13274	44
17	03923	16125	42672	21.09838	38118	79.94968	43
18	06849	20427	49611	22852	70835	81.85315	42
19	09792	24761	56614	36027	34.04199	83.84947	41
20	12752	29125	63679	49368	38232	85.94561	40
21	10.15730	12.33521	15.70810	21.62876	34.72952	88.14924	39
22	18725	37948	78005	76555	35.08380	90.46886	38
23	21739	42408	85268	90409	44539	92.91387	37
24	24770	46900	92597	22.04440	81452	95.49471	36
25	27819	51424	99995	18653	36.19141	98.22303	35
26	30887	55982	16.07462	33050	57633	101.11185	34
27	33973	60572	14999	47635	96953	104.17574	33
28	37077	65197	22607	62413	37.37127	107.43114	32
29	40201	69856	30287	77386	78185	110.89656	31
30	43343	74550	38041	92559	38.20155	114.59301	30
31	10.46505	12.79278	16.45869	23.07935	38.63068	118.54440	29
32	49685	84042	53772	23520	39.06957	122.77803	28
33	52886	88841	61751	39316	51855	127.32526	27
34	56106	93677	69808	55329	97797	132.22229	26
35	59346	98549	77944	71563	40.44820	137.51108	25
36	62605	13.03458	86159	88022	92963	143.24061	24
37	65885	08040	94456	24.04712	41.42266	149.46837	23
38	69186	13388	17.02835	21637	92772	156.26228	22
39	72507	18411	11297	38802	42.44525	163.70325	21
40	75849	23472	19843	56212	97571	171.88831	20
41	10.79212	13.28572	17.28476	24.73873	43.51961	180.93496	19
42	82596	33712	37196	91790	44.07746	190.98680	18
43	86001	38891	46005	25.09969	64980	202.22122	17
44	89428	44112	54903	28414	45.23720	214.85995	16
45	92877	49378	63893	47134	84026	229.18385	15
46	96348	54676	72975	66132	46.45963	245.55402	14
47	99841	60021	82152	85417	47.09596	264.44269	13
48	11.03356	65408	91424	26.04994	74997	286.47948	12
49	06894	70838	18.00794	24869	48.42241	312.52297	11
50	10455	76312	10262	45051	49.11406	343.77516	10
51	11.14039	13.81829	18.19930	26.65546	49.82576	381.97230	9
52	17646	87391	29501	86360	50.55840	429.71873	8
53	21277	92999	39274	27.07503	51.31290	491.10702	7
54	24932	98651	49153	28981	52.09027	572.95809	6
55	28610	14.04350	59139	50804	89156	687.54960	5
56	32313	10096	69233	72978	53.71790	859.43689	4
57	36040	15889	79438	95513	54.57046	1145.9157	3
58	39792	21730	89755	28.18417	55.45053	1718.8785	2
59	43569	27620	19.00185	41700	56.35946	3437.7468	1
60	47371	33559	10732	65371	57.29869	∞	0
′	5°	4°	3°	2°	1°	0°	′
	COSECANTS.						

′	0°		1°		2°		3°		′
	Tang	Cotang	Tang	Cotang	Tang	Cotang	Tang	Cotang	
0	.00000	Infinite.	.01746	57.2900	.03492	28.6363	.05241	19.0811	60
1	.00029	3437.75	.01775	56.3506	.03521	28.3994	.05270	18.9755	59
2	.00058	1718.87	.01804	55.4415	.03550	28.1664	.05299	18.8711	58
3	.00087	1145.92	.01833	54.5613	.03579	27.9372	.05328	18.7678	57
4	.00116	859.436	.01862	53.7086	.03609	27.7117	.05357	18.6656	56
5	.00145	687.549	.01891	52.8821	.03638	27.4899	.05387	18.5645	55
6	.00175	572.957	.01920	52.0807	.03667	27.2715	.05416	18.4645	54
7	.00204	491.106	.01949	51.3032	.03696	27.0566	.05445	18.3655	53
8	.00233	429.718	.01978	50.5485	.03725	26.8450	.05474	18.2677	52
9	.00262	381.971	.02007	49.8157	.03754	26.6367	.05503	18.1708	51
10	.00291	343.774	.02036	49.1039	.03783	26.4316	.05533	18.0750	50
11	.00320	312.521	.02066	48.4121	.03812	26.2296	.05562	17.9802	49
12	.00349	286.478	.02095	47.7395	.03842	26.0307	.05591	17.8863	48
13	.00378	264.441	.02124	47.0853	.03871	25.8348	.05620	17.7934	47
14	.00407	245.552	.02153	46.4489	.03900	25.6418	.05649	17.7015	46
15	.00436	229.182	.02182	45.8294	.03929	25.4517	.05678	17.6106	45
16	.00465	214.858	.02211	45.2261	.03958	25.2644	.05708	17.5205	44
17	.00495	202.219	.02240	44.6386	.03987	25.0798	.05737	17.4314	43
18	.00524	190.984	.02269	44.0661	.04016	24.8978	.05766	17.3432	42
19	.00553	180.932	.02298	43.5081	.04046	24.7185	.05795	17.2558	41
20	.00582	171.885	.02328	42.9641	.04075	24.5418	.05824	17.1693	40
21	.00611	163.700	.02357	42.4335	.04104	24.3675	.05854	17.0837	39
22	.00640	156.259	.02386	41.9158	.04133	24.1957	.05883	16.9990	38
23	.00669	149.465	.02415	41.4106	.04162	24.0263	.05912	16.9150	37
24	.00698	143.237	.02444	40.9174	.04191	23.8593	.05941	16.8319	36
25	.00727	137.507	.02473	40.4358	.04220	23.6945	.05970	16.7496	35
26	.00756	132.219	.02502	39.9655	.04250	23.5321	.05999	16.6681	34
27	.00785	127.321	.02531	39.5059	.04279	23.3718	.06029	16.5874	33
28	.00815	122.774	.02560	39.0568	.04308	23.2137	.06058	16.5075	32
29	.00844	118.540	.02589	38.6177	.04337	23.0577	.06087	16.4283	31
30	.00873	114.589	.02619	38.1885	.04366	22.9038	.06116	16.3499	30
31	.00902	110.892	.02648	37.7686	.04395	22.7519	.06145	16.2722	29
32	.00931	107.426	.02677	37.3579	.04424	22.6020	.06175	16.1952	28
33	.00960	104.171	.02706	36.9560	.04454	22.4541	.06204	16.1190	27
34	.00989	101.107	.02735	36.5627	.04483	22.3081	.06233	16.0435	26
35	.01018	98.2179	.02764	36.1776	.04512	22.1640	.06262	15.9687	25
36	.01047	95.4895	.02793	35.8006	.04541	22.0217	.06291	15.8945	24
37	.01076	92.9085	.02822	35.4313	.04570	21.8813	.06321	15.8211	23
38	.01105	90.4633	.02851	35.0695	.04599	21.7426	.06350	15.7483	22
39	.01135	88.1436	.02881	34.7151	.04628	21.6056	.06379	15.6762	21
40	.01164	85.9398	.02910	34.3678	.04658	21.4704	.06408	15.6048	20
41	.01193	83.8435	.02939	34.0273	.04687	21.3369	.06437	15.5340	19
42	.01222	81.8470	.02968	33.6935	.04716	21.2049	.06467	15.4638	18
43	.01251	79.9434	.02997	33.3662	.04745	21.0747	.06496	15.3943	17
44	.01280	78.1263	.03026	33.0452	.04774	20.9460	.06525	15.3254	16
45	.01309	76.3900	.03055	32.7303	.04803	20.8188	.06554	15.2571	15
46	.01338	74.7292	.03084	32.4213	.04833	20.6932	.06584	15.1893	14
47	.01367	73.1390	.03114	32.1181	.04862	20.5691	.06613	15.1222	13
48	.01396	71.6151	.03143	31.8205	.04891	20.4465	.06642	15.0557	12
49	.01425	70.1533	.03172	31.5284	.04920	20.3253	.06671	14.9898	11
50	.01455	68.7501	.03201	31.2416	.04949	20.2056	.06700	14.9244	10
51	.01484	67.4019	.03230	30.9599	.04978	20.0872	.06730	14.8596	9
52	.01513	66.1055	.03259	30.6833	.05007	19.9702	.06759	14.7954	8
53	.01542	64.8580	.03288	30.4116	.05037	19.8546	.06788	14.7317	7
54	.01571	63.6567	.03317	30.1446	.05066	19.7403	.06817	14.6685	6
55	.01600	62.4992	.03346	29.8823	.05095	19.6273	.06847	14.6059	5
56	.01629	61.3829	.03376	29.6245	.05124	19.5156	.06876	14.5438	4
57	.01658	60.3058	.03405	29.3711	.05153	19.4051	.06905	14.4823	3
58	.01687	59.2659	.03434	29.1220	.05182	19.2959	.06934	14.4212	2
59	.01716	58.2612	.03463	28.8771	.05212	19.1879	.06963	14.3607	1
60	.01746	57.2900	.03492	28.6363	.05241	19.0811	.06993	14.3007	0
	Cotang	Tang	Cotang	Tang	Cotang	Tang	Cotang	Tang	
′	89°		88°		87°		86°		′

′	4°		5°		6°		7°		′
	Tang	Cotang	Tang	Cotang	Tang	Cotang	Tang	Cotang	
0	.06993	14.3007	.08749	11.4301	.10510	9.51436	.12278	8.14435	60
1	.07022	14.2411	.08778	11.3919	.10540	9.48781	.12308	8.12481	59
2	.07051	14.1821	.08807	11.3540	.10569	9.46141	.12338	8.10536	58
3	.07080	14.1235	.08837	11.3163	.10599	9.43515	.12367	8.08600	57
4	.07110	14.0655	.08866	11.2789	.10628	9.40904	.12397	8.06674	56
5	.07139	14.0079	.08895	11.2417	.10657	9.38307	.12426	8.04756	55
6	.07168	13.9507	.08925	11.2048	.10687	9.35724	.12456	8.02848	54
7	.07197	13.8940	.08954	11.1681	.10716	9.33155	.12485	8.00948	53
8	.07227	13.8378	.08983	11.1316	.10746	9.30599	.12515	7.99058	52
9	.07256	13.7821	.09013	11.0954	.10775	9.28058	.12544	7.97176	51
10	.07285	13.7267	.09042	11.0594	.10805	9.25530	.12574	7.95302	50
11	.07314	13.6719	.09071	11.0237	.10834	9.23016	.12603	7.93438	49
12	.07344	13.6174	.09101	10.9882	.10863	9.20516	.12633	7.91582	48
13	.07373	13.5634	.09130	10.9529	.10893	9.18028	.12662	7.89734	47
14	.07402	13.5098	.09159	10.9178	.10922	9.15554	.12692	7.87895	46
15	.07431	13.4566	.09189	10.8829	.10952	9.13093	.12722	7.86064	45
16	.07461	13.4039	.09218	10.8483	.10981	9.10646	.12751	7.84242	44
17	.07490	13.3515	.09247	10.8139	.11011	9.08211	.12781	7.82428	43
18	.07519	13.2996	.09277	10.7797	.11040	9.05789	.12810	7.80622	42
19	.07548	13.2480	.09306	10.7457	.11070	9.03379	.12840	7.78825	41
20	.07578	13.1969	.09335	10.7119	.11099	9.00983	.12869	7.77035	40
21	.07607	13.1461	.09365	10.6783	.11128	8.98598	.12899	7.75254	39
22	.07636	13.0958	.09394	10.6450	.11158	8.96227	.12929	7.73480	38
23	.07665	13.0458	.09423	10.6118	.11187	8.93867	.12958	7.71715	37
24	.07695	12.9962	.09453	10.5789	.11217	8.91520	.12988	7.69957	36
25	.07724	12.9469	.09482	10.5462	.11246	8.89185	.13017	7.68208	35
26	.07753	12.8981	.09511	10.5136	.11276	8.86802	.13047	7.66466	34
27	.07782	12.8496	.09541	10.4813	.11305	8.84551	.13076	7.64732	33
28	.07812	12.8014	.09570	10.4491	.11335	8.82252	.13106	7.63005	32
29	.07841	12.7536	.09600	10.4172	.11364	8.79964	.13136	7.61287	31
30	.07870	12.7062	.09629	10.3854	.11394	8.77689	.13165	7.59575	30
31	.07899	12.6591	.09658	10.3538	.11423	8.75425	.13195	7.57872	29
32	.07929	12.6124	.09688	10.3224	.11452	8.73172	.13224	7.56176	28
33	.07958	12.5660	.09717	10.2913	.11482	8.70931	.13254	7.54487	27
34	.07987	12.5199	.09746	10.2602	.11511	8.68701	.13284	7.52806	26
35	.08017	12.4742	.09776	10.2294	.11541	8.66482	.13313	7.51132	25
36	.08046	12.4288	.09805	10.1988	.11570	8.64275	.13343	7.49465	24
37	.08075	12.3838	.09834	10.1683	.11600	8.62078	.13372	7.47806	23
38	.08104	12.3390	.09864	10.1381	.11629	8.59893	.13402	7.46154	22
39	.08134	12.2946	.09893	10.1080	.11[illegible]	8.57718	.13432	7.44509	21
40	.08163	12.2505	.09923	10.0780	.11688	8.55555	.13461	7.42871	20
41	.08192	12.2067	.09952	10.0483	.11718	8.53402	.13491	7.41240	19
42	.08221	12.1632	.09981	10.0187	.11747	8.51259	.13521	7.39616	18
43	.08251	12.1201	.10011	9.98931	.11777	8.49128	.13550	7.37999	17
44	.08280	12.0772	.10040	9.96007	.11806	8.47007	.13580	7.36389	16
45	.08309	12.0346	.10069	9.93101	.11836	8.44896	.13609	7.34786	15
46	.08339	11.9923	.10099	9.90211	.11865	8.42795	.13639	7.33190	14
47	.08368	11.9504	.10128	9.87338	.11895	8.40705	.13669	7.31600	13
48	.08397	11.9087	.10158	9.84482	.11924	8.38625	.13698	7.30018	12
49	.08427	11.8673	.10187	9.81641	.11954	8.36555	.13728	7.28442	11
50	.08456	11.8262	.10216	9.78817	.11983	8.34496	.13758	7.26873	10
51	.08485	11.7853	.10246	8.76009	.12013	8.32446	.13787	7.25310	9
52	.08514	11.7448	.10275	9.73217	.12042	8.30406	.13817	7.23754	8
53	.08544	11.7045	.10305	9.70441	.12072	8.28376	.13846	7.22204	7
54	.08573	11.6645	.10334	9.67680	.12101	8.26355	.13876	7.20661	6
55	.08602	11.6248	.10363	9.64935	.12131	8.24345	.13906	7.19125	5
56	.08632	11.5853	.10393	9.62205	.12160	8.22344	.13935	7.17594	4
57	.08661	11.5461	.10422	9.59490	.12190	8.20352	.13965	7.16071	3
58	.08690	11.5072	.10452	9.56791	.12219	8.18370	.13995	7.14553	2
59	.08720	11.4685	.10481	9.54106	.12249	8.16398	.14024	7.13042	1
60	.08749	11.4301	.10510	9.51436	.12278	8.14435	.14054	7.11537	0
′	Cotang	Tang	Cotang	Tang	Cotang	Tang	Cotang	Tang	′
	85°		84°		83°		82°		

′	8°		9°		10°		11°		′
	Tang	Cotang	Tang	Cotang	Tang	Cotang	Tang	Cotang	
0	.14054	7.11537	.15838	6.31375	.17633	5.67128	.19438	5.14455	60
1	.14084	7.10038	.15868	6.30189	.17663	5.66165	.19468	5.13658	59
2	.14113	7.08546	.15898	6.29007	.17693	5.65205	.19498	5.12862	58
3	.14143	7.07059	.15928	6.27829	.17723	5.64248	.19529	5.12069	57
4	.14173	7.05579	.15958	6.26655	.17753	5.63295	.19559	5.11279	56
5	.14202	7.04105	.15988	6.25486	.17783	5.62344	.19589	5.10490	55
6	.14232	7.02637	.16017	6.24321	.17813	5.61397	.19619	5.09704	54
7	.14262	6.91174	.16047	6.23160	.17843	5.60452	.19649	5.08921	53
8	.14291	6.99718	.16077	6.22003	.17873	5.59511	.19680	5.08139	52
9	.14321	6.98268	.16107	6.20851	.17903	5.58573	.19710	5.07360	51
10	.14351	6.96823	.16137	6.19703	.17933	5.57638	.19740	5.06584	50
11	.14381	6.95385	.16167	6.18559	.17963	5.56706	.19770	5.05809	49
12	.14410	6.93952	.16196	6.17419	.17993	5.55777	.19801	5.05037	48
13	.14440	6.92525	.16226	6.16283	.18023	5.54851	.19831	5.04267	47
14	.14470	6.91104	.16256	6.15151	.18053	5.53927	.19861	5.03499	46
15	.14499	6.89688	.16286	6.14023	.18083	5.53007	.19891	5.02734	45
16	.14529	6.88278	.16316	6.12899	.18113	5.52090	.19921	5.01971	44
17	.14559	6.86874	.16346	6.11779	.18143	5.51176	.19952	5.01210	43
18	.14588	6.85475	.16376	6.10664	.18173	5.50264	.19982	5.00451	42
19	.14618	6.84082	.16405	6.09552	.18203	5.49356	.20012	4.99695	41
20	.14648	6.82694	.16435	6.08444	.18233	5.48451	.20042	4.98940	40
21	.14678	6.81312	.16465	6.07340	.18263	5.47548	.20073	4.98188	39
22	.14707	6.79936	.16495	6.06240	.18293	5.46648	.20103	4.97438	38
23	.14737	6.78564	.16525	6.05143	.18323	5.45751	.20133	4.96690	37
24	.14767	6.77199	.16555	6.04051	.18353	5.44857	.20164	4.95945	36
25	.14796	6.75838	.16585	6.02962	.18384	5.43966	.20194	4.95201	35
26	.14826	6.74483	.16615	6.01878	.18414	5.43077	.20224	4.94460	34
27	.14856	6.73133	.16645	6.00797	.18444	5.42192	.20254	4.93721	33
28	.14886	6.71789	.16674	5.99720	.18474	5.41309	.20285	4.92984	32
29	.14915	6.70450	.16704	5.98646	.18504	5.40429	.20315	4.92249	31
30	.14945	6.69116	.16734	5.97576	.18534	5.39552	.20345	4.91516	30
31	.14975	6.67787	.16764	5.96510	.18564	5.38677	.20376	4.90785	29
32	.15005	6.66463	.16794	5.95448	.18594	5.37805	.20406	4.90056	28
33	.15034	6.65144	.16824	5.94390	.18624	5.36936	.20436	4.89330	27
34	.15064	6.63831	.16854	5.93335	.18654	5.36070	.20466	4.88605	26
35	.15094	6.62523	.16884	5.92283	.18684	5.35206	.20497	4.87882	25
36	.15124	6.61219	.16914	5.91236	.18714	5.34345	.20527	4.87162	24
37	.15153	6.59921	.16944	5.90191	.18745	5.33487	.20557	4.86444	23
38	.15183	6.58627	.16974	5.89151	.18775	5.32631	.20588	4.85727	22
39	.15213	6.57339	.17004	5.88114	.18805	5.31778	.20618	4.85013	21
40	.15243	6.56055	.17033	5.87080	.18835	5.30928	.20648	4.84300	20
41	.15272	6.54777	.17063	5.86051	.18865	5.30080	.20679	4.83590	19
42	.15302	6.53503	.17093	5.85024	.18895	5.29235	.20709	4.82882	18
43	.15332	6.52234	.17123	5.84001	.18925	5.28393	.20739	4.82175	17
44	.15362	6.50970	.17153	5.82982	.18955	5.27553	.20770	4.81471	16
45	.15391	6.49710	.17183	5.81966	.18986	5.26715	.20800	4.80769	15
46	.15421	6.48456	.17213	5.80953	.19016	5.25880	.20830	4.80068	14
47	.15451	6.47206	.17243	5.79944	.19046	5.25048	.20861	4.79370	13
48	.15481	6.45961	.17273	5.78938	.19076	5.24218	.20891	4.78673	12
49	.15511	6.44720	.17303	5.77936	.19106	5.23391	.20921	4.77978	11
50	.15540	6.43484	.17333	5.76937	.19136	5.22566	.20952	4.77286	10
51	.15570	6.42253	.17363	5.75941	.19166	5.21744	.20982	4.76595	9
52	.15600	6.41026	.17393	5.74949	.19197	5.20925	.21013	4.75906	8
53	.15630	6.39804	.17423	5.73960	.19227	5.20107	.21043	4.75219	7
54	.15600	6.38587	.17453	5.72974	.19257	5.19293	.21073	4.74534	6
55	.15689	6.37374	.17483	5.71992	.19287	5.18480	.21104	4.73851	5
56	.15719	6.36165	.17513	5.71013	.19317	5.17671	.21134	4.73170	4
57	.15749	6.34961	.17543	5.70037	.19347	5.16863	.21164	4.72490	3
58	.15779	6.33761	.17573	5.69064	.19378	5.16058	.21195	4.71813	2
59	.15809	6.32566	.17603	5.68094	.19408	5.15256	.21225	4.71137	1
60	.15838	6.31375	.17633	5.67128	.19438	5.14455	.21256	4.70463	0
′	Cotang	Tang	Cotang	Tang	Cotang	Tang	Cotang	Tang	′
	81°		80°		79°		78°		

′	12°		13°		14°		15°		′
	Tang	Cotang	Tang	Cotang	Tang	Cotang	Tang	Cotang	
0	.21256	4.70463	.23087	4.33148	.24933	4.01078	.26795	3.73205	60
1	.21286	4.69791	.23117	4.32573	.24964	4.00582	.26826	3.72771	59
2	.21316	4.69121	.23148	4.32001	.24995	4.00086	.26857	3.72338	58
3	.21347	4.68452	.23179	4.31430	.25026	3.99592	.26888	3.71907	57
4	.21377	4.67786	.23209	4.30860	.25056	3.99099	.26920	3.71476	56
5	.21408	4.67121	.23240	4.30291	.25087	3.98607	.26951	3.71046	55
6	.21438	4.66458	.23271	4.29724	.25118	3.98117	.26982	3.70616	54
7	.21469	4.65797	.23301	4.29159	.25149	3.97627	.27013	3.70188	53
8	.21499	4.65138	.23332	4.28595	.25180	3.97139	.27044	3.69761	52
9	.21529	4.64480	.23363	4.28032	.25211	3.96651	.27076	3.69335	51
10	.21560	4.63825	.23393	4.27471	.25242	3.96165	.27107	3.68909	50
11	.21590	4.63171	.23424	4.26911	.25273	3.95680	.27138	3.68485	49
12	.21621	4.62518	.23455	4.26352	.25304	3.95196	.27169	3.68061	48
13	.21651	4.61868	.23485	4.25795	.25335	3.94713	.27201	3.67638	47
14	.21682	4.61219	.23516	4.25239	.25366	3.94232	.27232	3.67217	46
15	.21712	4.60572	.23547	4.24685	.25397	3.93751	.27263	3.66796	45
16	.21743	4.59927	.23578	4.24132	.25428	3.93271	.27294	3.66376	44
17	.21773	4.59283	.23608	4.23580	.25459	3.92793	.27326	3.65957	43
18	.21804	4.58641	.23639	4.23030	.25490	3.92316	.27357	3.65538	42
19	.21834	4.58001	.23670	4.22481	.25521	3.91839	.27388	3.65121	41
20	.21864	4.57363	.23700	4.21933	.25552	3.91364	.27419	3.64705	40
21	.21895	4.56726	.23731	4.21387	.25583	3.90890	.27451	3.64289	39
22	.21925	4.56091	.23762	4.20842	.25614	3.90417	.27482	3.63874	38
23	.21956	4.55458	.23793	4.20298	.25645	3.89945	.27513	3.63461	37
24	.21986	4.54826	.23823	4.19756	.25676	3.89474	.27545	3.63048	36
25	.22017	4.54196	.23854	4.19215	.25707	3.89004	.27576	3.62636	35
26	.22047	4.53568	.23885	4.18675	.25738	3.88536	.27607	3.62224	34
27	.22078	4.52941	.23916	4.18137	.25769	3.88068	.27638	3.61814	33
28	.22108	4.52316	.23946	4.17600	.25800	3.87601	.27670	3.61405	32
29	.22139	4.51693	.23977	4.17064	.25831	3.87136	.27701	3.60996	31
30	.22169	4.51071	.24008	4.16530	.25862	3.86671	.27732	3.60588	30
31	.22200	4.50451	.24039	4.15997	.25893	3.86208	.27764	3.60181	29
32	.22231	4.49832	.24069	4.15465	.25924	3.85745	.27795	3.59775	28
33	.22261	4.49215	.24100	4.14934	.25955	3.85284	.27826	3.59370	27
34	.22292	4.48600	.24131	4.14405	.25986	3.84824	.27858	3.58966	26
35	.22322	4.47986	.24162	4.13877	.26017	3.84364	.27889	3.58562	25
36	.22353	4.47374	.24193	4.13350	.26048	3.83906	.27921	3.58160	24
37	.22383	4.46764	.24223	4.12825	.26079	3.83449	.27952	3.57758	23
38	.22414	4.46155	.24254	4.12301	.26110	3.82992	.27983	3.57357	22
39	.22444	4.45548	.24285	4.11778	.26141	3.82537	.28015	3.56957	21
40	.22475	4.44942	.24316	4.11256	.26172	3.82083	.28046	3.56557	20
41	.22505	4.44338	.24347	4.10736	.26203	3.81630	.28077	3.56159	19
42	.22536	4.43735	.24377	4.10216	.26235	3.81177	.28109	3.55761	18
43	.22567	4.43134	.24408	4.09699	.26266	3.80726	.28140	3.55364	17
44	.22597	4.42534	.24439	4.09182	.26297	3.80276	.28172	3.54968	16
45	.22628	4.41936	.24470	4.08666	.26328	3.79827	.28203	3.54573	15
46	.22658	4.41340	.24501	4.08152	.26359	3.79378	.28234	3.54179	14
47	.22689	4.40745	.24532	4.07639	.26390	3.78931	.28266	3.53785	13
48	.22719	4.40152	.24562	4.07127	.26421	3.78485	.28297	3.53393	12
49	.22750	4.39560	.24593	4.06616	.26452	3.78040	.28329	3.53001	11
50	.22781	4.38969	.24624	4.06107	.26483	3.77595	.28360	3.52609	10
51	.22811	4.38381	.24655	4.05599	.26515	3.77152	.28391	3.52219	9
52	.22842	4.37793	.24686	4.05092	.26546	3.76709	.28423	3.51829	8
53	.22872	4.37207	.24717	4.04586	.26577	3.76268	.28454	3.51441	7
54	.22903	4.36623	.24747	4.04081	.26608	3.75828	.28486	3.51053	6
55	.22934	4.36040	.24778	4.03578	.26639	3.75388	.28517	3.50666	5
56	.22964	4.35459	.24809	4.03076	.26670	3.74950	.28549	3.50279	4
57	.22995	4.34879	.24840	4.02574	.26701	3.74512	.28580	3.49894	3
58	.23026	4.34300	.24871	4.02074	.26733	3.74075	.28612	3.49509	2
59	.23056	4.33723	.24902	4.01576	.26764	3.73640	.28643	3.49125	1
60	.23087	4.33148	.24933	4.01078	.26795	3.73205	.28675	3.48741	0
′	Cotang	Tang	Cotang	Tang	Cotang	Tang	Cotang	Tang	′
	77°		76°		75°		74°		

′	16° Tang	16° Cotang	17° Tang	17° Cotang	18° Tang	18° Cotang	19° Tang	19° Cotang	′
0	.28675	3.48741	.30573	3.27085	.32492	3.07768	.34433	2.90421	60
1	.28706	3.48359	.30605	3.26745	.32524	3.07464	.34465	2.90147	59
2	.28738	3.47977	.30637	3.26406	.32556	3.07160	.34498	2.89873	58
3	.28769	3.47596	.30669	3.26067	.32588	3.06857	.34530	2.89600	57
4	.28800	3.47216	.30700	3.25729	.32621	3.06554	.34563	2.89327	56
5	.28832	3.46837	.30732	3.25392	.32653	3.06252	.34596	2.89055	55
6	.28864	3.46458	.30764	3.25055	.32685	3.05950	.34628	2.88783	54
7	.28895	3.46080	.30796	3.24719	.32717	3.05649	.34661	2.88511	53
8	.28927	3.45703	.30828	3.24383	.32749	3.05349	.34693	2.88240	52
9	.28958	3.45327	.30860	3.24049	.32782	3.05049	.34726	2.87970	51
10	.28990	3.44951	.30891	3.23714	.32814	3.04749	.34758	2.87700	50
11	.29021	3.44576	.30923	3.23381	.32846	3.04450	.34791	2.87430	49
12	.29053	3.44202	.30955	3.23048	.32878	3.04152	.34824	2.87161	48
13	.29084	3.43829	.30987	3.22715	.32911	3.03854	.34856	2.86892	47
14	.29116	3.43456	.31019	3.22384	.32943	3.03556	.34889	2.86624	46
15	.29147	3.43084	.31051	3.22053	.32975	3.03260	.34922	2.86356	45
16	.29179	3.42713	.31083	3.21722	.33007	3.02963	.34954	2.86089	44
17	.29210	3.42343	.31115	3.21392	.33040	3.02667	.34987	2.85822	43
18	.29242	3.41973	.31147	3.21063	.33072	3.02372	.35020	2.85555	42
19	.29274	3.41604	.31178	3.20734	.33104	3.02077	.35052	2.85289	41
20	.29305	3.41236	.31210	3.20406	.33136	3.01783	.35085	2.85023	40
21	.29337	3.40869	.31242	3.20079	.33169	3.01489	.35118	2.84758	39
22	.29368	3.40502	.31274	3.19752	.33201	3.01196	.35150	2.84494	38
23	.29400	3.40136	.31306	3.19426	.33233	3.00903	.35183	2.84229	37
24	.29432	3.39771	.31338	3.19100	.33266	3.00611	.35216	2.83965	36
25	.29463	3.39406	.31370	3.18775	.33298	3.00319	.35248	2.83702	35
26	.29495	3.39042	.31402	3.18451	.33330	3.00028	.35281	2.83439	34
27	.29526	3.38679	.31434	3.18127	.33363	2.99738	.35314	2.83176	33
28	.29558	3.38317	.31466	3.17804	.33395	2.99447	.35346	2.82914	32
29	.29590	3.37955	.31498	3.17481	.33427	2.99158	.35379	2.82653	31
30	.29621	3.37594	.31530	3.17159	.33460	2.98868	.35412	2.82391	30
31	.29653	3.37234	.31562	3.16828	.33492	2.98580	.35445	2.82130	29
32	.29685	3.36875	.31594	3.16517	.33524	2.98292	.35477	2.81870	28
33	.29716	3.36516	.31626	3.16197	.33557	2.98004	.35510	2.81610	27
34	.29748	3.36158	.31658	3.15877	.33589	2.97717	.35543	2.81350	26
35	.29780	3.35800	.31690	3.15558	.33621	2.97430	.35576	2.81091	25
36	.29811	3.35443	.31722	3.15240	.33654	2.97144	.35608	2.80833	24
37	.29843	3.35087	.31754	3.14922	.33686	2.96858	.35641	2.80574	23
38	.29875	3.34732	.31786	3.14605	.33718	2.96573	.35674	2.80316	22
39	.29906	3.34377	.31818	3.14288	.33751	2.96288	.35707	2.80059	21
40	.29938	3.34023	.31850	3.13972	.33783	2.96004	.35740	2.79802	20
41	.29970	3.33670	.31882	3.13656	.33816	2.95721	.35772	2.79545	19
42	.30001	3.33317	.31914	3.13341	.33848	2.95437	.35805	2.79289	18
43	.30033	3.32965	.31946	3.13027	.33881	2.95155	.35838	2.79033	17
44	.30065	3.32614	.31978	3.12713	.33913	2.94872	.35871	2.78778	16
45	.30097	3.32264	.32010	3.12400	.33945	2.94591	.35904	2.78523	15
46	.30128	3.31914	.32042	3.12087	.33978	2.94309	.35937	2.78269	14
47	.30160	3.31565	.32074	3.11775	.34010	2.94028	.35969	2.78014	13
48	.30192	3.31216	.32106	3.11464	.34043	2.93748	.36002	2.77761	12
49	.30224	3.30868	.32139	3.11153	.34075	2.93468	.36035	2.77507	11
50	.30255	3.30521	.32171	3.10842	.34108	2.93189	.36068	2.77254	10
51	.30287	3.30174	.32203	3.10532	.34140	2.92910	.36101	2.77002	9
52	.30319	3.29829	.32235	3.10223	.34173	2.92632	.36134	2.76750	8
53	.30351	3.29483	.32267	3.09914	.34205	2.92354	.36167	2.76498	7
54	.30382	3.29139	.32299	3.09606	.34238	2.92076	.36199	2.76247	6
55	.30414	3.28795	.32331	3.09298	.34270	2.91799	.36232	2.75996	5
56	.30446	3.28452	.32363	3.08991	.34303	2.91523	.36265	2.75746	4
57	.30478	3.28109	.32396	3.08685	.34335	2.91246	.36298	2.75496	3
58	.30509	3.27767	.32428	3.08379	.34368	2.90971	.36331	2.75246	2
59	.30541	3.27426	.32460	3.08073	.34400	2.90696	.36364	2.74997	1
60	.30573	3.27085	.32492	3.07768	.34433	2.90421	.36397	2.74748	0
′	Cotang	Tang	Cotang	Tang	Cotang	Tang	Cotang	Tang	′
	73°		72°		71°		70°		

′	20° Tang	20° Cotang	21° Tang	21° Cotang	22° Tang	22° Cotang	23° Tang	23° Cotang	′
0	.36397	2.74748	.38386	2.60509	.40403	2.47509	.42447	2.35585	60
1	.33430	2.74499	.38420	2.60283	.40436	2.47302	.42482	2.35395	59
2	.36463	2.74251	.38453	2.60057	.40470	2.47095	.42516	2.35205	58
3	.36496	2.74004	.38487	2.59831	.40504	2.46888	.42551	2.35015	57
4	.36529	2.73756	.38520	2.59606	.40538	2.46682	.42585	2.34825	56
5	.36562	2.73509	.38553	2.59381	.40572	2.46476	.42619	2.34636	55
6	.36595	2.73263	.38587	2.59156	.40606	2.46270	.42654	2.34447	54
7	.36628	2.73017	.38620	2.58932	.40640	2.46065	.42688	2.34258	53
8	.36661	2.72771	.38654	2.58708	.40674	2.45860	.42722	2.34069	52
9	.36694	2.72526	.38687	2.58484	.40707	2.45655	.42757	2.33881	51
10	.36727	2.72281	.38721	2.58261	.40741	2.45451	.42791	2.33693	50
11	.36760	2.72036	.38754	2.58038	.40775	2.45246	.42826	2.33505	49
12	.36793	2.71792	.38787	2.57815	.40809	2.45043	.42860	2.33317	48
13	.36826	2.71548	.38821	2.57593	.40843	2.44839	.42894	2.33130	47
14	.36859	2.71305	.38854	2.57371	.40877	2.44636	.42929	2.32943	46
15	.36892	2.71062	.38888	2.57150	.40911	2.44433	.42963	2.32756	45
16	.36925	2.70819	.38921	2.56928	.40945	2.44230	.42998	2.32570	44
17	.36958	2.70577	.38955	2.56707	.40979	2.44027	.43032	2.32383	43
18	.36991	2.70335	.38988	2.56487	.41013	2.43825	.43067	2.32197	42
19	.37024	2.70094	.39022	2.56266	.41047	2.43623	.43101	2.32012	41
20	.37057	2.69853	.39055	2.56046	.41081	2.43422	.43136	2.31826	40
21	.37090	2.69612	.39089	2.55827	.41115	2.43220	.43170	2.31641	39
22	.37123	2.69371	.39122	2.55608	.41149	2.43019	.43205	2.31456	38
23	.37157	2.69131	.39156	2.55389	.41183	2.42819	.43239	2.31271	37
24	.37190	2.68892	.39190	2.55170	.41217	2.42618	.43274	2.31086	36
25	.37223	2.68653	.39223	2.54952	.41251	2.42418	.43308	2.30902	35
26	.37256	2.68414	.39257	2.54734	.41285	2.42218	.43343	2.30718	34
27	.37289	2.68175	.39290	2.54516	.41319	2.42019	.43378	2.30534	33
28	.37322	2 67937	.39324	2.54299	.41353	2.41819	.43412	2.30351	32
29	.37355	2.67700	.39357	2.54082	.41387	2.41620	.43447	2.30167	31
30	.37388	2.67462	.39391	2.53865	.41421	2.41421	.43481	2.29984	30
31	.37422	2.67225	.39425	2.53648	.41455	2.41223	.43516	2.29801	29
32	.37455	2.66989	.39458	2.53432	.41490	2.41025	.43550	2.29619	28
33	.37488	2.66752	.39492	2.53217	.41524	2.40827	.43585	2.29437	27
34	.37521	2.66516	.39526	2.53001	.41558	2.40629	.43620	2.29254	26
35	.37554	2.66281	.39559	2.52786	.41592	2.40432	.43654	2.29073	25
36	.37588	2.66046	.39593	2.52571	.41626	2.40235	.43689	2.28891	24
37	.37621	2.65811	.39626	2.52357	.41660	2.40038	.43724	2.28710	23
38	.37654	2.65576	.39660	2.52142	.41694	2.39841	.43758	2.28528	22
39	.37687	2.65342	.39694	2.51929	.41728	2.39645	.43793	2.28348	21
40	.37720	2.65109	.39727	2.51715	.41763	2.39449	.43828	2.28167	20
41	.37754	2.64875	.39761	2.51502	.41797	2.39253	.43862	2.27987	19
42	.37787	2.64642	.39795	2.51289	.41831	2.39058	.43897	2.27806	18
43	.37820	2.64410	.39829	2.51076	.41865	2.38863	.43932	2.27626	17
44	.37853	2.64177	.39862	2.50864	41899	2.38668	.43966	2.27447	16
45	.37887	2.63945	.39896	2.50652	.41933	2.38473	.44001	2.27267	15
46	.37920	2.63714	.39930	2.50440	.41968	2.38279	.44036	2.27088	14
47	.37953	2.63483	.39963	2.50229	.42002	2.38084	.44071	2.26909	13
48	.37986	2.63252	.39997	2.50018	.42036	2.37891	.44105	2.26730	12
49	.38020	2.63021	.40031	2.49807	.42070	2.37697	.44140	2.26552	11
50	.38053	2.62791	.40065	2.49597	.42105	2.37504	.44175	2.26374	10
51	.38086	2.62561	.40098	2.49386	.42139	2.37311	.44210	2.26196	9
52	.38120	2.62332	.40132	2.49177	.42173	2.37118	.44244	2.26018	8
53	.38153	2.62103	.40166	2.48967	.42207	2.36925	.44279	2.25840	7
54	.38186	2.61874	.40200	2.48758	.42242	2.36733	.44314	2.25663	6
55	.38220	2.61646	.40234	2.48549	.42276	2.36541	.44349	2.25486	5
56	.38253	2.61418	.40267	2 48340	.42310	2.36349	.44384	2.25309	4
57	.38286	2.61190	.40301	2.48132	.42345	2.36158	.44418	2.25132	3
58	.38320	2.60963	.40335	2.47924	.42379	2.35967	.44453	2.24956	2
59	.38353	2.60736	.40369	2.47716	.42413	2.35776	.44488	2 24780	1
60	.38386	2 60509	.40403	2.47509	.42447	2.35585	.44523	2.24604	0
′	Cotang	Tang	Cotang	Tang	Cotang	Tang	Cotang	Tang	′
	69°		68°		67°		66°		

′	24° Tang	24° Cotang	25° Tang	25° Cotang	26° Tang	26° Cotang	27° Tang	27° Cotang	′
0	.44523	2.24604	.46631	2.14451	.48773	2.05030	.50953	1.96261	60
1	.44558	2.24428	.46666	2.14288	.48809	2.04879	.50989	1.96120	59
2	.44593	2.24252	.46702	2.14125	.48845	2.04728	.51026	1.95979	58
3	.44627	2.24077	.46737	2.13963	.48881	2.04577	.51063	1.95838	57
4	.44662	2.23902	.46772	2.13801	.48917	2.04426	.51099	1.95698	56
5	.44697	2.23727	.46808	2.13639	.48953	2.04276	.51136	1.95557	55
6	.44732	2.23553	.46843	2.13477	.48989	2.04125	.51173	1.95417	54
7	.44767	2.23378	.46879	2.13316	.49026	2.03975	.51209	1.95277	53
8	.44802	2.23204	.46914	2.13154	.49062	2.03825	.51246	1.95137	52
9	.44837	2.23030	.46950	2.12993	.49098	2.03675	.51283	1.94997	51
10	.44872	2.22857	.46985	2.12832	.49134	2.03526	.51319	1.94858	50
11	.44907	2.22683	.47021	2.12671	.49170	2.03376	.51356	1.94718	49
12	.44942	2.22510	.47056	2.12511	.49206	2.03227	.51393	1.94579	48
13	.44977	2.22337	.47092	2.12350	.49242	2.03078	.51430	1.94440	47
14	.45012	2.22164	.47128	2.12190	.49278	2.02929	.51467	1.94301	46
15	.45047	2.21992	.47163	2.12030	.49315	2.02780	.51503	1.94162	45
16	.45082	2.21819	.47199	2.11871	.49351	2.02631	.51540	1.94023	44
17	.45117	2.21647	.47234	2.11711	.49387	2.02483	.51577	1.93885	43
18	.45152	2.21475	.47270	2.11552	.49423	2.02335	.51614	1.93746	42
19	.45187	2.21304	.47305	2.11392	.49459	2.02187	.51651	1.93608	41
20	.45222	2.21132	.47341	2.11233	.49495	2.02039	.51688	1.93470	40
21	.45257	2.20961	.47377	2.11075	.49532	2.01891	.51724	1.93332	39
22	.45292	2.20790	.47412	2.10916	.49568	2.01743	.51761	1.93195	38
23	.45327	2.20619	.47448	2.10758	.49604	2.01596	.51798	1.93057	37
24	.45362	2.20449	.47483	2.10600	.49640	2.01449	.51835	1.92920	36
25	.45397	2.20278	.47519	2.10442	.49677	2.01302	.51872	1.92782	35
26	.45432	2.20108	.47555	2.10284	.49713	2.01155	.51909	1.92645	34
27	.45467	2.19938	.47590	2.10126	.49749	2.01008	.51946	1.92508	33
28	.45502	2.19769	.47626	2.09969	.49786	2.00862	.51983	1.92371	32
29	.45538	2.19599	.47662	2.09811	.49822	2.00715	.52020	1.92235	31
30	.45573	2.19430	.47698	2.09654	.49858	2.00569	.52057	1.92098	30
31	.45608	2.19261	.47733	2.09498	.49894	2.00423	.52094	1.91962	29
32	.45643	2.19092	.47769	2.09341	.49931	2.00277	.52131	1.91826	28
33	.45678	2.18923	.47805	2.09184	.49967	2.00131	.52168	1.91690	27
34	.45713	2.18755	.47840	2.09028	.50004	1.99986	.52205	1.91554	26
35	.45748	2.18587	.47876	2.08872	.50040	1.99841	.52242	1.91418	25
36	.45784	2.18419	.47912	2.08716	.50076	1.99695	.52279	1.91282	24
37	.45819	2.18251	.47948	2.08560	.50113	1.99550	.52316	1.91147	23
38	.45854	2.18084	.47984	2.08405	.50149	1.99406	.52353	1.91012	22
39	.45889	2.17916	.48019	2.08250	.50185	1.99261	.52390	1.90876	21
40	.45924	2.17749	.48055	2.08094	.50222	1.99116	.52427	1.90741	20
41	.45960	2.17582	.48091	2.07939	.50258	1.98972	.52464	1.90607	19
42	.45995	2.17416	.48127	2.07785	.50295	1.98828	.52501	1.90472	18
43	.46030	2.17249	.48163	2.07630	.50331	1.98684	.52538	1.90337	17
44	.46065	2.17083	.48198	2.07476	.50368	1.98540	.52575	1.90203	16
45	.46101	2.16917	.48234	2.07321	.50404	1.98396	.52613	1.90069	15
46	.46136	2.16751	.48270	2.07167	.50441	1.98253	.52650	1.89935	14
47	.46171	2.16585	.48306	2.07014	.50477	1.98110	.52687	1.89801	13
48	.46206	2.16420	.48342	2.06860	.50514	1.97966	.52724	1.89667	12
49	.46242	2.16255	.48378	2.06706	.50550	1.97823	.52761	1.89533	11
50	.46277	2.16090	.48414	2.06553	.50587	1.97681	.52798	1.89400	10
51	.46312	2.15925	.48450	2.06400	.50623	1.97538	.52836	1.89266	9
52	.46348	2.15760	.48486	2.06247	.50660	1.97395	.52873	1.89133	8
53	.46383	2.15596	.48521	2.06094	.50696	1.97253	.52910	1.89000	7
54	.46418	2.15432	.48557	2.05942	.50733	1.97111	.52947	1.88867	6
55	.46454	2.15268	.48593	2.05790	.50769	1.96969	.52985	1.88734	5
56	.46489	2.15104	.48629	2.05637	.50806	1.96827	.53022	1.88602	4
57	.46525	2.14940	.48665	2.05485	.50843	1.96685	.53059	1.88469	3
58	.46560	2.14777	.48701	2.05333	.50879	1.96544	.53096	1.88337	2
59	.46595	2.14614	.48737	2.05182	.50916	1.96402	.53134	1.88205	1
60	.46631	2.14451	.48773	2.05030	.50953	1.96261	.53171	1.88073	0
′	65° Cotang	65° Tang	64° Cotang	64° Tang	63° Cotang	63° Tang	62° Cotang	62° Tang	′

′	28° Tang	28° Cotang	29° Tang	29° Cotang	30° Tang	30° Cotang	31° Tang	31° Cotang	′
0	.53171	1.88073	.55431	1.80405	.57735	1.73205	.60086	1.66428	60
1	.53208	1.87941	.55469	1.80281	.57774	1.73089	.60126	1.66318	59
2	.53246	1.87809	.55507	1.80158	.57813	1.72973	.60165	1.66209	58
3	.53283	1.87677	.55545	1.80034	.57851	1.72857	.60205	1.66099	57
4	.53320	1.87546	.55583	1.79911	.57890	1.72741	.60245	1.65990	56
5	.53358	1.87415	.55621	1.79788	.57929	1.72625	.60284	1.65881	55
6	.53395	1.87283	.55659	1.79665	.57968	1.72509	.60324	1.65772	54
7	.53432	1.87152	.55697	1.79542	.58007	1.72393	.60364	1.65663	53
8	.53470	1.87021	.55736	1.79419	.58046	1.72278	.60403	1.65554	52
9	.53507	1.86891	.55774	1.79296	.58085	1.72163	.60443	1.65445	51
10	.53545	1.86760	.55812	1.79174	.58124	1.72047	.60483	1.65337	50
11	.53582	1.86630	.55850	1.79051	.58162	1.71932	.60522	1.65228	49
12	.53620	1.86499	.55888	1.78929	.58201	1.71817	.60562	1.65120	48
13	.53657	1.86369	.55926	1.78807	.58240	1.71702	.60602	1.65011	47
14	.53694	1.86239	.55964	1.78685	.58279	1.71588	.60642	1.64903	46
15	.53732	1.86109	.56003	1.78563	.58318	1.71473	.60681	1.64795	45
16	.53769	1.85979	.56041	1.78441	.58357	1.71358	.60721	1.64687	44
17	.53807	1.85850	.56079	1.78319	.58396	1.71244	.60761	1.64579	43
18	.53844	1.85720	.56117	1.78198	.58435	1.71129	.60801	1.64471	42
19	.53882	1.85591	.56156	1.78077	.58474	1.71015	.60841	1.64363	41
20	.53920	1.85462	.56194	1.77955	.58513	1.70901	.60881	1.64256	40
21	.53957	1.85333	.56232	1.77834	.58552	1.70787	.60921	1.64148	39
22	.53995	1.85204	.56270	1.77713	.58591	1.70673	.60960	1.64041	38
23	.54032	1.85075	.56309	1.77592	.58631	1.70560	.61000	1.63934	37
24	.54070	1.84946	.56347	1.77471	.58670	1.70446	.61040	1.63826	36
25	.54107	1.84818	.56385	1.77351	.58709	1.70332	.61080	1.63719	35
26	.54145	1.84689	.56424	1.77230	.58748	1.70219	.61120	1.63612	34
27	.54183	1.84561	.56462	1.77110	.58787	1.70106	.61160	1.63505	33
28	.54220	1.84433	.56501	1.76990	.58826	1.69992	.61200	1.63398	32
29	.54258	1.84305	.56539	1.76869	.58865	1.69879	.61240	1.63292	31
30	.54296	1.84177	.56577	1.76749	.58905	1.69766	.61280	1.63185	30
31	.54333	1.84049	.56616	1.76629	.58944	1.69653	.61320	1.63079	29
32	.54371	1.83922	.56654	1.76510	.58983	1.69541	.61360	1.62972	28
33	.54409	1.83794	.56693	1.76390	.59022	1.69428	.61400	1.62866	27
34	.54446	1.83667	.56731	1.76271	.59061	1.69316	.61440	1.62760	26
35	.54484	1.83540	.56769	1.76151	.59101	1.69203	.61480	1.62654	25
36	.54522	1.83413	.56808	1.76032	.59140	1.69091	.61520	1.62548	24
37	.54560	1.83286	.56846	1.75913	.59179	1.68979	.61561	1.62442	23
38	.54597	1.83159	.56885	1.75794	.59218	1.68866	.61601	1.62336	22
39	.54635	1.83033	.56923	1.75675	.59258	1.68754	.61641	1.62230	21
40	.54673	1.82906	.56962	1.75556	.59297	1.68643	.61681	1.62125	20
41	.54711	1.82780	.57000	1.75437	.59336	1.68531	.61721	1.62019	19
42	.54748	1.82654	.57039	1.75319	.59376	1.68419	.61761	1.61914	18
43	.54786	1.82528	.57078	1.75200	.59415	1.68308	.61801	1.61808	17
44	.54824	1.82402	.57116	1.75082	.59454	1.68196	.61842	1.61703	16
45	.54862	1.82276	.57155	1.74964	.59494	1.68085	.61882	1.61598	15
46	.54900	1.82150	.57193	1.74846	.59533	1.67974	.61922	1.61493	14
47	.54938	1.82025	.57232	1.74728	.59573	1.67863	61962	1.61388	13
48	.54975	1.81899	.57271	1.74610	.59612	1.67752	.62003	1.61283	12
49	.55013	1.81774	.57309	1.74492	.59651	1.67641	.62043	1.61179	11
50	.55051	1.81649	.57348	1.74375	.59691	1.67530	.62083	1.61074	10
51	.55089	1.81524	.57386	1.74257	.59730	1.67419	.62124	1.60970	9
52	.55127	1.81399	.57425	1.74140	.59770	1.67309	.62164	1.60865	8
53	.55165	1.81274	.57464	1.74022	.59809	1.67198	.62204	1.60761	7
54	.55203	1.81150	.57503	1.73905	.59849	1.67088	.62245	1.60657	6
55	.55241	1.81025	.57541	1.73788	.59888	1.66978	.62285	1.60553	5
56	.55279	1.80901	.57580	1.73671	.59928	1.66867	.62325	1.60449	4
57	.55317	1.80777	.57619	1.73555	.59967	1.66757	.62366	1.60345	3
58	.55355	1.80653	.57657	1.73438	.60007	1.66647	.62406	1.60241	2
59	.55393	1.80529	.57696	1.73321	.60046	1.66538	.62446	1.60137	1
60	.55431	1.80405	.57735	1.73205	.60086	1.66428	.62487	1.60033	0
′	Cotang	Tang	Cotang	Tang	Cotang	Tang	Cotang	Tang	′
	61°		60°		59°		58°		

′	32°		33°		34°		35°		′
	Tang	Cotang	Tang	Cotang	Tang	Cotang	Tang	Cotang	
0	.62487	1.60033	.64941	1.53986	.67451	1.48256	.70021	1.42815	60
1	.62527	1.59930	.64982	1.53888	.67493	1.48163	.70064	1.42726	59
2	.62568	1.59826	.65024	1.53791	.67536	1.48070	.70107	1.42638	58
3	.62608	1.59723	.65065	1.53693	.67578	1.47977	.70151	1.42550	57
4	.62649	1.59620	.65106	1.53595	.67620	1.47885	.70194	1.42462	56
5	.62689	1.59517	.65148	1.53497	.67663	1.47792	.70238	1.42374	55
6	.62730	1.59414	.65189	1.53400	.67705	1.47699	.70281	1.42286	54
7	.62770	1.59311	.65231	1.53302	.67748	1.47607	.70325	1.42198	53
8	.62811	1.59208	.65272	1.53205	.67790	1.47514	.70368	1.42110	52
9	.62852	1.59105	.65314	1.53107	.67832	1.47422	.70412	1.42022	51
10	.62892	1.59002	.65355	1.53010	.67875	1.47330	.70455	1.41934	50
11	.62933	1.58900	.65397	1.52913	.67917	1.47238	.70499	1.41847	49
12	.62973	1.58797	.65438	1.52816	.67960	1.47146	.70542	1.41759	48
13	.63014	1.58695	.65480	1.52719	.68002	1.47053	.70586	1.41672	47
14	.63055	1.58593	.65521	1.52622	.68045	1.46962	.70629	1.41584	46
15	.63095	1.58490	.65563	1.52525	.68088	1.46870	.70673	1.41497	45
16	.63136	1.58388	.65604	1.52429	.68130	1.46778	.70717	1.41409	44
17	.63177	1.58286	.65646	1.52332	.68173	1.46686	.70760	1.41322	43
18	.63217	1.58184	.65688	1.52235	.68215	1.46595	.70804	1.41235	42
19	.63258	1.58083	.65729	1.52139	.68258	1.46503	.70848	1.41148	41
20	.63299	1.57981	.65771	1.52043	.68301	1.46411	.70891	1.41061	40
21	.63340	1.57879	.65813	1.51946	.68343	1.46320	.70935	1.40974	39
22	.63380	1.57778	.65854	1.51850	.68386	1.46229	.70979	1.40887	38
23	.63421	1.57676	.65896	1.51754	.68429	1.46137	.71023	1.40800	37
24	.63462	1.57575	.65938	1.51658	.68471	1.46046	.71066	1.40714	36
25	.63503	1.57474	.65980	1.51562	.68514	1.45955	.71110	1.40627	35
26	.63544	1.57372	.66021	1.51466	.68557	1.45864	.71154	1.40540	34
27	.63584	1.57271	.66063	1.51370	.68600	1.45773	.71198	1.40454	33
28	.63625	1.57170	.66105	1.51275	.68642	1.45682	.71242	1.40367	32
29	.63666	1.57069	.66147	1.51179	.68685	1.45592	.71285	1.40281	31
30	.63707	1.56969	.66189	1.51084	.68728	1.45501	.71329	1.40195	30
31	.63748	1.56868	.66230	1.50988	.68771	1.45410	.71373	1.40109	29
32	.63789	1.56767	.66272	1.50893	.68814	1.45320	.71417	1.40022	28
33	.63830	1.56667	.66314	1.50797	.68857	1.45229	.71461	1.39936	27
34	.63871	1.56566	.66356	1.50702	.68900	1.45139	.71505	1.39850	26
35	.63912	1.56466	.66398	1.50607	.68942	1.45049	.71549	1.39764	25
36	.63953	1.56366	.66440	1.50512	.68985	1.44958	.71593	1.39679	24
37	.63994	1.56265	.66482	1.50417	.69028	1.44868	.71637	1.39593	23
38	.64035	1.56165	.66524	1.50322	.69071	1.44778	.71681	1.39507	22
39	.64076	1.56065	.66566	1.50228	.69114	1.44688	.71725	1.39421	21
40	.64117	1.55966	.66608	1.50133	.69157	1.44598	.71769	1.39336	20
41	.64158	1.55866	.66650	1.50038	.69200	1.44508	.71813	1.39250	19
42	.64199	1.55766	.66692	1.49944	.69243	1.44418	.71857	1.39165	18
43	.64240	1.55666	.66734	1.49849	.69286	1.44329	.71901	1.39079	17
44	.64281	1.55567	.66776	1.49755	.69329	1.44239	.71946	1.38994	16
45	.64322	1.55467	.66818	1.49661	.69372	1.44149	.71990	1.38909	15
46	.64363	1.55368	.66860	1.49566	.69416	1.44060	.72034	1.38824	14
47	.64404	1.55269	.66902	1.49472	.69459	1.43970	.72078	1.38738	13
48	.64446	1.55170	.66944	1.49378	.69502	1.43881	.72122	1.38653	12
49	.64487	1.55071	.66986	1.49284	.69545	1.43792	.72167	1.38568	11
50	.64528	1.54972	.67028	1.49190	.69588	1.43703	.72211	1.38484	10
51	.64569	1.54873	.67071	1.49097	.69631	1.43614	.72255	1.38399	9
52	.64610	1.54774	.67113	1.49003	.69675	1.43525	.72299	1.38314	8
53	.64652	1.54675	.67155	1.48909	.69718	1.43436	.72344	1.38229	7
54	.64693	1.54576	.67197	1.48816	.69761	1.43347	.72388	1.38145	6
55	.64734	1.54478	.67239	1.48722	.69804	1.43258	.72432	1.38060	5
56	.64775	1.54379	.67282	1.48629	.69847	1.43169	.72477	1.37976	4
57	.64817	1.54281	.67324	1.48536	.69891	1.43080	.72521	1.37891	3
58	.64858	1.54183	.67366	1.48442	.69934	1.42992	.72565	1.37807	2
59	.64899	1.54085	.67409	1.48349	.69977	1.42903	.72610	1.37722	1
60	.64941	1.53986	.67451	1.48256	.70021	1.42815	.72654	1.37638	0
	Cotang	Tang	Cotang	Tang	Cotang	Tang	Cotang	Tang	
′	57°		56°		55°		54°		′

′	36° Tang	36° Cotang	37° Tang	37° Cotang	38° Tang	38° Cotang	39° Tang	39° Cotang	′
0	.72654	1.37638	.75355	1.32704	.78129	1.27994	.80978	1.23490	60
1	.72699	1.37554	.75401	1.32624	.78175	1.27917	.81027	1.23416	59
2	.72743	1.37470	.75447	1.32544	.78222	1.27841	.81075	1.23343	58
3	.72788	1.37386	.75492	1.32464	.78269	1.27764	.81123	1.23270	57
4	.72832	1.37302	.75538	1.32384	.78316	1.27688	.81171	1.23196	56
5	.72877	1.37218	.75584	1.32304	.78363	1.27611	.81220	1.23123	55
6	.72921	1.37134	.75629	1.32224	.78410	1.27535	.81268	1.23050	54
7	.72966	1.37050	.75675	1.32144	.78457	1.27458	.81316	1.22977	53
8	.73010	1.36967	.75721	1.32064	.78504	1.27382	.81364	1.22904	52
9	.73055	1.36883	.75767	1.31984	.78551	1.27306	.81413	1.22831	51
10	.73100	1.36800	.75812	1.31904	.78598	1.27230	.81461	1.22758	50
11	.73144	1.36716	.75858	1.31825	.78645	1.27153	.81510	1.22685	49
12	.73189	1.36633	.75904	1.31745	.78692	1.27077	.81558	1.22612	48
13	.73234	1.36549	.75950	1.31666	.78739	1.27001	.81606	1.22539	47
14	.73278	1.36466	.75996	1.31586	.78786	1.26925	.81655	1.22467	46
15	.73323	1.36383	.76042	1.31507	.78834	1.26849	.81703	1.22394	45
16	.73368	1.36300	.76088	1.31427	.78881	1.26774	.81752	1.22321	44
17	.73413	1.36217	.76134	1.31348	.78928	1.26698	.81800	1.22249	43
18	.73457	1.36134	.76180	1.31269	.78975	1.26622	.81849	1.22176	42
19	.73502	1.36051	.76226	1.31190	.79022	1.26546	.81898	1.22104	41
20	.73547	1.35968	.76272	1.31110	.79070	1.26471	.81946	1.22031	40
21	.73592	1.35885	.76318	1.31031	.79117	1.26395	.81995	1.21959	39
22	.73637	1.35802	.76364	1.30952	.79164	1.26319	.82044	1.21886	38
23	.73681	1.35719	.76410	1.30873	.79212	1.26244	.82092	1.21814	37
24	.73726	1.35637	.76456	1.30795	.79259	1.26169	.82141	1.21742	36
25	.73771	1.35554	.76502	1.30716	.79306	1.26093	.82190	1.21670	35
26	.73816	1.35472	.76548	1.30637	.79354	1.26018	.82238	1.21598	34
27	.73861	1.35389	.76594	1.30558	.79401	1.25943	.82287	1.21526	33
28	.73906	1.35307	.76640	1.30480	.79449	1.25867	.82336	1.21454	32
29	.73951	1.35224	.76686	1.30401	.79496	1.25792	.82385	1.21382	31
30	.73996	1.35142	.76733	1.30323	.79544	1.25717	.82434	1.21310	30
31	.74041	1.35060	.76779	1.30244	.79591	1.25642	.82483	1.21238	29
32	.74086	1.34978	.76825	1.30166	.79639	1.25567	.82531	1.21166	28
33	.74131	1.34896	.76871	1.30087	.79686	1.25492	.82580	1.21094	27
34	.74176	1.34814	.76918	1.30009	.79734	1.25417	.82629	1.21023	26
35	.74221	1.34732	.76964	1.29931	.79781	1.25343	.82678	1.20951	25
36	.74267	1.34650	.77010	1.29853	.79829	1.25268	.82727	1.20879	24
37	.74312	1.34568	.77057	1.29775	.79877	1.25193	.82776	1.20808	23
38	.74357	1.34487	.77103	1.29696	.79924	1.25118	.82825	1.20736	22
39	.74402	1.34405	.77149	1.29618	.79972	1.25044	.82874	1.20665	21
40	.74447	1.34323	.77196	1.29541	.80020	1.24969	.82923	1.20593	20
41	.74492	1.34242	.77242	1.29463	.80067	1.24895	.82972	1.20522	19
42	.74538	1.34160	.77289	1.29385	.80115	1.24820	.83022	1.20451	18
43	.74583	1.34079	.77335	1.29307	.80163	1.24746	.83071	1.20379	17
44	.74628	1.33998	.77382	1.29229	.80211	1.24672	.83120	1.20308	16
45	.74674	1.33916	.77428	1.29152	.80258	1.24597	.83169	1.20237	15
46	.74719	1.33835	.77475	1.29074	.80306	1.24523	.83218	1.20166	14
47	.74764	1.33754	.77521	1.28997	.80354	1.24449	.83268	1.20095	13
48	.74810	1.33673	.77568	1.28919	.80402	1.24375	.83317	1.20024	12
49	.74855	1.33592	.77615	1.28842	.80450	1.24301	.83366	1.19953	11
50	.74900	1.33511	.77661	1.28764	.80498	1.24227	.83415	1.19882	10
51	.74946	1.33430	.77708	1.28687	.80546	1.24153	.83465	1.19811	9
52	.74991	1.33349	.77754	1.28610	.80594	1.24079	.83514	1.19740	8
53	.75037	1.33268	.77801	1.28533	.80642	1.24005	.83564	1.19669	7
54	.75082	1.33187	.77848	1.28456	.80690	1.23931	.83613	1.19599	6
55	.75128	1.33107	.77895	1.28379	.80738	1.23858	.83662	1.19528	5
56	.75173	1.33026	.77941	1.28302	.80786	1.23784	.83712	1.19457	4
57	.75219	1.32946	.77988	1.28225	.80834	1.23710	.83761	1.19387	3
58	.75264	1.32865	.78035	1.28148	.80882	1.23637	.83811	1.19316	2
59	.75310	1.32785	.78082	1.28071	.80930	1.23563	.83860	1.19246	1
60	.75355	1.32704	.78129	1.27994	.80978	1.23490	.83910	1.19175	0
′	Cotang	Tang	Cotang	Tang	Cotang	Tang	Cotang	Tang	′
	53°		52°		51°		50°		

′	40°		41°		42°		43°		′
	Tang	Cotang	Tang	Cotang	Tang	Cotang	Tang	Cotang	
0	.83910	1.19175	.86929	1.15037	.90040	1.11061	.93252	1.07237	60
1	.83960	1.19105	.86980	1.14969	.90093	1.10996	.93306	1.07174	59
2	.84009	1.19035	.87031	1.14902	.90146	1.10931	.93360	1.07112	58
3	.84059	1.18964	.87082	1.14834	.90199	1.10867	.93415	1.07049	57
4	.84108	1.18894	.87133	1.14767	.90251	1.10802	.93469	1.06987	56
5	.84158	1.18824	.87184	1.14699	.90304	1.10737	.93524	1.06925	55
6	.84208	1.18754	.87236	1.14632	.90357	1.10672	.93578	1.06862	54
7	.84258	1.18684	.87287	1.14565	.90410	1.10607	.93633	1.06800	53
8	.84307	1.18614	.87338	1.14498	.90463	1.10543	.93688	1.06738	52
9	.84357	1.18544	.87389	1.14430	.90516	1.10478	.93742	1.06676	51
10	.84407	1.18474	.87441	1.14363	.90569	1.10414	.93797	1.06613	50
11	.84457	1.18404	.87492	1.14296	.90621	1.10349	.93852	1.06551	49
12	.84507	1.18334	.87543	1.14229	.90674	1.10285	.93906	1.06489	48
13	.84556	1.18264	.87595	1.14162	.90727	1.10220	.93961	1.06427	47
14	.84606	1.18194	.87646	1.14095	.90781	1.10156	.94016	1.06365	46
15	.84656	1.18125	.87698	1.14028	.90834	1.10091	.94071	1.06303	45
16	.84706	1.18055	.87749	1.13961	.90887	1.10027	.94125	1.06241	44
17	.84756	1.17986	.87801	1.13894	.90940	1.09963	.94180	1.06179	43
18	.84806	1.17916	.87852	1.13828	.90993	1.09899	.94235	1.06117	42
19	.84856	1.17846	.87904	1.13761	.91046	1.09834	.94290	1.06056	41
20	.84906	1.17777	.87955	1.13694	.91099	1.09770	.94345	1.05994	40
21	.84956	1.17708	.88007	1.13627	.91153	1.09706	.94400	1.05932	39
22	.85006	1.17638	.88059	1.13561	.91206	1.09642	.94455	1.05870	38
23	.85057	1.17569	.88110	1.13494	.91259	1.09578	.94510	1.05809	37
24	.85107	1.17500	.88162	1.13428	.91313	1.09514	.94565	1.05747	36
25	.85157	1.17430	.88214	1.13361	.91366	1.09450	.94620	1.05685	35
26	.85207	1.17361	.88265	1.13295	.91419	1.09386	.94676	1.05624	34
27	.85257	1.17292	.88317	1.13228	.91473	1.09322	.94731	1.05562	33
28	.85308	1.17223	.88369	1.13162	.91526	1.09258	.94786	1.05501	32
29	.85358	1.17154	.88421	1.13096	.91580	1.09195	.94841	1.05439	31
30	.85408	1.17085	.88473	1.13029	.91633	1.09131	.94896	1.05378	30
31	.85458	1.17016	.88524	1.12963	.91687	1.09067	.94952	1.05317	29
32	.85509	1.16947	.88576	1.12897	.91740	1.09003	.95007	1.05255	28
33	.85559	1.16878	.88628	1.12831	.91794	1.08940	.95062	1.05194	27
34	.85609	1.16809	.88680	1.12765	.91847	1.08876	.95118	1.05133	26
35	.85660	1.16741	.88732	1.12699	.91901	1.08813	.95173	1.05072	25
36	.85710	1.16672	.88784	1.12633	91955	1.08749	.95229	1.05010	24
37	.85761	1.16603	.88836	1.12567	.92008	1.08686	.95284	1.04949	23
38	.85811	1.16535	.88888	1.12501	.92062	1.08622	.95340	1.04888	22
39	.85862	1.16466	.88940	1.12435	.92116	1.08559	.95395	1.04827	21
40	.85912	1.16398	.88992	1.12369	.92170	1.08496	.95451	1.04766	20
41	.85963	1.16329	.89045	1.12303	.92224	1.08432	.95506	1.04705	19
42	.86014	1.16261	.89097	1.12238	.92277	1.08369	.95562	1.04644	18
43	.86064	1.16192	.89149	1.12172	.92331	1.08306	.95618	1.04583	17
44	.86115	1.16124	.89201	1.12106	.92385	1.08243	.95673	1.04522	16
45	.86166	1.16056	.89253	1.12041	.92439	1.08179	.95729	1.04461	15
46	.86216	1.15987	.89306	1.11975	.92493	1.08116	.95785	1.04401	14
47	.86267	1.15919	.89358	1.11909	.92547	1.08053	.95841	1.04340	13
48	.86318	1.15851	.89410	1.11844	.92601	1.07990	.95897	1.04279	12
49	.86368	1.15783	.89463	1 11778	.92655	1.07927	.95952	1.04218	11
50	.86419	1.15715	.89515	1.11713	.92709	1.07864	.96008	1.04158	10
51	.86470	1.15647	.89567	1.11648	.92763	1.07801	.96064	1.04097	9
52	.86521	1.15579	.89620	1.11582	.92817	1.07738	.96120	1.04036	8
53	.86572	1.15511	.89672	1.11517	.92872	1.07676	.96176	1.03976	7
54	.86623	1.15443	.89725	1.11452	.92926	1.07613	.96232	1.03915	6
55	.86674	1.15375	.89777	1.11387	.92980	1.07550	.96288	1.03855	5
56	.86725	1.15308	.89830	1.11321	.93034	1.07487	.96344	1.03794	4
57	.86776	1.15240	.89883	1.11256	.93088	1.07425	.96400	1.03734	3
58	.86827	1.15172	.89935	1.11191	.93143	1.07362	.96457	1.03674	2
59	.86878	1.15104	.89988	1.11126	.93197	1.07299	.96513	1.03613	1
60	.86929	1.15037	.90040	1.11061	.93252	1.07237	.96569	1.03553	0
	Cotang	Tang	Cotang	Tang	Cotang	Tang	Cotang	Tang	
′	49°		48°		47°		46°		′

′	44°		′	′	44°		′	′	44°		′
	Tang	Cotang			Tang	Cotang			Tang	Cotang	
0	.96569	1.03553	60	20	.97700	1.02355	40	40	.98843	1.01170	20
1	.96625	1.03493	59	21	.97756	1.02295	39	41	.98901	1.01112	19
2	.96681	1.03433	58	22	.97813	1.02236	38	42	.98958	1.01053	18
3	.96738	1.03372	57	23	.97870	1.02176	37	43	.99016	1.00994	17
4	.96794	1.03312	56	24	.97927	1.02117	36	44	.99073	1.00935	16
5	.96850	1.03252	55	25	.97984	1.02057	35	45	.99131	1.00876	15
6	.96907	1.03192	54	26	.98041	1.01998	34	46	.99189	1.00818	14
7	.96963	1.03132	53	27	.98098	1.01939	33	47	.99247	1.00759	13
8	.97020	1.03072	52	28	.98155	1.01879	32	48	.99304	1.00701	12
9	.97076	1.03012	51	29	.98213	1.01820	31	49	.99362	1.00642	11
10	.97133	1.02952	50	30	.98270	1.01761	30	50	.99420	1.00583	10
11	.97189	1.02892	49	31	.98327	1.01702	29	51	.99478	1.00525	9
12	.97246	1.02832	48	32	.98384	1.01642	28	52	.90536	1.00467	8
13	.97302	1.02772	47	33	.98441	1.01583	27	53	.99594	1.00408	7
14	.97359	1.02713	46	34	.98499	1.01524	26	54	.99652	1.00350	6
15	.97416	1.02653	45	35	.98556	1.01465	25	55	.99710	1.00291	5
16	.97472	1.02593	44	36	.98613	1.01406	24	56	.99768	1.00233	4
17	.97529	1.02533	43	37	.98671	1.01347	23	57	.99826	1.00175	3
18	.97586	1.02474	42	38	.98728	1.01288	22	58	.99884	1.00116	2
19	.97643	1.02414	41	39	.98786	1.01229	21	59	.99942	1.00058	1
20	.97700	1.02355	40	40	.98843	1.01170	20	60	1.00000	1.00000	0
′	Cotang	Tang	′	′	Cotang	Tang	′	′	Cotang	Tang	′
	45°				45°				45°		

′	0°		1°		2°		3°		′
	Vers.	Exsec.	Vers.	Exsec.	Vers.	Exsec.	Vers.	Exsec.	
0	.00000	.00000	.00015	.00015	.00061	.00061	.00137	.00137	0
1	.00000	.00000	.00016	.00016	.00062	.00062	.00139	00139	1
2	.00000	.00000	.00016	.00016	.00063	.00063	.00140	.00140	2
3	.00000	.00000	.00017	.00017	.00064	.00064	.00142	.00142	3
4	.00000	.00000	.00017	.00017	.00065	.00065	.00143	.00143	4
5	.00000	.00000	.00018	.00018	.00066	.00066	.00145	.00145	5
6	.00000	.00000	.00018	.00018	.00067	.00067	.00146	.00147	6
7	.00000	.00000	.00019	.00019	.00068	.00068	.00148	.00148	7
8	.00000	.00000	.00020	.00020	.00069	.00069	.00150	.00150	8
9	.00000	.00000	.00020	.00020	.00070	.00070	.00151	.00151	9
10	.00000	.00000	.00021	.00021	.00071	.00072	.00153	.00153	10
11	.00001	.00001	.00021	.00021	.00073	.00073	.00154	.00155	11
12	.00001	.00001	.00022	.00022	.00074	.00074	.00156	.00156	12
13	.00001	.00001	.00023	.00023	.00075	.00075	.00158	.00158	13
14	.00001	00001	.00023	.00023	.00076	.00076	.00159	.00159	14
15	.00001	.00001	.00024	.00024	.00077	.00077	.00161	.00161	15
16	.00001	.00001	.00024	.00024	.00078	.00078	.00162	.00163	16
17	.00001	.00001	.00025	.00025	.00079	.00079	.00164	.00164	17
18	.00001	.00001	.00026	.00026	.00081	.00081	.00166	.00166	18
19	.00002	.00002	.00026	.00026	.00082	.00082	.00168	.00168	19
20	.00002	.00002	.00027	.00027	.00083	.00083	.00169	.00169	20
21	.00002	.00002	.00028	.00028	.00084	.00084	.00171	.00171	21
22	.00002	.00002	.00028	.00028	.00085	.00085	.00173	.00173	22
23	.00002	.00002	.00029	.00029	.00087	.00087	.00174	.00175	23
24	.00002	.00002	.00030	.00030	.00088	.00088	.00176	.00176	24
25	.00003	.00003	.00031	.00031	.00089	.00089	.00178	.00178	25
26	.00003	.00003	.00031	.00031	.00090	.00090	.00179	.00180	26
27	.00003	.00003	.00032	.00032	.00091	.00091	.00181	.00182	27
28	.00003	.00003	.00033	.00033	.00093	.00093	.00183	.00183	28
29	.00004	.00004	.00034	.00034	.00094	.00094	.00185	.00185	29
30	.00004	.00004	.00034	.00034	.00095	.00095	.00187	.00187	30
31	.00004	.00004	.00035	.00035	.00096	.00097	.00188	.00189	31
32	.00004	.00004	.00036	.00036	.00098	.00098	.00190	.00190	32
33	.00005	.00005	.00037	.00037	.00099	.00099	.00192	.00192	33
34	.00005	.00005	.00037	.00037	.00100	.00100	.00194	.00194	34
35	.00005	.00005	.00038	.00038	.00102	.00102	.00196	.00196	35
36	.00005	.00005	.00039	.00039	.00103	.00103	.00197	.00198	36
37	.00006	.00006	.00040	.00040	.00104	.00104	.00199	.00200	37
38	.00006	.00006	.00041	.00041	.00106	.00106	.00201	.00201	38
39	.00006	.00006	.00041	.00041	.00107	.00107	.00203	.00203	39
40	.00007	.00007	.00042	.00042	.00108	.00108	.00205	.00205	40
41	.00007	.00007	.00043	.00043	.00110	.00110	.00207	.00207	41
42	.00007	.00007	.00044	.00044	.00111	.00111	.00208	.00209	42
43	.00008	.00008	.00045	.00045	.00112	.00113	.00210	.00211	43
44	.00008	.00008	.00046	.00046	.00114	.00114	.00212	.00213	44
45	.00009	.00009	.00047	.00047	.00115	.00115	.00214	.00215	45
46	.00009	.00009	.00048	.00048	.00117	.00117	.00216	.00216	46
47	.00009	.00009	.00048	.00048	.00118	.00118	.00218	.00218	47
48	.00010	.00010	.00049	.00049	.00119	.00120	.00220	.00220	48
49	.00010	.00010	.00050	.00050	.00121	.00121	.00222	.00222	49
50	.00011	.00011	.00051	.00051	.00122	.00122	.00224	.00224	50
51	.00011	.00011	.00052	.00052	.00124	.00124	.00226	.00226	51
52	.00011	.00011	.00053	.00053	.00125	.00125	.00228	.00228	52
53	.00012	.00012	.00054	.00054	.00127	.00127	.00230	.00230	53
54	.00012	.00012	.00055	.00055	.00128	.00128	.00232	.00232	54
55	.00013	.00013	.00056	.00056	.00130	.00130	.00234	.00234	55
56	.00013	.00013	.00057	.00057	.00131	.00131	.00236	.00236	56
57	.00014	.00014	.00058	.00058	.00133	.00133	.00238	.00238	57
58	.00014	.00014	.00059	.00059	.00134	.00134	.00240	.00240	58
59	.00015	.00015	.00060	.00060	.00136	.00136	.00242	.00242	59
60	.00015	.00015	.00061	.00061	.00137	.00137	.00244	.00244	60

′	4° Vers.	4° Exsec.	5° Vers.	5° Exsec.	6° Vers.	6° Exsec.	7° Vers.	7° Exsec.	′
0	.00244	.00244	.00381	.00382	.00548	.00551	.00745	.00751	0
1	.00246	.00246	.00383	.00385	.00551	.00554	.00749	.00755	1
2	.00248	.00248	.00386	.00387	.00554	.00557	.00752	.00758	2
3	.00250	.00250	.00388	.00390	.00557	.00560	.00756	.00762	3
4	.00252	.00252	.00391	.00392	.00560	.00563	.00760	.00765	4
5	.00254	.00254	.00393	.00395	.00563	.00566	.00763	.00769	5
6	.00256	.00257	.00396	.00397	.00566	.00569	.00767	.00773	6
7	.00258	.00259	.00398	.00400	.00569	.00573	.00770	.00776	7
8	.00260	.00261	.00401	.00403	.00572	.00576	.00774	.00780	8
9	.00262	.00263	.00404	.00405	.00576	.00579	.00778	.00784	9
10	.00264	.00265	.00406	.00408	.00579	.00582	.00781	.00787	10
11	.00266	.00267	.00409	.00411	.00582	.00585	.00785	.00791	11
12	.00269	.00269	.00412	.00413	.00585	.00588	.00789	.00795	12
13	.00271	.00271	.00414	.00416	.00588	.00592	.00792	.00799	13
14	.00273	.00274	.00417	.00419	.00591	.00595	.00796	.00802	14
15	.00275	.00276	.00420	.00421	.00594	.00598	.00800	.00806	15
16	.00277	.00278	.00422	.00424	.00598	.00601	.00803	.00810	16
17	.00279	.00280	.00425	.00427	.00601	.00604	.00807	.00813	17
18	.00281	.00282	.00428	.00429	.00604	.00608	.00811	.00817	18
19	.00284	.00284	.00430	.00432	.00607	.00611	.00814	.00821	19
20	.00286	.00287	.00433	.00435	.00610	.00614	.00818	.00825	20
21	.00288	.00289	.00436	.00438	.00614	.00617	.00822	.00828	21
22	.00290	.00291	.00438	.00440	.00617	.00621	.00825	.00832	22
23	.00293	.00293	.00441	.00443	.00620	.00624	.00829	.00836	23
24	.00295	.00296	.00444	.00446	.00623	.00627	.00833	.00840	24
25	.00297	.00298	.00447	.00449	.00626	.00630	.00837	.00844	25
26	.00299	.00300	.00449	.00451	.00630	.00634	.00840	.00848	26
27	.00301	.00302	.00452	.00454	.00633	.00637	.00844	.00851	27
28	.00304	.00305	.00455	.00457	.00636	.00640	.00848	.00855	28
29	.00306	.00307	.00458	.00460	.00640	.00644	.00852	.00859	29
30	.00308	.00309	.00460	.00463	.00643	.00647	.00856	.00863	30
31	.00311	.00312	.00463	.00465	.00646	.00650	.00859	.00867	31
32	.00313	.00314	.00466	.00468	.00649	.00654	.00863	.00871	32
33	.00315	.00316	.00469	.00471	.00653	.00657	.00867	.00875	33
34	.00317	.00318	.00472	.00474	.00656	.00660	.00871	.00878	34
35	.00320	.00321	.00474	.00477	.00659	.00664	.00875	.00882	35
36	.00322	.00323	.00477	.00480	.00663	.00667	.00878	.00886	36
37	.00324	.00326	.00480	.00482	.00666	.00671	.00882	.00890	37
38	.00327	.00328	.00483	.00485	.00669	.00674	.00886	.00894	38
39	.00329	.00330	.00486	.00488	.00673	.00677	.00890	.00898	39
40	.00332	.00333	.00489	.00491	.00676	.00681	.00894	.00902	40
41	.00334	.00335	.00492	.00494	.00680	.00684	.00898	.00906	41
42	.00336	.00337	.00494	.00497	.00683	.00688	.00902	.00910	42
43	.00339	.00340	.00497	.00500	.00686	.00691	.00906	.00914	43
44	.00341	.00342	.00500	.00503	.00690	.00695	.00909	.00918	44
45	.00343	.00345	.00503	.00506	.00693	.00698	.00913	.00922	45
46	.00346	.00347	.00506	.00509	.00697	.00701	.00917	.00926	46
47	.00348	.00350	.00509	.00512	.00700	.00705	.00921	.00930	47
48	.00351	.00352	.00512	.00515	.00703	.00708	.00925	.00934	48
49	.00353	.00354	.00515	.00518	.00707	.00712	.00929	.00938	49
50	.00356	.00357	.00518	.00521	.00710	.00715	.00933	.00942	50
51	.00358	.00359	.00521	.00524	.00714	.00719	.00937	.00946	51
52	.00361	.00362	.00524	.00527	.00717	.00722	.00941	.00950	52
53	.00363	.00364	.00527	.00530	.00721	.00726	.00945	.00954	53
54	.00365	.00367	.00530	.00533	.00724	.00730	.00949	.00958	54
55	.00368	.00369	.00533	.00536	.00728	.00733	.00953	.00962	55
56	.00370	.00372	.00536	.00539	.00731	.00737	.00957	.00966	56
57	.00373	.00374	.00539	.00542	.00735	.00740	.00961	.00970	57
58	.00375	.00377	.00542	.00545	.00738	.00744	.00965	.00975	58
59	.00378	.00379	.00545	.00548	.00742	.00747	.00969	.00979	59
60	.00381	.00382	.00548	.00551	.00745	.00751	.00973	.00983	60

′	8° Vers.	8° Exsec.	9° Vers.	9° Exsec.	10° Vers.	10° Exsec.	11° Vers.	11° Exsec.	′
0	.00973	.00983	.01231	.01247	.01519	.01543	.01837	.01872	0
1	.00977	.00987	.01236	.01251	.01524	.01548	.01843	.01877	1
2	.00981	.00991	.01240	.01256	.01529	.01553	.01848	.01883	2
3	.00985	.00995	.01245	.01261	.01534	.01558	.01854	.01889	3
4	.00989	.00999	.01249	.01265	.01540	.01564	.01860	.01895	4
5	.00994	.01004	.01254	.01270	.01545	.01569	.01865	.01901	5
6	.00998	.01008	.01259	.01275	.01550	.01574	.01871	.01906	6
7	.01002	.01012	.01263	.01279	.01555	.01579	.01876	.01912	7
8	.01006	.01016	.01268	.01284	.01560	.01585	.01882	.01918	8
9	.01010	.01020	.01272	.01289	.01565	.01590	.01888	.01924	9
10	.01014	.01024	.01277	.01294	.01570	.01595	.01893	.01930	10
11	.01018	.01029	.01282	.01298	.01575	.01601	.01899	.01936	11
12	.01022	.01033	.01286	.01303	.01580	.01606	.01904	.01941	12
13	.01027	.01037	.01291	.01308	.01586	.01611	.01910	.01947	13
14	.01031	.01041	.01296	.01313	.01591	.01616	.01916	.01953	14
15	.01035	.01046	.01300	.01318	.01596	.01622	.01921	.01959	15
16	.01039	.01050	.01305	.01322	.01601	.01627	.01927	.01965	16
17	.01043	.01054	.01310	.01327	.01606	.01633	.01933	.01971	17
18	.01047	.01059	.01314	.01332	.01612	.01638	.01939	.01977	18
19	.01052	.01063	.01319	.01337	.01617	.01643	.01944	.01983	19
20	.01056	.01067	.01324	.01342	.01622	.01649	.01950	.01989	20
21	.01060	.01071	.01329	.01346	.01627	.01654	.01956	.01995	21
22	.01064	.01076	.01333	.01351	.01632	.01659	.01961	.02001	22
23	.01069	.01080	.01338	.01356	.01638	.01665	.01967	.02007	23
24	.01073	.01084	.01343	.01361	.01643	.01670	.01973	.02013	24
25	.01077	.01089	.01348	.01366	.01648	.01676	.01979	.02019	25
26	.01081	.01093	.01352	.01371	.01653	.01681	.01984	.02025	26
27	.01086	.01097	.01357	.01376	.01659	.01687	.01990	.02031	27
28	.01090	.01102	.01362	.01381	.01664	.01692	.01996	.02037	28
29	.01094	.01106	.01367	.01386	.01669	.01698	.02002	.02043	29
30	.01098	.01111	.01371	.01391	.01675	.01703	.02008	.02049	30
31	.01103	.01115	.01376	.01395	.01680	.01709	.02013	.02055	31
32	.01107	.01119	.01381	.01400	.01685	.01714	.02019	.02061	32
33	.01111	.01124	.01386	.01405	.01690	.01720	.02025	.02067	33
34	.01116	.01128	.01391	.01410	.01696	.01725	.02031	.02073	34
35	.01120	.01133	.01396	.01415	.01701	.01731	.02037	.02079	35
36	.01124	.01137	.01400	.01420	.01706	.01736	.02042	.02085	36
37	.01129	.01142	.01405	.01425	.01712	.01742	.02048	.02091	37
38	.01133	.01146	.01410	.01430	.01717	.01747	.02054	.02097	38
39	.01137	.01151	.01415	.01435	.01723	.01753	.02060	.02103	39
40	.01142	.01155	.01420	.01440	.01728	.01758	.02066	.02110	40
41	.01146	.01160	.01425	.01445	.01733	.01764	.02072	.02116	41
42	.01151	.01164	.01430	.01450	.01739	.01769	.02078	.02122	42
43	.01155	.01169	.01435	.01455	.01744	.01775	.02084	.02128	43
44	.01159	.01173	.01439	.01461	.01750	.01781	.02090	.02134	44
45	.01164	.01178	.01444	.01466	.01755	.01786	.02095	.02140	45
46	.01168	.01182	.01449	.01471	.01760	.01792	.02101	.02146	46
47	.01173	.01187	.01454	.01476	.01766	.01798	.02107	.02153	47
48	.01177	.01191	.01459	.01481	.01771	.01803	.02113	.02159	48
49	.01182	.01196	.01464	.01486	.01777	.01809	.02119	.02165	49
50	.01186	.01200	.01469	.01491	.01782	.01815	.02125	.02171	50
51	.01191	.01205	.01474	.01496	.01788	.01820	.02131	.02178	51
52	.01195	.01209	.01479	.01501	.01793	.01826	.02137	.02184	52
53	.01200	.01214	.01484	.01506	.01799	.01832	.02143	.02190	53
54	.01204	.01219	.01489	.01512	.01804	.01837	.02149	.02196	54
55	.01209	.01223	.01494	.01517	.01810	.01843	.02155	.02203	55
56	.01213	.01228	.01499	.01522	.01815	.01849	.02161	.02209	56
57	.01218	.01233	.01504	.01527	.01821	.01854	.02167	.02215	57
58	.01222	.01237	.01509	.01532	.01826	.01860	.02173	.02221	58
59	.01227	.01242	.01514	.01537	.01832	.01866	.02179	.02228	59
60	.01231	.01247	.01519	.01543	.01837	.01872	.02185	.02234	60

′	12°		13°		14°		15°		′
	Vers.	Exsec.	Vers.	Exsec.	Vers.	Exsec.	Vers.	Exsec.	
0	.02185	.02234	.02563	.02630	.02970	.03061	.03407	.03528	0
1	.02191	.02240	.02570	.02637	.02977	.03069	.03415	.03536	1
2	.02197	.02247	.02576	.02644	.02985	.03076	.03422	.03544	2
3	.02203	.02253	.02583	.02651	.02992	.03084	.03430	.03552	3
4	.02210	.02259	.02589	.02658	.02999	.03091	.03438	.03560	4
5	.02216	.02266	.02596	.02665	.03006	.03099	.03445	.03568	5
6	.02222	.02272	.02602	.02672	.03013	.03106	.03453	.03576	6
7	.02228	.02279	.02609	.02679	.03020	.03114	.03460	.03584	7
8	.02234	.02285	.02616	.02686	.03027	.03121	.03468	.03592	8
9	.02240	.02291	.02622	.02693	.03034	.03129	.03476	.03601	9
10	.02246	.02298	.02629	.02700	.03041	.03137	.03483	.03609	10
11	.02252	.02304	.02635	.02707	.03048	.03144	.03491	.03617	11
12	.02258	.02311	.02642	.02714	.03055	.03152	.03498	.03625	12
13	.02265	.02317	.02649	.02721	.03063	.03159	.03506	03633	13
14	.02271	.02323	.02655	.02728	.03070	.03167	.03514	.03642	14
15	.02277	.02330	.02662	.02735	.03077	.03175	.03521	.03650	15
16	.02283	.02336	.02669	.02742	.03084	.03182	.03529	.03658	16
17	.02289	.02343	.02675	.02749	.03091	.03190	.03537	.03666	17
18	.02295	.02349	.02682	.02756	.03098	.03198	.03544	.03674	18
19	.02302	.02356	.02689	.02763	.03106	.03205	.03552	.03683	19
20	.02308	.02362	.02696	.02770	.03113	.03213	.03560	.03691	20
21	.02314	.02369	.02702	.02777	.03120	.03221	.03567	.03699	21
22	.02320	.02375	.02709	.02784	.03127	.03228	.03575	.03708	22
23	.02327	.02382	.02716	.02791	.03134	.03236	.03583	.03716	23
24	.02333	.02388	.02722	.02799	.03142	.03244	.03590	.03724	24
25	.02339	.02395	.02729	.02806	.03149	.03251	.03598	.03732	25
26	.02345	.02402	.02736	.02813	.03156	.03259	.03606	.03741	26
27	.02352	.02408	.02743	.02820	.03163	.03267	.03614	.03749	27
28	.02358	.02415	.02749	.02827	.03171	.03275	.03621	.03758	28
29	.02364	.02421	.02756	.02834	.03178	.03282	.03629	.03766	29
30	.02370	.02428	.02763	.02842	.03185	.03290	.03637	.03774	30
31	.02377	.02435	.02770	.02849	.03193	.03298	.03645	.03783	31
32	.02383	.02441	.02777	.02856	.03200	.03306	.03653	.03791	32
33	.02389	.02448	.02783	.02863	.03207	.03313	.03660	.03799	33
34	.02396	.02454	.02790	.02870	.03214	.03321	.03668	.03808	34
35	.02402	.02461	.02797	.02878	.03222	.03329	.03676	.03816	35
36	.02408	.02468	.02804	.02885	.03229	.03337	.03684	.03825	36
37	.02415	.02474	.02811	.02892	.03236	.03345	.03692	.03833	37
38	.02421	.02481	.02818	.02899	.03244	.03353	.03699	.03842	38
39	.02427	.02488	.02824	.02907	.03251	.03360	.03707	.03850	39
40	.02434	.02494	.02831	.02914	.03258	.03368	.03715	.03858	40
41	.02440	.02501	.02838	.02921	.03266	.03376	.03723	.03867	41
42	.02447	.02508	.02845	.02928	.03273	.03384	.03731	.03875	42
43	.02453	.02515	.02852	.02936	.03281	.03392	.03739	.03884	43
44	.02459	.02521	.02859	.02943	.03288	.03400	.03747	.03892	44
45	.02466	.02528	.02866	.02950	.03295	.03408	.03754	.03901	45
46	.02472	.02535	.02873	.02958	.03303	.03416	.03762	.03909	46
47	.02479	.02542	.02880	.02965	.03310	.03424	.03770	.03918	47
48	.02485	.02548	.02887	.02972	.03318	.03432	.03778	.03927	48
49	.02492	.02555	.02894	.02980	.03325	.03439	.03786	.03935	49
50	.02498	.02562	.02900	.02987	.03333	.03447	.03794	.03944	50
51	.02504	.02569	.02907	.02994	.03340	.03455	.03802	.03952	51
52	.02511	.02576	.02914	.03002	.03347	.03463	.03810	.03961	52
53	.02517	.02582	.02921	.03009	.03355	.03471	.03818	.03969	53
54	.02524	.02589	.02928	.03017	.03362	.03479	.03826	.03978	54
55	.02530	.02596	.02935	.03024	.03370	.03487	.03834	.03987	55
56	.02537	.02603	.02942	.03032	.03377	.03495	.03842	.03995	56
57	.02543	.02610	.02949	.03039	.03385	.03503	.03850	.04004	57
58	.02550	.02617	.02956	.03046	.03392	.03512	.03858	.04013	58
59	.02556	.02624	.02963	.03054	.03400	.03520	.03866	.04021	59
60	.02563	.02630	.02970	.03061	.03407	.03528	.03874	.04030	60

′	16°		17°		18°		19°		′
	Vers.	Exsec.	Vers.	Exsec.	Vers.	Exsec.	Vers.	Exsec.	
0	.03874	.04030	.04370	.04569	.04894	.05146	.05448	.05762	0
1	.03882	.04039	.04378	.04578	.04903	.05156	.05458	.05773	1
2	.03890	.04047	.04387	.04588	.04912	.05166	.05467	.05783	2
3	.03898	.04056	.04395	.04597	.04921	.05176	.05477	.05794	3
4	.03906	.04065	.04404	.04606	.04930	.05186	.05486	.05805	4
5	.03914	.04073	.04412	.04616	.04939	.05196	.05496	.05815	5
6	.03922	.04082	.04421	.04625	.04948	.05206	.05505	.05826	6
7	.03930	.04091	.04429	.04635	.04957	.05216	.05515	.05836	7
8	.03938	.04100	.04438	.04644	.04967	.05226	.05524	.05847	8
9	.03946	.04108	.04446	.04653	.04976	.05236	.05534	.05858	9
10	.03954	.04117	.04455	.04663	.04985	.05246	.05543	.05869	10
11	.03963	.04126	.04464	.04672	.04994	.05256	.05553	.05879	11
12	.03971	.04135	.04472	.04682	.05003	.05266	.05562	.05890	12
13	.03979	.04144	.04481	.04691	.05012	.05276	.05572	.05901	13
14	.03987	.04152	.04489	.04700	.05021	.05286	.05582	.05911	14
15	.03995	.04161	.04498	.04710	.05030	.05297	.05591	.05922	15
16	.04003	.04170	.04507	.04719	.05039	.05307	.05601	.05933	16
17	.04011	.04179	.04515	.04729	.05048	.05317	.05610	.05944	17
18	.04019	.04188	.04524	.04738	.05057	.05327	.05620	.05955	18
19	.04028	.04197	.04533	.04748	.05067	.05337	.05630	.05965	19
20	.04036	.04206	.04541	.04757	.05076	.05347	.05639	.05976	20
21	.04044	.04214	.04550	.04767	.05085	.05357	.05649	.05987	21
22	.04052	.04223	.04559	.04776	.05094	.05367	.05658	.05998	22
23	.04060	.04232	.04567	.04786	.05103	.05378	.05668	.06009	23
24	.04069	.04241	.04576	.04795	.05112	.05388	.05678	.06020	24
25	.04077	.04250	.04585	.04805	.05122	.05398	.05687	.06030	25
26	.04085	.04259	.04593	.04815	.05131	.05408	.05697	.06041	26
27	.04093	.04268	.04602	.04824	.05140	.05418	.05707	.06052	27
28	.04102	.04277	.04611	.04834	.05149	.05429	.05716	.06063	28
29	.04110	.04286	.04620	.04843	.05158	.05439	.05726	.06074	29
30	.04118	.04295	.04628	.04853	.05168	.05449	.05736	.06085	30
31	.04126	.04304	.04637	.04863	.05177	.05460	.05746	.06096	31
32	.04135	.04313	.04646	.04872	.05186	.05470	.05755	.06107	32
33	.04143	.04322	.04655	.04882	.05195	.05480	.05765	.06118	33
34	.04151	.04331	.04663	.04891	.05205	.05490	.05775	.06129	34
35	.04159	.04340	.04672	.04901	.05214	.05501	.05785	.06140	35
36	.04168	.04349	.04681	.04911	.05223	.05511	.05794	.06151	36
37	.04176	.04358	.04690	.04920	.05232	.05521	.05804	.06162	37
38	.04184	.04367	.04699	.04930	.05242	.05532	.05814	.06173	38
39	.04193	.04376	.04707	.04940	.05251	.05542	.05824	.06184	39
40	.04201	.04385	.04716	.04950	.05260	.05552	.05833	.06195	40
41	.04209	.04394	.04725	.04959	.05270	.05563	.05843	.06206	41
42	.04218	.04403	.04734	.04969	.05279	.05573	.05853	.06217	42
43	.04226	.04413	.04743	.04979	.05288	.05584	.05863	.06228	43
44	.04234	.04422	.04752	.04989	.05298	.05594	.05873	.06239	44
45	.04243	.04431	.04760	.04998	.05307	.05604	.05882	.06250	45
46	04251	.04440	.04769	.05008	.05316	.05615	.05892	.06261	46
47	.04260	.04449	.04778	.05018	.05326	.05625	.05902	.06272	47
48	.04268	.04458	.04787	.05028	.05335	.05636	.05912	.06283	48
49	.04276	.04468	.04796	.05038	.05344	.05646	.05922	.06295	49
50	.04285	.04477	.04805	.05047	.05354	.05657	.05932	.06306	50
51	.04293	.04486	.04814	.05057	.05363	.05667	.05942	.06317	51
52	.04302	.04495	.04823	.05067	.05373	.05678	.05951	.06328	52
53	.04310	.04504	.04832	.05077	.05382	.05688	.05961	.06339	53
54	.04319	.04514	.04841	.05087	.05391	.05699	.05971	.06350	54
55	.04327	.04523	.04850	.05097	.05401	.05709	.05981	.06362	55
56	.04336	.04532	.04858	.05107	.05410	.05720	.05991	.06373	56
57	.04344	.04541	.04867	.05116	.05420	.05730	.06001	.06384	57
58	.04353	.04551	.04876	.05126	.05429	.05741	.06011	.06395	58
59	.04361	.04560	.04885	.05136	.05439	.05751	.06021	.06407	59
60	.04370	.04569	.04894	.05146	.05448	.05762	.06031	.06418	60

′	20° Vers.	20° Exsec.	21° Vers.	21° Exsec.	22° Vers.	22° Exsec.	23° Vers.	23° Exsec.	′
0	.06031	.06418	.06642	.07115	.07282	.07853	.07950	.08636	0
1	.06041	.06429	.06652	.07126	.07293	.07866	.07961	.08649	1
2	.06051	.06440	.06663	.07138	.07303	.07879	.07972	.08663	2
3	.06061	.06452	.06673	.07150	.07314	.07892	.07984	.08676	3
4	.06071	.06463	.06684	.07162	.07325	.07904	.07995	.08690	4
5	.06081	.06474	.06694	.07174	.07336	.07917	.08006	.08703	5
6	.06091	.06486	.06705	.07186	.07347	.07930	.08018	.08717	6
7	.06101	.06497	.06715	.07199	.07358	.07943	.08029	.08730	7
8	.06111	.06508	.06726	.07211	.07369	.07955	.08041	.08744	8
9	.06121	.06520	.06736	.07223	.07380	.07968	.08052	.08757	9
10	.06131	.06531	.06747	.07235	.07391	.07981	.08064	.08771	10
11	.06141	.06542	.06757	.07247	.07402	.07994	.08075	.08784	11
12	.06151	.06554	.06768	.07259	.07413	.08006	.08086	.08798	12
13	.06161	.06565	.06778	.07271	.07424	.08019	.08098	.08811	13
14	.06171	.06577	.06789	.07283	.07435	.08032	.08109	.08825	14
15	.06181	.06588	.06799	.07295	.07446	.08045	.08121	.08839	15
16	.06191	.06600	.06810	.07307	.07457	.08058	.08132	.08852	16
17	.06201	.06611	.06820	.07320	.07468	.08071	.08144	.08866	17
18	.06211	.06622	.06831	.07332	.07479	.08084	.08155	.08880	18
19	.06221	.06634	.06841	.07344	.07490	.08097	.08167	.08893	19
20	.06231	.06645	.06852	.07356	.07501	.08109	.08178	.08907	20
21	.06241	.06657	.06863	.07368	.07512	.08122	.08190	.08921	21
22	.06252	.06668	.06873	.07380	.07523	.08135	.08201	.08934	22
23	.06262	.06680	.06884	.07393	.07534	.08148	.08213	.08948	23
24	.06272	.06691	.06894	.07405	.07545	.08161	.08225	.08962	24
25	.06282	.06703	.06905	.07417	.07556	.08174	.08236	.08975	25
26	.06292	.06715	.06916	.07429	.07568	.08187	.08248	.08989	26
27	.06302	.06726	.06926	.07442	.07579	.08200	.08259	.09003	27
28	.06312	.06738	.06937	.07454	.07590	.08213	.08271	.09017	28
29	.06323	.06749	.06948	.07466	.07601	.08226	.08282	.09030	29
30	.06333	.06761	.06958	.07479	.07612	.08239	.08294	.09044	30
31	.06343	.06773	.06969	.07491	.07623	.08252	.08306	.09058	31
32	.06353	.06784	.06980	.07503	.07634	.08265	.08317	.09072	32
33	.06363	.06796	.06990	.07516	.07645	.08278	.08329	.09086	33
34	.06374	.06807	.07001	.07528	.07657	.08291	.08340	.09099	34
35	.06384	.06819	.07012	.07540	.07668	.08305	.08352	.09113	35
36	.06394	.06831	.07022	.07553	.07679	.08318	.08364	.09127	36
37	.06404	.06843	.07033	.07565	.07690	.08331	.08375	.09141	37
38	.06415	.06854	.07044	.07578	.07701	.08344	.08387	.09155	38
39	.06425	.06866	.07055	.07590	.07713	.08357	.08399	.09169	39
40	.06435	.06878	.07065	.07602	.07724	.08370	.08410	.09183	40
41	.06445	.06889	.07076	.07615	.07735	.08383	.08422	.09197	41
42	.06456	.06901	.07087	.07627	.07746	.08397	.08434	.09211	42
43	.06466	.06913	.07098	.07640	.07757	.08410	.08445	.09224	43
44	.06476	.06925	.07108	.07652	.07769	.08423	.08457	.09238	44
45	.06486	.06936	.07119	.07665	.07780	.08436	.08469	.09252	45
46	.06497	.06948	.07130	.07677	.07791	.08449	.08481	.09266	46
47	.06507	.06960	.07141	.07690	.07802	.08463	.08492	.09280	47
48	.06517	.06972	.07151	.07702	.07814	.08476	.08504	.09294	48
49	.06528	.06984	.07162	.07715	.07825	.08489	.08516	.09308	49
50	.06538	.06995	.07173	.07727	.07836	.08503	.08528	.09323	50
51	.06548	.07007	.07184	.07740	.07848	.08516	.08539	.09337	51
52	.06559	.07019	.07195	.07752	.07859	.08529	.08551	.09351	52
53	.06569	.07031	.07206	.07765	.07870	.08542	.08563	.09365	53
54	.06580	.07043	.07216	.07778	.07881	.08556	.08575	.09379	54
55	.06590	.07055	.07227	.07790	.07893	.08569	.08586	.09393	55
56	.06600	.07067	.07238	.07803	.07904	.08582	.08598	.09407	56
57	.06611	.07079	.07249	.07816	.07915	.08596	.08610	.09421	57
58	.06621	.07091	.07260	.07828	.07927	.08609	.08622	.09435	58
59	.06632	.07103	.07271	.07841	.07938	.08623	.08634	.09449	59
60	.06642	.07115	.07282	.07853	.07950	.08636	.08645	.09464	60

′	24° Vers.	24° Exsec.	25° Vers.	25° Exsec.	26° Vers.	26° Exsec.	27° Vers.	27° Exsec.	′
0	.08645	.09464	.09369	.10338	.10121	.11260	.10899	.12233	0
1	.08657	.09478	.09382	.10353	.10133	.11276	.10913	.12249	1
2	.08669	.09492	.09394	.10368	.10146	.11292	.10926	.12266	2
3	.08681	.09506	.09406	.10383	.10159	.11308	.10939	.12283	3
4	.08693	.09520	.09418	.10398	.10172	.11323	.10952	.12299	4
5	.08705	.09535	.09431	.10413	.10184	.11339	.10965	.12316	5
6	.08717	.09549	.09443	.10428	.10197	.11355	.10979	.12333	6
7	.08728	.09563	.09455	.10443	.10210	.11371	.10992	.12349	7
8	.08740	.09577	.09468	.10458	.10223	.11387	.11005	.12366	8
9	.08752	.09592	.09480	.10473	.10236	.11403	.11019	.12383	9
10	.08764	.09606	.09493	.10488	.10248	.11419	.11032	.12400	10
11	.08776	.09620	.09505	.10503	.10261	.11435	.11045	.12416	11
12	.08788	.09635	.09517	.10518	.10274	.11451	.11058	.12433	12
13	.08800	.09649	.09530	.10533	.10287	.11467	.11072	.12450	13
14	.08812	.09663	.09542	.10549	.10300	.11483	.11085	.12467	14
15	.08824	.09678	.09554	.10564	.10313	.11499	.11098	.12484	15
16	.08836	.09692	.09567	.10579	.10326	.11515	.11112	.12501	16
17	.08848	.09707	.09579	.10594	.10338	.11531	.11125	.12518	17
18	.08860	.09721	.09592	.10609	.10351	.11547	.11138	.12534	18
19	.08872	.09735	.09604	.10625	.10364	.11563	.11152	.12551	19
20	.08884	.09750	.09617	.10640	.10377	.11579	.11165	.12568	20
21	.08896	.09764	.09629	.10655	.10390	.11595	.11178	.12585	21
22	.08908	.09779	.09642	.10670	.10403	.11611	.11192	.12602	22
23	.08920	.09793	.09654	.10686	.10416	.11627	.11205	.12619	23
24	.08932	.09808	.09666	.10701	.10429	.11643	.11218	.12636	24
25	.08944	.09822	.09679	.10716	.10442	.11659	.11232	.12653	25
26	.08956	.09837	.09691	.10731	.10455	.11675	.11245	.12670	26
27	.08968	.09851	.09704	.10747	.10468	.11691	.11259	.12687	27
28	.08980	.09866	.09716	.10762	.10481	.11708	.11272	.12704	28
29	.08992	.09880	.09729	.10777	.10494	.11724	.11285	.12721	29
30	.09004	.09895	.09741	.10793	.10507	.11740	.11299	.12738	30
31	.09016	.09909	.09754	.10808	.10520	.11756	.11312	.12755	31
32	.09028	.09924	.09767	.10824	.10533	.11772	.11326	.12772	32
33	.09040	.09939	.09779	.10839	.10546	.11789	.11339	.12789	33
34	.09052	.09953	.09792	.10854	.10559	.11805	.11353	.12807	34
35	.09064	.09968	.09804	.10870	.10572	.11821	.11366	.12824	35
36	.09076	.09982	.09817	.10885	.10585	.11838	.11380	.12841	36
37	.09089	.09997	.09829	.10901	.10598	.11854	.11393	.12858	37
38	.09101	.10012	.09842	.10916	.10611	.11870	.11407	.12875	38
39	.09113	.10026	.09854	.10932	.10624	.11886	.11420	.12892	39
40	.09125	.10041	.09867	.10947	.10637	.11903	.11434	.12910	40
41	.09137	.10055	.09880	.10963	.10650	.11919	.11447	.12927	41
42	.09149	.10071	.09892	.10978	.10663	.11936	.11461	.12944	42
43	.09161	.10085	.09905	.10994	.10676	.11952	.11474	.12961	43
44	.09174	.10100	.09918	.11009	.10689	.11968	.11488	.12979	44
45	.09186	.10115	.09930	.11025	.10702	.11985	.11501	.12996	45
46	.09198	.10130	.09943	.11041	.10715	.12001	.11515	.13013	46
47	.09210	.10144	.09955	.11056	.10728	.12018	.11528	.13031	47
48	.09222	.10159	.09968	.11072	.10741	.12034	.11542	.13048	48
49	.09234	.10174	.09981	.11087	.10755	.12051	.11555	.13065	49
50	.09247	.10189	.09993	.11103	.10768	.12067	.11569	.13083	50
51	.09259	.10204	.10006	.11119	.10781	.12084	.11583	.13100	51
52	.09271	.10218	.10019	.11134	.10794	.12100	.11596	.13117	52
53	.09283	.10233	.10032	.11150	.10807	.12117	.11610	.13135	53
54	.09296	.10248	.10044	.11166	.10820	.12133	.11623	.13152	54
55	.09308	.10263	.10057	.11181	.10833	.12150	.11637	.13170	55
56	.09320	.10278	.10070	.11197	.10847	.12166	.11651	.13187	56
57	.09332	.10293	.10082	.11213	.10860	.12183	.11664	.13205	57
58	.09345	.10308	.10095	.11229	.10873	.12199	.11678	.13222	58
59	.09357	.10323	.10108	.11244	.10886	.12216	.11692	.13240	59
60	.09369	.10338	.10121	.11260	.10899	.12233	.11705	.13257	60

′	28°		29°		30°		31°		′
	Vers.	Exsec.	Vers.	Exsec.	Vers.	Exsec.	Vers.	Exsec.	
0	.11705	.13257	.12538	.14335	.13397	.15470	.14283	.16663	0
1	.11719	.13275	.12552	.14354	.13412	.15489	.14298	.16684	1
2	.11733	.13292	.12566	.14372	.13427	.15509	.14313	.16704	2
3	.11746	.13310	.12580	.14391	.13441	.15528	.14328	.16725	3
4	.11760	.13327	.12595	.14409	.13456	.15548	.14343	.16745	4
5	.11774	.13345	.12609	.14428	.13470	.15567	.14358	.16766	5
6	.11787	.13362	.12623	.14446	.13485	.15587	.14373	.16786	6
7	.11801	.13380	.12637	.14465	.13499	.15606	.14388	.16806	7
8	.11815	.13398	.12651	.14483	.13514	.15626	.14403	.16827	8
9	.11828	.13415	.12665	.14502	.13529	.15645	.14418	.16848	9
10	.11842	.13433	.12679	.14521	.13543	.15665	.14433	.16868	10
11	.11856	.13451	.12694	.14539	.13558	.15684	.14449	.16889	11
12	.11870	.13468	.12708	.14558	.13573	.15704	.14464	.16909	12
13	.11883	.13486	.12722	.14576	.13587	.15724	.14479	.16930	13
14	.11897	.13504	.12736	.14595	.13602	.15743	.14494	.16950	14
15	.11911	.13521	.12750	.14614	.13616	.15763	.14509	.16971	15
16	.11925	.13539	.12765	.14632	.13631	.15782	.14524	.16992	16
17	.11938	.13557	.12779	.14651	.13646	.15802	.14539	.17012	17
18	.11952	.13575	.12793	.14670	.13660	.15822	.14554	.17033	18
19	.11966	.13593	.12807	.14689	.13675	.15841	.14569	.17054	19
20	.11980	.13610	.12822	.14707	.13690	.15861	.14584	.17075	20
21	.11994	.13628	.12836	.14726	.13705	.15881	.14599	.17095	21
22	.12007	.13646	.12850	.14745	.13719	.15901	.14615	.17116	22
23	.12021	.13664	.12864	.14764	.13734	.15920	.14630	.17137	23
24	.12035	.13682	.12879	.14782	.13749	.15940	.14645	.17158	24
25	.12049	.13700	.12893	.14801	.13763	.15960	.14660	.17178	25
26	.12063	.13718	.12907	.14820	.13778	.15980	.14675	.17199	26
27	.12077	.13735	.12921	.14839	.13793	.16000	.14690	.17220	27
28	.12091	.13753	.12936	.14858	.13808	.16019	.14706	.17241	28
29	.12104	.13771	.12950	.14877	.13822	.16039	.14721	.17262	29
30	.12118	.13789	.12964	.14896	.13837	.16059	.14736	.17283	30
31	.12132	.13807	.12979	.14914	.13852	.16079	.14751	.17304	31
32	.12146	.13825	.12993	.14933	.13867	.16099	.14766	.17325	32
33	.12160	.13843	.13007	.14952	.13881	.16119	.14782	.17346	33
34	.12174	.13861	.13022	.14971	.13896	.16139	.14797	.17367	34
35	.12188	.13879	.13036	.14990	.13911	.16159	.14812	.17388	35
36	.12202	.13897	.13051	.15009	.13926	.16179	.14827	.17409	36
37	.12216	.13916	.13065	.15028	.13941	.16199	.14843	.17430	37
38	.12230	.13934	.13079	.15047	.13955	.16219	.14858	.17451	38
39	.12244	.13952	.13094	.15066	.13970	.16239	.14873	.17472	39
40	.12257	.13970	.13108	.15085	.13985	.16259	.14888	.17493	40
41	.12271	.13988	.13122	.15105	.14000	.16279	.14904	.17514	41
42	.12285	.14006	.13137	.15124	.14015	.16299	.14919	.17535	42
43	.12299	.14024	.13151	.15143	.14030	.16319	.14934	.17556	43
44	.12313	.14042	.13166	.15162	.14044	.16339	.14949	.17577	44
45	.12327	.14061	.13180	.15181	.14059	.16359	.14965	.17598	45
46	.12341	.14079	.13195	.15200	.14074	.16380	.14980	.17620	46
47	.12355	.14097	.13209	.15219	.14089	.16400	.14995	.17641	47
48	.12369	.14115	.13223	.15239	.14104	.16420	.15011	.17662	48
49	.12383	.14134	.13238	.15258	.14119	.16440	.15026	.17683	49
50	.12397	.14152	.13252	.15277	.14134	.16460	.15041	.17704	50
51	.12411	.14170	.13267	.15296	.14149	.16481	.15057	.17726	51
52	.12425	.14188	.13281	.15315	.14164	.16501	.15072	.17747	52
53	.12439	.14207	.13296	.15335	.14179	.16521	.15087	.17768	53
54	.12454	.14225	.13310	.15354	.14194	.16541	.15103	.17790	54
55	.12468	.14243	.13325	.15373	.14208	.16562	.15118	.17811	55
56	.12482	.14262	.13339	.15393	.14223	.16582	.15134	.17832	56
57	.12496	.14280	.13354	.15412	.14238	.16602	.15149	.17854	57
58	.12510	.14299	.13368	.15431	.14253	.16623	.15164	.17875	58
59	.12524	.14317	.13383	.15451	.14268	.16643	.15180	.17896	59
60	.12538	.14335	.13397	.15470	.14283	.16663	.15195	.17918	60

′	32° Vers.	32° Exsec.	33° Vers.	33° Exsec.	34° Vers.	34° Exsec.	35° Vers.	35° Exsec.	′
0	.15195	.17918	.16133	.19236	.17096	.20622	.18085	.22077	0
1	.15211	.17939	.16149	.19259	.17113	.20645	.18101	.22102	1
2	.15226	.17961	.16165	.19281	.17129	.20669	.18118	.22127	2
3	.15241	.17982	.16181	.19304	.17145	.20693	.18135	.22152	3
4	.15257	.18004	.16196	.19327	.17161	.20717	.18152	.22177	4
5	.15272	.18025	.16212	.19349	.17178	.20740	.18168	.22202	5
6	.15288	.18047	.16228	.19372	.17194	.20764	.18185	.22227	6
7	.15303	.18068	.16244	.19394	.17210	.20788	.18202	.22252	7
8	.15319	.18090	.16260	.19417	.17227	.20812	.18218	.22277	8
9	.15334	.18111	.16276	.19440	.17243	.20836	.18235	.22302	9
10	.15350	.18133	.16292	.19463	.17259	.20859	.18252	.22327	10
11	.15365	.18155	.16308	.19485	.17276	.20883	.18269	.22352	11
12	.15381	.18176	.16324	.19508	.17292	.20907	.18286	.22377	12
13	.15396	.18198	.16340	.19531	.17308	.20931	.18302	.22402	13
14	.15412	.18220	.16355	.19554	.17325	.20955	.18319	.22428	14
15	.15427	.18241	.16371	.19576	.17341	.20979	.18336	.22453	15
16	.15443	.18263	.16387	.19599	.17357	.21003	.18353	.22478	16
17	.15458	.18285	.16403	.19622	.17374	.21027	.18369	.22503	17
18	.15474	.18307	.16419	.19645	.17390	.21051	.18386	.22528	18
19	.15489	.18328	.16435	.19668	.17407	.21075	.18403	.22554	19
20	.15505	.18350	.16451	.19691	.17423	.21099	.18420	.22579	20
21	.15520	.18372	.16467	.19713	.17439	.21123	.18437	.22604	21
22	.15536	.18394	.16483	.19736	.17456	.21147	.18454	.22629	22
23	.15552	.18416	.16499	.19759	.17472	.21171	.18470	.22655	23
24	.15567	.18437	.16515	.19782	.17489	.21195	.18487	.22680	24
25	.15583	.18459	.16531	.19805	.17505	.21220	.18504	.22706	25
26	.15598	.18481	.16547	.19828	.17522	.21244	.18521	.22731	26
27	.15614	.18503	.16563	.19851	.17538	.21268	.18538	.22756	27
28	.15630	.18525	.16579	.19874	.17554	.21292	.18555	.22782	28
29	.15645	.18547	.16595	.19897	.17571	.21316	.18572	.22807	29
30	.15661	.18569	.16611	.19920	.17587	.21341	.18588	.22833	30
31	.15676	.18591	.16627	.19944	.17604	.21365	.18605	.22858	31
32	.15692	.18613	.16644	.19967	.17620	.21389	.18622	.22884	32
33	.15708	.18635	.16660	.19990	.17637	.21414	.18639	.22909	33
34	.15723	.18657	.16676	.20013	.17653	.21438	.18656	.22935	34
35	.15739	.18679	.16692	.20036	.17670	.21462	.18673	.22960	35
36	.15755	.18701	.16708	.20059	.17686	.21487	.18690	.22986	36
37	.15770	.18723	.16724	.20083	.17703	.21511	.18707	.23012	37
38	.15786	.18745	.16740	.20106	.17719	.21535	.18724	.23037	38
39	.15802	.18767	.16756	.20129	.17736	.21560	.18741	.23063	39
40	.15818	.18790	.16772	.20152	.17752	.21584	.18758	.23089	40
41	.15833	.18812	.16788	.20176	.17769	.21609	.18775	.23114	41
42	.15849	.18834	.16805	.20199	.17786	.21633	.18792	.23140	42
43	.15865	.18856	.16821	.20222	.17802	.21658	.18809	.23166	43
44	.15880	.18878	.16837	.20246	.17819	.21682	.18826	.23192	44
45	.15896	.18901	.16853	.20269	.17835	.21707	.18843	.23217	45
46	.15912	.18923	.16869	.20292	.17852	.21731	.18860	.23243	46
47	.15928	.18945	.16885	.20316	.17868	.21756	.18877	.23269	47
48	.15943	.18967	.16902	.20339	.17885	.21781	.18894	.23295	48
49	.15959	.18990	.16918	.20363	.17902	.21805	.18911	.23321	49
50	.15975	.19012	.16934	.20386	.17918	.21830	.18928	.23347	50
51	.15991	.19034	.16950	.20410	.17935	.21855	.18945	.23373	51
52	.16006	.19057	.16966	.20433	.17952	.21879	.18962	.23399	52
53	.16022	.19079	.16983	.20457	.17968	.21904	.18979	.23424	53
54	.16038	.19102	.16999	.20480	.17985	.21929	.18996	.23450	54
55	.16054	.19124	.17015	.20504	.18001	.21953	.19013	.23476	55
56	.16070	.19146	.17031	.20527	.18018	.21978	.19030	.23502	56
57	.16085	.19169	.17047	.20551	.18035	.22003	.19047	.23529	57
58	.16101	.19191	.17064	.20575	.18051	.22028	.19064	.23555	58
59	.16117	.19214	.17080	.20598	.18068	.22053	.19081	.23581	59
60	.16133	.19236	.17096	.20622	.18085	.22077	.19098	.23607	60

′	36°		37°		38°		39°		′
	Vers.	Exsec.	Vers.	Exsec.	Vers.	Exsec.	Vers.	Exsec.	
0	.19098	.23607	.20136	.25214	.21199	.26902	.22285	.28676	0
1	.19115	.23633	.20154	.25241	.21217	.26931	.22304	.28706	1
2	.19133	.23659	.20171	.25269	.21235	.26960	.22322	.28737	2
3	.19150	.23685	.20189	.25296	.21253	.26988	.22340	.28767	3
4	.19167	.23711	.20207	.25324	.21271	.27017	.22359	.28797	4
5	.19184	.23738	.20224	.25351	.21289	.27046	.22377	.28828	5
6	.19201	.23764	.20242	.25379	.21307	.27075	.22395	.28858	6
7	.19218	.23790	.20259	.25406	.21324	.27104	.22414	.28889	7
8	.19235	.23816	.20277	.25434	.21342	.27133	.22432	.28919	8
9	.19252	.23843	.20294	.25462	.21360	.27162	.22450	.28950	9
10	.19270	.23869	.20312	.25489	.21378	.27191	.22469	.28980	10
11	.19287	.23895	.20329	.25517	.21396	.27221	.22487	.29011	11
12	.19304	.23922	.20347	.25545	.21414	.27250	.22506	.29042	12
13	.19321	.23948	.20365	.25572	.21432	.27279	.22524	.29072	13
14	.19338	.23975	.20382	.25600	.21450	.27308	.22542	.29103	14
15	.19356	.24001	.20400	.25628	.21468	.27337	.22561	.29133	15
16	.19373	.24028	.20417	.25656	.21486	.27366	.22579	.29164	16
17	.19390	.24054	.20435	.25683	.21504	.27396	.22598	.29195	17
18	.19407	.24081	.20453	.25711	.21522	.27425	.22616	.29226	18
19	.19424	.24107	.20470	.25739	.21540	.27454	.22634	.29256	19
20	.19442	.24134	.20488	.25767	.21558	.27483	.22653	.29287	20
21	.19459	.24160	.20506	.25795	.21576	.27513	.22671	.29318	21
22	.19476	.24187	.20523	.25823	.21595	.27542	.22690	.29349	22
23	.19493	.24213	.20541	.25851	.21613	.27572	.22708	.29380	23
24	.19511	.24240	.20559	.25879	.21631	.27601	.22727	.29411	24
25	.19528	.24267	.20576	.25907	.21649	.27630	.22745	.29442	25
26	.19545	.24293	.20594	.25935	.21667	.27660	.22764	.29473	26
27	.19562	.24320	.20612	.25963	.21685	.27689	.22782	.29504	27
28	.19580	.24347	.20629	.25991	.21703	.27719	.22801	.29535	28
29	.19597	.24373	.20647	.26019	.21721	.27748	.22819	.29566	29
30	.19614	.24400	.20665	.26047	.21739	.27778	.22838	.29597	30
31	.19632	.24427	.20682	.26075	.21757	.27807	.22856	.29628	31
32	.19649	.24454	.20700	.26104	.21775	.27837	.22875	.29659	32
33	.19666	.24481	.20718	.26132	.21794	.27867	.22893	.29690	33
34	.19684	.24508	.20736	.26160	.21812	.27896	.22912	.29721	34
35	.19701	.24534	.20753	.26188	.21830	.27926	.22930	.29752	35
36	.19718	.24561	.20771	.26216	.21848	.27956	.22949	.29784	36
37	.19736	.24588	.20789	.26245	.21866	.27985	.22967	.29815	37
38	.19753	.24615	.20807	.26273	.21884	.28015	.22986	.29846	38
39	.19770	.24642	.20824	.26301	.21902	.28045	.23004	.29877	39
40	.19788	.24669	.20842	.26330	.21921	.28075	.23023	.29909	40
41	.19805	.24696	.20860	.26358	.21939	.28105	.23041	.29940	41
42	.19822	.24723	.20878	.26387	.21957	.28134	.23060	.29971	42
43	.19840	.24750	.20895	.26415	.21975	.28164	.23079	.30003	43
44	.19857	.24777	.20913	.26443	.21993	.28194	.23097	.30034	44
45	.19875	.24804	.20931	.26472	.22012	.28224	.23116	.30066	45
46	.19892	.24832	.20949	.26500	.22030	.28254	.23134	.30097	46
47	.19909	.24859	.20967	.26529	.22048	.28284	.23153	.30129	47
48	.19927	.24886	.20985	.26557	.22066	.28314	.23172	.30160	48
49	.19944	.24913	.21002	.26586	.22084	.28344	.23190	.30192	49
50	.19962	.24940	.21020	.26615	.22103	.28374	.23209	.30223	50
51	.19979	.24967	.21038	.26643	.22121	.28404	.23228	.30255	51
52	.19997	.24995	.21056	.26672	.22139	.28434	.23246	.30287	52
53	.20014	.25022	.21074	.26701	.22157	.28464	.23265	.30318	53
54	.20032	.25049	.21092	.26729	.22176	.28495	.23283	.30350	54
55	.20049	.25077	.21109	.26758	.22194	.28525	.23302	.30382	55
56	.20066	.25104	.21127	.26787	.22212	.28555	.23321	.30413	56
57	.20084	.25131	.21145	.26815	.22231	.28585	.23339	.30445	57
58	.20101	.25159	.21163	.26844	.22249	.28615	.23358	.30477	58
59	.20119	.25186	.21181	.26873	.22267	.28646	.23377	.30509	59
60	.20136	.25214	.21199	.26902	.22285	.28676	.23396	.30541	60

′	40°		41°		42°		43°		′
	Vers.	Exsec.	Vers.	Exsec.	Vers.	Exsec.	Vers.	Exsec.	
0	.23396	.30541	.24529	.32501	.25686	.34563	.26865	.36733	0
1	.23414	.30573	.24548	.32535	.25705	.34599	.26884	.36770	1
2	.23433	.30605	.24567	.32568	.25724	.34634	.26904	.36807	2
3	.23452	.30636	.24586	.32602	.25744	.34669	.26924	.36844	3
4	.23470	.30668	.24605	.32636	.25763	.34704	.26944	.36881	4
5	.23489	.30700	.24625	.32669	.25783	.34740	.26964	.36919	5
6	.23508	.30732	.24644	.32703	.25802	.34775	.26984	.36956	6
7	.23527	.30764	.24663	.32737	.25822	.34811	.27004	.36993	7
8	.23545	.30796	.24682	.32770	.25841	.34846	.27024	.37030	8
9	.23564	.30829	.24701	.32804	.25861	.34882	.27043	.37068	9
10	.23583	.30861	.24720	.32838	.25880	.34917	.27063	.37105	10
11	.23602	.30893	.24739	.32872	.25900	.34953	.27083	.37143	11
12	.23620	.30925	.24759	.32905	.25920	.34988	.27103	.37180	12
13	.23639	.30957	.24778	.32939	.25939	.35024	.27123	.37218	13
14	.23658	.30989	.24797	.32973	.25959	.35060	.27143	.37255	14
15	.23677	.31022	.24816	.33007	.25978	.35095	.27163	.37293	15
16	.23696	.31054	.24835	.33041	.25998	.35131	.27183	.37330	16
17	.23714	.31086	.24854	.33075	.26017	.35167	.27203	.37368	17
18	.23733	.31119	.24874	.33109	.26037	.35203	.27223	.37406	18
19	.23752	.31151	.24893	.33143	.26056	.35238	.27243	.37443	19
20	.23771	.31183	.24912	.33177	.26076	.35274	.27263	.37481	20
21	.23790	.31216	.24931	.33211	.26096	.35310	.27283	.37519	21
22	.23808	.31248	.24950	.33245	.26115	.35346	.27303	.37556	22
23	.23827	.31281	.24970	.33279	.26135	.35382	.27323	.37594	23
24	.23846	.31313	.24989	.33314	.26154	.35418	.27343	.37632	24
25	.23865	.31346	.25008	.33348	.26174	.35454	.27363	.37670	25
26	.23884	.31378	.25027	.33382	.26194	.35490	.27383	.37708	26
27	.23903	.31411	.25047	.33416	.26213	.35526	.27403	.37746	27
28	.23922	.31443	.25066	.33451	.26233	.35562	.27423	.37784	28
29	.23941	.31476	.25085	.33485	.26253	.35598	.27443	.37822	29
30	.23959	.31509	.25104	.33519	.26272	.35634	.27463	.37860	30
31	.23978	.31541	.25124	.33554	.26292	.35670	.27483	.37898	31
32	.23997	.31574	.25143	.33588	.26312	.35707	.27503	.37936	32
33	.24016	.31607	.25162	.33622	.26331	.35743	.27523	.37974	33
34	.24035	.31640	.25182	.33657	.26351	.35779	.27543	.38012	34
35	.24054	.31672	.25201	.33691	.26371	.35815	.27563	.38051	35
36	.24073	.31705	.25220	.33726	.26390	.35852	.27583	.38089	36
37	.24092	.31738	.25240	.33760	.26410	.35888	.27603	.38127	37
38	.24111	.31771	.25259	.33795	.26430	.35924	.27623	.38165	38
39	.24130	.31804	.25278	.33830	.26449	.35961	.27643	.38204	39
40	.24149	.31837	.25297	.33864	.26469	.35997	.27663	.38242	40
41	.24168	.31870	.25317	.33899	.26489	.36034	.27683	.38280	41
42	.24187	.31903	.25336	.33934	.26509	.36070	.27703	.38319	42
43	.24206	.31936	.25356	.33968	.26528	.36107	.27723	.38357	43
44	.24225	.31969	.25375	.34003	.26548	.36143	.27743	.38396	44
45	.24244	.32002	.25394	.34038	.26568	.36180	.27764	.38434	45
46	.24262	.32035	.25414	.34073	.26588	.36217	.27784	.38473	46
47	.24281	.32068	.25433	.34108	.26607	.36253	.27804	.38512	47
48	.24300	.32101	.25452	.34142	.26627	.36290	.27824	.38550	48
49	.24320	.32134	.25472	.34177	.26647	.36327	.27844	.38589	49
50	.24339	.32168	.25491	.34212	.26667	.36363	.27864	.38628	50
51	.24358	.32201	.25511	.34247	.26686	.36400	.27884	.38666	51
52	.24377	.32234	.25530	.34282	.26706	.36437	.27905	.38705	52
53	.24396	.32267	.25549	.34317	.26726	.36474	.27925	.38744	53
54	.24415	.32301	.25569	.34352	.26746	.36511	.27945	.38783	54
55	.24434	.32334	.25588	.34387	.26766	.36548	.27965	.38822	55
56	.24453	.32368	.25608	.34423	.26785	.36585	.27985	.38860	56
57	.24472	.32401	.25627	.34458	.26805	.36622	.28005	.38899	57
58	.24491	.32434	.25647	.34493	.26825	.36659	.28026	.38938	58
59	.24510	.32468	.25666	.34528	.26845	.36696	.28046	.38977	59
60	.24529	.32501	.25686	.34563	.26865	.36733	.28066	.39016	60

′	44°		45°		46°		47°		′
	Vers.	Exsec.	Vers.	Exsec.	Vers.	Exsec.	Vers.	Exsec.	
0	.28066	.39016	.29289	.41421	.30534	.43956	.31800	.46628	0
1	.28086	.39055	.29310	.41463	.30555	.43999	.31821	.46674	1
2	.28106	.39095	.29330	.41504	.30576	.44042	.31843	.46719	2
3	.28127	.39134	.29351	.41545	.30597	.44086	.31864	.46765	3
4	.28147	.39173	.29372	.41586	.30618	.44129	.31885	.46811	4
5	.28167	.39212	.29392	.41627	.30639	.44173	.31907	.46857	5
6	.28187	.39251	.29413	.41669	.30660	.44217	.31928	.46903	6
7	.28208	.39291	.29433	.41710	.30681	.44260	.31949	.46949	7
8	.28228	.39330	.29454	.41752	.30702	.44304	.31971	.46995	8
9	.28248	.39369	.29475	.41793	.30723	.44347	.31992	.47041	9
10	.28268	.39409	.29495	.41835	.30744	.44391	.32013	.47087	10
11	.28289	.39448	.29516	.41876	.30765	.44435	.32035	.47134	11
12	.28309	.39487	.29537	.41918	.30786	.44479	.32056	.47180	12
13	.28329	.39527	.29557	.41959	.30807	.44523	.32077	.47226	13
14	.28350	.39566	.29578	.42001	.30828	.44567	.32099	.47272	14
15	.28370	.39606	.29599	.42042	.30849	.44610	.32120	.47319	15
16	.28390	.39646	.29619	.42084	.30870	.44654	.32141	.47365	16
17	.28410	.39685	.29640	.42126	.30891	.44698	.32163	.47411	17
18	.28431	.39725	.29661	.42168	.30912	.44742	.32184	.47458	18
19	.28451	.39764	.29681	.42210	.30933	.44787	.32205	.47504	19
20	.28471	.39804	.29702	.42251	.30954	.44831	.32227	.47551	20
21	.28492	.39844	.29723	.42293	.30975	.44875	.32248	.47598	21
22	.28512	.39884	.29743	.42335	.30996	.44919	.32270	.47644	22
23	.28532	.39924	.29764	.42377	.31017	.44963	.32291	.47691	23
24	.28553	.39963	.29785	.42419	.31038	.45007	.32312	.47738	24
25	.28573	.40003	.29805	.42461	.31059	.45052	.32334	.47784	25
26	.28593	.40043	.29826	.42503	.31080	.45096	.32355	.47831	26
27	.28614	.40083	.29847	.42545	.31101	.45141	.32377	.47878	27
28	.28634	.40123	.29868	.42587	.31122	.45185	.32398	.47925	28
29	.28655	.40163	.29888	.42630	.31143	.45229	.32420	.47972	29
30	.28675	.40203	.29909	.42672	.31165	.45274	.32441	.48019	30
31	.28695	.40243	.29930	.42714	.31186	.45319	.32462	.48066	31
32	.28716	.40283	.29951	.42756	.31207	.45363	.32484	.48113	32
33	.28736	.40324	.29971	.42799	.31228	.45408	.32505	.48160	33
34	.28757	.40364	.29992	.42841	.31249	.45452	.32527	.48207	34
35	.28777	.40404	.30013	.42883	.31270	.45497	.32548	.48254	35
36	.28797	.40444	.30034	.42926	.31291	.45542	.32570	.48301	36
37	.28818	.40485	.30054	.42968	.31312	.45587	.32591	.48349	37
38	.28838	.40525	.30075	.43011	.31334	.45631	.32613	.48396	38
39	.28859	.40565	.30096	.43053	.31355	.45676	.32634	.48443	39
40	.28879	.40606	.30117	.43096	.31376	.45721	.32656	.48491	40
41	.28900	.40646	.30138	.43139	.31397	.45766	.32677	.48538	41
42	.28920	.40687	.30158	.43181	.31418	.45811	.32699	.48586	42
43	.28941	.40727	.30179	.43224	.31439	.45856	.32720	.48633	43
44	.28961	.40768	.30200	.43267	.31461	.45901	.32742	.48681	44
45	.28981	.40808	.30221	.43310	.31482	.45946	.32763	.48728	45
46	.29002	.40849	.30242	.43352	.31503	.45992	.32785	.48776	46
47	.29022	.40890	.30263	.43395	.31524	.46037	.32806	.48824	47
48	.29043	.40930	.30283	.43438	.31545	.46082	.32828	.48871	48
49	.29063	.40971	.30304	.43481	.31567	.46127	.32849	.48919	49
50	.29084	.41012	.30325	.43524	.31588	.46173	.32871	.48967	50
51	.29104	.41053	.30346	.43567	.31609	.46218	.32893	.49015	51
52	.29125	.41093	.30367	.43610	.31630	.46263	.32914	.49063	52
53	.29145	.41134	.30388	.43653	.31651	.46309	.32936	.49111	53
54	.29166	.41175	.30409	.43696	.31673	.46354	.32957	.49159	54
55	.29187	.41216	.30430	.43739	.31694	.46400	.32979	.49207	55
56	.29207	.41257	.30451	.43783	.31715	.46445	.33001	.49255	56
57	.29228	.41298	.30471	.43826	.31736	.46491	.33022	.49303	57
58	.29248	.41339	.30492	.43869	.31758	.46537	.33044	.49351	58
59	.29269	.41380	.30513	.43912	.31779	.46582	.33065	.49399	59
60	.29289	.41421	.30534	.43956	.31800	.46628	.33087	.49448	60

′	48°		49°		50°		51°		′
	Vers.	Exsec.	Vers.	Exsec.	Vers.	Exsec.	Vers.	Exsec.	
0	.33087	.49448	.34394	.52425	.35721	.55572	.37068	.58902	0
1	.33109	.49496	.34416	.52476	.35744	.55626	.37091	.58959	1
2	.33130	.49544	.34438	.52527	.35766	.55680	.37113	.59016	2
3	.33152	.49593	.34460	.52579	.35788	.55734	.37136	.59073	3
4	.33173	.49641	.34482	.52630	.35810	.55789	.37158	.59130	4
5	.33195	.49690	.34504	.52681	.35833	.55843	.37181	.59188	5
6	.33217	.49738	.34526	.52732	.35855	.55897	.37204	.59245	6
7	.33238	.49787	.34548	.52784	.35877	.55951	.37226	.59302	7
8	.33260	.49835	.34570	.52835	.35900	.56005	.37249	.59360	8
9	.33282	.49884	.34592	.52886	.35922	.56060	.37272	.59418	9
10	.33303	.49933	.34614	.52938	.35944	.56114	.37294	.59475	10
11	.33325	.49981	.34636	.52989	.35967	.56169	.37317	.59533	11
12	.33347	.50030	.34658	.53041	.35989	.56223	.37340	.59590	12
13	.33368	.50079	.34680	.53092	.36011	.56278	.37362	.59648	13
14	.33390	.50128	.34702	.53144	.36034	.56332	.37385	.59706	14
15	.33412	.50177	.34724	.53196	.36056	.56387	.37408	.59764	15
16	.33434	.50226	.34746	.53247	.36078	.56442	.37430	.59822	16
17	.33455	.50275	.34768	.53299	.36101	.56497	.37453	.59880	17
18	.33477	.50324	.34790	.53351	.36123	.56551	.37476	.59938	18
19	.33499	.50373	.34812	.53403	.36146	.56606	.37498	.59996	19
20	.33520	.50422	.34834	.53455	.36168	.56661	.37521	.60054	20
21	.33542	.50471	.34856	.53507	.36190	.56716	.37544	.60112	21
22	.33564	.50521	.34878	.53559	.36213	.56771	.37567	.60171	22
23	.33586	.50570	.34900	.53611	.36235	.56826	.37589	.60229	23
24	.33607	.50619	.34923	.53663	.36258	.56881	.37612	.60287	24
25	.33629	.50669	.34945	.53715	.36280	.56937	.37635	.60346	25
26	.33651	.50718	.34967	.53768	.36302	.56992	.37658	.60404	26
27	.33673	.50767	.34989	.53820	.36325	.57047	.37680	.60463	27
28	.33694	.50817	.35011	.53872	.36347	.57103	.37703	.60521	28
29	.33716	.50866	.35033	.53924	.36370	.57158	.37726	.60580	29
30	.33738	.50916	.35055	.53977	.36392	.57213	.37749	.60639	30
31	.33760	.50966	.35077	.54029	.36415	.57269	.37771	.60698	31
32	.33782	.51015	.35099	.54082	.36437	.57324	.37794	.60756	32
33	.33803	.51065	.35122	.54134	.36460	.57380	.37817	.60815	33
34	.33825	.51115	.35144	.54187	.36482	.57436	.37840	.60874	34
35	.33847	.51165	.35166	.54240	.36504	.57491	.37862	.60933	35
36	.33869	.51215	.35188	.54292	.36527	.57547	.37885	.60992	36
37	.33891	.51265	.35210	.54345	.36549	.57603	.37908	.61051	37
38	.33912	.51314	.35232	.54398	.36572	.57659	.37931	.61111	38
39	.33934	.51364	.35254	.54451	.36594	.57715	.37954	.61170	39
40	.33956	.51415	.35277	.54504	.36617	.57771	.37976	.61229	40
41	.33978	.51465	.35299	.54557	.36639	.57827	.37999	.61288	41
42	.34000	.51515	.35321	.54610	.36662	.57883	.38022	.61348	42
43	.34022	.51565	.35343	.54663	.36684	.57939	.38045	.61407	43
44	.34044	.51615	.35365	.54716	.36707	.57995	.38068	.61467	44
45	.34065	.51665	.35388	.54769	.36729	.58051	.38091	.61526	45
46	.34087	.51716	.35410	.54822	.36752	.58108	.38113	.61586	46
47	.34109	.51766	.35432	.54876	.36775	.58164	.38136	.61646	47
48	.34131	.51817	.35454	.54929	.36797	.58221	.38159	.61705	48
49	.34153	.51867	.35476	.54982	.36820	.58277	.38182	.61765	49
50	.34175	.51918	.35499	.55036	.36842	.58333	.38205	.61825	50
51	.34197	.51968	.35521	.55089	.36865	.58390	.38228	.61885	51
52	.34219	.52019	.35543	.55143	.36887	.58447	.38251	.61945	52
53	.34241	.52069	.35565	.55196	.36910	.58503	.38274	.62005	53
54	.34262	.52120	.35588	.55250	.36932	.58560	.38296	.62065	54
55	.34284	.52171	.35610	.55303	.36955	.58617	.38319	.62125	55
56	.34306	.52222	.35632	.55357	.36978	.58674	.38342	.62185	56
57	.34328	.52273	.35654	.55411	.37000	.58731	.38365	.62246	57
58	.34350	.52323	.35677	.55465	.37023	.58788	.38388	.62306	58
59	.34372	.52374	.35699	.55518	.37045	.58845	.38411	.62366	59
60	.34394	.52425	.35721	.55572	.37068	.58902	.38434	.62427	60

′	52° Vers.	52° Exsec.	53° Vers.	53° Exsec.	54° Vers.	54° Exsec.	55° Vers.	55° Exsec.	′
0	.38434	.62427	.39819	.66164	.41221	.70130	.42642	.74345	0
1	.38457	.62487	.39842	.66228	.41245	.70198	.42666	.74417	1
2	.38480	.62548	.39865	.66292	.41269	.70267	.42690	.74490	2
3	.38503	.62609	.39888	.66357	.41292	.70335	.42714	.74562	3
4	.38526	.62669	.39911	.66421	.41316	.70403	.42738	.74635	4
5	.38549	.62730	.39935	.66486	.41339	.70472	.42762	.74708	5
6	.38571	.62791	.39958	.66550	.41363	.70540	.42785	.74781	6
7	.38594	.62852	.39981	.66615	.41386	.70609	.42809	.74854	7
8	.38617	.62913	.40005	.66679	.41410	.70677	.42833	.74927	8
9	.38640	.62974	.40028	.66744	.41433	.70746	.42857	.75000	9
10	.38663	.63035	.40051	.66809	.41457	.70815	.42881	.75073	10
11	.38686	.63096	.40074	.66873	.41481	.70884	.42905	.75146	11
12	.38709	.63157	.40098	.66938	.41504	.70953	.42929	.75219	12
13	.38732	.63218	.40121	.67003	.41528	.71022	.42953	.75293	13
14	.38755	.63279	.40144	.67068	.41551	.71091	.42976	.75366	14
15	.38778	.63341	.40168	.67133	.41575	.71160	.43000	.75440	15
16	.38801	.63402	.40191	.67199	.41599	.71229	.43024	.75513	16
17	.38824	.63464	.40214	.67264	.41622	.71298	.43048	.75587	17
18	.38847	.63525	.40237	.67329	.41646	.71368	.43072	.75661	18
19	.38870	.63587	.40261	.67394	.41670	.71437	.43096	.75734	19
20	.38893	.63648	.40284	.67460	.41693	.71506	.43120	.75808	20
21	.38916	.63710	.40307	.67525	.41717	.71576	.43144	.75882	21
22	.38939	.63772	.40331	.67591	.41740	.71646	.43168	.75956	22
23	.38962	.63834	.40354	.67656	.41764	.71715	.43192	.76031	23
24	.38985	.63895	.40378	.67722	.41788	.71785	.43216	.76105	24
25	.39009	.63957	.40401	.67788	.41811	.71855	.43240	.76179	25
26	.39032	.64019	.40424	.67853	.41835	.71925	.43264	.76253	26
27	.39055	.64081	.40448	.67919	.41859	.71995	.43287	.76328	27
28	.39078	.64144	.40471	.67985	.41882	.72065	.43311	.76402	28
29	.39101	.64206	.40494	.68051	.41906	.72135	.43335	.76477	29
30	.39124	.64268	.40518	.68117	.41930	.72205	.43359	.76552	30
31	.39147	.64330	.40541	.68183	.41953	.72275	.43383	.76626	31
32	.39170	.64393	.40565	.68250	.41977	.72346	.43407	.76701	32
33	.39193	.64455	.40588	.68316	.42001	.72416	.43431	.76776	33
34	.39216	.64518	.40611	.68382	.42024	.72487	.43455	.76851	34
35	.39239	.64580	.40635	.68449	.42048	.72557	.43479	.76926	35
36	.39262	.64643	.40658	.68515	.42072	.72628	.43503	.77001	36
37	.39286	.64705	.40682	.68582	.42096	.72698	.43527	.77077	37
38	.39309	.64768	.40705	.68648	.42119	.72769	.43551	.77152	38
39	.39332	.64831	.40728	.68715	.42143	.72840	.43575	.77227	39
40	.39355	.64894	.40752	.68782	.42167	.72911	.43599	.77303	40
41	.39378	.64957	.40775	.68848	.42191	.72982	.43623	.77378	41
42	.39401	.65020	.40799	.68915	.42214	.73053	.43647	.77454	42
43	.39424	.65083	.40822	.68982	.42238	.73124	.43671	.77530	43
44	.39447	.65146	.40846	.69049	.42262	.73195	.43695	.77606	44
45	.39471	.65209	.40869	.69116	.42285	.73267	.43720	.77681	45
46	.39494	.65272	.40893	.69183	.42309	.73338	.43744	.77757	46
47	.39517	.65336	.40916	.69250	.42333	.73409	.43768	.77833	47
48	.39540	.65399	.40939	.69318	.42357	.73481	.43792	.77910	48
49	.39563	.65462	.40963	.69385	.42381	.73552	.43816	.77986	49
50	.39586	.65526	.40986	.69452	.42404	.73624	.43840	.78062	50
51	.39610	.65589	.41010	.69520	.42428	.73696	.43864	.78138	51
52	.39633	.65653	.41033	.69587	.42452	.73768	.43888	.78215	52
53	.39656	.65717	.41057	.69655	.42476	.73840	.43912	.78291	53
54	.39679	.65780	.41080	.69723	.42499	.73911	.43936	.78368	54
55	.39702	.65844	.41104	.69790	.42523	.73983	.43960	.78445	55
56	.39726	.65908	.41127	.69858	.42547	.74056	.43984	.78521	56
57	.39749	.65972	.41151	.69926	.42571	.74128	.44008	.78598	57
58	.39772	.66036	.41174	.69994	.42595	.74200	.44032	.78675	58
59	.39795	.66100	.41198	.70062	.42619	.74272	.44057	.78752	59
60	.39819	.66164	.41221	.70130	.42642	.74345	.44081	.78829	60

′	56° Vers.	56° Exsec.	57° Vers.	57° Exsec.	58° Vers.	58° Exsec.	59° Vers.	59° Exsec.	′
0	.44081	.78829	.45536	.83608	.47008	.88708	.48496	.94160	0
1	.44105	.78906	.45560	.83690	.47033	.88796	.48521	.94254	1
2	.44129	.78984	.45585	.83773	.47057	.88884	.48546	.94349	2
3	.44153	.79061	.45609	.83855	.47082	.88972	.48571	.94443	3
4	.44177	.79138	.45634	.83938	.47107	.89060	.48596	.94537	4
5	.44201	.79216	.45658	.84020	.47131	.89148	.48621	.94632	5
6	.44225	.79293	.45683	.84103	.47156	.89237	.48646	.94726	6
7	.44250	.79371	.45707	.84186	.47181	.89325	.48671	.94821	7
8	.44274	.79449	.45731	.84269	.47206	.89414	.48696	.94916	8
9	.44298	.79527	.45756	.84352	.47230	.89503	.48721	.95011	9
10	.44322	.79604	.45780	.84435	.47255	.89591	.48746	.95106	10
11	.44346	.79682	.45805	.84518	.47280	.89680	.48771	.95201	11
12	.44370	.79761	.45829	.84601	.47304	.89769	.48796	.95296	12
13	.44395	.79839	.45854	.84685	.47329	.89858	.48821	.95392	13
14	.44419	.79917	.45878	.84768	.47354	.89948	.48846	.95487	14
15	.44443	.79995	.45903	.84852	.47379	.90037	.48871	.95583	15
16	.44467	.80074	.45927	.84935	.47403	.90126	.48896	.95678	16
17	.44491	.80152	.45951	.85019	.47428	.90216	.48921	.95774	17
18	.44516	.80231	.45976	.85103	.47453	.90305	.48946	.95870	18
19	.44540	.80309	.46000	.85187	.47478	.90395	.48971	.95966	19
20	.44564	.80388	.46025	.85271	.47502	.90485	.48996	.96062	20
21	.44588	.80467	.46049	.85355	.47527	.90575	.49021	.96158	21
22	.44612	.80546	.46074	.85439	.47552	.90665	.49046	.96255	22
23	.44637	.80625	.46098	.85523	.47577	.90755	.49071	.96351	23
24	.44661	.80704	.46123	.85608	.47601	.90845	.49096	.96448	24
25	.44685	.80783	.46147	.85692	.47626	.90935	.49121	.96544	25
26	.44709	.80862	.46172	.85777	.47651	.91026	.49146	.96641	26
27	.44734	.80942	.46196	.85861	.47676	.91116	.49171	.96738	27
28	.44758	.81021	.46221	.85946	.47701	.91207	.49196	.96835	28
29	.44782	.81101	.46246	.86031	.47725	.91297	.49221	.96932	29
30	.44806	.81180	.46270	.86116	.47750	.91388	.49246	.97029	30
31	.44831	.81260	.46295	.86201	.47775	.91479	.49271	.97127	31
32	.44855	.81340	.46319	.86286	.47800	.91570	.49296	.97224	32
33	.44879	.81419	.46344	.86371	.47825	.91661	.49321	.97322	33
34	.44903	.81499	.46368	.86457	.47849	.91752	.49346	.97420	34
35	.44928	.81579	.46393	.86542	.47874	.91844	.49372	.97517	35
36	.44952	.81659	.46417	.86627	.47899	.91935	.49397	.97615	36
37	.44976	.81740	.46442	.86713	.47924	.92027	.49422	.97713	37
38	.45001	.81820	.46466	.86799	.47949	.92118	.49447	.97811	38
39	.45025	.81900	.46491	.86885	.47974	.92210	.49472	.97910	39
40	.45049	.81981	.46516	.86990	.47998	.92302	.49497	.98008	40
41	.45073	.82061	.46540	.87056	.48023	.92394	.49522	.98107	41
42	.45098	.82142	.46565	.87142	.48048	.92486	.49547	.98205	42
43	.45122	.82222	.46589	.87229	.48073	.92578	.49572	.98304	43
44	.45146	.82303	.46614	.87315	.48098	.92670	.49597	.98403	44
45	.45171	.82384	.46639	.87401	.48123	.92762	.49623	.98502	45
46	.45195	.82465	.46663	.87488	.48148	.92855	.49648	.98601	46
47	.45219	.82546	.46688	.87574	.48172	.92947	.49673	.98700	47
48	.45244	.82627	.46712	.87661	.48197	.93040	.49698	.98799	48
49	.45268	.82709	.46737	.87748	.48222	.93133	.49723	.98899	49
50	.45292	.82790	.46762	.87834	.48247	.93226	.49748	.98998	50
51	.45317	.82871	.46786	.87921	.48272	.93319	.49773	.99098	51
52	.45341	.82953	.46811	.88008	.48297	.93412	.49799	.99198	52
53	.45365	.83034	.46836	.88095	.48322	.93505	.49824	.99298	53
54	.45390	.83116	.46860	.88183	.48347	.93598	.49849	.99398	54
55	.45414	.83198	.46885	.88270	.48372	.93692	.49874	.99498	55
56	.45439	.83280	.46909	.88357	.48396	.93785	.49899	.99598	56
57	.45463	.83362	.46934	.88445	.48421	.93879	.49924	.99698	57
58	.45487	.83444	.46959	.88532	.48446	.93973	.49950	.99799	58
59	.45512	.83526	.46983	.88620	.48471	.94066	.49975	.99899	59
60	.45536	.83608	.47008	.88708	.48496	.94160	.50000	1.00000	60

′	60° Vers.	60° Exsec.	61° Vers.	61° Exsec.	62° Vers.	62° Exsec.	63° Vers.	63° Exsec.	′
0	.50000	1.00000	.51519	1.06267	.53053	1.13005	.54601	1.20269	0
1	.50025	1.00101	.51544	1.06375	.53079	1.13122	.54627	1.20395	1
2	.50050	1.00202	.51570	1.06483	.53104	1.13239	.54653	1.20521	2
3	.50076	1.00303	.51595	1.06592	.53130	1.13356	.54679	1.20647	3
4	.50101	1.00404	.51621	1.06701	.53156	1.13473	.54705	1.20773	4
5	.50126	1.00505	.51646	1.06809	.53181	1.13590	.54731	1.20900	5
6	.50151	1.00607	.51672	1.06918	.53207	1.13707	.54757	1.21026	6
7	.50176	1.00708	.51697	1.07027	.53233	1.13825	.54782	1.21153	7
8	.50202	1.00810	.51723	1.07137	.53258	1.13942	.54808	1.21280	8
9	.50227	1.00912	.51748	1.07246	.53284	1.14060	.54834	1.21407	9
10	.50252	1.01014	.51774	1.07356	.53310	1.14178	.54860	1.21535	10
11	.50277	1.01116	.51799	1.07465	.53336	1.14296	.54886	1.21662	11
12	.50303	1.01218	.51825	1.07575	.53361	1.14414	.54912	1.21790	12
13	.50328	1.01320	.51850	1.07685	.53387	1.14533	.54938	1.21918	13
14	.50353	1.01422	.51876	1.07795	.53413	1.14651	.54964	1.22045	14
15	.50378	1.01525	.51901	1.07905	.53439	1.14770	.54990	1.22174	15
16	.50404	1.01628	.51927	1.08015	.53464	1.14889	.55016	1.22302	16
17	.50429	1.01730	.51952	1.08126	.53490	1.15008	.55042	1.22430	17
18	.50454	1.01833	.51978	1.08236	.53516	1.15127	.55068	1.22559	18
19	.50479	1.01936	.52003	1.08347	.53542	1.15246	.55094	1.22688	19
20	.50505	1.02039	.52029	1.08458	.53567	1.15366	.55120	1.22817	20
21	.50530	1.02143	.52054	1.08569	.53593	1.15485	.55146	1.22946	21
22	.50555	1.02246	.52080	1.08680	.53619	1.15605	.55172	1.23075	22
23	.50581	1.02349	.52105	1.08791	.53645	1.15725	.55198	1.23205	23
24	.50606	1.02453	.52131	1.08903	.53670	1.15845	.55224	1.23334	24
25	.50631	1.02557	.52156	1.09014	.53696	1.15965	.55250	1.23464	25
26	.50656	1.02661	.52182	1.09126	.53722	1.16085	.55276	1.23594	26
27	.50682	1.02765	.52207	1.09238	.53748	1.16206	.55302	1.23724	27
28	.50707	1.02869	.52233	1.09350	.53774	1.16326	.55328	1.23855	28
29	.50732	1.02973	.52259	1.09462	.53799	1.16447	.55354	1.23985	29
30	.50758	1.03077	.52284	1.09574	.53825	1.16568	.55380	1.24116	30
31	.50783	1.03182	.52310	1.09686	.53851	1.16689	.55406	1.24247	31
32	.50808	1.03286	.52335	1.09799	.53877	1.16810	.55432	1.24378	32
33	.50834	1.03391	.52361	1.09911	.53903	1.16932	.55458	1.24509	33
34	.50859	1.03496	.52386	1.10024	.53928	1.17053	.55484	1.24640	34
35	.50884	1.03601	.52412	1.10137	.53954	1.17175	.55510	1.24772	35
36	.50910	1.03706	.52438	1.10250	.53980	1.17297	.55536	1.24903	36
37	.50935	1.03811	.52463	1.10363	.54006	1.17419	.55563	1.25035	37
38	.50960	1.03916	.52489	1.10477	.54032	1.17541	.55589	1 25167	38
39	.50986	1.04022	.52514	1.10590	.54058	1.17663	.55615	1.25300	39
40	.51011	1.04128	.52540	1.10704	.54083	1.17786	.55641	1.25432	40
41	.51036	1.04233	.52566	1.10817	.54109	1.17909	.55667	1.25565	41
42	.51062	1.04339	.52591	1.10931	.54135	1.18031	.55693	1.25697	42
43	.51087	1.04445	.52617	1.11045	.54161	1.18154	.55719	1.25830	43
44	.51113	1.04551	.52642	1.11159	.54187	1.18277	.55745	1.25963	44
45	.51138	1.04658	.52668	1.11274	.54213	1.18401	.55771	1.26097	45
46	.51163	1.04764	.52694	1.11388	.54238	1.18524	.55797	1.26230	46
47	.51189	1.04870	.52719	1.11503	.54264	1.18648	.55823	1.26364	47
48	.51214	1.04977	.52745	1.11617	.54290	1.18772	.55849	1.26498	48
49	.51239	1.05084	.52771	1.11732	.54316	1.18895	.55876	1.26632	49
50	.51265	1.05191	.52796	1.11847	.54342	1.19019	.55902	1.26766	50
51	.51290	1.05298	.52822	1.11963	.54368	1.19144	.55928	1.26900	51
52	.51316	1.05405	.52848	1.12078	.54394	1.19268	.55954	1.27035	52
53	.51341	1.05512	.52873	1.12193	.54420	1.19393	.55980	1.27169	53
54	.51366	1.05619	.52899	1.12309	.54446	1.19517	.56006	1.27304	54
55	.51392	1.05727	.52924	1.12425	.54471	1.19642	.56032	1.27439	55
56	.51417	1.05835	.52950	1.12540	.54497	1.19767	.56058	1.27574	56
57	.51443	1.05942	.52976	1.12657	.54523	1.19892	.56084	1.27710	57
58	.51468	1.06050	.53001	1.12773	.54549	1.20018	.56111	1.27845	58
59	.51494	1.06158	.53027	1.12889	.54575	1.20143	.56137	1.27981	59
60	.51519	1.06267	.53053	1.13005	.54601	1.20269	.56163	1.28117	60

′	64° Vers.	64° Exsec.	65° Vers.	65° Exsec.	66° Vers.	66° Exsec.	67° Vers.	67° Exsec.	′
0	.56163	1.28117	.57738	1.36620	.59326	1.45859	.60927	1.55930	0
1	.56189	1.28253	.57765	1.36768	.59353	1.46020	.60954	1.56106	1
2	.56215	1.28390	.57791	1.36916	.59379	1.46181	.60980	1.56282	2
3	.56241	1.28526	.57817	1.37064	.59406	1.46342	.61007	1.56458	3
4	.56267	1.28663	.57844	1.37212	.59433	1.46504	.61034	1.56634	4
5	.56294	1.28800	.57870	1.37361	.59459	1.46665	.61061	1.56811	5
6	.56320	1.28937	.57896	1.37509	.59486	1.46827	.61088	1.56988	6
7	.56346	1.29074	.57923	1.37658	.59512	1.46989	.61114	1.57165	7
8	.56372	1.29211	.57949	1.37808	.59539	1.47152	.61141	1.57342	8
9	.56398	1.29349	.57976	1.37957	.59566	1.47314	.61168	1.57520	9
10	.56425	1.29487	.58002	1.38107	.59592	1.47477	.61195	1.57698	10
11	.56451	1.29625	.58028	1.38256	.59619	1.47640	.61222	1.57876	11
12	.56477	1.29763	.58055	1.38406	.59645	1.47804	.61248	1.58054	12
13	.56503	1.29901	.58081	1.38556	.59672	1.47967	.61275	1.58233	13
14	.56529	1.30040	.58108	1.38707	.59699	1.48131	.61302	1.58412	14
15	.56555	1.30179	.58134	1.38857	.59725	1.48295	.61329	1.58591	15
16	.56582	1.30318	.58160	1.39008	.59752	1.48459	.61356	1.58771	16
17	.56608	1.30457	.58187	1.39159	.59779	1.48624	.61383	1.58950	17
18	.56634	1.30596	.58213	1.39311	.59805	1.48789	.61409	1.59130	18
19	.56660	1.30735	.58240	1.39462	.59832	1.48954	.61436	1.59311	19
20	.56687	1.30875	.58266	1.39614	.59859	1.49119	.61463	1.59491	20
21	.56713	1.31015	.58293	1.39766	.59885	1.49284	.61490	1.59672	21
22	.56739	1.31155	.58319	1.39918	.59912	1.49450	.61517	1.59853	22
23	.56765	1.31295	.58345	1.40070	.59938	1.49616	.61544	1.60035	23
24	.56791	1.31436	.58372	1.40222	.59965	1.49782	.61570	1.60217	24
25	.56818	1.31576	.58398	1.40375	.59992	1.49948	.61597	1.60399	25
26	.56844	1.31717	.58425	1.40528	.60018	1.50115	.61624	1.60581	26
27	.56870	1.31858	.58451	1.40681	.60045	1.50282	.61651	1.60763	27
28	.56896	1.31999	.58478	1.40835	.60072	1.50449	.61678	1.60946	28
29	.56923	1.32140	.58504	1.40988	.60098	1.50617	.61705	1.61129	29
30	.56949	1.32282	.58531	1.41142	.60125	1.50784	.61732	1.61313	30
31	.56975	1.32424	.58557	1.41296	.60152	1.50952	.61759	1.61496	31
32	.57001	1.32566	.58584	1.41450	.60178	1.51120	.61785	1.61680	32
33	.57028	1.32708	.58610	1.41605	.60205	1.51289	.61812	1.61864	33
34	.57054	1.32850	.58637	1.41760	.60232	1.51457	.61839	1.62049	34
35	.57080	1.32993	.58663	1.41914	.60259	1.51626	.61866	1.62234	35
36	.57106	1.33135	.58690	1.42070	.60285	1.51795	.61893	1.62419	36
37	.57133	1.33278	.58716	1.42225	.60312	1.51965	.61920	1.62604	37
38	.57159	1.33422	.58743	1.42380	.60339	1.52134	.61947	1.62790	38
39	.57185	1.33565	.58769	1.42536	.60365	1.52304	.61974	1.62976	39
40	.57212	1.33708	.58796	1.42692	.60392	1.52474	.62001	1.63162	40
41	.57238	1.33852	.58822	1.42848	.60419	1.52645	.62027	1.63348	41
42	.57264	1.33996	.58849	1.43005	.60445	1.52815	.62054	1.63535	42
43	.57291	1.34140	.58875	1.43162	.60472	1.52986	.62081	1.63722	43
44	.57317	1.34284	.58902	1.43318	.60499	1.53157	.62108	1.63909	44
45	.57343	1.34429	.58928	1.43476	.60526	1.53329	.62135	1.64097	45
46	.57369	1.34573	.58955	1.43633	.60552	1.53500	.62162	1.64285	46
47	.57396	1.34718	.58981	1.43790	.60579	1.53672	.62189	1.64473	47
48	.57422	1.34863	.59008	1.43948	.60606	1.53845	.62216	1.64662	48
49	.57448	1.35009	.59034	1.44106	.60633	1.54017	.62243	1.64851	49
50	.57475	1.35154	.59061	1.44264	.60659	1.54190	.62270	1.65040	50
51	.57501	1.35300	.59087	1.44423	.60686	1.54363	.62297	1.65229	51
52	.57527	1.35446	.59114	1.44582	.60713	1.54536	.62324	1.65419	52
53	.57554	1.35592	.59140	1.44741	.60740	1.54709	.62351	1.65609	53
54	.57580	1.35738	.59167	1.44900	.60766	1.54883	.62378	1.65799	54
55	.57606	1.35885	.59194	1.45059	.60793	1.55057	.62405	1.65989	55
56	.57633	1.36031	.59220	1.45219	.60820	1.55231	.62431	1.66180	56
57	.57659	1.36178	.59247	1.45378	.60847	1.55405	.62458	1.66371	57
58	.57685	1.36325	.59273	1.45539	.60873	1.55580	.62485	1.66563	58
59	.57712	1.36473	.59300	1.45699	.60900	1.55755	.62512	1.66755	59
60	.57738	1.36620	.59326	1.45859	.60927	1.55930	.62539	1.66947	60

′	68°		69°		70°		71°		′
	Vers.	Exsec.	Vers.	Exsec.	Vers.	Exsec.	Vers.	Exsec.	
0	.62539	1.66947	.64163	1.79043	.65798	1.92380	.67443	2.07155	0
1	.62566	1.67139	.64190	1.79254	.65825	1.92614	.67471	2.07415	1
2	.62593	1.67332	.64218	1.79466	.65853	1.92849	.67498	2.07675	2
3	.62620	1.67525	.64245	1.79679	.65880	1.93083	.67526	2.07936	3
4	.62647	1.67718	.64272	1.79891	.65907	1.93318	.67553	2.08197	4
5	.62674	1.67911	.64299	1.80104	.65935	1.93554	.67581	2.08459	5
6	.62701	1.68105	.64326	1.80318	.65962	1.93790	.67608	2.08721	6
7	.62728	1.68299	.64353	1.80531	.65989	1.94026	.67636	2.08983	7
8	.62755	1.68494	.64381	1.80746	.66017	1.94263	.67663	2.09246	8
9	.62782	1.68689	.64408	1.80960	.66044	1.94500	.67691	2.09510	9
10	.62809	1.68884	.64435	1.81175	.66071	1.94737	.67718	2.09774	10
11	.62836	1 69079	.64462	1.81390	.66099	1.94975	.67746	2.10038	11
12	.62863	1.69275	.64489	1.81605	.66126	1.95213	.67773	2.10303	12
13	.62890	1.69471	.64517	1.81821	.66154	1.95452	.67801	2.10568	13
14	.62917	1.69667	.64544	1.82037	.66181	1.95691	.67829	2.10834	14
15	.62944	1.69864	.64571	1.82254	.66208	1.95931	.67856	2.11101	15
16	.62971	1.70061	.64598	1.82471	.66236	1.96171	.67884	2.11367	16
17	.62998	1.70258	.64625	1.82688	.66263	1.96411	.67911	2.11635	17
18	.63025	1.70455	.64653	1.82906	.66290	1.96652	.67939	2.11903	18
19	.63052	1.70653	.64680	1.83124	.66318	1.96893	.67966	2.12171	19
20	.63079	1.70851	.64707	1.83342	.66345	1.97135	.67994	2.12440	20
21	.63106	1.71050	.64734	1.83561	.66373	1.97377	.68021	2.12709	21
22	.63133	1.71249	.64761	1.83780	.66400	1.97619	.68049	2.12979	22
23	.63161	1.71448	.64789	1.83999	.66427	1.97862	.68077	2.13249	23
24	.63188	1.71647	.64816	1.84219	.66455	1.98106	.68104	2.13520	24
25	.63215	1.71847	.64843	1.84439	.66482	1.98349	.68132	2.13791	25
26	.63242	1.72047	.64870	1.84659	.66510	1.98594	.68159	2.14063	26
27	.63269	1.72247	.64898	1.84880	.66537	1.98838	.68187	2.14335	27
28	.63296	1.72448	.64925	1.85102	.66564	1.99083	.68214	2.14608	28
29	.63323	1.72649	.64952	1.85323	.66592	1.99329	.68242	2.14881	29
30	.63350	1.72850	.64979	1.85545	.66619	1.99574	.68270	2.15155	30
31	.63377	1.73052	.65007	1.85767	.66647	1.99821	.68297	2.15429	31
32	.63404	1.73254	.65034	1.85990	.66674	2.00067	.68325	2.15704	32
33	.63431	1.73456	.65061	1.86213	.66702	2.00315	.68352	2.15979	33
34	.63458	1.73659	.65088	1.86437	.66729	2.00562	.68380	2.16255	34
35	.63485	1.73862	.65116	1.86661	.66756	2.00810	.68408	2.16531	35
36	.63512	1.74065	.65143	1.86885	.66784	2.01059	.68435	2.16808	36
37	.63539	1.74269	.65170	1.87109	.66811	2.01308	.68463	2.17085	37
38	.63566	1.74473	.65197	1.87334	.66839	2.01557	.68490	2.17363	38
39	.63594	1.74677	.65225	1.87560	.66866	2.01807	.68518	2.17641	39
40	.63621	1.74881	.65252	1.87785	.66894	2.02057	.68546	2.17920	40
41	.63648	1.75086	.65279	1.88011	.66921	2.02308	.68573	2.18199	41
42	.63675	1.75292	.65306	1.88238	.66949	2.02559	.68601	2.18479	42
43	.63702	1.75497	.65334	1.88465	.66976	2.02810	.68628	2.18759	43
44	.63729	1.75703	.65361	1.88692	.67003	2.03062	.68656	2.19040	44
45	.63756	1.75909	.65388	1.88920	.67031	2.03315	.68684	2.19322	45
46	.63783	1.76116	.65416	1.89148	.67058	2.03568	.68711	2.19604	46
47	.63810	1.76323	.65443	1.89376	.67086	2.03821	.68739	2.19886	47
48	.63838	1.76530	.65470	1.89605	.67113	2.04075	.68767	2.20169	48
49	.63865	1.76737	.65497	1.89834	.67141	2.04329	.68794	2.20453	49
50	.63892	1.76945	.65525	1.90063	.67168	2.04584	.68822	2.20737	50
51	.63919	1.77154	.65552	1.90293	.67196	2.04839	.68849	2.21021	51
52	.63946	1.77362	.65579	1.90524	.67223	2.05094	.68877	2.21306	52
53	.63973	1.77571	.65607	1.90754	.67251	2.05350	.68905	2.21592	53
54	.64000	1 77780	.65634	1.90986	.67278	2.05607	.68932	2.21878	54
55	.64027	1.77990	.65661	1.91217	.67306	2.05864	.68960	2.22165	55
56	.64055	1.78200	.65689	1.91449	.67333	2.06121	.68988	2.22452	56
57	.64082	1.78410	.65716	1.91681	.67361	2.06379	.69015	2.22740	57
58	.64109	1.78621	.65743	1.91914	.67388	2.06637	.69043	2.23028	58
59	.64136	1.78832	.65771	1.92147	.67416	2.06896	.69071	2.23317	59
60	.64163	1.79043	.65798	1.92380	.67443	2.07155	.69098	2.23607	60

′	72°		73°		74°		75°		′
	Vers.	Exsec.	Vers.	Exsec.	Vers.	Exsec.	Vers.	Exsec.	
0	.69098	2.23607	.70763	2.42030	.72436	2.62796	.74118	2.86370	0
1	.69126	2.23897	.70791	2.42356	.72464	2.63164	.74146	2.86790	1
2	.69154	2.24187	.70818	2.42683	.72492	2.63533	.74174	2.87211	2
3	.69181	2.24478	.70846	2.43010	.72520	2.63903	.74202	2.87633	3
4	.69209	2.24770	.70874	2.43337	.72548	2.64274	.74231	2.88056	4
5	.69237	2.25062	.70902	2.43666	.72576	2.64645	.74259	2.88479	5
6	.69264	2.25355	.70930	2.43995	.72604	2.65018	.74287	2.88904	6
7	.69292	2.25648	.70958	2.44324	.72632	2.65391	.74315	2.89330	7
8	.69320	2.25942	.70985	2.44655	.72660	2.65765	.74343	2.89756	8
9	.69347	2 26237	.71013	2.44986	.72688	2.66140	.74371	2.90184	9
10	.69375	2.26531	.71041	2.45317	.72716	2.66515	.74399	2.90613	10
11	.69403	2.26827	.71069	2.45650	.72744	2.66892	.74427	2.91042	11
12	.69430	2.27123	.71097	2.45983	.72772	2.67269	.74455	2.91473	12
13	.69458	2.27420	.71125	2.46316	.72800	2.67647	.74484	2.91904	13
14	.69486	2.27717	.71153	2.46651	.72828	2.68025	.74512	2.92337	14
15	.69514	2.28015	.71180	2.46986	.72856	2.68405	.74540	2.92770	15
16	.69541	2.28313	.71208	2.47321	.72884	2.68785	.74568	2.93204	16
17	.69569	2.28612	.71236	2.47658	.72912	2.69167	.74596	2.93640	17
18	.69597	2.28912	.71264	2.47995	.72940	2.69549	.74624	2.94076	18
19	.69624	2.29212	.71292	2.48333	.72968	2.69931	.74652	2.94514	19
20	.69652	2.29512	.71320	2.48671	.72996	2.70315	.74680	2.94952	20
21	.69680	2.29814	.71348	2.49010	.73024	2.70700	.74709	2.95392	21
22	.69708	2.30115	.71375	2.49350	.73052	2.71085	.74737	2.95832	22
23	.69735	2.30418	.71403	2.49691	.73080	2.71471	.74765	2.96274	23
24	.69763	2.30721	.71431	2.50032	.73108	2.71858	.74793	2.96716	24
25	.69791	2.31024	.71459	2.50374	.73136	2.72246	.74821	2.97160	25
26	.69818	2.31328	.71487	2.50716	.73164	2.72635	.74849	2.97604	26
27	.69846	2.31633	.71515	2.51060	.73192	2.73024	.74878	2.98050	27
28	.69874	2.31939	.71543	2.51404	.73220	2.73414	.74906	2.98497	28
29	.69902	2.32244	.71571	2.51748	.73248	2.73806	.74934	2.98944	29
30	.69929	2.32551	.71598	2.52094	.73276	2.74198	.74962	2.99393	30
31	.69957	2.32858	.71626	2.52440	.73304	2.74591	.74990	2.99843	31
32	.69985	2.33166	.71654	2.52787	.73332	2.74984	.75018	3.00293	32
33	.70013	2.33474	.71682	2.53134	.73360	2.75379	.75047	3.00745	33
34	.70040	2.33783	.71710	2.53482	.73388	2.75775	.75075	3.01198	34
35	.70068	2.34092	.71738	2.53831	.73416	2.76171	.75103	3.01652	35
36	.70096	2.34403	.71766	2.54181	.73444	2.76568	.75131	3.02107	36
37	.70124	2.34713	.71794	2.54531	.73472	2.76966	.75159	3.02563	37
38	.70151	2.35025	.71822	2.54883	.73500	2.77365	.75187	3.03020	38
39	.70179	2.35336	.71850	2.55235	.73529	2.77765	.75216	3.03479	39
40	.70207	2.35649	.71877	2.55587	.73557	2.78166	.75244	3.03938	40
41	.70235	2.35962	.71905	2.55940	.73585	2.78568	.75272	3.04398	41
42	.70263	2.36276	.71933	2.56294	.73613	2.78970	.75300	3.04860	42
43	.70290	2.36590	.71961	2.56649	.73641	2.79374	.75328	3.05322	43
44	.70318	2.36905	.71989	2.57005	.73669	2.79778	.75356	3.05786	44
45	.70346	2.37221	.72017	2.57361	.73697	2.80183	.75385	3.06251	45
46	.70374	2.37537	.72045	2.57718	.73725	2.80589	.75413	3.06717	46
47	.70401	2.37854	.72073	2.58076	.73753	2.80996	.75441	3.07184	47
48	.70429	2.38171	.72101	2.58434	.73781	2.81404	.75469	3.07652	48
49	.70457	2.38489	.72129	2.58794	.73809	2.81813	.75497	3.08121	49
50	.70485	2.38808	.72157	2.59154	.73837	2.82223	.75526	3.08591	50
51	.70513	2.39128	.72185	2.59514	.73865	2.82633	.75554	3.09063	51
52	.70540	2.39448	.72213	2.59876	.73893	2.83045	.75582	3.09535	52
53	.70568	2.39768	.72241	2.60238	.73921	2.83457	.75610	3.10009	53
54	.70596	2.40089	.72269	2.60601	.73950	2.83871	.75639	3.10484	54
55	.70624	2.40411	.72296	2.60965	.73978	2.84285	.75667	3.10960	55
56	.70652	2.40734	.72324	2.61330	.74006	2.84700	.75695	3.11437	56
57	.70679	2.41057	.72352	2.61695	.74034	2.85116	.75723	3.11915	57
58	.70707	2.41381	.72380	2.62061	.74062	2.85533	.75751	3.12394	58
59	.70735	2.41705	.72408	2.62428	.74090	2.85951	.75780	3.12875	59
60	.70763	2.42030	.72436	2.62796	.74118	2.86370	.75808	3.13357	60

′	76°		77°		78°		79°		′
	Vers.	Exsec.	Vers.	Exsec.	Vers.	Exsec.	Vers.	Exsec.	
0	.75808	3.13357	.77505	3.44541	.79209	3.80973	.80919	4.24084	0
1	.75836	3.13839	.77533	3.45102	.79237	3.81633	.80948	4.24870	1
2	.75864	3.14323	.77562	3.45664	.79266	3.82294	.80976	4.25658	2
3	.75892	3.14809	.77590	3.46228	.79294	3.82956	.81005	4.26448	3
4	.75921	3.15295	.77618	3.46793	.79323	3.83621	.81033	4.27241	4
5	.75949	3.15782	.77647	3.47360	.79351	3.84288	.81062	4.28036	5
6	.75977	3.16271	.77675	3.47928	.79380	3.84956	.81090	4.28833	6
7	.76005	3.16761	.77703	3.48498	.79408	3.85627	.81119	4.29634	7
8	.76034	3.17252	.77732	3.49069	.79437	3.86299	.81148	4.30436	8
9	.76062	3.17744	.77760	3.49642	.79465	3.86973	.81176	4.31241	9
10	.76090	3.18238	.77788	3.50216	.79493	3.87649	.81205	4.32049	10
11	.76118	3.18733	.77817	3.50791	.79522	3.88327	.81233	4.32859	11
12	.76147	3.19228	.77845	3.51368	.79550	3.89007	.81262	4.33671	12
13	.76175	3.19725	.77874	3.51947	.79579	3.89689	.81290	4.34486	13
14	.76203	3.20224	.77902	3.52527	.79607	3.90373	.81319	4.35304	14
15	.76231	3.20723	.77930	3.53109	.79636	3.91058	.81348	4.36124	15
16	.76260	3.21224	.77959	3.53692	.79664	3.91746	.81376	4.36947	16
17	.76288	3.21726	.77987	3.54277	.79693	3.92436	.81405	4.37772	17
18	.76316	3.22229	.78015	3.54863	.79721	3.93128	.81433	4.38600	18
19	.76344	3.22734	.78044	3.55451	.79750	3.93821	.81462	4.39430	19
20	.76373	3.23239	.78072	3.56041	.79778	3.94517	.81491	4.40263	20
21	.76401	3.23746	.78101	3.56632	.79807	3.95215	.81519	4.41099	21
22	.76429	3.24255	.78129	3.57224	.79835	3.95914	.81548	4.41937	22
23	.76458	3.24764	.78157	3.57819	.79864	3.96616	.81576	4.42778	23
24	.76486	3.25275	.78186	3.58414	.79892	3.97320	.81605	4.43622	24
25	.76514	3.25787	.78214	3.59012	.79921	3.98025	.81633	4.44468	25
26	.76542	3.26300	.78242	3.59611	.79949	3.98733	.81662	4.45317	26
27	.76571	3.26814	.78271	3.60211	.79978	3.99443	.81691	4.46169	27
28	.76599	3.27330	.78299	3.60813	.80006	4.00155	.81719	4.47023	28
29	.76627	3.27847	.78328	3.61417	.80035	4.00869	.81748	4.47881	29
30	.76655	3.28366	.78356	3.62023	.80063	4.01585	.81776	4.48740	30
31	.76684	3.28885	.78384	3.62630	.80092	4.02303	.81805	4.49603	31
32	.76712	3.29406	.78413	3.63238	.80120	4.03024	.81834	4.50468	32
33	.76740	3.29929	.78441	3.63849	.80149	4.03746	.81862	4.51337	33
34	.76769	3.30452	.78470	3.64461	.80177	4.04471	.81891	4.52208	34
35	.76797	3.30977	.78498	3.65074	.80206	4.05197	.81919	4.53081	35
36	.76825	3.31503	.78526	3.65690	.80234	4.05926	.81948	4.53958	36
37	.76854	3.32031	.78555	3.66307	.80263	4.06657	.81977	4.54837	37
38	.76882	3.32560	.78583	3.66925	.80291	4.07390	.82005	4.55720	38
39	.76910	3.33090	.78612	3.67545	.80320	4.08125	.82034	4.56605	39
40	.76938	3.33622	.78640	3.68167	.80348	4.08863	.82063	4.57493	40
41	.76967	3.34154	.78669	3.68791	.80377	4.09602	.82091	4.58383	41
42	.76995	3.34689	.78697	3.69417	.80405	4.10344	.82120	4.59277	42
43	.77023	3.35224	.78725	3.70044	.80434	4.11088	.82148	4.60174	43
44	.77052	3.35761	.78754	3.70673	.80462	4.11835	.82177	4.61073	44
45	.77080	3.36299	.78782	3.71303	.80491	4.12583	.82206	4.61976	45
46	.77108	3.36839	.78811	3.71935	.80520	4.13334	.82234	4.62881	46
47	.77137	3.37380	.78839	3.72569	.80548	4.14087	.82263	4.63790	47
48	.77165	3.37923	.78868	3.73205	.80577	4.14842	.82292	4.64701	48
49	.77193	3.38466	.78896	3.73843	.80605	4.15599	.82320	4.65616	49
50	.77222	3.39012	.78924	3.74482	.80634	4.16359	.82349	4.66533	50
51	.77250	3.39558	.78953	3.75123	.80662	4.17121	.82377	4.67454	51
52	.77278	3.40106	.78981	3.75766	.80691	4.17886	.82406	4.68377	52
53	.77307	3.40656	.79010	3.76411	.80719	4.18652	.82435	4.69304	53
54	.77335	3.41206	.79038	3.77057	.80748	4.19421	.82463	4.70234	54
55	.77363	3.41759	.79067	3.77705	.80776	4.20193	.82492	4.71166	55
56	.77392	3.42312	.79095	3.78355	.80805	4.20966	.82521	4.72102	56
57	.77420	3.42867	.79123	3.79007	.80833	4.21742	.82549	4.73041	57
58	.77448	3.43424	.79152	3.79661	.80862	4.22521	.82578	4.73983	58
59	.77477	3.43982	.79180	3.80316	.80891	4.23301	.82607	4.74929	59
60	.77505	3.44541	.79209	3.80973	.80919	4.24084	.82635	4.75877	60

′	80° Vers.	80° Exsec.	81° Vers.	81° Exsec.	82° Vers.	82° Exsec.	83° Vers.	83° Exsec.	′
0	.82635	4.75877	.84357	5.39245	.86083	6.18530	.87813	7.20551	0
1	.82664	4.76829	.84385	5.40422	.86112	6.20020	.87842	7.22500	1
2	.82692	4.77784	.84414	5.41602	.86140	6.21517	.87871	7.24457	2
3	.82721	4.78742	.84443	5.42787	.86169	6.23019	.87900	7.26425	3
4	.82750	4.79703	.84471	5.43977	.86198	6.24529	.87929	7.28402	4
5	.82778	4.80667	.84500	5.45171	.86227	6.26044	.87957	7.30388	5
6	.82807	4.81635	.84529	5.46369	.86256	6.27566	.87986	7.32384	6
7	.82836	4.82606	.84558	5.47572	.86284	6.29095	.88015	7.34390	7
8	.82864	4.83581	.84586	5.48779	.86313	6.30630	.88044	7.36405	8
9	.82893	4.84558	.84615	5.49991	.86342	6.32171	.88073	7.38431	9
10	.82922	4.85539	.84644	5.51208	.86371	6.33719	.88102	7.40466	10
11	.82950	4.86524	.84673	5.52429	.86400	6.35274	.88131	7.42511	11
12	.82979	4.87511	.84701	5.53655	.86428	6.36835	.88160	7.44566	12
13	.83008	4.88502	.84730	5.54886	.86457	6.38403	.88188	7.46632	13
14	.83036	4.89497	.84759	5.56121	.86486	6.39978	.88217	7.48707	14
15	.83065	4.90495	.84788	5.57361	.86515	6.41560	.88246	7.50793	15
16	.83094	4.91496	.84816	5.58606	.86544	6.43148	.88275	7.52889	16
17	.83122	4.92501	.84845	5.59855	.86573	6.44743	.88304	7.54996	17
18	.83151	4.93509	.84874	5.61110	.86601	6.46346	.88333	7.57113	18
19	.83180	4.94521	.84903	5.62369	.86630	6.47955	.88362	7.59241	19
20	.83208	4.95536	.84931	5.63633	.86659	6.49571	.88391	7.61379	20
21	.83237	4.96555	.84960	5.64902	.86688	6.51194	.88420	7.63528	21
22	.83266	4.97577	.84989	5.66176	.86717	6.52825	.88448	7.65688	22
23	.83294	4.98603	.85018	5.67454	.86746	6.54462	.88477	7.67859	23
24	.83323	4.99633	.85046	5.68738	.86774	6.56107	.88506	7.70041	24
25	.83352	5.00666	.85075	5.70027	.86803	6.57759	.88535	7.72234	25
26	.83380	5.01703	.85104	5.71321	.86832	6.59418	.88564	7.74438	26
27	.83409	5.02743	.85133	5.72620	.86861	6.61085	.88593	7.76653	27
28	.83438	5.03787	.85162	5.73924	.86890	6.62759	.88622	7.78880	28
29	.83467	5.04834	.85190	5.75233	.86919	6.64441	.88651	7.81118	29
30	.83495	5.05886	.85219	5.76547	.86947	6.66130	.88680	7.83367	30
31	.83524	5.06941	.85248	5.77866	.86976	6.67826	.88709	7.85628	31
32	.83553	5.08000	.85277	5.79191	.87005	6.69530	.88737	7.87901	32
33	.83581	5.09062	.85305	5.80521	.87034	6.71242	.88766	7.90186	33
34	.83610	5.10129	.85334	5.81856	.87063	6.72962	.88795	7.92482	34
35	.83639	5.11199	.85363	5.83196	.87092	6.74689	.88824	7.94791	35
36	.83667	5.12273	.85392	5.84542	.87120	6.76424	.88853	7.97111	36
37	.83696	5.13350	.85420	5.85893	.87149	6.78167	.88882	7.99444	37
38	.83725	5.14432	.85449	5.87250	.87178	6.79918	.88911	8.01788	38
39	.83754	5.15517	.85478	5.88612	.87207	6.81677	.88940	8.04146	39
40	.83782	5.16607	.85507	5.89979	.87236	6.83443	.88969	8.06515	40
41	.83811	5.17700	.85536	5.91352	.87265	6.85218	.88998	8.08897	41
42	.83840	5.18797	.85564	5.92731	.87294	6.87001	.89027	8.11292	42
43	.83868	5.19898	.85593	5.94115	.87322	6.88792	.89055	8.13699	43
44	.83897	5.21004	.85622	5.95505	.87351	6.90592	.89084	8.16120	44
45	.83926	5.22113	.85651	5.96900	.87380	6.92400	.89113	8.18553	45
46	.83954	5.23226	.85680	5.98301	.87409	6.94216	.89142	8.20999	46
47	.83983	5.24343	.85708	5.99708	.87438	6.96040	.89171	8.23459	47
48	.84012	5.25464	.85737	6.01120	.87467	6.97873	.89200	8.25931	48
49	.84041	5.26590	.85766	6.02538	.87496	6.99714	.89229	8.28417	49
50	.84069	5.27719	.85795	6.03962	.87524	7.01565	.89258	8.30917	50
51	.84098	5.28853	.85823	6.05392	.87553	7.03423	.89287	8.33430	51
52	.84127	5.29991	.85852	6.06828	.87582	7.05291	.89316	8.35957	52
53	.84155	5.31133	.85881	6.08269	.87611	7.07167	.89345	8.38497	53
54	.84184	5.32279	.85910	6.09717	.87640	7.09052	.89374	8.41052	54
55	.84213	5.33429	.85939	6.11171	.87669	7.10946	.89403	8.43620	55
56	.84242	5.34584	.85967	6.12630	.87698	7.12849	.89431	8.46203	56
57	.84270	5.35743	.85996	6.14096	.87726	7.14760	.89460	8.48800	57
58	.84299	5.36906	.86025	6.15568	.87755	7.16681	.89489	8.51411	58
59	.84328	5.38073	.86054	6.17046	.87784	7.18612	.89518	8.54037	59
60	.84357	5.39245	.86083	6.18530	.87813	7.20551	.89547	8.56677	60

'	84°		85°		86°		'
	Vers.	Exsec.	Vers.	Exsec.	Vers.	Exsec.	
0	.89547	8.56677	.91284	10.47371	.93024	13.33559	0
1	.89576	8.59332	.91313	10.51199	.93053	13.39547	1
2	.89605	8.62002	.91342	10.55052	.93082	13.45586	2
3	.89634	8.64687	.91371	10.58932	.93111	13.51676	3
4	.89663	8.67387	.91400	10.62837	.93140	13.57817	4
5	.89692	8.70103	.91429	10.66769	.93169	13.64011	5
6	.89721	8.72833	.91458	10.70728	.93198	13.70258	6
7	.89750	8.75579	.91487	10.74714	.93227	13.76558	7
8	.89779	8.78341	.91516	10.78727	.93257	13.82913	8
9	.89808	8.81119	.91545	10.82768	.93286	13.89323	9
10	.89836	8.83912	.91574	10.86837	.93315	13.95788	10
11	.89865	8.86722	.91603	10.90934	.93344	14.02310	11
12	.89894	8.89547	.91632	10.95060	.93373	14.08890	12
13	.89923	8.92389	.91661	10.99214	.93402	14.15527	13
14	.89952	8.95248	.91690	11.03397	.93431	14.22223	14
15	.89981	8.98123	.91719	11.07610	.93460	14.28979	15
16	.90010	9.01015	.91748	11.11852	.93489	14.35795	16
17	.90039	9.03923	.91777	11.16125	.93518	14.42672	17
18	.90068	9.06849	.91806	11.20427	.93547	14.49611	18
19	.90097	9.09792	.91835	11.24761	.93576	14.56614	19
20	.90126	9.12752	.91864	11.29125	.93605	14.63679	20
21	.90155	9.15730	.91893	11.33521	.93634	14.70810	21
22	.90184	9.18725	.91922	11.37948	.93663	14.78005	22
23	.90213	9.21739	.91951	11.42408	.93692	14.85268	23
24	.90242	9.24770	.91980	11.46900	.93721	14.92597	24
25	.90271	9.27819	.92009	11.51424	.93750	14.99995	25
26	.90300	9.30887	.92038	11.55982	.93779	15.07462	26
27	.90329	9.33973	.92067	11.60572	.93808	15.14999	27
28	.90358	9.37077	.92096	11.65197	.93837	15.22607	28
29	.90386	9.40201	.92125	11.69856	.93866	15.30287	29
30	.90415	9.43343	.92154	11.74550	.93895	15.38041	30
31	.90444	9.46505	.92183	11.79278	.93924	15.45869	31
32	.90473	9.49685	.92212	11.84042	.93953	15.53772	32
33	.90502	9.52886	.92241	11.88841	.93982	15.61751	33
34	.90531	9.56106	.92270	11.93677	.94011	15.69808	34
35	.90560	9.59346	.92299	11.98549	.94040	15.77944	35
36	.90589	9.62605	.92328	12.03458	.94069	15.86159	36
37	.90618	9.65885	.92357	12.08040	.94098	15.94456	37
38	.90647	9.69186	.92386	12.13388	.94127	16.02835	38
39	.90676	9.72507	.92415	12.18411	.94156	16.11297	39
40	.90705	9.75849	.92444	12.23472	.94186	16.19843	40
41	.90734	9.79212	.92473	12.28572	.94215	16.28476	41
42	.90763	9.82596	.92502	12.33712	.94244	16.37196	42
43	.90792	9.86001	.92531	12.38891	.94273	16.46005	43
44	.90821	9.89428	.92560	12.44112	.94302	16.54903	44
45	.90850	9.92877	.92589	12.49373	.94331	16.63893	45
46	.90879	9.96348	.92618	12.54676	.94360	16.72975	46
47	.90908	9.99841	.92647	12.60021	.94389	16.82152	47
48	.90937	10.03356	.92676	12.65408	.94418	16.91424	48
49	.90966	10.06894	.92705	12.70838	.94447	17.00794	49
50	.90995	10.10455	.92734	12.76312	.94476	17.10262	50
51	.91024	10.14039	.92763	12.81829	.94505	17.19830	51
52	.91053	10.17646	.92792	12.87391	.94534	17.29501	52
53	.91082	10.21277	.92821	12.92999	.94563	17.39274	53
54	.91111	10.24932	.92850	12.98651	.94592	17.49153	54
55	.91140	10.28610	.92879	13.04350	.94621	17.59139	55
56	.91169	10.32313	.92908	13.10096	.94650	17.69233	56
57	.91197	10.36040	.92937	13.15889	.94679	17.79438	57
58	.91226	10.39792	.92966	13.21730	.94708	17.89755	58
59	.91255	10.43569	.92995	13.27620	.94737	18.00185	59
60	.91284	10.47371	.93024	13.33559	.94766	18.10732	60

′	87° Vers.	87° Exsec.	88° Vers.	88° Exsec.	89° Vers.	89° Exsec.	′
0	.94766	18.10732	.96510	27.65371	.98255	56.29869	0
1	.94795	18.21397	.96539	27.89440	.98284	57.26976	1
2	.94825	18.32182	.96568	28.13917	.98313	58.27431	2
3	.94854	18.43088	.96597	28.38812	.98342	59.31411	3
4	.94883	18.54119	.96626	28.64137	.98371	60.39105	4
5	.94912	18.65275	.96655	28.89903	.98400	61.50715	5
6	.94941	18.76560	.96684	29.16120	.98429	62.66460	6
7	.94970	18.87976	.96714	29.42802	.98458	63.86572	7
8	.94999	18.99524	.96743	29.69960	.98487	65.11304	8
9	.95028	19.11208	.96772	29.97607	.98517	66.40927	9
10	.95057	19.23028	.96801	30.25758	.98546	67.75736	10
11	.95086	19.34989	.96830	30.54425	.98575	69.16047	11
12	.95115	19.47093	.96859	30.83623	.98604	70.62285	12
13	.95144	19.59341	.96888	31.13366	.98633	72.14583	13
14	.95173	19.71737	.96917	31.43671	.98662	73.73586	14
15	.95202	19.84283	.96946	31.74554	.98691	75.39655	15
16	.95231	19.96982	.96975	32.06030	.98720	77.13274	16
17	.95260	20.09838	.97004	32.38118	.98749	78.94968	17
18	.95289	20.22852	.97033	32.70835	.98778	80.85315	18
19	.95318	20.36027	.97062	33.04199	.98807	82.84947	19
20	.95347	20.49368	.97092	33.38232	.98836	84.94561	20
21	.95377	20.62876	.97121	33.72952	.98866	87.14924	21
22	.95406	20.76555	.97150	34.08380	.98895	89.46886	22
23	.95435	20.90409	.97179	34.44539	.98924	91.91387	23
24	.95464	21.04440	.97208	34.81452	.98953	94.49471	24
25	.95493	21.18653	.97237	35.19141	.98982	97.22303	25
26	.95522	21.33050	.97266	35.57633	.99011	100.1119	26
27	.95551	21.47635	.97295	35.96953	.99040	103.1757	27
28	.95580	21.62413	.97324	36.37127	.99069	106.4311	28
29	.95609	21.77386	.97353	36.78185	.99098	109.8966	29
30	.95638	21.92559	.97382	37.20155	.99127	113.5930	30
31	.95667	22.07935	.97411	37.63068	.99156	117.5444	31
32	.95696	22.23520	.97440	38.06957	.99186	121.7780	32
33	.95725	22.39316	.97470	38.51855	.99215	126.3253	33
34	.95754	22.55329	.97499	38.97797	.99244	131.2223	34
35	.95783	22.71563	.97528	39.44820	.99273	136.5111	35
36	.95812	22.88022	.97557	39.92963	.99302	142.2406	36
37	.95842	23.04712	.97586	40.42266	.99331	148.4684	37
38	.95871	23.21637	.97615	40.92772	.99360	155.2623	38
39	.95900	23.38802	.97644	41.44525	.99389	162.7033	39
40	.95929	23.56212	.97673	41.97571	.99418	170.8883	40
41	.95958	23.73873	.97702	42.51961	.99447	179.9350	41
42	.95987	23.91790	.97731	43.07746	.99476	189.9868	42
43	.96016	24.09969	.97760	43.64980	.99505	201.2212	43
44	.96045	24.28414	.97789	44.23720	.99535	213.8600	44
45	.96074	24.47134	.97819	44.84026	.99564	228.1839	45
46	.96103	24.66132	.97848	45.45963	.99593	244.5540	46
47	.96132	24.85417	.97877	46.09596	.99622	263.4427	47
48	.96161	25.04994	.97906	46.74997	.99651	285.4795	48
49	.96190	25.24869	.97935	47.42241	.99680	311.5230	49
50	.96219	25.45051	.97964	48.11406	.99709	342.7752	50
51	.96248	25.65546	.97993	48.82576	.99738	380.9723	51
52	.96277	25.86360	.98022	49.55840	.99767	428.7187	52
53	.96307	26.07503	.98051	50.31290	.99796	490.1070	53
54	.96336	26.28981	.98080	51.09027	.99825	571.9581	54
55	.96365	26.50804	.98109	51.89156	.99855	686.5496	55
56	.96394	26.72978	.98138	52.71790	.99884	858.4369	56
57	.96423	26.95513	.98168	53.57046	.99913	1144.916	57
58	.96452	27.18417	.98197	54.45053	.99942	1717.874	58
59	.96481	27.41700	.98226	55.35946	.99971	3436.747	59
60	.96510	27.65371	.98255	56.29869	1.00000	Infinite	60

$\phi°$	$y = lC$		$x = l(1 - E)$		$\phi°$	$y = lC$		$x = l(1 - E)$	
	C	Dif.	E	Dif.		C	Dif.	E	Dif.
0° 10′	0.00097	97	0.00000	0	10° 10′	0.05901	97	0.00314	11
20	194	97	0	1	20	5998	96	325	10
30	291	97	1	0	30	6094	96	335	11
40	388	97	1	1	40	6190	96	346	11
50	485	97	2	1	50	6286	97	357	11
1	0.00582	97	0.00003	1	11	0.06383	96	0.00368	11
10	679	97	4	1	10	6479	96	379	12
20	776	97	5	2	20	6575	96	391	11
30	873	97	7	1	30	6671	96	402	12
40	970	97	8	2	40	6767	96	414	12
50	1067	97	10	2	50	6863	96	426	12
2	0.01164	96	0.00012	2	12	0.06959	96	0.00438	12
10	1260	97	14	3	10	7055	96	450	12
20	1357	97	17	2	20	7151	96	462	13
30	1454	97	19	3	30	7247	96	475	13
40	1551	97	22	2	40	7343	96	488	14
50	1648	97	24	3	50	7439	96	502	12
3	0.01745	97	0.00027	4	13	0.07535	96	0.00514	13
10	1842	97	31	3	10	7631	96	527	13
20	1939	97	34	3	20	7727	96	540	14
30	2036	97	37	4	30	7823	96	554	13
40	2133	96	41	4	40	7919	96	567	14
50	2229	97	45	4	50	8015	95	581	14
4	0.02326	97	0.00049	4	14	0.08110	96	0.00595	15
10	2423	97	53	4	10	8206	96	610	14
20	2520	97	57	5	20	8302	95	624	15
30	2617	97	62	4	30	8397	96	639	14
40	2714	96	66	5	40	8493	95	653	15
50	2810	97	71	5	50	8588	96	668	15
5	0.02907	97	0.00076	5	15	0.08684	96	0.00683	15
10	3004	97	81	6	10	8780	95	698	16
20	3101	97	87	5	20	8875	95	714	15
30	3198	96	92	6	30	8970	96	729	16
40	3294	97	98	6	40	9066	95	745	16
50	3391	97	104	6	50	9161	96	761	16
6	0.03488	97	0.00110	6	16	0.09257	95	0.00777	16
10	3585	96	116	6	10	9352	95	793	17
20	3681	97	122	7	20	9447	96	810	16
30	3778	97	129	6	30	9543	95	826	17
40	3875	96	135	7	40	9638	95	843	17
50	3971	97	142	7	50	9733	95	860	17
7	0.04068	97	0.00149	7	17	0.09828	95	0.00877	17
10	4165	96	156	8	10	9923	95	894	17
20	4261	97	164	7	20	10018	95	911	18
30	4358	97	171	8	30	10113	95	929	18
40	4455	96	179	8	40	10208	95	947	17
50	4551	97	187	8	50	10303	95	964	18
8	0.04648	96	0.00195	8	18	0.10398	95	0.00982	19
10	4744	97	203	8	10	10493	95	1001	18
20	4841	96	211	9	20	10588	95	1019	19
30	4937	97	220	9	30	10683	95	1038	18
40	5034	96	229	8	40	10778	95	1056	19
50	5130	97	237	9	50	10873	94	1075	19
9	0.05227	96	0.00246	10	19	0.10967	95	0.01094	19
10	5323	97	256	9	10	11062	95	1113	20
20	5420	96	265	10	20	11157	94	1133	19
30	5516	96	275	9	30	11251	95	1152	20
40	5612	97	284	10	40	11346	94	1172	20
50	5709	96	294	10	50	11440	95	1192	20
10	5805	96	304	10	20	11535	188	1212	40

$\phi°$	$y = lC$: C	Dif.	$x = l(1 - E)$: E	Dif.	$\phi°$	$y = lC$: C	Dif.	$x = l(1 - E)$: E	Dif.
20° 20′	0.11723	189	0.01252	41	30° 30′	0.17388	273	0.02797	91
40	11912	189	1293	42	31	17661	273	2888	93
21	12101	188	1335	42	30	17934	272	2981	94
20	12289	188	1377	44	32	18206	272	3075	95
40	12477	188	1421	43	30	18478	271	3170	97
22	0.12665	188	0.01464	45	33	0.18749	270	0.03267	98
20	12853	187	1509	45	30	19019	269	3365	99
40	13040	188	1554	45	34	19288	269	3464	101
23	13228	187	1599	47	30	19557	269	3565	102
20	13415	187	1646	47	35	19826	268	3667	104
40	13602	187	1693	47	30	20094	267	3771	105
24	0.13789	186	0.01740	49	36	0.20361	266	0.03876	107
20	13975	187	1789	49	30	20627	266	3983	107
40	14162	186	1838	49	37	20893	265	4090	109
25	14348	186	1887	50	30	21158	265	4199	111
20	14534	186	1937	51	38	21423	263	4310	112
40	14720	185	1988	52	30	21686	263	4422	113
26	0.14905	186	0.02040	52	39	0.21949	263	0.04535	114
20	15091	185	2092	52	30	22212	262	4649	116
40	15276	185	2144	54	40	22474	521	4765	236
27	15461	184	2198	54	41	22995	518	5001	240
20	15645	185	2252	55	42	23513	515	5241	246
40	15830	184	2307	55	43	24028	512	5487	252
28	0.16014	184	0.02362	56	44	24540	509	5739	256
20	16198	184	2418	56	45	0.25049	505	0.05995	261
40	16382	183	2474	57	46	25554	503	6256	267
29	16565	184	2531	58	47	26057	499	6523	271
20	16749	183	2589	59	48	26556	496	6794	276
40	16932	182	2648	59	49	27052	492	7070	282
30	17114	274	2707	90	50	27544		7352	

TABLE XV.—DEFLECTION-ANGLES FOR TRANSITION-CURVES.

Transit at $P.T.C.$, $n'' = 0$. $(\delta_0°) = \frac{I_1°}{3} A_0 - B_0$

Tr. at quarter-point, $n'' = \frac{1}{4}$. $(\delta_{\frac{1}{4}}°) = \frac{I_1°}{3} A_{\frac{1}{4}} - B_{\frac{1}{4}}$

n	ϕ for $\frac{I_1}{3} = 1$	A_0	B_0 for $\frac{I}{3} =$ 4°	6°	8°	10°	12°	14°	16°	$A_{\frac{1}{4}}$	$B_{\frac{1}{4}}$ for $\frac{I_1}{3} =$ 4°	8°	12°	14°	16°	n
.0	.00	.00								.0625						.0
.05	.0075	.0025								.0775						.05
.1	.03	.01								.0975						.1
.15	.0675	.0225								.1225						.15
.2	.12	.04								.1525						.2
.25	.1875	.0625								.1875						.25
.3	.27	.09								.2275						.3
.35	.3675	.1225								.2725						.35
.4	.48	.16								.3225						.4
.45	.6075	.2025						1	1	.3775						.45
.5	.75	.25					1	1	1	.4375						.5
.55	.9075	.3025				1	1	2	3	.5025						.55
.6	1.08	.36			1	1	2	3	4	.5725					1	.6
.65	1.2675	.4225			1	2	3	5	7	.6475			1	1	2	.65
.7	1.47	.49		1	1	3	5	7	11	.7275		1	2	3	4	.7
.75	1.6875	.5625		1	2	4	7	11	17	.8125		1	3	4	6	.75
.8	1.92	.64		1	3	6	11	17	25	.9025		1	4	6	9	.8
.85	2.1675	.7225	1	2	4	9	15	24	36	.9975		2	6	10	15	.85
.9	2.43	.81	1	3	6	12	21	34	51	1.0975		3	9	15	22	.9
.95	2.7075	.9025	1	4	9	17	30	47	71	1.2025	1	4	14	22	33	.95
1.	3.	1.	1	5	12	23	41	64	97	1.3125	1	6	20	31	47	1.

TABLE XV.—DEFLECTION ANGLES FOR TRANSITION CURVES.

Transit at mid-point, $n'' = \frac{1}{2}$.

$$(\delta_{\frac{1}{2}}°) = \frac{I_1°}{3}A_{\frac{1}{2}} - B_{\frac{1}{2}}$$

n	ϕ for $\frac{I_1}{3} = 1$	$A_{\frac{1}{2}}$	$B_{\frac{1}{2}}$ for $\frac{I_1°}{3}=$ 8°	12°	16°
.0	.00	.25		1	1
.05	.0075	.2775			1
.1	.03	.31			1
.15	.0675	.3475			1
.2	.12	.39			
25	.1875	.4375			
.3	.27	.49			
.35	.3675	.5475			
.4	.48	.61			
.45	.6075	.6775			
.5	.75	.75			
.55	.9075	.8275			
.6	1.08	.91			
.65	1.2675	.9975			
.7	1.47	1.09			
.75	1.6875	1.1875			1
.8	1.92	1.29		1	1
.85	2.1675	1.3975		1	3
.9	2.43	1.51	1	2	5
.95	2.7075	1.6275	1	3	8
1.	3.	1.75	2	6	14

Tr. at three-quarter point, $n'' = \frac{3}{4}$.

$$(\delta_{\frac{3}{4}}°) = \frac{I_1°}{3}A_{\frac{3}{4}} - B_{\frac{3}{4}}$$

$A_{\frac{3}{4}}$	$B_{\frac{3}{4}}$ for $\frac{I_1°}{3}=$ 6°	10°	14°	16°	n
.5625	1	4	11	17	.0
.6025	1	4	10	15	.05
.6475	1	3	8	12	.1
.6975	1	2	7	10	.15
.7525		2	6	8	.2
.8125		1	4	6	.25
.8775		1	3	4	.3
.9475		1	2	3	.35
1.0225			1	2	.4
1.1025			1	1	.45
1.1875				1	.5
1.2775					.55
1.3725					.6
1.4725					.65
1.5775					.7
1.6875					.75
1.8025					.8
1.9225					.85
2.0475					.9
2.1775					.95
2.3125			1	1	1.

Transit at $P.C._1$, $n'' = 1$.

$$(\delta_1°) = \frac{I_1°}{3}A_1 - B_1$$

n	ϕ for $\frac{I_1}{3} = 1$	A_1	B_1 for $\frac{I_1°}{3}=$ 4°	6°	8°	10°	12°	14°	16°
.0	.00	1.	1	5	12	23	41	64	97
.05	.0075	1.0525	1	4	11	20	36	58	86
.1	.03	1.11	1	4	9	18	32	51	76
.15	.0675	1.1725	1	3	8	16	28	44	66
.2	.12	1.24	1	3	7	13	23	37	56
.25	.1875	1.3125	1	2	6	11	20	31	47
.3	.27	1.39	1	2	5	9	16	26	39
.35	.3675	1.4725		2	4	7	13	21	31
.4	.48	1.56		1	3	6	10	16	24
.45	.6075	1.6525		1	2	4	8	12	18
.5	.75	1.75		1	2	3	6	9	14
.55	.9075	1.8525			1	2	4	6	9
.6	1.08	1.96			1	1	3	4	6
.65	1.2675	2.0725				1	2	3	4
.7	1.47	2.19					1	1	2
.75	1.6875	2.3125						1	1
.8	1.92	2.44							
.85	2.1675	2.5725							
.9	2.43	2.71							
.95	2.7075	2.8525							
1.	3.	3.							

Transit at $P.T.C._1$, or $P.C._1$, deflections from tangent to circular curve.

$$(\delta_O°) = \frac{I_1°}{3}A_O + B_1$$

A_O	B_1 for $\frac{I_1°}{3}=$ 4°	6°	8°	10°	12°	14°	16°	n
2.	1	5	12	23	41	64	97	.0
1.9475	1	4	11	20	36	58	86	.05
1.89	1	4	9	18	32	51	76	.1
1.8275	1	3	8	16	28	44	66	.15
1.76	1	3	7	13	23	37	56	.2
1.6875	1	2	6	11	20	31	47	.25
1.61	1	2	5	9	16	26	39	.3
1.5275		2	4	7	13	21	31	.35
1.44		1	3	6	10	16	24	.4
1.3475		1	2	4	8	12	18	.45
1.25		1	2	3	6	9	14	.5
1.1475			1	2	4	6	9	.55
1.04			1	1	3	4	6	.6
.9275				1	2	3	4	.65
.81					1	1	2	.7
.6875						1	1	.75
.56								.8
.4275								.85
.29								.9
.1475								.95
.00								1.

l_1	2° Curve. $I°$	x_1	y_1	e	e'	F	x'	y'	3° Curve. $I°$	x_1	y_1	e	e'	F	x'	y'	l_1
40	.40	40	.09			.02	20	.01	.60	40	.13			.03	20	.02	40
60	.60	60	.21			5	30	3	.90	60	.31			8	30	4	60
80	.80	80	.37			9	40	5	1.20	80	.56			.14	40	7	80
100	1.00	100	.58			.14	50	7	1.50	100	.88			.22	50	.11	100
120	1.20	120	.84			.21	60	.10	1.80	120	1.26			.32	60	.16	120
140	1.40	140	1.14			.28	70	.14	2.10	140	1.71			.43	70	.21	140
160	1.60	160	1.49			.37	80	.19	2.40	160	2.23			.56	80	.28	160
180	1.80	180	1.89			.47	90	.24	2.70	180	2.82		.01	.71	90	.35	180
200	2.00	200	2.33			.58	100	.29	3.00	200	3.49		1	.87	100	.44	200
220	2.20	220	2.84			.70	110	.35	3.30	220	4.23		1	1.06	110	.53	220
240	2.40	240	3.35			.84	120	.42	3.60	239.9	5.03		1	1.26	120	.63	240
260	2.60	259.9	3.93		.01	.98	130	.49	3.90	259.9	5.90		2	1.48	130	.74	260
280	2.80	279.9	4 56		1	1.14	140	.57	4.20	279.8	6.84		2	1.71	140	.86	280
300	3.00	299.9	5.24		1	1.31	150	.65	4.50	299.8	7.85	.01	2	1.96	150	.98	300
320	3.20	319.9	5.96		2	1.49	160	.74	4.80	319.7	8.93	1	2	2.23	160	1.12	320
340	3.40	339.9	6.73		2	1.68	170	.84	5.10	339.7	10.08	1	3	2.52	170	1.26	340
360	3.60	359.9	7.54		2	1.88	180	.94	5.40	359.7	11.30	1	3	2.83	179.9	1.41	360
380	3.80	379.8	8.40	.01	2	2.10	190	1.05	5.70	379.6	12.59	1	4	3.15	189.9	1.57	380
400	4.00	399.8	9.30	1	3	2.32	200	1.16	6.00	399.6	13.95	1	5	3.49	199.9	1.74	400
420	4.20	419.8	10.25	1	4	2.56	210	1.28	6.30	419.5	15.38	2	7	3.85	209.9	1.92	420
440	4.40	439.7	11.26	1	4	2.81	219.9	1.41	6.60	439.4	16.88	2	8	4.22	219.9	2.11	440
460	4.60	459.7	12.31	1	4	3.08	229.9	1.54	6.90	459.4	18.45	2	9	4.61	229.9	2.31	460
480	4.80	479.7	13 40	1	5	3.35	239.9	1.68	7.20	479.3	20.08	2	.10	5.02	239.9	2.51	480
500	5.00	499.6	14.54	1	6	3.63	249.9	1.82	7.50	499.2	21.78	3	.11	5.45	249.9	2.72	500
520	5.20	519.6	15.72	1	7	3.93	259.9	1.96	7.80	519.0	23.56	3	.12	5.89	259.8	2.94	520
540	5.40	539.5	16.95	2	7	4.24	269.9	2.12	8.10	538.9	25.41	3	.14	6.35	269.8	3.18	540
560	5.60	559.5	18.23	2	8	4.56	279.9	2.28	8.40	558.8	27.32	4	.16	6.83	279.8	3.42	560
580	5.80	579.4	19.56	2	9	4.89	289.9	2.44	8.70	578.7	29.30	4	.18	7.33	289.8	3.66	580
600	6.00	599.3	20.93	2	9	5.23	299.9	2.61	9.00	598.5	31.36	5	.20	7.84	299.7	3.92	600
620	6.20	619.3	22.35	2	.10	5.59	309.9	2.79	9.30	618.4	33.48	6	.22	8.37	309.7	4.19	620
640	6.40	639.2	23.81	2	.11	5.95	319.8	2.97	9.60	638.2	35.67	6	.24	8.92	319.7	4.46	640
660	6.60	659.1	25.32	3	.12	6.33	329.8	3.16	9.90	658.0	37.93	6	.27	9.48	329.6	4.74	660
680	6.80	679.0	26.87	3	.13	6.72	339.8	3.36	10.20	677.8	40.26	7	.30	10.07	339.6	5.03	680
700	7.00	699.0	28.47	4	.14	7.12	319.8	3.56	10.50	697.6	42.66	7	.33	10.67	349.5	5.33	700

l_1	4° Curve.							
	$I°$	x_1	y_1	e	e'	F	x'	y'
40	0.80	40	.18			.04	20	.02
60	1.20	60	.42			.10	30	5
80	1.60	80	.74			.18	40	9
100	2.00	100	1.16			.29	50	.14
120	2.40	120	1.67			.42	60	.21
140	2.80	140	2.28			.57	70	.28
160	3.20	160	2.98			.75	80	.37
180	3.60	179.9	3.77		.01	.94	90	.47
200	4.00	199.9	4.65		1	1.16	100	.58
220	4 40	219.9	5.62		1	1.41	110	.70
240	4.80	239.8	6.69	.01	2	1.67	120	.84
260	5.20	259.8	7.85	1	2	1.96	130	.98
280	5.60	279.7	9.11	1	3	2.28	139.9	1.14
300	6.00	299.7	10.46	1	3	2.62	149.9	1.31
320	6.40	319.6	11.90	1	4	2.98	159.9	1.49
340	6.80	339.5	13.43	1	5	3.36	169.9	1.68
360	7.20	359.4	15.06	2	6	3.77	179.9	1.88
380	7.60	379.3	16.78	2	7	4.20	189.9	2.10
400	8.00	399.2	18.59	2	9	4.65	199.9	2.32
420	8.40	419.1	20.49	3	.11	5.12	209.8	2.56
440	8.80	439.0	22.49	3	.14	5.62	219.8	2.81
460	9.20	458.9	24.58	3	.15	6.15	229.8	3.07
480	9.60	478.7	26.75	4	.17	6.69	239.8	3.34
500	10.00	498.5	29.02	5	.20	7.26	249.7	3.63
520	10.40	518.3	31.38	6	.23	7.85	259.7	3.92
540	10.80	538.1	33.84	6	.25	8.47	269.6	4.23
560	11.20	557.9	36.39	7	.26	9.11	279.6	4.55
580	11.60	577.6	39.03	7	.29	9.77	289.6	4.88
600	12.00	597.4	41.75	8	.32	10.45	299.5	5.22
620	12.40	617.1	44.57	9	.36	11.15	309.5	5.57
640	12.80	636.8	47.48	.1	.4	11.88	319.4	5.93
660	13.20	656.5	50.48	.1	.5	12.64	329.4	6.30
680	13.60	676.2	53.58	.1	.5	13.42	339.3	6.69
700	14 00	695.8	56.77	.1	.6	14.22	349.2	7.08

5° Curve.								l_1
$I°$	x_1	y_1	e	e'	F	x'	y'	
1.00	40	.23		−.01	.06	20	.03	40
1.50	60	.52		1	.13	30	6	60
2.00	80	.93		1	.23	40	.12	80
2.50	100	1.45		1	.36	50	.18	100
3.00	120	2.09		1	.52	60	.26	120
3.50	140	2.85		1	.71	70	.36	140
4.00	159.9	3.72		−.01	.93	80	.47	160
4.50	179.9	4.71			1.18	90	.59	180
5.00	199.8	5.81			1.45	100	.73	200
5.50	219.8	7.08	.01		1.76	110	.88	220
6.00	239.7	8.37	1	+.01	2.09	119.9	1.05	240
6.50	259.7	9.82	1	1	2.46	129.9	1.23	260
7.00	279.6	11.39	1	2	2.85	139.9	1.42	280
7.50	299.5	13.07	2	3	3.27	149.9	1.63	300
8.00	319.4	14.87	2	5	3.72	159.9	1.86	320
8.50	339.3	16.78	2	6	4.20	169.9	2.10	340
9.00	359.1	18.81	3	8	4.70	179.8	2.35	360
9.50	379.0	20.96	3	.10	5.24	189.8	2.62	380
10.00	398.8	23.22	3	.12	5.81	199.8	2.90	400
10.50	418.6	25.59	4	.2	6.40	209.7	3.20	420
11.00	438.4	28.08	5	.2	7.03	219.7	3.51	440
11.50	458.1	30.69	6	.2	7.68	229.7	3.83	460
12.00	477.9	33.40	7	.3	8.36	239.6	4.17	480
12.50	497.7	36.23	7	.3	9.07	249.6	4.53	500
13.00	517.3	39.18	8	.4	9.81	259.5	4.90	520
13.50	537.0	42.24	9	.4	10.58	269.5	5.26	540
14.00	556.7	45.41	.1	.4	11.37	279.4	5.67	560
14.50	576.3	48.70	.1	.5	12.20	289.3	6.08	580
15.00	595.9	52.10	.1	.5	13.05	299.3	6.51	600
15.50	615.4	55.61	.1	.6	13.93	309.2	6.95	620
16.00	635.0	59.24	.2	.6	14.85	319.1	7.39	640
16.50	654.5	62.98	.2	.7	15.79	329.0	7.86	660
17.00	674.0	66.83	.2	.7	16.76	339.0	8.34	680
17.50	693.5	70.79	.2	.8	17.75	348.9	8.84	700

l_1	6° Curve.								7° Curve.								l_1
	$I°$	x_1	y_1	e	e'	F	x'	y'	$I°$	x_1	y_1	e	e'	F	x'	y'	
40	1.20	40	.28		−.01	.07	20	.03	1.40	40	.32		−.01	.08	20	.04	40
60	1.80	60	.63		1	.16	30	.08	2.10	60	.73		2	.18	30	.09	60
80	2.40	80	1.11		1	.28	40	.14	2.80	80	1.30		2	.33	40	.16	80
100	3.00	100	1.74		1	.44	50	.22	3.50	100	2.04		2	.51	50	.25	100
120	3.60	120	2.51		1	.63	60	.31	4.20	119.9	2.93		2	.73	60	.36	120
140	4.20	139.9	3.42		1	.86	70	.43	4.90	139.9	3.99		2	1.00	70	.50	140
160	4.80	159.9	4.47		1	1.12	80	.56	5.60	159.8	5.21		2	1.30	80	.65	160
180	5.40	179.8	5.66		1	1.42	90	.71	6.30	179.8	6.59	.01	2	1.65	90	.82	180
200	6.00	199.8	6.98	.01	−.01	1.75	100	.87	7.00	199.7	8.14	1	−.01	2.04	90.9	1.02	200
220	6.60	219.7	8.44	1		2.11	109.9	1.05	7.70	219.6	9.85	1		2.46	109.9	1.23	220
240	7.20	239.6	10.04	1	+.01	2.51	119.9	1.25	8.40	239.5	11.72	2		2.93	119.9	1.46	240
260	7.80	259.5	11.78	1	2	2.95	129.9	1.47	9.10	259.4	13.75	2	+.01	3.44	129.9	1.72	260
280	8.40	279.4	13.66	2	3	3.42	139.9	1.71	9.80	279.2	15.93	3	1	3.99	139.9	1.99	280
300	9.00	299.3	15.68	2	4	3.92	149.9	1.96	10.50	299.0	18.28	3	4	4.57	149.8	2.28	300
320	9.60	319.1	17.83	3	6	4.46	159.8	2.23	11.20	318.8	20.79	4	8	5.20	159.8	2.60	320
340	10.20	338.9	20.12	3	8	5.03	169.8	2.52	11.90	338.5	23.46	5	.11	5.87	169.7	2.93	340
360	10.80	358.7	22.56	4	.11	5.64	179.8	2.82	12.60	358.3	26.29	5	.14	6.58	179.7	3.28	360
380	11.40	378.5	25.13	5	.14	6.28	189.7	3.14	13.30	378.0	29.28	6	.17	7.33	189.7	3.66	380
400	12.00	398.2	27.83	5	.18	6.96	199.7	3.48	14.00	397.6	32.43	7	.22	8.12	199.6	4.05	400
420	12.60	418.0	30.68	6	.2	7.68	209.7	3.83	14.70	417.2	35.74	9	.3	8.95	209.5	4.46	420
440	13.20	437.7	33.06	7	.2	8.43	219.6	4.20	15.40	436.8	39.21	.10	.4	9.82	219.4	4.89	440
460	13.80	457.3	36.78	8	.3	9.21	229.5	4.59	16.10	456.4	42.84	.11	.4	10.73	229.4	5.35	460
480	14.40	477.0	40.03	9	.3	10.03	239.5	5.00	16.80	475.9	46.62	.13	.4	11.68	239.3	5.82	480
500	15.00	496.7	43.42	.10	.4	10.88	249.4	5.42	17.50	495.4	50.56	.14	.5	12.67	249.2	6.31	500
520	15.60	516.2	46.94	.1	.5	11.76	259.3	5.86	18.20	514.8	54.66	.2	.5	13.70	259.1	6.82	520
540	16.20	535.7	50.60	.1	.5	12.68	269.2	6.32	18.90	534.2	58.91	.2	.6	14.78	269.0	7.35	540
560	16.80	555.2	54.39	.2	.5	13.63	279.2	6.79	19.60	553.5	63.32	.2	.7	15.89	278.9	7.90	560
580	17.40	574.7	58.32	.2	.6	14.62	289.1	7.28	20.30	572.8	67.88	.2	.8	17.04	288.8	8.47	580
600	18.00	594.1	62.38	.2	.8	15.64	299.0	7.79	21.00	592.0	72.60	.3	.9	18.23	298.6	9.06	600
620	18.60	613.5	66.58	.2	.9	16.70	308.9	8.31	21.70	611.2	77.47	.3	1.0	19.46	308.5	9.67	620
640	19.20	632.8	70.91	.2	1.0	17.79	318.7	8.85	22.40	630.3	82.49	.3	1.1	20.73	318.4	10.29	640
660	19.80	652.1	75.37	.2	1.0	18.91	328.6	9.40	23.10	649.4	87.67	.3	1.2	22.04	328.3	10.94	660
680	20.40	671.4	79.97	.3	1.1	20.07	338.5	9.97	23.80	668.4	93.00	.4	1.3	23.39	338.1	11.60	680
700	21.00	690.6	84.70	.3	1.1	21.27	348.4	10.56	24.50	687.3	98.48	.4	1.5	24.78	347.9	12.28	700

l_1	8° Curve.								9° Curve.								l_1
	I°	x_1	y_1	e	e'	F	x'	y'	I°	x_1	y_1	e	e'	F	x'	y'	
40	1.60	40	.37			.09	20	.05	1.80	40	.42			.11	20	.05	40
60	2.40	60	.84			.21	30	.11	2.70	60	.94			.24	30	.12	60
80	3.20	80	1.49			.37	40	.19	3.60	80	1.67			.42	40	.21	80
100	4.00	100	2.33			.58	50	.29	4.50	99.9	2.62			.66	50	.33	100
120	4.80	119.9	3.35			.84	60	.42	5.40	119.9	3.77		.01	.94	60	.47	120
140	5.60	139.9	4.56		.01	1.14	70	.57	6.30	139.8	5.13	.01	1	1.28	70	.64	140
160	6.40	159.8	5.95	.01	2	1.49	80	.74	7.20	159.7	6.70	1	2	1.68	80	.84	160
180	7.20	179.7	7.53	1	3	1.88	89.9	.94	8.10	179.6	8.47	1	3	2.12	89.9	1.06	180
200	8.00	199.6	9.29	1	4	2.32	99.9	1.16	9.00	199.5	10.45	2	4	2.61	99.9	1.30	200
220	8.80	219.5	11.24	2	6	2.81	109.9	1.40	9.90	219.3	12.64	2	6	3.16	109.9	1.58	220
240	9.60	239.3	13.38	2	7	3.35	119.9	1.67	10.80	239.2	15.04	3	9	3.76	119.9	1.88	240
260	10.40	259.1	15.70	3	.10	3.93	129.8	1.96	11.70	258.9	17.64	3	.12	4.42	129.8	2.20	260
280	11.20	278.9	18.20	3	.13	4.56	139.8	2.27	12.60	278.7	20.45	4	.16	5.12	139.8	2.55	280
300	12.00	298.7	20.88	4	.16	5.23	149.8	2.61	13.50	298.3	23.47	5	.20	5.88	149.7	2.93	300
320	12.80	318.4	23.74	5	.20	5.95	159.7	2.97	14.40	318.0	26.69	6	.24	6.69	159.7	3.33	320
340	13.60	338.1	26.79	6	.24	6.71	169.7	3.35	15.30	337.6	30.11	8	.30	7.55	169.6	3.76	340
360	14.40	357.7	30.02	7	.29	7.52	179.6	3.75	16.20	357.1	33.73	9	.36	8.45	179.5	4.21	360
380	15.20	377.3	33.43	8	.34	8.37	189.5	4.17	17.10	376.6	37.56	.11	.43	9.41	189.4	4.69	380
400	16.00	396.9	37.02	.10	.41	9.27	199.5	4.62	18.00	396.1	41.59	.12	.51	10.43	199.3	5.19	400
420	16.80	416.4	40.79	.1	.5	10.22	209.4	5.09	18.90	415.4	45.82	.1	.6	11.49	209.2	5.72	420
440	17.60	435.9	44.74	.1	.6	11.21	219.3	5.58	19.80	434.7	50.25	.2	.6	12.61	219.1	6.27	440
460	18.40	455.3	48.87	.1	.7	12.25	229.2	6.10	20.70	454.0	54.88	.2	.7	13.78	229.0	6.84	460
480	19.20	474.6	53.18	.2	.7	13.34	239.1	6.64	21.60	473.2	59.70	.2	.9	14.99	238.8	7.44	480
500	20.00	493.9	57.66	.2	.8	14.47	249.0	7.19	22.50	492.3	64.72	.2	1.0	16.26	248.7	8.07	500
520	20.80	513.2	62.32	.2	.9	15.64	258.8	7.77	23.40	511.4	69.94	.3	1.1	17.58	258.5	8.72	520
540	21.60	532.4	67.16	.2	1.0	16.86	268.7	8.37	24.30	530.4	75.35	.3	1.3	18.95	268.3	9.39	540
560	22.40	551.5	72.18	.3	1.0	18.13	278.6	9.00	25.20	549.2	80.96	.3	1.5	20.37	278.1	10.08	560
580	23.20	570.6	77.37	.3	1.2	19.45	288.4	9.65	26.10	568.0	86.76	.4	1.6	21.84	287.9	10.80	580
600	24.00	589.6	82.73	.3	1.4	20.81	298.2	10.31	27.00	586.7	92.75	.4	1.8	23.36	297.7	11.55	600
620	24.80	608.5	88.26	.4	1.4	22.21	308.1	11.00	27.90	605.8	98.93	.5	2.0	24.93	307.4	12.31	620
640	25.60	627.3	93.96	.4	1.6	23.66	317.9	11.71	28.80	624.0	105.30	.5	2.2	26.55	317.2	13.10	640
660	26.40	646.1	99.84	.4	1.9	25.15	327.6	12.44	29.70	642.5	111.86	.6	2.3	28.22	327.0	13.92	660
680	27.20	664.8	105.88	.5	2.0	26.68	337.4	13.19	30.60	660.9	118.60	.6	2.5	29.94	336.8	14.76	680
700	28.00	683.5	112.09	.5	2.2	28.25	347.2	13.96	31.50	679.1	125.53	.7	2.7	31.71	346.5	15.62	700

l_1	10° Curve.								11° Curve.								l_1
	$I°$	x_1	y_1	e	e'	F	x'	y'	$I°$	x_1	y_1	e	e'	F	x'	y'	
40	2.00	40	.47			.12	20	.06	2.20	40	.51			.13	20	.06	40
60	3.00	60	1.05			.26	30	.13	3.30	60	1.15			.29	30	.14	60
80	4.00	80	1.86			.47	40	.23	4.40	80	2.05			.51	40	.26	80
100	5.00	99.9	2.91			.73	50	.36	5.50	99.9	3.20			.80	50	.40	100
120	6.00	119.9	4.19		.01	1.05	60	.52	6.60	119.8	4.60	.01		1.15	60	.57	120
140	7.00	139.8	5.70	.01	1	1.43	70	.71	7.70	139.7	6.25	1	.01	1.56	70	.78	140
160	8.00	159.7	7.44	1	2	1.86	79.9	93	8.80	159.6	8.16	1	3	2.04	79.9	1.02	160
180	9.00	179.6	9.41	1	4	2.36	89.9	1.18	9.90	179.5	10.33	1	5	2.58	89.9	1.29	180
200	10.00	199.4	11.61	2	6	2.91	99.9	1.45	11.00	199.3	12.75	2	7	3.19	99.9	1.59	200
220	11.00	219.2	14.04	3	8	3.51	109.9	1.75	12.10	219.0	15.43	3	.10	3.86	109.8	1.93	220
240	12.00	239.0	16.70	3	.11	4.18	119.8	2.09	13.20	238.7	18.36	4	.13	4.60	119.8	2.30	240
260	13.00	258.7	19.59	4	.15	4.90	129.8	2.45	14.30	258.4	21.54	5	.18	5.40	129.7	2.69	260
280	14.00	278.3	22.71	5	.19	5.68	139.7	2.84	15.40	278.0	24.96	6	.23	6.25	139.7	3.12	280
300	15.00	297.9	26.05	6	.24	6.52	149.7	3.26	16.50	297.5	28.63	8	.28	7.17	149.6	3.58	300
320	16.00	317.5	29.62	8	.29	7.42	159.6	3.70	17.60	317.0	32.55	.10	.35	8.15	159.5	4.07	320
340	17.00	337.0	33.41	9	.37	8.37	169.5	4.17	18.70	336.4	36.71	.11	.44	9.20	169.4	4.59	340
360	18.00	356.5	37.43	.11	.44	9.38	179.4	4.67	19.80	355.7	41.11	.13	.52	10.31	179.3	5.13	360
380	19.00	375.9	41.67	.13	.52	10.45	189.3	5.20	20.90	375.0	45.76	.15	.62	11.48	189.2	5.70	380
400	20.00	395.2	46.13	.15	.61	11.57	199.2	5.76	22.00	394.2	50.65	.18	.73	12.72	199.0	6.31	400
420	21.00	414.5	50.81	.2	.7	12.75	209.1	6.34	23.10	413.3	55.78	.2	.8	14.01	208.9	6.95	420
440	22.00	433.6	55.72	.2	.8	13.99	218.9	6.95	24.20	432.2	61.15	.2	1.0	15.37	218.7	7.62	440
460	23.00	452.6	60.84	.2	.9	15.29	228.7	7.59	25.30	451.1	66.76	.3	1.1	16.79	228.5	8.33	460
480	24.00	471.7	66.18	.3	1.1	16.64	238.6	8.25	26.40	469.9	72.61	.3	1.3	18.28	238.3	9.06	480
500	25.00	490.6	71.73	.3	1.2	18.05	248.4	8.94	27.50	488.6	78.69	.4	1.4	19.83	248.1	9.81	500
520	26.00	509.4	77.50	.3	1.3	19.51	258.2	9.66	28.60	507.2	85.00	.4	1.6	21.44	257.9	10.59	520
540	27.00	528.1	83.48	.4	1.5	21.08	268.0	10.40	29.70	525.7	91.53	.4	1.8	23.11	267.6	11.40	540
560	28.00	546.8	89.67	.4	1.7	22.60	277.8	11.17	30.80	544.1	98.29	.5	2.0	24.83	277.4	12.24	560
580	29.00	565.3	96.07	.5	2.0	24.23	287.5	11.97	31.90	562.4	105.27	.6	2.3	26.60	287.1	13.11	580
600	30.00	583.8	102.68	.5	2.2	25.91	297.3	12.79	33.00	580.5	112.48	.6	2.5	28.44	296.8	14.01	600
620	31.00	602.1	109.49	.6	2.4	27.64	307.0	13.64	34.10	598.5	119.91	.7	2.7	30.34	306.5	14.93	620
640	32.00	620.3	116.51	.6	2.6	29.44	316.7	14.51	35.20	616.4	127.56	.7	3.0	32.31	316.1	15.87	640
660	33.00	638.4	123.73	.7	2.9	31.29	326.4	15.40	36.30	634.1	135.43	.8	3.3	34.34	325.7	16.84	660
680	34.00	656.5	131.15	.7	3.2	33.19	336.0	16.31	37.40	651.7	143.51	.9	3.7	36.42	335.3	17.84	680
700	35.00	674.3	138.77	.8	3.4	35.14	346.7	17.25	38.50	669.2	151.79	.9	4.0	38.56	341.9	18.87	700

l_1	12° Curve.								13° Curve.								l_1
	$I°$	x_1	y_1	e	e'	F	x'	y'	$I°$	x_1	y_1	e	e'	F	x'	y'	
40	2.40	40	.56			.14	20	.07	2.60	40	.60			.15	20	.07	40
60	3.60	60	1.26			.32	30	.16	3.90	60	1.36			.34	30	.17	60
80	4.80	79.9	2.23			.56	40	.28	5.20	79.9	2.42			.61	40	.30	80
100	6.00	99.9	3.49			.87	50	.44	6.50	99.9	3.78			.95	50	.47	100
120	7.20	119.8	5.02	.01		1.26	60	.63	7.80	119.8	5.44	.01		1.36	60	.68	120
140	8.40	139.7	6.83	1	.01	1.71	69.9	.85	9.10	139.6	7.40	1	.02	1.85	69.9	.92	140
160	9.60	159.6	8.92	1	3	2.23	79.9	1.11	10.40	159.5	9.66	2	4	2.42	79.9	1.21	160
180	10.80	179.4	11.28	2	5	2.82	89.9	1.41	11.70	179.3	12.22	2	6	3.06	89.8	1.53	180
200	12.00	199.1	13.92	3	8	3.49	99.8	1.74	13.00	199.0	15.07	3	.10	3.77	99.8	1.88	200
220	13.20	218.8	16.83	4	.11	4.22	109.8	2.10	14.30	218.6	18.22	4	.13	4.56	109.8	2.28	220
240	14.40	238.5	20.01	5	.15	5.01	119.7	2.50	15.60	238.2	21.66	6	.18	5.42	119.7	2.71	240
260	15.60	258.1	23.47	6	.20	5.88	129.7	2.93	16.90	257.7	25.39	7	.24	6.36	129.6	3.17	260
280	16.80	277.6	27.20	8	.27	6.82	139.6	3.40	18.20	277.2	29.43	9	.31	7.38	139.5	3.68	280
300	18.00	297.0	31.20	9	.34	7.82	149.5	3.90	19.50	296.5	33.76	.11	.39	8.48	149.4	4.22	300
320	19.20	316.4	35.46	.11	.42	8.89	159.4	4.43	20.80	315.8	38.37	.13	.49	9.64	159.3	4.79	320
340	20.40	335.7	39.98	.13	.52	10.03	169.3	4.99	22.10	335.0	43.26	.16	.60	10.87	169.2	5.40	340
360	21.60	354.9	44.77	.15	.62	11.24	179.1	5.59	23.40	354.1	48.43	.19	.72	12.18	179.0	6.04	360
380	22.80	374.0	49.88	.18	.75	12.52	189.0	6.22	24.70	373.0	53.88	.22	.86	13.56	188.8	6.72	380
400	24.00	393.0	55.15	.21	.88	13.86	198.8	6.87	26.00	391.8	59.61	.25	1.03	15.01	198.6	7.43	400
420	25.20	411.9	60.73	.2	1.0	15.27	208.6	7.57	27.30	410.5	65.62	.3	1.2	16.53	208.4	8.18	420
440	26.40	430.8	66.56	.3	1.2	16.75	218.4	8.29	28.60	429.2	71.91	.3	1.4	18.13	218.2	8.96	440
460	27.60	449.4	72.64	.3	1.4	18.30	228.2	9.05	29.90	447.6	78.47	.4	1.6	19.80	227.9	9.77	460
480	28.80	468.0	78.98	.4	1.6	19.92	238.0	9.84	31.20	466.0	85.29	.4	1.8	21.54	237.6	10.62	480
500	30.00	486.4	85.57	.4	1.8	21.60	247.7	10.65	32.50	484.2	92.38	.5	2.1	23.35	247.3	11.50	500
520	31.20	504.8	92.40	.5	2.0	23.34	257.4	11.49	33.80	502.2	99.78	.5	2.3	25.23	257.0	12.41	520
540	32.40	523.0	99.48	.5	2.2	25.15	267.1	12.37	35.10	520.1	107.34	.6	2.6	27.18	266.7	13.35	540
560	33.60	541.0	106.80	.6	2.5	27.02	276.8	13.28	36.40	537.8	115.21	.7	2.9	29.21	276.3	14.32	560
580	34.80	559.0	114.36	.6	2.8	28.96	286.4	14.21	37.70	555.4	123.33	.7	3.2	31.31	285.9	15.32	580
600	36.00	576.8	122.16	.7	3.1	30.96	296.1	15.17	39.00	572.8	131.70	.8	3.6	33.47	295.4	16.35	600
620	37.20	594.4	130.19	.8	3.4	33.03	305.7	16.16	40.30	590.0	140.31	.9	4.0	35.70	304.9	17.41	620
640	38.40	611.9	138.45	.8	3.8	35.16	315.2	17.18	41.60	607.1	149.16	1.0	4.4	38.00	314.4	18.50	640
660	39.60	629.2	146.93	.9	4.1	37.35	324.8	18.23	42.90	624.0	158.23	1.1	4.8	40.36	323.9	19.62	660
680	40.80	646.3	155.64	1.0	4.5	39.60	334.3	19.30	44.20	640.6	167.56	1.2	5.3	42.79	333.3	20.77	680
700	42.00	663.3	164.58	1.1	5.0	41.92	343.8	20.40	45.50	657.1	177.11	1.3	5.8	45.28	342.8	21.95	700

l_1	14° Curve.								16° Curve.								l_1
	$I°$	x_1	y_1	e	e'	F	x'	y'	$I°$	x_1	y_1	e	e'	F	x'	y'	
40	2.8	40	.65			.16	20	.08	3.2	40	.74			.19	20	.09	40
60	4.2	60	1.47			.37	30	.18	4.8	60	1.67			.42	30	.21	60
80	5.6	79.9	2.60			.65	40	.32	6.4	79.9	2.98			.75	40	.37	80
100	7.0	99.8	4.07			1.02	50	.51	8.0	99.8	4.65	.01	.02	1.16	50	.58	100
120	8.4	119.7	5.86	.01		1.47	60	.73	9.6	119.7	6.69	1	4	1.67	59.9	.83	120
140	9.8	139.6	7.97	1	.02	1.99	69 9	.99	11.2	139.5	9.10	2	7	2.27	69.9	1.13	140
160	11.2	159.4	10.40	2	4	2.60	79.9	1.30	12.8	159.2	11.87	2	.10	2.97	79.9	1.48	160
180	12.6	179.1	13.15	3	6	3.29	89.8	1.64	14.4	178.9	15.01	4	.14	3.76	89.8	1.87	180
200	14.0	198.8	16.22	4	.10	4.06	99.8	2.03	16.0	198.5	18.51	5	.20	4.64	99.7	2.31	200
220	15.4	218.4	19.61	5	.15	4.92	109.7	2.45	17.6	217.9	22.37	6	.27	5.61	109.7	2.79	220
240	16.8	237.9	23.31	6	.22	5.85	119.7	2.91	19.2	237.3	26.59	8	.35	6.68	119.5	3.32	240
260	18.2	257.4	27.33	8	.29	6.86	129.5	3.41	20.8	256.6	31.16	.10	.42	7.83	129.4	3.89	260
280	19.6	276.7	31.66	.11	.36	7.95	139.4	3.95	22.4	275.8	36.09	.13	.50	9.07	139.3	4.50	280
300	21.0	296.0	36.30	.13	.46	9.12	149.3	4.53	24.0	294.8	41.37	.16	.63	10.40	149.1	5.15	300
320	22.4	315.1	41.24	.2	.6	10.36	159.2	5.14	25.6	313.7	46.99	.2	.8	11.82	158.9	5.85	320
340	23.8	334.2	46.49	.2	.7	11.68	169.0	5.79	27.2	332.4	52.95	.2	1.0	13.33	168.7	6.59	340
360	25.2	353.1	52.01	.2	.8	13.09	178.8	6.48	28.8	351.0	59.24	.3	1.2	14.93	178.5	7.38	360
380	26.6	371.9	57.89	.2	1.0	14.58	188.6	7.21	30.4	369.4	65.86	.3	1.4	16.62	188.2	8.20	380
400	28.0	390.6	64.05	.3	1.2	16.14	198.4	7.98	32.0	387.7	72.82	.4	1.7	18.40	197.9	9.06	400
420	29.4	409.1	70.50	.3	1.4	17.78	208.2	8.78	33.6	405.8	80.10	.4	1.8	20.26	207.6	9.96	420
440	30.8	427.5	77.23	.4	1.6	19.50	217.9	9.62	35.2	423.7	87.70	.5	2.2	22.21	217.3	10.90	440
460	32.2	445.7	84.24	.4	1.8	21.29	227.6	10.49	36.8	441.4	95.61	.6	2.5	24.25	226.9	11.88	460
480	33.6	463.8	91.54	.5	2.1	23.16	237.3	11.39	38.4	458.9	103.84	.6	2.9	26.37	236.4	12.90	480
500	35.0	481.7	99.12	.6	2.4	25.11	246.9	12.32	40.0	476.2	112.37	.7	3.3	28.58	246.0	13.95	500
520	36.4	499.4	106.98	.6	2.7	27.13	256.5	13.29	41.6	493.3	121.19	.8	3.7	30.87	255.5	15.04	520
540	37.8	517.0	115.11	.7	3.0	29.22	266.1	14.29	43.2	510.1	130.30	.9	4.1	33.24	265.0	16.16	540
560	39.2	534.4	123.50	.8	3.4	31.39	275.7	15.33	44.8	526.7	139.70	1.0	4.6	35.69	274.4	17.32	560
580	40.6	551.6	132.15	.8	3.8	33.63	285.2	16.40	46.4	543.1	149.38	1.1	5.1	38.23	283.8	18.51	580
600	42.0	568.6	141.07	.9	4.2	35.94	294.7	17.50	48.0	559.2	159.33	1.2	5.6	40.84	293.1	19.72	600
620	43.4	585.4	150.24	1.0	4.6	38.33	304.2	18.63	49.6	575.1	169.55	1.3	6.2	43.53	302.4	20.96	620
640	44.8	602.0	159.66	1.1	5.1	40.79	313.6	19.79	51.2	590.8	180.03	1.4	6.8	46.30	311.7	22.24	640
660	46.2	618 4	169.32	1.3	5.5	43.32	323.0	20.97	52.8	606.1	190.76	1.6	7.4	49.15	320.9	23.55	660
680	47.6	634.6	179.22	1.4	6.1	45.92	332.3	22.18	54.4	621.2	201.73	1.7	8.2	52.08	330.0	24.89	680
700	49.0	650.5	189.35	1.5	6.7	48.58	341.6	23.42	56.0	636.0	212.95	1.8	9.0	55.09	339.1	26.25	700

l_1	18° Curve.								20° Curve.								l_1
	$I°$	x_1	y_1	e	e'	F	x'	y'	$I°$	x_1	y_1	e	e'	F	x'	y'	
40	3.6	40	.84			.21	20	.10	4.0	40	.93			.23	20	.12	40
60	5.4	59.9	1.88			.47	30	.24	6.0	59.9	2.09			.52	30	.26	60
80	7.2	79.9	3.35			.84	40	.42	8.0	79.8	3.72		.01	.93	40	0.46	80
100	9.0	99.8	5.23	.01	.02	1.31	50	.65	10.0	99.7	5.81	.01	2	1.45	49.9	0.72	100
120	10.8	119.6	7.52	1	5	1.88	59.9	.94	12.0	119.5	8.35	2	6	2.09	59.9	1.04	120
140	12.6	139.3	10.22	2	8	2.56	69.9	1.28	14.0	139.2	11.35	3	.10	2.84	69.8	1.42	140
160	14.4	159.0	13.34	3	.13	3.34	79.8	1.67	16.0	158.8	14.81	4	.15	3.71	79.8	1.85	160
180	16.2	178.6	16.87	4	.19	4.23	89.8	2.11	18.0	178.2	18.72	6	.22	4.70	89.7	2.34	180
200	18.0	198.0	20.80	6	.25	5.22	99.7	2.60	20.0	197.6	23.07	8	.30	5.79	99.6	2.88	200
220	19.8	217.4	25.13	8	.33	6.31	109.6	3.14	22.0	216.8	27.86	.10	.41	7.00	109.5	3.48	220
240	21.6	236.6	29.85	.10	.44	7.50	119.4	3.73	24.0	235.8	33.09	.13	.54	8.31	119.3	4.13	240
260	23.4	255.7	34.97	.13	.57	8.79	129.3	4.37	26.0	254.7	38.75	.16	.70	9.75	129.1	4.83	260
280	25.2	274.6	40.48	.17	.71	10.18	139.1	5.05	28.0	273.4	44.84	.20	.88	11.30	138.9	5.58	280
300	27.0	293.4	46.37	.20	.87	11.67	148.9	5.78	30.0	291.9	51.35	.25	1.07	12.96	148.6	6.39	300
320	28.8	312.0	52.65	.2	1.1	13.27	158.7	6.56	32.0	310.2	58.27	.3	1.3	14.73	158.3	7.25	320
340	30.6	330.4	59.30	.3	1.3	14.97	168.4	7.38	34.0	328.2	65.58	.4	1.6	16.60	168.0	8.16	340
360	32.4	348.7	66.32	.3	1.5	16.76	178.1	8.25	36.0	346.0	73.30	.4	1.9	18.58	177.6	9.11	360
380	34.2	366.7	73.70	.4	1.8	18.65	187.8	9.16	38.0	363.6	81.40	.5	2.2	20.67	187.2	10.11	380
400	36.0	384.5	81.43	.5	2.1	20.63	197.4	10.12	40.0	380.9	89.89	.6	2.6	22.86	196.8	11.15	400
420	37.8	402.1	89.51	.5	2.4	22.71	207.0	11.12	42.0	398.0	98.75	.7	3.0	25.16	206.3	12.21	420
440	39.6	419.5	97.94	.6	2.8	24.89	216.5	12.16	44.0	414.8	107.97	.7	3.5	27.56	215.7	13.38	440
460	41.4	436.6	106.72	.7	3.2	27.17	226.1	13.21	46.0	431.2	117.54	.8	3.9	30.06	225.1	14.56	460
480	43.2	453.4	115.88	.8	3.6	29.54	235.5	14.36	48.0	447.4	127.46	1.0	4.5	32.67	234.5	15.77	480
500	45.0	470.0	125.24	.9	4.1	32.00	245.0	15.52	50.0	463.2	137.71	1.1	5.1	35.37	243.8	17.02	500
520	46.8	486.4	134.97	1.0	4.6	34.55	254.3	[illegible]	52.0	478.8	148.29	1.2	5.7	38.17	253.0	18.31	520
540	48.6	502.4	145.01	1.1	5.2	37.19	263.7	17.94	54.0	494.0	159.17	1.3	6.4	41.07	262.2	19.63	540
560	50.4	518.2	155.34	1.2	5.8	39.92	272.9	19.20	56.0	508.8	170.36	1.5	7.2	44.07	271.3	20.99	560
580	52.2	533.7	165.96	1.3	6.4	42.74	282.2	20.50	58.0	523.3	181.83	1.6	8.0	47.16	280.4	22.39	580
600	54.0	548.9	176.86	1.5	7.1	45.64	291.4	21.83	60.0	537.5	193.58	1.7	8.8	50.33	289.4	23.81	600
620	55.8	563.8	188.03	1.6	7.8	48.63	300.5	23.20	62.0	551.2	205.60	1.8	9.8	53.60	298.3	25.26	620
640	57.6	578.4	199.46	1.7	8.6	51.70	309.6	24.59	64.0	564.6	217.87	2.1	10.7	56.96	307.1	26.74	640
660	59.4	592.6	211.15	1.9	9.4	54.85	318.6	26.00	66.0	577.6	230.38	2.3	11.7	60.41	315.9	28.25	660
680	61.2	606.5	223.09	2.1	10.3	58.09	327.6	27.45	68.0	590.3	243.12	2.4	12.8	63.94	324.6	29.77	680
700	63.0	620.1	235.28	2.2	11.2	61.40	336.5	28.93	70.0	602.5	256.07	2.6	14.0	67.56	333.3	31.32	700

l_1	22° CURVE.								24° CURVE.								l_1
	$I°$	x_1	y_1	e	e'	F	x'	y'	$I°$	x_1	y_1	e	e'	F	x'	y'	
40	4.4	40	1.02			.26	20	.13	4.8	40	1.12			.28	20	.14	40
60	6.6	59.9	2.30			.58	30	.29	7.2	59.9	2.51			.63	30	.31	60
80	8.8	79.8	4.09		.01	1.02	40	51	9.6	79.8	4.46	.01	.02	1.12	40	56	80
100	11.0	99.6	6.38	.01	2	1.60	49.9	80	12.0	99.6	6.96	1	4	1.74	49.9	.87	100
120	13.2	119.4	9.18	2	7	2.30	59.9	1.15	14.4	119.2	10.01	2	8	2.51	59.9	1.25	120
140	15.4	139.0	12.48	3	.12	3.13	69.8	1.56	16.8	138.8	13.69	4	14	3.41	69.8	1.70	140
160	17.6	158.5	16.27	4	.18	4.08	79.7	2.03	19.2	158.2	17.73	5	.21	4.45	79.7	2.21	160
180	19.8	177.9	20.56	6	.27	5.16	89.6	2.57	21.6	177.5	22.39	8	.31	5.62	89.6	2.79	180
200	22.0	197.1	25.33	9	.37	6.37	99.5	3.16	24.0	196.5	27.57	.11	.42	6.93	99.4	3.44	200
220	24.2	216.1	30.58	.12	.50	7.69	109.3	3.81	26.4	215.4	33.27	.15	.58	8.38	109.2	4.15	220
240	26.4	235.0	36.30	.15	.65	9.13	119.1	4.53	28.8	234.0	39.49	.18	.77	9.97	119.0	4.92	240
260	28.6	253.6	42.49	.19	.83	10.71	128.9	5.30	31.2	252.4	46.20	.23	.99	11.67	128.7	5.75	260
280	30.8	272.0	49.14	.24	1.05	12.42	138.6	6.12	33.6	270.5	53.40	.29	1.25	13.51	138.4	6.64	280
300	33.0	290.2	56.24	.29	1.30	14.22	148.3	7.00	36.0	288.4	61.08	.35	1.54	15.48	148.0	7.59	300
320	35.2	308.1	63.78	.4	1.6	16.15	158.0	7.93	38.4	305.9	69.22	.4	1.9	17.58	157.6	8.59	320
340	37.4	325.8	71.75	.4	1.9	18.21	167.6	8.92	40.8	323.2	77.82	.5	2.3	19.81	167.2	9.65	340
360	39.6	343.2	80.15	.5	2.3	20.38	177.2	9.95	43.2	340.1	86.86	.6	2.7	22.15	176.6	10.77	360
380	41.8	360.3	88.95	.6	2.7	22.66	186.7	11.03	45.6	356.6	96.33	.7	3.2	24.63	186.0	11.93	380
400	44.0	377.0	98.15	.7	3.1	25.05	196.1	12.17	48.0	372.8	106.22	.8	3.7	27.22	195.4	13.14	400
420	46.2	393.5	107.75	.8	3.6	27.56	205.5	13.34	50.4	388.6	116.50	.9	4.3	29.94	204.7	14.40	420
440	48.4	409.6	117.72	.9	4.2	30.18	214.9	14.56	52.8	404.1	127.17	1.0	4.9	32.77	213.9	15.70	440
460	50.6	425.4	128.05	1.0	4.8	32.92	224.1	15.82	55.2	419.1	138.21	1.2	5.6	35.72	223.1	17.05	460
480	52.8	440.8	138.73	1.1	5.5	35.75	233.3	17.12	57.6	433.7	149.60	1.3	6.4	38.77	232.1	18.44	480
500	55.0	455.9	149.75	1.2	6.2	38.69	242.5	18.46	60.0	447.9	161.32	1.5	7.8	41.93	241.1	19.85	500
520	57.2	470.5	161.10	1.4	7.0	41.74	251.6	19.84	62.4	461.6	173.36	1.6	8.2	45.22	250.0	21.31	520
540	59.4	484.8	172.76	1.5	7.8	44.89	260.6	21.25	64.8	474.9	185.70	1.8	9.2	48.60	258.9	22.80	540
560	61.6	498.6	184.71	1.7	8.7	48.14	269.5	22.70	67.2	487.7	198.32	2.0	10.2	52.09	267.6	24.31	560
580	63.8	512.1	196.94	1.8	9.7	51.48	278.4	24.17	69.6	500.1	211.21	2.2	11.4	55.68	276.3	25.84	580
600	66.0	525.1	209.44	2.0	10.7	54.93	287.2	25.67	72.0	511.9	224.34	2.4	12 6	59.37	284.9	27.40	600
620	68.2	537.7	222.19	2.2	11.8	58.46	295.9	27.19	74.4	523.8	237.70	2.6	13.9	63.15	293.4	28.99	620
640	70.4	549.9	235.17	2 4	13.0	62.09	304.5	28.74	76.8	534.2	251.26	2.8	15 3	67.03	301.7	30.60	640
660	72.6	561.6	248.37	2 6	14.2	65.80	313.1	30.30	79.2	544.6	265.00	3.0	16.8	70.98	310.0	32.23	660
680	74.8	572.9	261.77	2.8	15.5	69.61	321.6	31.89	81.6	554.4	278.91	3.2	18.4	75.03	318.2	33.86	680
700	77.0	583.7	275.36	3.0	17.0	73.49	329.9	33.49	84.0	563.8	292.97	3.4	20.1	79.18	326.3	35.51	700

l_1	26° CURVE. $I°$	x_1	y_1	e	e'	F	x'	y'	28° CURVE. $I°$	x_1	y_1	e	e'	F	x'	y'	l_1
40	5.2	40.0	1.21			.30	20	.15	5.6	39.9	1.30			.33	20	.16	40
50	6.5	49.9	1.89			.47	25	.24	7.0	49.9	2.08			.51	25	.25	50
60	7.8	59.9	2.72			.68	30	.34	8.4	59.8	2.93			.74	30	.37	60
70	9.1	69.8	3.70		.01	.93	35	.46	9.8	69.8	3 99		.01	1.00	34.9	.50	70
80	10.4	79.7	4.83	.01	2	1.21	39.9	.60	11.2	79.7	5.20	.01	2	1.30	39.9	.65	80
90	11.7	89.6	6.11	1	3	1.53	44.9	.76	12.6	89.5	6.57	1	3	1.64	44.9	.82	90
100	13.0	99.5	7.54	2	5	1.89	49.9	.94	14.0	99.4	8.11	2	5	2.03	49.9	1.01	100
110	14.3	109.3	9.11	2	6	2.28	54.9	1.14	15.4	109.2	9.80	2	8	2.46	54.8	1.22	110
120	15.6	119.1	10.83	3	9	2.71	59.8	1.35	16.8	119.0	11.65	3	.11	2.93	59.8	1.45	120
130	16.9	128.9	12.70	3	.12	3.18	64.8	1.58	18.2	128.7	13.66	4	.14	3.43	64.7	1.70	130
140	18.2	138.6	14.72	4	.15	3.69	69.7	1.84	19.6	138 4	15.83	5	.18	3.97	69.7	1.97	140
150	19.5	148.3	16.88	5	.19	4.24	74.7	2.11	21.0	148.0	18.15	6	.23	4.56	74.6	2.26	150
160	20.8	157.9	19.18	6	.24	4.82	79.6	2.40	22.4	157.6	20.62	.1	.3	5.18	79.6	2.57	160
170	22.1	167.5	21.63	8	.30	5.44	84.6	2.70	23.8	167.1	23.24	.1	.3	5.84	84.5	2.90	170
180	23.4	177.0	24.21	9	.36	6.09	89.5	3.02	25.2	176.5	26.02	.1	.4	6.54	89.4	3.24	180
190	24.7	186.5	26.94	.11	.43	6.78	94.4	3.36	26.6	185.9	28.95	.1	.5	7.29	94.3	3.61	190
200	26.0	195.9	29.81	.13	.51	7.50	99.3	3.71	28.0	195.3	32.02	.1	.6	8.07	99.2	3.99	200
210	27.3	205.3	32.81	.2	.6	8.26	104.2	4.09	29.4	204.6	35.25	.2	.7	8.89	104.1	4.39	210
220	28.6	214.6	35.95	.2	.7	9.06	109.1	4.48	30.8	213.7	38.61	.2	.8	9.75	108.9	4.81	220
230	29.9	223.8	39.23	.2	.8	9.90	113.9	4.89	32.2	222.8	42.12	.2	.9	10.64	113.8	5.24	230
240	31.2	233.0	42.61	.2	.9	10.77	118.8	5.31	33.6	231.9	45.77	.3	1.0	11.58	118.6	5.69	240
250	32.5	242.1	46.19	.2	1.0	11.67	123.7	5.75	35.0	240.8	49.56	.3	1.2	12.55	123.4	6.16	250
260	33.8	251.1	49.87	.3	1.1	12.61	128.5	6.20	36.4	249.7	53.49	.3	1.4	13.56	128.2	6.64	260
270	35.1	260.0	53.67	.3	1.3	13.59	133.3	6.67	37.8	258.5	57.55	.4	1.5	14.61	133.0	7.14	270
280	36.4	268.9	57.60	.3	1.5	14.60	138.1	7.16	39.2	267.2	61.75	.4	1.7	15.69	137.8	7.66	280
290	37.7	277.7	61.66	.4	1.6	15.65	142.9	7.66	40.6	275.8	66.07	.4	1.9	16.81	142.6	8.20	290
300	39.0	286.4	65.85	.4	1.8	16.73	147.7	8.17	42.0	284.3	70.53	.5	2.1	17.97	147.3	8.75	300
310	40.3	295.0	70.16	.5	2.0	17.85	152.4	8.70	43.4	292.7	75.12	.5	2.3	19.16	152.1	9.31	310
320	41.6	303.5	74.58	.5	2.2	19.00	157.2	9.25	44.8	301.0	79.83	.5	2.6	20.39	156.8	9.89	320
330	42.9	312.0	79.12	.6	2.4	20.18	161.9	9.81	46.2	309.2	84.66	.6	2.8	21.66	161.5	10.48	330
340	44.2	320.3	83.78	.6	2.6	21.39	166.7	10.38	47.6	317.3	89 61	.6	3.1	22.96	166.1	11.09	340
350	45.5	328.5	88.55	.7	2.9	22 64	171.4	10.97	49.0	325.2	94.68	.7	3.4	24.29	170.8	11.71	350
360	46.8	336.7	93.41	.7	3.1	23.92	176.1	11.57	50.4	333.1	99 86	.8	3.7	25.66	175.4	12.34	360
370	48.1	344.8	98.44	.7	3.4	25.23	180.8	12.18	51.8	340.9	105.16	.9	4.0	27.07	180.0	12.99	370

l_1	32° CURVE.								36° CURVE.								l_1
	$I°$	x_1	y_1	e	e'	F	x'	y'	$I°$	x_1	y_1	e	e'	F	x'	y'	
30	4.8	30	.84			.21	15	.10	5.4	30	.94			.24	15	.12	30
40	6.4	39.9	1.49			.37	20	.18	7.2	39.9	1.67			.42	20	.21	40
50	8.0	49.9	2.32		.01	.58	25	.29	9.0	49.9	2.61		.01	.65	25	.32	50
60	9.6	59.8	3.35	.01	2	.84	30	.42	10.8	59.8	3.76	.01	2	.94	30	.47	60
70	11.2	69.7	4.55	1	3	1.14	34.9	.57	12.6	69.7	5.11	1	4	1.28	34.9	.64	70
80	12.8	79.6	5.93	1	5	1.49	39.9	.74	14.4	79.5	6.67	1	6	1.67	39.9	.83	80
90	14.4	89.4	7.50	2	7	1.88	44.9	.94	16.2	89.3	8.43	2	9	2.11	44.9	1.05	90
100	16.0	99.2	9.25	2	.10	2.32	49.9	1.16	18.0	99.0	10.40	3	.13	2.61	49.8	1.30	100
110	17.6	109.0	11.18	3	.13	2.80	54.8	1.39	19.8	108.7	12.56	4	.16	3.15	54.8	1.57	110
120	19.2	118.7	13.29	4	.17	3.34	59.8	1.66	21.6	118.3	14.92	5	.22	3.75	59.7	1.86	120
130	20.8	128.3	15.58	5	.21	3.91	64.7	1.94	23.4	127.8	17.48	6	.28	4.39	64.6	2.18	130
140	22.4	137.9	18.04	7	.25	4.53	69.6	2.25	25.2	137.3	20.24	8	.35	5.09	69.5	2.52	140
150	24.0	147.4	20.68	8	.31	5.20	74.5	2.58	27.0	146.7	23.19	.10	.44	5.84	74.4	2.99	150
160	25.6	156.8	23.49	.1	.4	5.91	79.4	2.93	28.8	156.0	26.33	.1	.5	6.63	79.3	3.28	160
170	27.2	166.2	26.47	.1	.5	6.67	84.3	3.30	30.6	165.2	29.65	.1	.6	7.48	84.2	3.69	170
180	28.8	175.5	29.62	.1	.6	7.46	89.2	3.69	32.4	174.3	33.16	.2	.7	8.38	89.0	4.12	180
190	30.4	184.7	32.93	.2	.7	8.31	94.1	4.10	34.2	183.3	36.85	.2	.9	9.32	93.9	4.58	190
200	32.0	193.8	36.41	.2	.8	9.20	98.9	4.53	36.0	192.2	40.72	.2	1.0	10.31	98.7	5.06	200
210	33.6	202.9	40.05	.2	.9	10.13	103.8	4.98	37.8	201.0	44.75	.3	1.2	11.35	103.5	5.56	210
220	35.2	211.8	43.85	.3	1.1	11.10	108.6	5.45	39.6	209.7	48.97	.3	1.4	12.44	108.2	6.08	220
230	36.8	220.7	47.81	.3	1.2	12.12	113.4	5.94	41.4	218.3	53.36	.3	1.6	13.58	113.0	6.62	230
240	38.4	229.4	51.92	.3	1.4	13.19	118.2	6.45	43.2	226.7	57.92	.4	1.8	14.77	117.7	7.18	240
250	40.0	238.1	56.19	.4	1.6	14.29	123.0	6.97	45.0	235.0	62.62	.4	2.0	16.00	122.5	7.76	250
260	41.6	246.6	60.60	.4	1.8	15.43	127.7	7.52	46.8	243.2	67.48	.5	2.3	17.27	127.1	8.36	260
270	43.2	255.0	65.15	.4	2.0	16.62	132.5	8.08	48.6	251.2	72.50	.5	2.6	18.59	131.8	8.97	270
280	44.8	263.3	69.85	.5	2.3	17.84	137.2	8.66	50.4	259.1	77.67	.6	2.9	19.96	136.4	9.60	280
290	46.4	271.5	74.69	.5	2.5	19.11	141.9	9.25	52.2	266.8	82.98	.6	3.2	21.37	141.1	10.25	290
300	48.0	279.6	79.67	.6	2.8	20.42	146.5	9.86	54.0	274.4	88.43	.7	3.5	22.82	145.7	10.92	300
310	49.6	287.5	84.78	.6	3.1	21.76	151.2	10.48	55.8	281.9	94.02	.8	3.9	24.31	150.2	11.60	310
320	51.2	295.4	90.01	.7	3.4	23.15	155.8	11.12	57.6	289.2	99.73	.8	4.3	25.85	154.8	12.30	320
330	52.8	303.1	95.38	.8	3.7	24.57	160.4	11.77	59.4	296.3	105.57	.9	4.7	27.42	159.3	13.00	330
340	54.4	310.6	100.87	.8	4.1	26.04	165.0	12.44	61.2	303.2	111.54	1.0	5.1	29.04	163.8	13.72	340
350	56.0	318.0	106.48	.9	4.5	27.54	169.5	13.12	63.0	310.0	117.64	1.1	5.6	30.70	168.2	14.46	350
360	57.6	325.3	112.20	1.0	4.9	29.08	174.1	13.82	64.8	316.7	123.85	1.2	6.1	32.40	172.6	15.21	360

l_1	40° Curve.								44° Curve.								l_1
	I°	x_1	y_1	e	e'	F	x'	y'	I°	x_1	y_1	e	e'	F	x'	y'	
30	6.0	30	1.05			.26	15	.13	6.6	29.9	1.15			.29	15	.14	30
40	8.0	39.9	1.86			.46	20	.23	8.8	39.9	2.04			.51	20	.25	40
50	10.0	49.8	2.90		.01	.73	25	.36	11.0	49.8	3.19		.01	.80	25	.40	50
60	12.0	59.7	4.17	.01	3	1.04	29.9	.52	13.2	59.7	4.59	.01	3	1.15	29.9	.57	60
70	14.0	69.6	5.68	1	5	1.42	34.9	.71	15.4	69.5	6.24	1	6	1.56	34.9	.78	70
80	16.0	79.4	7.40	2	7	1.86	39.9	.93	17.6	79.2	8.14	2	9	2.04	39.9	1.02	80
90	18.0	89.1	9.36	3	.11	2.35	44.8	1.17	19.8	88.9	10.28	3	.13	2.58	44.8	1.28	90
100	20.0	98.8	11.53	4	.15	2.89	49.8	1.44	22.0	98.5	12.66	4	.18	3.18	49.7	1.58	100
110	22.0	108.4	13.93	5	.20	3.50	54.7	1.74	24.2	108.0	15.29	6	.25	3.84	54.7	1.90	110
120	24.0	117.9	16.54	6	.27	4.16	59.7	2.06	26.4	117.5	18.15	8	.32	4.56	59.5	2.26	120
130	26.0	127.3	19.37	8	.35	4.87	64.5	2.41	28.6	126.8	21.25	.10	.41	5.35	64.5	2.65	130
140	28.0	136.7	22.42	.10	.44	5.65	69.4	2.79	30.8	136.0	24.57	.12	.52	6.21	69.3	3.06	140
150	30.0	145.9	25.67	.13	.53	6.48	74.3	3.19	33.0	145.1	28.12	.14	.65	7.11	74.2	3.50	150
160	32.0	155.1	29.13	.2	.6	7.36	79.2	3.62	35.2	154.1	31.89	.2	.8	8.08	79.0	3.97	160
170	34.0	164.1	32.79	.2	.8	8.30	84.0	4.08	37.4	162.9	35.88	.2	.9	9.10	83.8	4.46	170
180	36.0	173.0	36.65	.2	1.0	9.29	88.8	4.55	39.6	171.6	40.07	.2	1.1	10.19	88.6	4.97	180
190	38.0	181.8	40.70	.2	1.1	10.33	93.6	5.05	41.8	180.1	44.47	.3	1.3	11.33	93.3	5.51	190
200	40.0	190.4	44.95	.3	1.3	11.43	98.4	5.58	44.0	188.5	49.07	.3	1.6	12.52	98.0	6.08	200
210	42.0	199.0	49.38	.3	1.5	12.58	103.1	6.12	46.2	196.7	53.87	.4	1.8	13.78	102.7	6.67	210
220	44.0	207.4	53.99	.4	1.7	13.78	107.8	6.69	48.4	204.8	58.86	.4	2.1	15.09	107.4	7.28	220
230	46.0	215.6	58.77	.4	2.0	15.03	112.5	7.28	50.6	212.7	64.02	.5	2.4	16.46	112.0	7.91	230
240	48.0	223.7	63.73	.5	2.2	16.33	117.2	7.88	52.8	220.4	69.36	.5	2.8	17.87	116.6	8.56	240
250	50.0	231.6	68.85	.5	2.5	17.68	121.9	8.51	55.0	227.9	74.87	.6	3.1	19.34	121.2	9.23	250
260	52.0	239.4	74.14	.6	2.8	19.08	126.5	9.15	57.2	235.2	80.55	.7	3.5	20.87	125.8	9.92	260
270	54.0	247.0	79.58	.6	3.2	20.53	131.1	9.81	59.4	242.4	86.38	.7	3.9	22.45	130.3	10.62	270
280	56.0	254.4	85.18	.7	3.6	22.03	135.6	10.49	61.6	249.3	92.35	.8	4.3	24.07	134.8	11.35	280
290	58.0	261.6	90.92	.8	4.0	23.58	140.2	11.19	63.8	256.0	98.47	.9	4.8	25.74	139.2	12.08	290
300	60.0	268.7	96.79	.9	4.4	25.17	144.7	11.90	66.0	262.5	104.72	1.0	5.3	27.46	143.6	12.83	300
310	62.0	275.6	102.80	.9	4.9	26.80	149.1	12.63	68.2	268.8	111.09	1.1	5.9	29.23	147.9	13.59	310
320	64.0	282.3	108.94	1.0	5.3	28.48	153.6	13.37	70.4	274.9	117.58	1.2	6.5	31.05	152.2	14.37	320
330	66.0	288.8	115.19	1.1	5.8	30.20	158.0	14.12	72.6	280.8	124.18	1.3	7.1	32.90	156.5	15.15	330
340	68.0	295.1	121.56	1.2	6.4	31.97	162.3	14.88	74.8	286.4	130.88	1.4	7.8	34.80	160.8	15.94	340
350	70.0	301.2	128.03	1.3	7.0	33.78	166.6	15.66	77.0	291.8	137.68	1.5	8.5	36.74	165.0	16.74	350
360	72.0	307.2	134.61	1.4	7.6	35.63	170.9	16.45	79.2	297.0	144.57	1.6	9.3	38.74	169.1	17.55	360

l_1	48° Curve.								52° Curve.								l_1
	$I°$	x_1	y_1	e	e'	F	x'	y'	$I°$	x_1	y_1	e	e'	F	x'	y'	
30	7.2	29.9	1.25			.32	15	.16	7.8	29.9	1.36			.34	15	0.17	30
40	9.6	39.9	2.23		.01	.56	20	.28	10.4	39.9	2.42		.01	.60	20	.30	40
50	12.0	49.8	3.48	.01	2	.87	25	.43	13.0	49.7	3.77	.01	2	.94	25	.47	50
60	14.4	59.6	5.00	1	4	1.25	29.9	.62	15.6	59.5	5.42	1	4	1.36	29.9	.68	60
70	16.8	69.4	6.80	2	7	1.70	34.9	.85	18.2	69.3	7.36	2	8	1.85	34.9	.92	70
80	19.2	79.1	8.87	3	.11	2.22	39.8	1.11	20.8	78.9	9.59	3	.12	2.41	39.8	1.20	80
90	21.6	88.7	11.19	4	.15	2.81	44.8	1.40	23.4	88.5	12.11	5	.18	3.04	44.7	1.51	90
100	24.0	98.2	13.79	5	.22	3.47	49.7	1.72	26.0	97.9	14.90	6	.26	3.75	49.6	1.86	100
110	26.4	107.7	16.64	7	.29	4.19	54.6	2.07	28.6	107.3	17.98	.1	.2	4.53	54.5	2.24	110
120	28.8	117.0	19.74	9	.38	4.98	59.5	2.46	31.2	116.5	21.32	.1	.4	5.38	59.4	2.65	120
130	31.2	126.2	23.10	.11	.49	5.83	64.3	2.87	33.8	125.5	24.93	.1	.6	6.31	64.2	3.10	130
140	33.6	135.2	26.70	.14	.62	6.75	69.2	3.32	36.4	134.4	28.80	.2	.7	7.30	69.1	3.58	140
150	36.0	144.2	30.54	.17	.77	7.74	74.0	3.79	39.0	143.2	32.92	.2	.9	8.37	73.8	4.09	150
160	38.4	153.0	34.61	.2	.9	8.79	78.8	4.30	41.6	151.8	37.29	.3	1.1	9.50	78.6	4.62	160
170	40.8	161.6	38.91	.2	1.1	9.90	83.6	4.83	44.2	160.1	41.89	.3	1.3	10.70	83.8	5.19	170
180	43.2	170.0	43.43	.3	1.3	11.07	88.3	5.39	46.8	168.3	46.72	.4	1.6	11.96	88.0	5.79	180

l_1	56° Curve.								60° Curve.								l_1
	$I°$	x_1	y_1	e	e'	F	x'	y'	$I°$	x_1	y_1	e	e'	F	x'	y'	
30	8.4	29.9	1.46			.37	15	.18	9.0	29.9	1.57		.01	.39	15	.20	30
40	11.2	39.8	2.60		.01	.65	20	.32	12.0	39.8	2.78	.01	2	.70	20	.35	40
50	14.0	49.7	4.06	.01	2	1.02	24.9	.51	15.0	49.7	4.34	1	4	1.09	24.9	.54	50
60	16.8	59.5	5.83	2	5	1.46	29.9	.73	18.0	59.4	6.24	2	7	1.67	29.9	.78	60
70	19.6	69.2	7.92	3	9	1.90	34.9	.99	21.0	69.1	8.47	3	.12	2.13	34.8	1.06	70
80	22.4	78.8	10.31	4	.15	2.59	39.8	1.29	24.0	78.6	11.03	4	.18	2.77	39.8	1.37	80
90	25.2	88.3	13.01	5	21	3.27	44.7	1.62	27.0	88.0	13.91	6	.25	3.50	44.7	1.73	90
100	28.0	97.6	16.01	6	.30	4.03	49.6	1.99	30.0	97.3	17.12	8	.35	4.32	49.5	2.13	100
110	30.8	106.9	19.31	.1	.4	4.87	54.5	2.40	33.0	106.4	20.62	.1	.5	5.21	54.4	2.57	110
120	33.6	116.0	22.89	.1	.5	5.79	59.3	2.85	36.0	115.3	24.43	.1	.6	6.19	59.2	3.04	120
130	36.4	124.9	26.75	.2	.7	6.78	64.1	3.32	39.0	124.1	28.53	.2	.8	7.25	64.0	3.54	130
140	39.2	133.6	30.88	.2	.9	7.85	68.9	3.83	42.0	132.7	32.92	.2	1.0	8.38	68.8	4.08	140
150	42.0	142.2	35.27	.2	1.2	8.98	73.7	4.37	45.0	141.0	37.58	.2	1.2	9.60	73.5	4.65	150
160	44.8	150.5	39.92	.3	1.3	10.20	78.4	4.95	48.0	149.1	42.49	.3	1.5	10.89	78.2	5.26	160
170	47.6	158.6	44.81	.3	1.5	11.48	83.1	5.55	51.0	157.0	47.65	.4	1.8	12.26	72.8	5.89	170
180	50.4	166.6	49.93	.4	1.8	12.83	87.7	6.17	54.0	164.7	53.06	.4	2.1	13.69	87.4	6.55	180

Base, $2b$ = 14 feet. Side slopes 1½ to 1.

C. H.	.0	.1	.2	.3	.4	.5	.6	.7	.8	.9
0	0.0	1.4	2.9	4.3	5.8	7.4	8.9	10.5	12.2	13.8
1	15.5	17.2	19.0	20.7	22.5	24.4	26.2	28.1	30.1	32.0
2	34.0	36.0	38.1	40.1	42.2	44.4	46.5	48.7	51.0	53.2
3	55.5	57.8	60.2	62.5	64.9	67.4	69.8	72.3	74.9	77.4
4	80.0	82.6	85.3	87.9	90.6	93.4	96.1	98.9	101.8	104.6
5	107.5	110.4	113.4	116.3	119.3	122.4	125.4	128.5	131.7	134.8
6	138.0	141.2	144.5	147.7	151.0	154.4	157.7	161.1	164.6	168.0
7	171.5	175.0	178.6	182.1	185.7	189.4	193.0	196.7	200.5	204.2
8	208.0	211.8	215.7	219.5	223.4	227.4	231.3	235.3	239.4	243.4
9	247.5	251.6	255.8	259.9	264.1	268.4	272.6	276.9	281.3	285.6
10	290.0	294.4	298.9	303.3	307.8	312.4	316.9	321.7	326.2	330.8
11	335.5	340.2	345.0	349.7	354.5	359.4	364.2	369.1	374.1	379.0
12	384.0	389.0	394.1	399.1	404.2	409.4	414.5	419.7	425.0	430.2
13	435.5	440.8	446.2	451.5	456.9	462.4	467.8	473.3	478.9	484.4
14	490.0	495.6	501.3	506.9	512.6	518.4	524.1	529.9	535.8	541.6
15	547.5	553.4	559.4	565.3	571.3	577.4	583.4	589.5	595.7	601.8
16	608.0	614.2	620.5	626.7	633.0	639.4	645.7	652.1	658.6	665.0
17	671.5	678.0	684.6	691.1	697.7	704.4	711.0	717.7	724.5	731.2
18	738.0	744.8	751.7	758.5	765.4	772.4	779.3	786.3	793.4	800.4
19	807.5	814.6	821.8	828.9	836.1	843.4	850.6	857.9	865.3	872.6
20	880.0	887.4	894.9	902.3	909.8	917.4	924.9	932.5	940.2	947.8
21	955.5	963.2	971.0	978.7	986.5	994.4	1002.2	1010.1	1018.1	1026.0
22	1034.0	1042.0	1050.1	1058.1	1066.2	1074.4	1082.5	1090.7	1099.0	1107.2
23	1115.5	1123.8	1132.2	1140.5	1148.9	1157.4	1165.8	1174.3	1182.9	1191.4
24	1200.0	1208.6	1217.3	1225.9	1234.6	1243.4	1252.1	1260.9	1269.8	1278.6
25	1287.5	1296.4	1305.4	1314.3	1323.3	1332.4	1341.4	1350.5	1359.7	1368.8
26	1378.0	1387.2	1396.5	1405.7	1415.0	1424.4	1433.7	1443.1	1452.6	1462.0
27	1471.5	1481.0	1490.6	1500.1	1509.7	1519.4	1529.0	1538.7	1548.5	1558.2
28	1568.0	1577.8	1587.7	1597.5	1607.4	1617.4	1627.3	1637.3	1647.4	1657.4
29	1667.5	1677.6	1687.8	1697.9	1708.1	1718.4	1728.6	1738.9	1749.3	1759.6

Base, $2b$ = 15 feet. Side slopes 1½ to 1.

C. H.	.0	1	.2	.3	.4	.5	.6	.7	.8	.9
0	0.0	1.5	3.1	4.6	6.2	7.9	9.5	11.2	13.0	14.7
1	16.5	18.3	20.2	22.0	23.9	25.9	27.8	29.8	31.9	33.9
2	36.0	38.1	40.3	42.4	44.6	46.9	49.1	51.4	53.8	56.1
3	58.5	60.9	63.4	65.8	68.3	70.9	73.4	76.0	78.7	81.3
4	84.0	86.7	89.5	92.2	95.0	97.9	100.7	103.6	106.6	109.5
5	112.5	115.5	118.6	121.6	124.7	127.9	131.0	134.2	137.5	140.7
6	144.0	147.3	150.7	154.0	157.4	160.9	164.3	167.8	171.4	174.9
7	178.5	182.1	185.8	189.4	193.1	196.9	200.6	204.4	208.3	212.1
8	216.0	219.9	223.9	227.8	231.8	235.9	239.9	244.0	248.2	252.3
9	256.5	260.7	265.0	269.2	273.5	277.9	282.2	286.6	291.1	295.5
10	300.0	304.5	309.1	313.6	318.2	322.9	327.5	332.2	337.0	341.7
11	346.5	351.3	356.2	361.0	365.9	370.9	375.8	380.8	385.9	390.9
12	396.0	401.1	406.3	411.4	416.6	421.9	427.1	432.4	437.8	443.1
13	448.5	453.9	459.4	464.8	470.3	475.9	481.4	487.0	492.7	498.3
14	504.0	509.7	515.5	521.2	527.0	532.9	538.7	544.6	550.6	556.5
15	562.5	568.5	574.6	580.6	586.7	592.9	599.0	605.2	611.5	617.7
16	624.0	630.3	636.7	643.0	649.4	655.9	662.3	668.8	675.4	681.9
17	688.5	695.1	701.8	708.4	715.1	721.9	728.6	735.4	742.3	749.1
18	756.0	762.9	769.9	776.8	783.8	790.9	797.9	805.0	812.2	819.3
19	826.5	833.7	841.0	848.2	855.5	862.9	870.2	877.6	885.1	892.5
20	900.0	907.5	915.1	922.6	930.2	937.9	945.5	953.2	961.0	968.7
21	976.5	984.3	992.2	1000.0	1007.9	1015.9	1023.8	1031.8	1039.9	1047.9
22	1056.0	1064.1	1072.3	1080.4	1088.6	1096.9	1105.1	1113.4	1121.8	1130.1
23	1138.5	1146.9	1155.4	1163.8	1172.3	1180.9	1189.4	1178.0	1206.7	1215.3
24	1224.0	1232.7	1241.5	1250.2	1259.0	1267.9	1276.7	1285.6	1294.6	1303.5
25	1312.5	1321.5	1330.6	1339.6	1348.7	1357.9	1367.0	1376.2	1385.5	1394.7
26	1404.0	1413.3	1422.7	1432.0	1441.4	1450.9	1460.3	1469.8	1479.4	1488.9
27	1498.5	1508.1	1517.8	1527.4	1537.1	1546.9	1556.6	1566.4	1576.3	1586.1
28	1596.0	1605.9	1615.9	1625.8	1635.8	1645.9	1655.9	1666.0	1676.2	1686.3
29	1696.5	1706.7	1717.0	1727.2	1737.5	1747.9	1758.2	1768.6	1779.1	1789.5

Base, $2b$ = 28 feet. Side slopes $1\frac{1}{4}$ to 1.

C. H.	.0	.1	.2	.3	.4	.5	.6	.7	.8	.9
0	0.0	2.8	5.7	8.5	11.4	14.4	17.3	20.3	23.4	26.4
1	29.5	32.6	35.8	38.9	42.1	45.4	48.6	51.9	55.3	58.6
2	62.0	65.4	68.9	72.3	75.8	79.4	82.9	86.5	90.2	93.8
3	97.5	101.2	105.0	108.7	112.5	116.4	120.2	124.1	128.1	132.0
4	136.0	140.0	144.1	148.1	152.2	156.4	160.5	164.7	169.0	173.2
5	177.5	181.8	186.2	190.5	194.9	199.4	203.8	208.3	212.9	217.4
6	222.0	226.6	231.3	235.9	240.6	245.4	250.1	254.9	259.8	264.6
7	269.5	274.4	279.4	284.3	289.3	294.4	299.4	304.5	309.7	314.8
8	320.0	325.2	330.5	335.7	341.0	346 4	351.7	357.1	362.6	368.0
9	373.5	379.0	384.6	390.1	395.7	401.4	407.0	412.7	418.5	424.2
10	430.0	435.8	441.7	447.5	453.4	459.4	465.3	471.3	477.4	483.4
11	489.5	495.6	501.8	507.9	514.1	520.4	526.6	532.9	539.3	545.6
12	552.0	558.4	564.9	571.3	577.8	584.4	590.9	597.5	604.2	610.8
13	617.5	624.2	631.0	637.7	644.5	651.4	658.2	665.1	672.1	679.0
14	686.0	693.0	700.1	707.1	714.2	721.4	728.5	735.7	743.0	750.2
15	757.5	764.8	772.2	779.5	786.9	794.4	801.8	809.3	816.9	824.4
16	832.0	839.6	847.3	854.9	862.6	870.4	878.1	885.9	893.8	901.6
17	909.5	917.4	925.4	933.3	941.3	949.4	957.4	965.5	973.7	981.8
18	990.0	998.2	1006.5	1014.7	1023.0	1031.4	1039.7	1048.1	1056.6	1065.0
19	1073.5	1082.0	1090.6	1099.1	1107.7	1116.4	1125.0	1133.7	1142.5	1151.2
20	1160.0	1168.8	1177.7	1186.5	1195.4	1204.4	1213.3	1222.3	1231.4	1240.4
21	1249.5	1258.6	1267.8	1276.9	1286.1	1295.4	1304.6	1313.9	1323.3	1332.6
22	1342.0	1351.4	1360.9	1370.3	1379.8	1389.4	1398.9	1408.5	1418.2	1427.8
23	1437.5	1447.2	1457.0	1466.7	1476.5	1486.4	1496.2	1506.1	1516.1	1526.0
24	1536.0	1546.0	1556.1	1566.1	1576.2	1586.4	1596.5	1606.7	1617.0	1627.2
25	1637.5	1647.8	1658.2	1668.5	1678.9	1689.4	1699.8	1710.3	1720.9	1731.4
26	1742.0	1752.6	1763.3	1773.9	1784.6	1795.4	1806.1	1816.9	1827.8	1838.6
27	1849.5	1860.4	1871.4	1882 3	1893.3	1904.4	1915.4	1926.5	1937.7	1948.8
28	1960 0	1971.2	1982.5	1993.7	2005.0	2016.4	2027.7	2039.1	2050.6	2062.0
29	2073.5	2085 0	2096.6	2108.1	2119.7	2131.4	2143.0	2154.7	2166.5	2178.2

Base, $2b$ = 18 feet. Side slopes 1 to 1.

C. H.	.0	.1	.2	.3	.4	.5	.6	.7	.8	.9
0	0.0	1.8	3.6	5.5	7.4	9.3	11.2	13.1	15.0	17.0
1	19.0	21.0	23.0	25.1	27.2	29.3	31.4	33.5	35.6	37.8
2	40.0	42.2	44.4	46.7	49.0	51.3	53.6	55.9	58.2	60.6
3	63.0	65.4	67.8	70.3	72.8	75.3	77.8	80.3	82.8	85.4
4	88.0	90.6	93.2	95.9	98.6	101.3	104.0	106.7	109.4	112.2
5	115.0	117.8	120.6	123.5	126.4	129.3	132.2	135.1	138.0	141.0
6	144.0	147.0	150.0	153.1	156.2	159.3	162.4	165.5	168.6	171.8
7	175.0	178.2	181.4	184.7	188.0	191.3	194.6	197.9	201.2	204.6
8	208.0	211.4	214.8	218.3	221 8	225.3	228.8	232.3	235.8	239.4
9	243.0	246.6	250.2	253.9	257.6	261.3	265.0	268.7	272.4	276.2
10	280.0	283.8	287.6	291.5	295.4	299.3	303.2	307.1	311.0	315.0
11	319.0	323.0	327.0	331.1	335.2	339.3	343.4	347.5	351.6	355.8
12	360.0	364.2	368.4	372.7	377.0	381.3	385.6	389.9	394.2	398.6
13	403.0	407.4	411.8	416.3	420.8	425.3	429.8	434.3	438.8	443.4
14	448.0	452.6	457.2	461.9	466.6	471.3	476.0	480.7	485.4	490.2
15	495.0	499.8	504.6	509.5	514.4	519.3	524.2	529.1	534.0	539.0
16	544.0	549.0	554.0	559.1	564.2	569.3	574.4	579.5	584.6	589.8
17	595.0	600.2	605.4	610.7	616.0	621.3	626.6	631.9	637.2	642.6
18	648.0	653 4	658.8	664.3	669.8	675.3	680.8	686.3	691.8	697.4
19	703.0	708.6	714.2	719.9	725.6	731.3	737.0	742.7	748.4	754.2
20	760.0	765.8	771.6	777.5	783.4	789.3	795.2	801.1	807.0	813.0
21	819.0	825.0	831.0	837.1	843.2	849.3	855.4	861.5	867.6	873.8
22	880.0	886.2	892.4	898.7	905.0	911.3	917.6	923.9	930.2	936.6
23	943.0	949.4	955.8	962.3	968.8	975.3	981.8	988.3	994.8	1001.4
24	1008.0	1014.6	1021.2	1027.9	1034.6	1041.3	1048.0	1054.7	1061.4	1068.2
25	1075 0	1081.8	1088.6	1095.5	1102.4	1109.3	1116.2	1123.1	1130.0	1137.0
26	1144.0	1151.0	1158.0	1165.1	1172.2	1179.3	1186.4	1193.5	1200.6	1207.8
27	1215.0	1222.2	1229.4	1236.7	1244.0	1251.3	1258.6	1265.9	1273.2	1280.6
28	1288.0	1295 4	1302.8	1310.3	1317.8	1325.3	1332.8	1340.3	1347.8	1355.4
29	1363.0	1370.6	1378.2	1385.9	1393.6	1401.3	1409.0	1416.7	1424 4	1432.2

Base, $2b$ = 20 feet. Side slopes 1 to 1.

C. H.	.0	.1	.2	.3	.4	.5	.6	.7	.8	.9
0	0.0	2.0	4.0	6.1	8.2	10.3	12.4	14.5	16.6	18.8
1	21.0	23.2	25.4	27.7	30.0	32.3	34.6	36.9	39.2	41.6
2	44.0	46.4	48.8	51.3	53.8	56.3	58.8	61.3	63.8	66.4
3	69.0	71.6	74.2	76.9	79.6	82.3	85.0	87.7	90.4	93.2
4	96.0	98.8	101.6	104.5	107.4	110.3	113.2	116.1	119.0	122.0
5	125.0	128.0	131.0	134.1	137.2	140.3	143.4	146.5	149.6	152.8
6	156.0	159.2	162.4	165.7	169.0	172.3	175.6	178.9	182.2	185.6
7	189.0	192.4	195.8	199.3	202.8	206.3	209.8	213.3	216.8	220.4
8	224.0	227.6	231.2	234.9	238.6	242.3	246.0	249.7	253.4	257.2
9	261.0	264.8	268.6	272.5	276.4	280.3	284.2	288.1	292.0	296.0
10	300.0	304.0	308.0	312.1	316.2	320.3	324.4	328.5	332.6	336.8
11	341.0	345.2	349.4	353.7	358.0	362.3	366.6	370.9	375.2	379.6
12	384.0	388.4	392.8	397.3	401.8	406.3	410.8	415.3	419.8	424.4
13	429.0	433.6	438.2	442.9	447.6	452.3	457.0	461.7	466.4	471.2
14	476.0	480.8	485.6	490.5	495.4	500.3	505.2	510.1	515.0	520.0
15	525.0	530.0	535.0	540.1	545.2	550.3	555.4	560.5	565.6	570.8
16	576.0	581.2	586.4	591.7	597.0	602.3	607.6	612.9	618.2	623.6
17	629.0	634.4	639.8	645.3	650.8	656.3	661.8	667.3	672.8	678.4
18	684.0	689.6	695.2	700.9	706.6	712.3	718.0	723.7	729.4	735.2
19	741.0	746.8	752.6	758.5	764.4	770.3	776.2	782.1	788.0	794.0
20	800.0	806.0	812.0	818.1	824.2	830.3	836.4	842.5	848.6	854.8
21	861.0	867.2	873.4	879.7	886.0	892.3	898.6	904.9	911.2	917.6
22	924.0	930.4	936.8	943.3	949.8	956.3	962.8	969.3	975.8	982.4
23	989.0	995.6	1002.2	1008.9	1015.6	1022.3	1029.0	1035.7	1042.4	1049.2
24	1056.0	1062.8	1069.6	1076.5	1083.4	1090.3	1097.2	1104.1	1111.0	1118.0
25	1125.0	1132.0	1139.0	1146.1	1153.2	1160.3	1167.4	1174.5	1181.6	1188.8
27	1196.0	1203.2	1210.4	1217.7	1225.0	1232.3	1239.6	1246.9	1254.2	1261.6
26	1269.0	1276.4	1283.8	1291.3	1298.8	1306.3	1313.8	1321.3	1328.8	1336.4
28	1344.0	1351.6	1359.2	1366.9	1374.6	1382.3	1390.0	1397.7	1405.4	1413.2
29	1421.0	1428.8	1436.6	1444.5	1452.4	1460.3	1468.2	1476.1	1484.0	1492.0

Base, $2b$ = 30 feet. Side slopes 1 to 1.

C. H.	.0	.1	.2	.3	.4	.5	.6	.7	.8	.9
0	0.0	3.0	6.0	9.1	12.2	15.3	18.4	21.5	24.6	27.8
1	31.0	34.2	37.4	40.7	44.0	47.3	50.6	53.9	57.2	60.6
2	64.0	67.4	70.8	74.3	77.8	81.3	84.8	88.3	91.8	95.4
3	99.0	102.6	106.2	109.9	113.6	117.3	121.0	124.7	128.4	132.2
4	136.0	139.8	143.6	147.5	151.4	155.3	159.2	163.1	167.0	171.0
5	175.0	179.0	183.0	187.1	191.2	195.3	199.4	203.5	207.6	211.8
6	216.0	220.2	224.4	228.7	233.0	237.3	241.6	245.9	250.2	254.6
7	259.0	263.4	267.8	272.3	276.8	281.3	285.8	290.3	294.8	299.4
8	304.0	308.6	313.2	317.9	322.6	327.3	332.0	336.7	341.4	346.2
9	351.0	355.8	360.6	365.5	370.4	375.3	380.2	385.1	390.0	395.0
10	400.0	405.0	410.0	415.1	420.2	425.3	430.4	435.5	440.6	445.8
11	451.0	456.2	461.4	466.7	472.0	477.3	482.6	487.9	493.2	498.6
12	504.0	509.4	514.8	520.3	525.8	531.3	536.8	542.3	547.8	553.4
13	559.0	564.6	570.2	575.9	581.6	587.3	593.0	598.7	604.4	610.2
14	616.0	621.8	627.6	633.5	639.4	645.3	651.2	657.1	663.0	669.0
15	675.0	681.0	687.0	693.1	699.2	705.3	711.4	717.5	723.6	729.8
16	736.0	742.2	748.4	754.7	761.0	767.3	773.6	779.9	786.2	792.6
17	799.0	805.4	811.8	818.3	824.8	831.3	837.8	844.3	850.8	857.4
18	864.0	870.6	877.2	883.9	890.6	897.3	904.0	910.7	917.4	924.2
19	931.0	937.8	944.6	951.5	958.4	965.3	972.2	979.1	986.0	993.0
20	1000.0	1007.0	1014.0	1021.1	1028.2	1035.3	1042.4	1049.5	1056.6	1063.8
21	1071.0	1078.2	1085.4	1092.7	1100.0	1107.3	1114.6	1121.9	1129.2	1136.6
22	1144.0	1151.4	1158.8	1166.3	1173.8	1181.3	1188.8	1196.3	1203.8	1211.4
23	1219.0	1226.6	1234.2	1241.9	1249.6	1257.3	1265.0	1272.7	1280.4	1288.2
24	1296.0	1303.8	1311.6	1319.5	1327.4	1335.3	1343.2	1351.1	1359.0	1367.0
25	1375.0	1383.0	1391.0	1399.1	1407.2	1415.3	1423.4	1431.5	1439.6	1447.8
26	1456.0	1464.2	1472.4	1480.7	1489.0	1497.3	1505.6	1513.9	1522.2	1530.6
27	1539.0	1547.4	1555.8	1564.3	1572.8	1581.3	1589.8	1598.3	1606.8	1615.4
28	1624.0	1632.6	1641.2	1649.9	1658.6	1667.3	1676.0	1684.7	1693.4	1702.2
29	1711.0	1719.8	1728.6	1737.5	1746.4	1755.3	1764.2	1773.1	1782.0	1791.0

Base, $2b$ = 16 feet. Side slopes ¼ to 1.

C. H.	.0	.1	.2	.3	.4	.5	.6	.7	.8	.9
0	0.0	1.6	3.2	4.8	6.4	8.1	9.7	11.3	13.0	14.6
1	16.3	17.9	19.6	21.2	22.9	24.6	26.2	27.9	29.6	31.3
2	33.0	34.7	36.4	38.1	39.8	41.6	43.3	45.0	46.8	48.5
3	50.3	52.0	53.8	55.5	57.3	59.1	60.8	62.6	64.4	66.2
4	68.0	69.8	71.6	73.4	75.2	77.1	78.9	80.7	82.6	84.4
5	86.3	88.1	90.0	91.8	93.7	95.6	97.4	99.3	101.2	103.1
6	105.0	106.9	108.8	110.7	112.6	114.6	116.5	118.4	120.4	122.3
7	124.3	126.2	128.2	130.1	132.1	134.1	136.0	138.0	140.0	142.0
8	144.0	146.0	148.0	150.0	152.0	154.1	156.1	158.1	160.2	162.2
9	164.3	166.3	168.4	170.4	172.5	174.6	176.6	178.7	180.8	182.9
10	185.0	187.1	189.2	191.3	193.4	195.6	197.7	199.8	202.0	204.1
11	206.3	208.4	210.6	212.7	214.9	217.1	219.2	221.4	223.6	225.8
12	228.0	230.2	232.4	234.6	236.8	239.1	241.3	243.5	245.8	248.0
13	250.3	252.5	254.8	257.0	259.3	261.6	263.8	266.1	268.4	270.7
14	273.0	275.3	277.6	279.9	282.2	284.6	286.9	289.2	291.6	293.9
15	296.3	298.6	301.0	303.3	305.7	308.1	310.4	312.8	315.2	317.6
16	320.0	322.4	324.8	327.2	329.6	332.1	334.5	336.9	339.4	341.8
17	344.3	346.7	349.2	351.6	354.1	356.6	359.0	361.5	364.0	366.5
18	369.0	371.5	374.0	376.5	379.0	381.6	384.1	386.6	389.2	391.7
19	394.3	396.8	399.4	401.9	404.5	407.1	409.6	412.2	414.8	417.4
20	420.0	422.6	425.2	427.8	430.4	433.1	435.7	438.3	441.0	443.6
21	446.3	448.9	451.6	454.2	456.9	450.6	462.2	464.9	467.6	470.3
22	473.0	475.7	478.4	481.1	483.8	486.6	489.3	492.0	494.8	497.5
23	500.3	503.0	505.8	508.5	511.3	514.1	516.8	519.6	522.4	525.2
24	528.0	530.8	533.6	536.4	539.2	542.1	544.9	547.7	550.6	553.4
25	556.3	559.1	562.0	564.8	567.7	570.6	573.4	576.3	579.2	582.1
26	585.0	587.9	590.8	593.7	596.6	599.6	602.5	605.4	608.4	611.3
27	614.3	617.2	620.2	623.1	626.1	629.1	632.0	635.0	638.0	641.0
28	644.0	647.0	650.0	653.0	656.0	659.1	662.1	665.1	668.2	671.2
29	674.3	677.3	680.4	683.4	686.5	689.6	692.6	695.7	698.8	701.9

Base, $2b$ = 18 feet. Side slopes ¼ to 1.

C. H.	.0	.1	.2	.3	.4	.5	.6	.7	.8	.9
0	0.0	1.8	3.6	5.4	7.2	9.1	10.9	12.7	14.6	16.4
1	18.3	20.1	22.0	23.8	25.7	27.6	29.4	31.3	33.2	35.1
2	37.0	38.9	40.8	42.7	44.6	46.6	48.5	50.4	52.4	54.3
3	56.3	58.2	60.2	62.1	64.1	66.1	68.0	70.0	.0	74.0
4	76.0	78.0	80.0	82.0	84.0	86.1	88.1	90.1	.2	94.2
5	96.3	98.3	100.4	102.4	104.5	106.6	108.6	110.7	112.8	114.9
6	117.0	119.1	121.2	123.3	125.4	127.6	129.7	131.8	134.0	136.1
7	138.3	140.4	142.6	144.7	146.9	149.1	151.2	153.4	155.6	157.8
8	160.0	162.2	164.4	166.6	168.8	171.1	173.3	175.5	177.8	180.0
9	182.3	184.5	186.8	189.0	191.3	193.6	195.8	198.1	200.4	202.7
10	205.0	207.3	209.6	211.9	214.2	216.6	218.9	221.2	223.6	225.9
11	228.3	230.6	233.0	235.3	237.7	240.1	212.4	244.8	247.2	249.6
12	252.0	254.4	256.8	259.2	261.6	264.1	266.5	268.9	271.4	273.8
13	276.3	278.7	281.2	283.6	286.1	288.6	291.0	293.5	296.0	298.5
14	301.0	303.5	306.0	308.5	311.0	313.6	316.1	318.6	321.2	323.7
15	326.3	328.8	331.4	333.9	336.5	339.1	341.6	344.2	346 8	349.4
16	352.0	354.6	357.2	359.8	362.4	365.1	367.7	370.3	373.0	375.6
17	378.3	380.9	383.6	386.2	388.9	391.6	394.2	396.9	399.6	402.3
18	405.0	407.7	410.4	413.1	415.8	418.6	421.3	424.0	426.8	429.5
19	432.3	435.0	437.8	440.5	443.3	446.1	448.8	451.6	454.4	457.2
20	460.0	462.8	465.6	468.4	471.2	474.1	476.9	479.7	482.6	485.4
21	488.3	491.1	494.0	496.8	499.7	502.6	505.4	508.3	511.2	514.1
22	517.0	519.9	522.8	525.7	528.6	531.6	534.5	537.4	540.4	543.3
23	546.3	549.2	552.2	555.1	558.1	561.1	564.0	567.0	570.0	573.0
24	576.0	579.0	582.0	585.0	588.0	591.1	594.1	597.1	600.2	603.2
25	606.3	609.3	612.4	615.4	618.5	621.6	624.6	627.7	630.8	633.9
26	637 0	640.1	643.2	646.3	649.4	652.6	655.7	658.8	662.0	665.1
27	668.3	671.4	674.6	677.7	680.9	684.1	687.2	600.4	693.6	696.8
28	700.0	703.2	706.4	709.6	712.8	716.1	719.3	722.5	725.8	729.0
29	732.3	735.5	738.8	742.0	745.3	748.6	751.8	755.1	758.4	761.7

Correction = $(h_m - h_0)^2 s$. (See **221.**)

SIDE SLOPES 1½ TO 1.

$h_m - h_0$	.0	.1	.2	.3	.4	.5	.6	.7	.8	.9
0	0.0	0.0	0.1	0.1	0.2	0.4	0.5	0.7	1.0	1.2
1	1.5	1.8	2.2	2.5	2.9	3.4	3.8	4.3	4.9	5.4
2	6.0	6.6	7.3	7.9	8.6	9.4	10.1	10.9	11.8	12.6
3	13.5	14.4	15.4	16.3	17.3	18.4	19.4	20.5	21.7	22.8
4	24.0	25.2	26.5	27.7	29.0	30.4	31.7	33.1	34.6	36.0
5	37.5	39.0	40.6	42.1	43.7	45.4	47.0	48.7	50.5	52.2
6	54.0	55.8	57.7	59.5	61.4	63.4	65.3	67.3	69.4	71.4
7	73.5	75.6	77.8	79.9	82.1	84.4	86.6	88.9	91.3	93.6
8	96.0	98.4	100.9	103.3	105.8	108.4	110.9	113.5	116.2	118.8
9	121.5	124.2	127.0	129.7	132.5	135.4	138.2	141.1	144.1	147.0
10	150.0	153.0	156.1	159.1	162.2	165.4	168.5	171.7	175.0	178.2
11	181.5	184.8	188.2	191.5	194.9	198.4	201.8	205.3	208.9	212.4

SIDE SLOPES 1 TO 1.

$h_m - h_0$	.0	.1	.2	.3	.4	.5	.6	.7	.8	.9
0	0.0	0.0	0.0	0.1	0.2	0.3	0.4	0.5	0.6	0.8
1	1.0	1.2	1.4	1.7	2.0	2.3	2.6	2.9	3.2	3.6
2	4.0	4.4	4.8	5.3	5.8	6.3	6.8	7.3	7.8	8.4
3	9.0	9.6	10.2	10.9	11.6	12.3	13.0	13.7	14.4	15.2
4	16.0	16.8	17.6	18.5	19.4	20.3	21.2	22.1	23.0	24.0
5	25.0	26.0	27.0	28.1	29.2	30.3	31.4	32.5	33.6	34.8
6	36.0	37.2	38.4	39.7	41.0	42.3	43.6	44.9	46.2	47.6
7	49.0	50.4	51.8	53.3	54.8	56.3	57.8	59.3	60.8	62.4
8	64.0	65.6	67.2	68.9	70.6	72.3	74.0	75.7	77.4	79.2
9	81.0	82.8	84.6	86.5	88.4	90.3	92.2	94.1	96.0	98.0
10	100.0	102.0	104.0	106.1	108.2	110.3	112.4	114.5	116.6	118.8
11	121.0	123.2	125.4	127.7	130.0	132.3	134.6	136.9	139.2	141.6

SIDE SLOPES ½ TO 1.

$h_m - h_0$	.0	.1	.2	.3	.4	.5	.6	.7	.8	.9
0	0.0	0.0	0.0	0.0	0.1	0.1	0.2	0.2	0.3	0.4
1	0.5	0.6	0.7	0.8	1.0	1.1	1.3	1.4	1.6	1.8
2	2.0	2.2	2.4	2.6	2.9	3.1	3.4	3.6	3.9	4.2
3	4.5	4.8	5.1	5.4	5.8	6.1	6.5	6.8	7.2	7.6
4	8.0	8.4	8.8	9.2	9.7	10.1	10.6	11.0	11.5	12.0
5	12.5	13.0	13.5	14.0	14.6	15.1	15.7	16.2	16.8	17.4
6	18.0	18.6	19.2	19.8	20.5	21.1	21.8	22.4	23.1	23.8
7	24.5	25.2	25.9	26.6	27.4	28.1	28.9	29.6	30.4	31.2
8	32.0	32.8	33.6	34.4	35.3	36.1	37.0	37.8	38.7	39.6
9	40.5	41.4	42.3	43.2	44.2	45.1	46.1	47.0	48.0	49.0
10	50.0	51.0	52.0	53.0	54.1	55.1	56.2	57.2	58.3	59.4
11	60.5	61.6	62.7	63.8	65.0	66.1	67.3	68.4	69.6	70.8

IDE SLOPES ¼ TO 1.

$h_m - h_0$	.0	.1	.2	.3	.4	.5	.6	.7	.8	.9
0	0.0	0.0	0.0	0.0	0.0	0.1	0.1	0.1	0.2	0.2
1	0.3	0.3	0.4	0.4	0.5	0.6	0.6	0.7	0.8	0.9
2	1.0	1.1	1.2	1.3	1.4	1.6	1.7	1.8	2.0	2.1
3	2.3	2.4	2.6	2.7	2.9	3.1	3.2	3.4	3.6	3.8
4	4.0	4.2	4.4	4.6	4.8	5.1	5.3	5.5	5.8	6.0
5	6 3	6.5	6.8	7.0	7.3	7.6	7.8	8.1	8.4	8.7
6	9.0	9 3	9.6	9.9	10.2	10.6	10.9	11.2	11.6	11.9
7	12.3	12.6	13.0	13.3	13.7	14.1	14.4	14.8	15.2	15.6
8	16.0	16.4	16.8	17.2	17.6	18.1	18.5	18.9	19.4	19.8
9	20.3	20.7	21.2	21.6	22.1	22.6	23.0	23.5	24.0	24.5
10	25.0	25.5	26.0	26.5	27.0	27.6	28.1	28.6	29.2	29.7
11	30.3	30.8	21.4	31.9	32.5	33.1	33.6	34.2	34.8	35.4

Depth	Base 12	Base 14	Base 16	Base 18	Base 22	Base 24	Base 26	Base 28
1	45	53	60	68	82	90	97	105
2	93	107	122	137	167	181	196	211
3	142	163	186	208	253	275	297	319
4	193	222	252	281	341	370	400	430
5	245	282	319	356	431	468	505	542
6	300	344	389	433	522	567	611	656
7	356	408	460	512	616	668	719	771
8	415	474	533	593	711	770	830	889
9	475	542	608	675	808	875	942	1008
10	537	611	685	759	907	981	1056	1130
11	601	682	764	845	1008	1090	1171	1253
12	667	756	844	933	1111	1200	1289	1378
13	734	831	926	1023	1216	1312	1408	1505
14	804	907	1010	1115	1322	1426	1530	1633
15	873	986	1096	1208	1431	1542	1653	1764
16	948	1067	1184	1304	1541	1659	1778	1896
17	1023	1149	1274	1401	1653	1779	1905	2031
18	1100	1233	1366	1500	1767	1900	2033	2167
19	1179	1319	1460	1601	1882	2023	2164	2305
20	1259	1407	1555	1704	2000	2148	2296	2444
21	1342	1497	1653	1808	2119	2275	2431	2586
22	1426	1589	1752	1915	2241	2404	2567	2730
23	1512	1682	1853	2023	2364	2534	2705	2875
24	1600	1778	1955	2133	2489	2667	2844	3022
25	1690	1875	2060	2245	2616	2801	2986	3171
26	1781	1974	2166	2359	2744	2937	3130	3322
27	1875	2075	2274	2475	2875	3075	3275	3475
28	1970	2178	2384	2593	3007	3215	3422	3630
29	2068	2282	2496	2712	3142	3356	3571	3786
30	2167	2389	2610	2833	3278	3500	3722	3944
31	2268	2497	2726	2956	3416	3645	3875	4105
32	2370	2607	2844	3081	3556	3793	4030	4267
33	2475	2719	2964	3208	3697	3942	4186	4431
34	2581	2833	3085	3337	3841	4093	4344	4596
35	2690	2949	3208	3468	3986	4245	4505	4764
36	2800	3067	3333	3600	4133	4400	4667	4933
37	2912	3186	3460	3734	4282	4556	4831	5105
38	3026	3307	3589	3870	4433	4715	4996	5278
39	3142	3431	3719	4008	4586	4875	5164	5453
40	3259	3556	3852	4148	4741	5037	5333	5630
41	3379	3682	3986	4290	4897	5201	5505	5808
42	3500	3811	4122	4433	5056	5367	5678	5989
43	3623	3942	4260	4579	5216	5534	5853	6171
44	3748	4074	4400	4726	5378	5704	6030	6356
45	3875	4208	4541	4875	5542	5875	6208	6542
46	4004	4344	4684	5026	5707	6048	6389	6730
47	4134	4482	4830	5179	5875	6223	6571	6919
48	4267	4622	4978	5333	6044	6400	6756	7111
49	4401	4764	5127	5490	6216	6579	6942	7305
50	4537	4907	5278	5648	6389	6759	7130	7500
51	4675	5053	5430	5808	6564	6942	7319	7697
52	4815	5200	5584	5970	6741	7126	7511	7896
53	4956	5349	5741	6134	6919	7312	7705	8097
54	5100	5500	5900	6300	7100	7500	7900	8300
55	5245	5653	6060	6468	7282	7690	8097	8505
56	5393	5807	6222	6637	7467	7881	8296	8711
57	5542	5964	6386	6808	7653	8075	8497	8919
58	5693	6122	6552	6981	7841	8270	8700	9130
59	5845	6282	6719	7156	8031	8468	8905	9342
60	6000	6444	6889	7333	8222	8667	9111	9556

Depth	Base 12	Base 14	Base 16	Base 18	Base 22	Base 24	Base 26	Base 28
1	46	54	61	69	83	91	98	106
2	96	111	126	141	170	185	200	215
3	150	172	194	217	261	283	306	328
4	207	237	267	296	356	385	415	444
5	269	306	343	380	454	491	528	565
6	333	378	422	467	556	600	644	689
7	402	454	506	557	661	713	765	817
8	474	533	593	652	770	830	889	948
9	550	617	683	750	883	950	1017	1083
10	630	704	778	852	1000	1074	1148	1222
11	713	794	876	957	1120	1202	1283	1365
12	800	889	978	1067	1244	1333	1422	1511
13	891	987	1083	1180	1372	1469	1565	1661
14	985	1089	1193	1296	1504	1607	1711	1815
15	1083	1194	1306	1417	1639	1750	1861	1972
16	1185	1304	1422	1541	1779	1896	2015	2133
17	1291	1417	1543	1669	1920	2046	2172	2298
18	1400	1533	1667	1800	2067	2200	2333	2467
19	1513	1654	1794	1935	2217	2357	2498	2639
20	1630	1778	1926	2074	2370	2519	2667	2815
21	1750	1906	2061	2217	2528	2683	2839	2994
22	1874	2037	2200	2363	2689	2852	3015	3178
23	2002	2172	2343	2513	2854	3024	3194	3365
24	2133	2311	2489	2667	3022	3200	3378	3556
25	2269	2454	2639	2824	3194	3380	3565	3750
26	2407	2600	2793	2985	3370	3563	3756	3948
27	2550	2750	2950	3150	3550	3750	3950	4151
28	2696	2904	3111	3319	3733	3941	4148	4356
29	2846	3061	3276	3491	3920	4135	4350	4565
30	3000	3222	3444	3667	4111	4333	4556	4778
31	3157	3387	3617	3846	4306	4535	4765	4994
32	3319	3556	3793	4030	4504	4741	4978	5215
33	3483	3728	3972	4217	4706	4950	5194	5439
34	3652	3904	4156	4407	4911	5163	5415	5667
35	3824	4083	4343	4602	5120	5380	5639	5898
36	4000	4267	4533	4800	5333	5600	5867	6133
37	4180	4454	4728	5002	5550	5824	6098	6372
38	4363	4644	4926	5207	5770	6052	6333	6615
39	4550	4839	5128	5417	5994	6283	6572	6861
40	4741	5037	5333	5630	6222	6519	6815	7111
41	4935	5239	5543	5846	6454	6757	7061	7365
42	5133	5444	5756	6067	6689	7000	7311	7622
43	5335	5654	5972	6291	6928	7246	7565	7883
44	5541	5867	6193	6519	7170	7496	7822	8148
45	5750	6083	6417	6750	7417	7750	8083	8417
46	5963	6304	6644	6985	7667	8007	8348	8689
47	6180	6528	6876	7224	7920	8269	8617	8965
48	6400	6756	7111	7467	8178	8533	8889	9244
49	6624	6987	7350	7713	8439	8802	9165	9528
50	6852	7222	7593	7963	8704	9074	9444	9815
51	7083	7461	7839	8217	8972	9350	9728	10106
52	7319	7704	8089	8474	9244	9630	10015	10400
53	7557	7950	8343	8735	9520	9913	10306	10698
54	7800	8200	8600	9000	9800	10200	10600	11000
55	8046	8454	8861	9269	10083	10491	10898	11306
56	8296	8711	9126	9541	10370	10785	11200	11615
57	8550	8972	9394	9817	10661	11083	11506	11928
58	8807	9237	9667	10096	10956	11385	11815	12244
59	9069	9506	9943	10380	11254	11691	12128	12565
60	9333	9778	10222	10667	11556	12000	12444	12889

Depth	Base 12	Base 14	Base 16	Base 18	Base 20	Base 28	Base 30	Base 32
1	48	56	63	70	78	107	115	122
2	104	119	133	148	163	222	237	252
3	167	189	211	233	256	344	367	389
4	237	267	296	326	356	474	504	533
5	315	352	389	426	463	611	648	685
6	400	444	489	533	578	756	800	844
7	493	544	596	648	700	907	959	1011
8	593	652	711	770	830	1067	1126	1185
9	700	767	833	900	967	1233	1300	1367
10	815	889	963	1037	1111	1407	1481	1556
11	937	1019	1100	1181	1263	1589	1670	1752
12	1067	1156	1244	1333	1422	1778	1867	1956
13	1204	1300	1396	1493	1589	1974	2070	2167
14	1348	1452	1556	1659	1763	2178	2281	2385
15	1500	1611	1722	1833	1944	2389	2500	2611
16	1659	1778	1896	2015	2133	2607	2726	2844
17	1826	1952	2078	2204	2330	2833	2959	3085
18	2000	2133	2267	2400	2533	3067	3200	3333
19	2181	2322	2463	2604	2744	3307	3448	3589
20	2370	2519	2667	2815	2963	3556	3704	3852
21	2567	2722	2878	3033	3189	3811	3967	4122
22	2770	2933	3096	3259	3422	4074	4237	4444
23	2981	3152	3322	3493	3663	4344	4515	4685
24	3200	3378	3556	3733	3911	4622	4800	4978
25	3426	3611	3796	3981	4167	4907	5093	5278
26	3659	3852	4044	4237	4430	5200	5393	5585
27	3900	4100	4300	4500	4700	5500	5700	5900
28	4148	4356	4563	4770	4978	5807	6015	6222
29	4404	4619	4833	5048	5263	6122	6337	6552
30	4667	4889	5111	5333	5556	6444	6667	6889
31	4937	5167	5396	5626	5856	6774	7004	7233
32	5215	5452	5689	5926	6163	7111	7348	7585
33	5500	5744	5989	6233	6478	7456	7700	7944
34	5793	6044	6296	6548	6800	7807	8059	8311
35	6093	6352	6611	6870	7130	8167	8426	8685
36	6400	6667	6933	7200	7467	8533	8800	9067
37	6715	6989	7263	7537	7811	8907	9181	9456
38	7037	7319	7600	7881	8163	9289	9570	9852
39	7367	7656	7944	8233	8522	9678	9967	10256
40	7704	8000	8296	8593	8889	10074	10370	10667
41	8048	8352	8656	8959	9263	10478	10781	11085
42	8400	8711	9022	9333	9644	10889	11200	11511
43	8759	9078	9396	9715	10033	11307	11626	11944
44	9126	9452	9778	10104	10430	11733	12059	12385
45	9500	9833	10167	10500	10833	12167	12500	12833
46	9881	10222	10563	10904	11244	12607	12948	13289
47	10270	10619	10967	11315	11663	13056	13404	13752
48	10667	11022	11378	11733	12089	13511	13867	14222
49	11070	11433	11796	12159	12522	13974	14337	14700
50	11481	11852	12222	12593	12963	14444	14815	15185
51	11900	12278	12656	13033	13411	14922	15300	15678
52	12326	12711	13096	13481	13867	15407	15793	16178
53	12759	13152	13544	13937	14330	15900	16293	16685
54	13200	13600	14000	14400	14800	16400	16800	17200
55	13648	14056	14463	14870	15278	16907	17315	17722
56	14104	14519	14933	15348	15763	17422	17837	18252
57	14567	14989	15411	15833	16256	17944	18367	18780
58	15037	15467	15896	16326	16756	18474	18904	19333
59	15515	15952	16389	16826	17263	19011	19448	19885
60	16000	16444	16889	17333	17778	19556	20000	20444

Depth	Base 12	Base 14	Base 16	Base 18	Base 20	Base 28	Base 30	Base 32
1	50	57	65	72	80	109	117	124
2	111	126	141	156	170	230	244	259
3	183	206	228	250	272	361	383	406
4	267	296	326	356	385	504	533	563
5	361	398	435	472	509	657	694	731
6	467	511	556	600	644	822	867	911
7	583	635	687	739	791	998	1050	1102
8	711	770	830	889	948	1185	1244	1304
9	850	917	983	1050	1116	1383	1450	1517
10	1000	1074	1148	1222	1296	1593	1667	1741
11	1161	1243	1324	1406	1487	1813	1894	1976
12	1333	1422	1511	1600	1689	2044	2133	2222
13	1517	1613	1709	1806	1902	2287	2383	2480
14	1711	1815	1919	2022	2126	2541	2644	2748
15	1917	2028	2139	2250	2361	2806	2917	3028
16	2133	2252	2370	2489	2607	3081	3200	3319
17	2361	2487	2613	2739	2865	3369	3494	3620
18	2600	2733	2867	3000	3133	3667	3800	3933
19	2850	2991	3131	3272	3413	3976	4117	4257
20	3111	3259	3407	3556	3704	4296	4444	4592
21	3383	3539	3694	3850	4005	4628	4783	4939
22	3667	3830	3993	4156	4318	4970	5133	5296
23	3961	4131	4302	4472	4642	5324	5494	5665
24	4267	4444	4622	4800	4978	5689	5867	6044
25	4583	4769	4954	5139	5324	6065	6250	6435
26	4911	5104	5296	5489	5681	6452	6644	6837
27	5250	5450	5650	5850	6050	6850	7050	7250
28	5600	5807	6015	6222	6430	7259	7467	7674
29	5961	6176	6391	6606	6820	7680	7894	8109
30	6333	6556	6778	7000	7222	8111	8333	8555
31	6717	6946	7176	7406	7635	8554	8783	9013
32	7111	7348	7585	7822	8059	9007	9244	9482
33	7517	7761	8006	8250	8494	9472	9717	9962
34	7933	8185	8437	8689	8941	9948	10200	10452
35	8361	8620	8880	9139	9398	10435	10694	10954
36	8800	9067	9333	9600	9867	10933	11200	11467
37	9250	9524	9798	10072	10346	11443	11717	11991
38	9711	9993	10274	10556	10837	11963	12244	12526
39	10183	10472	10761	11050	11339	12494	12783	13072
40	10667	10963	11259	11556	11852	13037	13333	13630
41	11161	11465	11769	12072	12376	13591	13894	14198
42	11667	11978	12289	12600	12911	14156	14467	14778
43	12183	12502	12820	13139	13457	14731	15050	15369
44	12711	13037	13363	13689	14015	15319	15644	15970
45	13250	13583	13917	14250	14583	15917	16250	16583
46	13800	14141	14481	14822	15163	16526	16867	17207
47	14361	14709	15057	15406	15754	17146	17494	17843
48	14933	15289	15644	16000	16356	17778	18133	18489
49	15517	15880	16243	16606	16968	18420	18783	19146
50	16111	16481	16852	17222	17592	19074	19444	19815
51	16717	17094	17472	17850	18228	19739	20117	20494
52	17333	17719	18104	18489	18874	20415	20800	21185
53	17961	18354	18746	19139	19531	21102	21494	21887
54	18600	19000	19400	19800	20200	21800	22200	22600
55	19250	19657	20065	20472	20880	22509	22917	23324
56	19911	20326	20741	21156	21570	23230	23644	24059
57	20583	21006	21428	21850	22272	23961	24383	24805
58	21267	21696	22126	22556	22985	24704	25133	25563
59	21961	22398	22835	23272	23709	25457	25894	26332
60	22667	23111	23556	24000	24444	26222	26667	27111

Depth	Base 12	Base 14	Base 16	Base 18	Base 20	Base 28	Base 30	Base 32
1	52	59	67	74	81	111	119	126
2	119	133	148	163	178	237	252	267
3	200	222	244	267	289	378	400	422
4	296	326	356	385	415	533	563	593
5	407	444	481	519	556	704	741	778
6	533	578	622	667	711	889	933	978
7	674	726	778	830	881	1089	1141	1193
8	830	889	948	1007	1067	1304	1363	1422
9	1000	1067	1133	1200	1267	1533	1600	1667
10	1185	1259	1333	1407	1481	1778	1852	1926
11	1385	1467	1548	1630	1711	2037	2119	2200
12	1600	1689	1778	1867	1956	2311	2400	2489
13	1820	1926	2022	2119	2215	2600	2696	2793
14	2074	2178	2281	2385	2489	2904	3007	3111
15	2333	2444	2556	2667	2778	3222	3333	3444
16	2607	2726	2844	2963	3081	3556	3674	3793
17	2896	3022	3148	3274	3400	3904	4030	4156
18	3200	3333	3467	3600	3733	4267	4400	4533
19	3519	3659	3800	3941	4081	4644	4785	4926
20	3852	4000	4148	4296	4444	5037	5185	5333
21	4200	4356	4511	4667	4822	5444	5600	5756
22	4563	4730	4889	5052	5215	5867	6030	6193
23	4941	5111	5281	5452	5622	6304	6474	6644
24	5333	5511	5689	5867	6044	6756	6933	7111
25	5741	5926	6111	6296	6481	7222	7407	7593
26	6163	6356	6548	6741	6933	7704	7896	8089
27	6600	6800	7000	7200	7400	8200	8400	8600
28	7052	7259	7467	7674	7881	8711	8919	9126
29	7519	7733	7948	8163	8378	9237	9452	9667
30	8000	8222	8444	8667	8889	9778	10000	10222
31	8496	8726	8956	9185	9415	10333	10563	10793
32	9007	9244	9481	9719	9956	10904	11141	11378
33	9533	9778	10022	10267	10511	11489	11733	11978
34	10074	10326	10578	10830	11081	12089	12341	12593
35	10630	10889	11148	11407	11667	12704	12963	13222
36	11200	11467	11733	12000	12267	13333	13600	13867
37	11785	12059	12333	12607	12881	13978	14252	14526
38	12385	12667	12948	13230	13511	14637	14919	15200
39	13000	13289	13578	13867	14156	15311	15600	15889
40	13630	13926	14222	14519	14815	16000	16296	16593
41	14274	14578	14881	15185	15489	16704	17007	17311
42	14933	15244	15556	15867	16178	17422	17733	18044
43	15607	15926	16224	16563	16881	18156	18474	18793
44	16296	16622	16948	17274	17600	18904	19230	19556
45	17000	17333	17667	18000	18333	19667	20000	20333
46	17719	18059	18400	18741	19081	20444	20785	21126
47	18452	18800	19148	19496	19844	21237	21585	21933
48	19200	19556	19911	20267	20622	22044	22400	22756
49	19963	20326	20689	21052	21415	22867	23230	23593
50	20741	20711	21481	21852	22222	23704	24074	24444
51	21533	21911	22289	22667	23044	24556	24933	25311
52	22341	22726	23111	23496	23881	25422	25807	26193
53	23163	23556	23948	24341	24733	26304	26696	27089
54	24000	24400	24800	25200	25600	27200	27600	28000
55	24852	25259	25667	26074	26481	28111	28519	28926
56	25719	26133	26548	26963	27378	29037	29452	29867
57	26600	27022	27444	27867	28289	29978	30400	30822
58	27496	27926	28356	28785	29215	30933	31363	31793
59	28407	28844	29281	29719	30156	31904	32341	32778
60	29333	29778	30222	30667	31111	32889	33333	33778

Depth	Base 12	Base 14	Base 16	Base 18	Base 20	Base 28	Base 30	Base 32
1	56	63	70	78	85	115	122	130
2	133	148	163	178	193	252	267	281
3	233	256	278	300	322	411	433	456
4	356	385	415	444	474	593	622	652
5	500	537	574	611	648	796	833	870
6	667	711	756	800	844	1022	1067	1111
7	856	907	959	1011	1063	1270	1322	1374
8	1067	1126	1185	1244	1304	1541	1600	1659
9	1300	1367	1433	1500	1567	1833	1900	1967
10	1556	1630	1704	1778	1852	2148	2222	2296
11	1833	1915	1996	2078	2159	2485	2567	2648
12	2133	2222	2311	2400	2489	2844	2933	3022
13	2456	2552	2648	2744	2841	3226	3322	3419
14	2800	2904	3007	3111	3215	3630	3733	3837
15	3167	3278	3389	3500	3611	4056	4167	4278
16	3556	3674	3793	3911	4030	4504	4622	4741
17	3967	4093	4219	4344	4470	4974	5100	5226
18	4400	4533	4667	4800	4933	5467	5600	5733
19	4856	4996	5137	5278	5419	5981	6122	6263
20	5333	5481	5630	5778	5926	6519	6667	6815
21	5833	5989	6144	6300	6456	7078	7233	7389
22	6356	6519	6681	6844	7007	7659	7822	7985
23	6900	7070	7241	7411	7581	8263	8433	8504
24	7467	7644	7822	8000	8178	8889	9067	9144
25	8056	8241	8426	8611	8796	9537	9722	9807
26	8667	8859	9052	9244	9437	10207	10400	10593
27	9300	9500	9700	9900	10100	10900	11100	11300
28	9956	10163	10370	10578	10785	11615	11822	12030
29	10633	10848	11063	11278	11493	12352	12567	12781
30	11333	11556	11778	12000	12222	13111	13333	13556
31	12056	12285	12515	12744	12974	13893	14122	14352
32	12800	13037	13274	13511	13748	14696	14933	15170
33	13567	13811	14056	14300	14544	15522	15767	16011
34	14356	14607	14859	15111	15363	16370	16622	16874
35	15167	15426	15685	15944	16204	17241	17500	17759
36	16000	16267	16533	16800	17067	18133	18400	18667
37	16856	17130	17404	17678	17952	19048	19322	19596
38	17733	18015	18296	18578	18859	19985	20267	20548
39	18633	18922	19211	19500	19789	20941	21233	21522
40	19556	19852	20148	20444	20741	21926	22222	22516
41	20500	20804	21107	21411	21715	22930	23233	23537
42	21467	21778	22089	22400	22711	23956	24267	24578
43	22456	22774	23093	23411	23730	25004	25322	25641
44	23467	23793	24119	24444	24770	26074	26400	26726
45	24500	24833	25167	25500	25833	27167	27500	27833
46	25556	25896	26237	26578	26919	28281	28622	28963
47	26633	26981	27330	27678	28026	29419	29767	30115
48	27733	28089	28444	28800	29156	30578	30933	31289
49	28856	29219	29581	29944	30307	31759	32122	32485
50	30000	30370	30741	31111	31481	32963	33333	33704
51	31167	31544	31922	32300	32678	34189	34567	34944
52	32356	32741	33126	33511	33896	35437	35822	36207
53	33567	33959	34352	34744	35137	36707	37100	37493
54	34800	35200	35600	36000	36400	38000	38400	38800
55	36056	36463	36870	37278	37685	39315	39722	40130
56	37333	37748	38163	38578	38993	40652	41067	41481
57	38633	39056	39478	39900	40322	42011	42433	42856
58	39956	40385	40815	41244	41674	43393	43822	44252
59	41300	41737	42174	42611	43048	44796	45233	45670
60	42667	43111	43556	44000	44444	46222	46667	47111

Area. Sq. Ft.	Cubic Yards.	Area. Sq. Ft.	Cubic Yards.	Area. Sq. Ft.	Cubic Yards.	Area. Sq. Ft.	Cubic Yards.	Area. Sq. Ft.	Cubic Yards.
1	3.7	51	188.9	101	374.1	151	559.3	201	744.4
2	7.4	52	192.6	102	377.8	152	563.0	202	748.2
3	11.1	53	196.3	103	381.5	153	566.7	203	751.9
4	14.8	54	200.0	104	385.2	154	570.4	204	755.6
5	18.5	55	203.7	105	388.9	155	574.1	205	759.3
6	22.2	56	207.4	106	392.6	156	577.8	206	763.0
7	25.9	57	211.1	107	396.3	157	581.5	207	766.7
8	29.6	58	214.8	108	400 0	158	585.2	208	770.4
9	33.3	59	218.5	109	403.7	159	588.9	209	774.1
10	37.0	60	222.2	110	407.4	160	592.6	210	777.8
11	40.7	61	225.9	111	411.1	161	596.3	211	781 5
12	44.4	62	229.6	112	414.8	162	600.0	212	785.2
13	48.1	63	233.3	113	418.5	163	603.7	213	788.9
14	51.9	64	237.0	114	422.2	164	607.4	214	792.6
15	55.6	65	240.7	115	425.9	165	611.1	215	796.3
16	59.3	66	244.4	116	429.6	166	614 8	216	800.0
17	63.0	67	248.2	117	433.3	167	618.5	217	803.7
18	66.7	68	251.9	118	437.0	168	622.2	218	807.4
19	70.4	69	255.6	119	440.7	169	625.9	219	811.1
20	74.1	70	259.3	120	444.4	170	629.6	220	814.8
21	77.8	71	263.0	121	448.2	171	633.3	221	818.5
22	81.5	72	266.7	122	451.9	172	637.0	222	822.2
23	85.2	73	270.4	123	455.6	173	640.7	223	825.9
24	88.9	74	274.1	124	459.3	174	644.4	224	829.6
25	92.6	75	277.8	125	463.0	175	648.2	225	833.3
26	96.3	76	281.5	126	466.7	176	651.9	226	837.0
27	100.0	77	285 2	127	470.4	177	655.6	227	840.7
28	103.7	78	288.9	128	474 1	178	659.3	228	844.4
29	107.4	79	292.6	129	477.8	179	663.0	229	848.2
30	111.1	80	296.3	130	481.5	180	666.7	230	851.9
31	114.8	81	300.0	131	485.2	181	670.4	231	855.6
32	118.5	82	303.7	132	488 9	182	674.1	232	859.3
33	122.2	83	307.4	133	492.6	183	677.8	233	863.0
34	125.9	84	311.1	134	496.3	184	681.5	234	866.7
35	129 6	85	314.8	135	500.0	185	685.2	235	870.4
36	133.3	86	318.5	136	503.7	186	688.9	236	874.1
37	137.0	87	322.2	137	507.4	187	692.6	237	877.8
38	140.7	88	325.9	138	511.1	188	696.3	238	881.5
39	144 4	89	329.6	139	514.8	189	700.0	239	885.2
40	148.2	90	333.3	140	518.5	190	703.7	240	888.9
41	151.9	91	337.0	141	522.2	191	707.4	241	892.6
42	155.6	92	340.7	142	525.9	192	711.1	242	896.3
43	159.3	93	344.4	143	529.6	193	714.8	243	900.0
44	163.0	94	348.2	144	533.3	194	718.5	244	903.7
45	166.7	95	351.9	145	537.0	195	722.2	245	907.4
46	170.4	96	355.6	146	540.7	196	725.9	246	911.1
47	174.1	97	359.3	147	544.4	197	729.6	247	914.8
48	177.8	98	363.0	148	548.2	198	733.3	248	918.5
49	181.5	99	366.7	149	551.9	199	737.0	249	922.2
50	185.2	100	370.4	150	555.6	200	740.7	250	925.9

Area. Sq. Ft.	Cubic Yards.	Area. Sq. Ft.	Cubic Yards.	Area. Sq. Ft.	Cubic Yards.	Area. Sq. Ft.	Cubic Yards.	Area. Sq. Ft.	Cubic Yards.
251	929.6	301	1114.8	351	1300.0	401	1485.2	451	1670.4
252	933.3	302	1118.5	352	1303.7	402	1488.9	452	1674.1
253	937.0	303	1122.2	353	1307.4	403	1492.6	453	1677.8
254	940.7	304	1125.9	354	1311.1	404	1496.3	454	1681.5
255	944.4	305	1129.6	355	1314.8	405	1500.0	455	1685.2
256	948.2	306	1133.3	356	1318.5	406	1503.7	456	1688.9
257	951.9	307	1137.0	357	1322.2	407	1507.4	457	1692.6
258	955.6	308	1140.7	358	1325.9	408	1511.1	458	1696.3
259	959.3	309	1144.4	359	1329.6	409	1514.8	459	1700.0
260	963.0	310	1148.2	360	1333.3	410	1518.5	460	1703.7
261	966.7	311	1151.9	361	1337.0	411	1522.2	461	1707.4
262	970.4	312	1155.6	362	1340.7	412	1525.9	462	1711.1
263	974.1	313	1159.3	363	1344.4	413	1529.6	463	1714.8
264	977.8	314	1163.0	364	1348.2	414	1533.3	464	1718.5
265	981.5	315	1166.7	365	1351.9	415	1537.0	465	1722.2
266	985.2	316	1170.4	366	1355.6	416	1540.7	466	1725.9
267	988.9	317	1174.1	367	1359.3	417	1544.4	467	1729.6
268	992.6	318	1177.8	368	1363.0	418	1548.2	468	1733.3
269	996.3	319	1181.5	369	1366.7	419	1551.9	469	1737.0
270	1000.0	320	1185.2	370	1370.4	420	1555.6	470	1740.7
271	1003.7	321	1188.9	371	1374.1	421	1559.3	471	1744.4
272	1007.4	322	1192.6	372	1377.8	422	1563.0	472	1748.2
273	1011.1	323	1196.3	373	1381.5	423	1566.7	473	1751.9
274	1014.8	324	1200.0	374	1385.2	424	1570.4	474	1755.6
275	1018.5	325	1203.7	375	1388.9	425	1574.1	475	1759.3
276	1022.2	326	1207.4	376	1392.6	426	1577.8	476	1763.0
277	1025.9	327	1211.1	377	1396.3	427	1581.5	477	1766.7
278	1029.6	328	1214.8	378	1400.0	428	1585.2	478	1770.4
279	1033.3	329	1218.5	379	1403.7	429	1588.9	479	1774.1
280	1037.0	330	1222.2	380	1407.4	430	1592.6	480	1777.8
281	1040.7	331	1225.9	381	1411.1	431	1596.3	481	1781.5
282	1044.4	332	1229.6	382	1414.8	432	1600.0	482	1785.2
283	1048.2	333	1233.3	383	1418.5	433	1603.7	483	1788.9
284	1051.9	334	1237.0	384	1422.2	434	1607.4	484	1792.6
285	1055.6	335	1240.7	385	1425.9	435	1611.1	485	1796.3
286	1059.3	336	1244.4	386	1429.6	436	1614.8	486	1800.0
287	1063.0	337	1248.2	387	1433.3	437	1618.5	487	1803.7
288	1066.7	338	1251.9	388	1437.0	438	1622.2	488	1807.4
289	1070.4	339	1255.6	389	1440.7	439	1625.9	489	1811.1
290	1074.1	340	1259.3	390	1444.4	440	1629.6	490	1814.8
291	1077.8	341	1263.0	391	1448.2	441	1633.3	491	1818.5
292	1081.5	342	1266.7	392	1451.9	442	1637.0	492	1822.2
293	1085.2	343	1270.4	393	1455.6	443	1640.7	493	1825.9
294	1088.9	344	1274.1	394	1459.3	444	1644.4	494	1829.6
295	1092.6	345	1277.8	395	1463.0	445	1648.2	495	1833.3
296	1096.3	346	1281.5	396	1466.7	446	1651.9	496	1837.0
297	1100.0	347	1285.2	397	1470.4	447	1655.6	497	1840.7
298	1103.7	348	1288.9	398	1474.1	448	1659.3	498	1844.4
299	1107.4	349	1292.6	399	1477.8	449	1663.0	499	1848.2
300	1111.1	350	1296.3	400	1481.5	450	1666.7	500	1851.9

Area. Sq. Ft.	Cubic Yards.	Area. Sq. Ft.	Cubic Yards.	Area. Sq. Ft.	Cubic Yards.	Area. Sq. Ft.	Cubic Yards.	Area. Sq. Ft.	Cubic Yards.
501	1855.6	551	2040.7	601	2225.9	651	2411.1	701	2596.3
502	1859.3	552	2044.4	602	2229.6	652	2414.8	702	2600.0
503	1863.0	553	2048.2	603	2233.3	653	2418.5	703	2603.7
504	1866.7	554	2051.9	604	2237.0	654	2422.2	704	2607.4
505	1870.4	555	2055.6	605	2240.7	655	2425.9	705	2611.1
506	1874.1	556	2059.3	606	2244.4	656	2429.6	706	2614.8
507	1877.8	557	2063.0	607	2248.2	657	2433.3	707	2618.5
508	1881.5	558	2066.7	608	2251.9	658	2437.0	708	2622.2
509	1885.2	559	2070.4	609	2255.6	659	2440.7	709	2625.9
510	1888.9	560	2074.1	610	2259.3	660	2444.4	710	2629.6
511	1892.6	561	2077.8	611	2263.0	661	2448.2	711	2633.3
512	1896.3	562	2081.5	612	2266.7	662	2451.9	712	2637.0
513	1900.0	563	2085.2	613	2270.4	663	2455.6	713	2640.7
514	1903.7	564	2088.9	614	2274.1	664	2459.3	714	2644.4
515	1907.4	565	2092.6	615	2277.8	665	2463.0	715	2648.2
516	1911.1	566	2096.3	616	2281.5	666	2466.7	716	2651.9
517	1914.8	567	2100.0	617	2285.2	667	2470.4	717	2655.6
518	1918.5	568	2103.7	618	2288.9	668	2474.1	718	2659.3
519	1922.2	569	2107.4	619	2292.6	669	2477.8	719	2663.0
520	1925.9	570	2111.1	620	2296.3	670	2481.5	720	2666.7
521	1929 6	571	2114.8	621	2300.0	671	2485.2	721	2670.4
522	1933.3	572	2118.5	622	2303.7	672	2488.9	722	2674.1
523	1937.0	573	2122.2	623	2307.4	673	2492 6	723	2677.8
524	1940.7	574	2125.9	624	2311.1	674	2496.3	724	2681.5
525	1944.4	575	2129.6	625	2314.8	675	2500.0	725	2685.2
526	1948.2	576	2133.3	626	2318.5	676	2503.7	726	2688.9
527	1951.9	577	2137.0	627	2322.2	677	2507.4	727	2692.6
528	1955.6	578	2140.7	628	2325.9	678	2511.1	728	2696.3
529	1959.3	579	2144.4	629	2329.6	679	2514.8	729	2700.0
530	1963.0	580	2148 2	630	2333.3	680	2518.5	730	2703.7
531	1966.7	581	2151.9	631	2337.0	681	2522.2	731	2707.4
532	1970.4	582	2155.6	632	2340.7	682	2525.9	732	2711.1
533	1974.1	583	2159.3	633	2344.4	683	2529.6	733	2714.8
534	1977.8	584	2163.0	634	2348.2	684	2533.3	734	2718.5
535	1981.5	585	2166.7	635	2351.9	685	2537.0	735	2722.2
536	1985.2	586	2170.4	636	2355.6	686	2540 7	736	2725.9
537	1988.9	587	2174.1	637	2359.3	687	2544.4	737	2729.6
538	1992.6	588	2177.8	638	2363 0	688	2548.2	738	2733.3
539	1996.3	589	2181.5	639	2366.7	689	2551.9	739	2737.0
540	2000.0	590	2185.2	640	2370.4	690	2555.6	740	2740.7
541	2003.7	591	2188.9	641	2374.1	691	2559.3	741	2744.4
542	2007.4	592	2192.6	642	2377.8	692	2563.0	742	2748.2
543	2011.1	593	2196.3	643	2381.5	693	2566.7	743	2751.9
544	2014.8	594	2200.0	644	2385.2	694	2570.4	744	2755.6
545	2018.5	595	2203.7	645	2388.9	695	2574.1	745	2759.3
546	2022.2	596	2207.4	646	2392.6	696	2577.8	746	2763.0
547	2025.9	597	2211.1	647	2396.3	697	2581.5	747	2766.7
548	2029.6	598	2214.8	648	2400.0	698	2585.2	748	2770.4
549	2033.3	599	2218.5	649	2403.7	699	2588.9	749	2774.1
550	2037.0	600	2222.2	650	2407.4	700	2592.6	750	2777.8

Area. Sq. Ft.	Cubic Yards.	Area. Sq. Ft.	Cubic Yards.	Area. Sq. Ft.	Cubic Yards.	Area. Sq. Ft.	Cubic Yards.	Area. Sq. Ft.	Cubic Yards.
751	2781.5	801	2966.7	851	3151.9	901	3337.0	951	3522.2
752	2785.2	802	2970.4	852	3155.6	902	3340.7	952	3525.9
753	2788.9	803	2974.1	853	3159.3	903	3344.4	953	3529.6
754	2792.6	804	2977.8	854	3163.0	904	3348.2	954	3533.3
755	2796.3	805	2981.5	855	3166.7	905	3351.9	955	3537.0
756	2800.0	806	2985.2	856	3170.4	906	3355.6	956	3540.7
757	2803.7	807	2988.9	857	3174.1	907	3359.3	957	3544.4
758	2807.4	808	2992.6	858	3177.8	908	3363.0	958	3548.2
759	2811.1	809	2996.3	859	3181.5	909	3366.7	959	3551.9
760	2814.8	810	3000.0	860	3185.2	910	3370.4	960	3555.6
761	2818.5	811	3003.7	861	3188.9	911	3374.1	961	3559.3
762	2822.2	812	3007.4	862	3192.6	912	3377.8	962	3563.0
763	2825.9	813	3011.1	863	3196.3	913	3381.5	963	3566.7
764	2829.6	814	3014.8	864	3200.0	914	3385.2	964	3570.4
765	2833.3	815	3018.5	865	3203.7	915	3388.9	965	3574.1
766	2837.0	816	3022.2	866	3207.4	916	3392.6	966	3577.8
767	2840.7	817	3025.9	867	3211.1	917	3396.3	967	3581.5
768	2844.4	818	3029.6	868	3214.8	918	3400.0	968	3585.2
769	2848.2	819	3033.3	869	3218.5	919	3403.7	969	3588.9
770	2851.9	820	3037.0	870	3222.2	920	3407.4	970	3592.6
771	2855.6	821	3040.7	871	3225.9	921	3411.1	971	3596.3
772	2859.3	822	3044.4	872	3229.6	922	3414.8	972	3600.0
773	2863.0	823	3048.2	873	3233.3	923	3418.5	973	3603.7
774	2866.7	824	3051.9	874	3237.0	924	3422.2	974	3607.4
775	2870.4	825	3055.6	875	3240.7	925	3425.9	975	3611.1
776	2874.1	826	3059.3	876	3244.4	926	3429.6	976	3614.8
777	2877.8	827	3063.0	877	3248.2	927	3433.3	977	3618.5
778	2881.5	828	3066.7	878	3251.9	928	3437.0	978	3622.2
779	2885.2	829	3070.4	879	3255.6	929	3440.7	979	3625.9
780	2888.9	830	3074.1	880	3259.3	930	3444.4	980	3629.6
781	2892.6	831	3077.8	881	3263.0	931	3448.2	981	3633.3
782	2896.3	832	3081.5	882	3266.7	932	3451.9	982	3637.0
783	2900.0	833	3085.2	883	3270.4	933	3455.6	983	3640.7
784	2903.7	834	3088.9	884	3274.1	934	3459.3	984	3644.4
785	2907.4	835	3092.6	885	3277.8	935	3463.0	985	3648.2
786	2911.1	836	3096.3	886	3281.5	936	3466.7	986	3651.9
787	2914.8	837	3100.0	887	3285.2	937	3470.4	987	3655.6
788	2918.5	838	3103.7	888	3288.9	938	3474.1	988	3659.3
789	2922.2	839	3107.4	889	3292.6	939	3477.8	989	3663.0
790	2925.9	840	3111.1	890	3296.3	940	3481.5	990	3666.7
791	2929.6	841	3114.8	891	3300.0	941	3485.2	991	3670.4
792	2933.3	842	3118.5	892	3303.7	942	3488.9	992	3674.1
793	2937.0	843	3122.2	893	3307.4	943	3492.6	993	3677.8
794	2940.7	844	3125.9	894	3311.1	944	3496.3	994	3681.5
795	2944.4	845	3129.6	895	3314.8	945	3500.0	995	3685.2
796	2948.2	846	3133.3	896	3318.5	946	3503.7	996	3688.9
797	2951.9	847	3137.0	897	3322.2	947	3507.4	997	3692.6
798	2955.6	848	3140.7	898	3325.9	948	3511.1	998	3696.3
799	2959.3	849	3144.4	899	3329.6	949	3514.8	999	3700.0
800	2963.0	850	3148.2	900	3333.3	950	3518.5	1000	3703.7

Rise per Cent.	Feet per Mile.	Rise per Cent.	Feet per Mile.	Rise per Cent.	Feet per Mile.	Rise per Cent.	Feet per Mile.
.01	.528	.61	32.208	1.21	63.888	1.81	95.568
.02	1.056	.62	32.736	1.22	64.416	1.82	96.096
.03	1.584	.63	33.264	1.23	64.944	1.83	96.624
.04	2.112	.64	33.792	1.24	65.472	1.84	97.152
.05	2.640	.65	34.320	1.25	66.000	1.85	97.680
.06	3.168	.66	34.848	1.26	66.528	1.86	98.208
.07	3.696	.67	35.376	1.27	67.056	1.87	98.736
.08	4.224	.68	35.904	1.28	67.584	1.88	99.264
.09	4.752	.69	36.432	1.29	68.112	1.89	99.792
.10	5.280	.70	36.960	1.30	68.640	1.90	100.320
.11	5.808	.71	37.488	1.31	69.168	1.91	100.848
.12	6.336	.72	38.016	1.32	69.696	1.92	101.376
.13	6.864	.73	38.544	1.33	70.224	1.93	101.904
.14	7.392	.74	39.072	1.34	70.752	1.94	102.432
.15	7.920	.75	39.600	1.35	71.280	1.95	102.960
.16	8.448	.76	40.128	1.36	71.808	1.96	103.488
.17	8.976	.77	40.656	1.37	72.336	1.97	104.016
.18	9.504	.78	41.184	1.38	72.864	1.98	104.544
.19	10.032	.79	41.712	1.39	73.392	1.99	105.072
.20	10.560	.80	42.240	1.40	73.920	2.00	105.600
.21	11.088	.81	42.768	1.41	74.448	2.10	110.880
.22	11.616	.82	43.296	1.42	74.976	2.20	116.160
.23	12.144	.83	43.824	1.43	75.504	2.30	121.440
.24	12.672	.84	44.352	1.44	76.032	2.40	126.720
.25	13.200	.85	44.880	1.45	76.560	2.50	132.000
.26	13.728	.86	45.408	1.46	77.088	2.60	137.280
.27	14.256	.87	45.936	1.47	77.616	2.70	142.560
.28	14.784	.88	46.464	1.48	78.144	2.80	147.840
.29	15.312	.89	46.992	1.49	78.672	2.90	153.120
.30	15.840	.90	47.520	1.50	79.200	3.00	158.400
.31	16.368	.91	48.048	1.51	79.728	3.10	163.680
.32	16.896	.92	48.576	1.52	80.256	3.20	168.960
.33	17.424	.93	49.104	1.53	80.784	3.30	174.240
.34	17.952	.94	49.632	1.54	81.312	3.40	179.520
.35	18.480	.95	50.160	1.55	81.840	3.50	184.800
.36	19.008	.96	50.688	1.56	82.368	3.60	190.080
.37	19.536	.97	51.[illegible]16	1.57	82.896	3.70	195.360
.38	20.064	.98	51.744	1.58	83.424	3.80	200.640
.39	20.592	.99	52.272	1.59	83.952	3.90	205.920
.40	21.120	1.00	52.800	1.60	84.480	4.00	211.200
.41	21.648	1.01	53.328	1.61	85.008	4.10	216.480
.42	22.176	1.02	53.856	1.62	85.536	4.20	221.760
.43	22.704	1.03	54.384	1.63	86.064	4.30	227.040
.44	23.232	1.04	54.912	1.64	86.592	4.40	232.320
.45	23.760	1.05	55.440	1.65	87.120	4.50	237.600
.46	24.288	1.06	55.968	1.66	87.648	4.60	242.880
.47	24.816	1.07	56.496	1.67	88.176	4.70	248.160
.48	25.344	1.08	57.024	1.68	88.704	4.80	253.440
.49	25.872	1.09	57.552	1.69	89.232	4.90	258.720
.50	26.400	1.10	58.080	1.70	89.760	5.00	264.000
.51	26.928	1.11	58.608	1.71	90.288	5.10	269.280
.52	27.456	1.12	59.136	1.72	90.816	5.20	274.560
.53	27.984	1.13	59.664	1.73	91.344	5.30	279.840
.54	28.512	1.14	60.192	1.74	91.872	5.40	285.120
.55	29.040	1.15	60.720	1.75	92.400	5.50	290.400
.56	29.568	1.16	61.248	1.76	92.928	5.60	295.680
.57	30.096	1.17	61.776	1.77	93.456	5.70	300.960
.58	30.624	1.18	62.304	1.78	93.984	5.80	306.240
.59	31.152	1.19	62.832	1.79	94.512	5.90	311.520
.60	31.680	1.20	63.360	1.80	95.040	6.00	316.800

Angle of Inclination.	Vertical Rise in 100 Horizontal.	Horizontal Distance to a Rise of 10.	Angle of Inclination.	Vertical Rise in 100 Horizontal.	Horizontal Distance to a Rise of 10.	Angle of Inclination.	Vertical Rise in 100 Horizontal.	Horizontal Distance to a Rise of 10.
0° 20′	.58	1718.9	7° 20′	12.87	77.7	16°	28.67	34.9
40	1.16	859.4	40	13.46	74.3	17	30.57	32.7
1	1.75	572.9	8	14.05	71.2	18	32.49	30.8
20	2.33	429.6	20	14.65	68.3	19	34.43	29.0
40	2.91	343.7	40	15.24	65.6	20	36.40	27.5
2	3.49	286.4	9	15.84	63.1	21	38.39	26.1
20	4.08	245.4	20	16.44	60.8	22	40.40	24.8
40	4.66	214.7	40	17.03	58.7	23	42.45	23.6
3	5.24	190.8	10	17.63	56.7	24	44.52	22.5
20	5.82	171.7	20	18.23	54.8	25	46.63	21.4
40	6.41	156.0	40	18.84	53.1	26	48.77	20.5
4	6.99	143.0	11	19.44	51.4	27	50.95	19.6
20	7.58	132.0	30	20.35	49.2	28	53.17	18.8
40	8.16	122.5	12	21.26	47.0	29	55.43	18.0
5	8.75	114.3	30	22.17	45.1	30	57.74	17.3
20	9.34	107.1	13	23.09	43.3	35	70.02	14.3
40	9.92	100.8	30	24.01	41.7	40	83.91	11.9
6	10.51	95.1	14	24.93	40.1	45	100.00	10.0
20	11.10	90.1	30	25.86	38.7	50	119.18	8.4
40	11.69	85.6	15	26.79	37.3	55	142.81	7.0
7	12.28	81.4	30	27.73	36.1	60	173.21	5.8

TABLE XXIII.—MATERIAL REQUIRED FOR ONE MILE OF TRACK.

RAIL WEIGHTS.			RAILROAD SPIKES.				
					Required for Ties 2 ft. Apart.		
Pounds per Yard.	Short Tons of 2000 lbs.	Long Tons of 2240 lbs.	Size under Head.	Average No. per Keg of 200 lbs.	Weight in Lbs.	No of Kegs.	For Rails Weighing
12	21.12	18.857	5½ × $\frac{9}{16}$	360	5870	29.3	45 to 70 lbs.
16	28.16	25.143	5 × $\frac{9}{16}$	400	5280	26.4	40 " 56 "
20	35.20	31.429	5 × ½	450	4690	23.5	35 " 40 "
25	44.00	39.286	4½ × ½	530	3980	19.9	28 " 35 "
30	52.80	47.143	4 × ½	600	3520	17.6	24 " 35 "
35	61.60	55.000	4½ × $\frac{7}{16}$	680	3110	15.5	20 " 30 "
40	70.40	62.857	4 × $\frac{7}{16}$	720	2930	14.7	20 " 30 "
45	79.20	70.714	3½ × $\frac{7}{16}$	900	2350	11.7	16 " 25 "
56	98.56	88.000	4 × ⅜	1000	2110	10.6	16 " 25 "
60	105.60	94.286	3½ × ⅜	1190	1770	8.9	16 " 20 "
70	123.20	110.000	3 × ⅜	1240	1700	8.5	16 " 20 "
80	140.80	125.714	2½ × ⅜	1342	1570	7.9	12 " 16 "

NUMBER OF CROSS-TIES.					NUMBER OF SPLICE-JOINTS. Two Bars with Four Bolts and Nuts to Each Joint.				
Distance apart, c. to c., in Feet.					Length of Rail in Feet.				
1.5	1.75	2.0	2.25	2.50	20	24	26	28	30
3520	3017	2640	2347	2112	528	440	406	377	352

CONVERSION OF ENGLISH INCHES INTO CENTIMETRES.

Ins.	0	1	2	3	4	5	6	7	8	9
	Cm.	Cm.	Cm.	Cm.	Cm.	Cm.	Cm.	Cm.	Cm.	Cm.
0	0.000	2.540	5.080	7.620	10.16	12.70	15.24	17.78	20.32	22.86
10	25.40	27.94	30.48	33.02	35.56	38.10	40.64	43.18	45.72	48.26
20	50.80	53.34	55.88	58.42	60.96	63.50	66.04	68.58	71.12	73.66
30	76.20	78.74	81.28	83.82	86.36	88.90	91.44	93.98	96.52	99.06
40	101.60	104.14	106.68	109.22	111.76	114.30	116.84	119.38	121.92	124.46
50	127.00	129.54	132.08	134.62	137.16	139.70	142.24	144.78	147.32	149.86
60	152.40	154.94	157.48	160.02	162.56	165.10	167.64	170.18	172.72	175.26
70	177.80	180.34	182.88	185.42	187.96	190.50	193.04	195.58	198.12	200.96
80	203.20	205.74	208.28	210.82	213.36	215.90	218.44	220.98	223.52	226.06
90	228.60	231.14	233.68	236.22	238.76	241.30	243.84	246.38	248.92	251.46
100	254.00	256.54	259.08	261.62	264.16	266.70	269.24	271.78	274.32	276.86

CONVERSION OF CENTIMETRES INTO ENGLISH INCHES.

Cm.	0	1	2	3	4	5	6	7	8	9
	Ins.	Ins.	Ins.	Ins.	Ins.	Ins.	Ins.	Ins.	Ins.	Ins.
0	0.000	0.394	0.787	1.181	1.575	1.969	2.302	2.756	3.150	3.543
10	3.937	4.331	4.742	5.118	5.512	5.906	6.299	6.693	7.087	7.480
20	7.874	8.268	8.662	9.055	9.449	9.843	10.236	10.630	11.024	11.418
30	11.811	12.205	12.599	12.992	13.386	13.780	14.173	14.567	14.961	15.355
40	15.748	16.142	16.536	16.929	17.323	17.717	18.111	18.504	18.898	19.292
50	19.685	20.079	20.473	20.867	21.260	21.654	22.048	22.441	22.835	23.229
60	23.622	24.016	24.410	24.804	25.197	25.591	25.985	26.378	26.772	27.166
70	27.560	27.953	28.347	28.741	29.134	29.528	29.922	30.316	30.709	31.103
80	31.497	31.890	32.284	32.678	33.071	33.465	33.859	34.253	34.646	35.040
90	35.434	35.827	36.221	36.615	37.009	37.402	37.796	38.190	38.583	38.977
100	39.370	39.764	40.158	40.552	40.945	41.339	41.733	42.126	42.520	42.914

CONVERSION OF ENGLISH FEET INTO METRES.

Feet.	0	1	2	3	4	5	6	7	8	9
	Met.	Met.	Met.	Met.	Met.	Met.	Met.	Met.	Met.	Met.
0	0.000	0.3048	0.6096	0.9144	1.2192	1.5239	1.8287	2.1335	2.4383	2.7431
10	3.0479	3.3527	3.6575	3.9623	4.2671	4.5719	4.8767	5.1815	5.4863	5.7911
20	6.0359	6.4006	6.7055	7.0102	7.3150	7.6198	7.9246	8.2294	8.5342	8.8390
30	9.1438	9.4486	9.7534	10.058	10.363	10.668	10.972	11.277	11.582	11.887
40	12.192	12.496	12.801	13.106	13.411	13.716	14.020	14.325	14.630	14.935
50	15.239	15.544	15.849	16.154	16.459	16.763	17.068	17.373	17.678	17.983
60	18.287	18.592	18.897	19.202	19.507	19.811	20.116	20.421	20.726	21.031
70	21.335	21.640	21.945	22.250	22.555	22.859	23.164	23.469	23.774	24.079
80	24.383	24.688	24.993	25.298	25.602	25.907	26.212	26.517	26.822	27.126
90	27.431	27.736	28.041	28.346	28.651	28.955	29.260	29.565	29.870	30.174
100	30.479	30.784	31.089	31.394	31.698	32.003	32.308	32.613	32.918	33.222

CONVERSION OF METRES INTO ENGLISH FEET.

Met.	0	1	2	3	4	5	6	7	8	9
	Feet.	Feet.	Feet.	Feet.	Feet.	Feet.	Feet.	Feet.	Feet.	Feet.
0	0.000	3.2809	6.5618	9.8427	13.123	16.404	19.685	22.966	26.247	29.528
10	32.809	36.090	39.371	42.651	45.932	49.213	52.494	55.775	59.056	62.337
20	65.618	68.899	72.179	75.461	78.741	82.022	85.303	88.584	91.865	95.146
30	98.427	101.71	104.99	108.27	111.55	114.83	118.11	121.39	124.67	127.96
40	131.24	134.52	137.80	141.08	144.36	147.64	150.92	154.20	157.48	160.76
50	164.04	167.33	170.61	173.89	177.17	180.45	183.73	187.01	190.29	193.57
60	196.85	200.13	203.42	206.70	209.98	213.26	216.54	219.82	223.10	226.38
70	229.66	232.94	236.22	239.51	242.79	246.07	249.35	252.63	255.91	259.19
80	262.47	265.75	269.03	272.31	275.60	278.88	282.16	285.44	288.72	292.00
90	295.28	298.56	391.84	305.12	308.40	311.69	314.97	318.25	321.53	324.81
100	328.09	331.37	334.65	337.93	341.21	344.49	347.78	351.06	354.34	357.62

CONVERSION OF ENGLISH STATUTE-MILES INTO KILOMETRES.

Miles.	0	1	2	3	4	5	6	7	8	9
	Kilo.	Kilo.	Kilo.	Kilo.	Kilo.	Kilo.	Kilo.	Kilo.	Kilo.	Kilo.
0	0.0000	1.6093	3.2186	4.8279	6.4372	8.0465	9.6558	11.2652	12.8745	14.4818
10	16.093	17.702	19.312	20.921	22.530	24.139	25.749	27.358	28.967	30.577
20	32.186	33.795	35.405	37.014	38.623	40.232	41.842	43.451	45.060	46.670
30	48.279	49.888	51.498	53.107	54.716	56.325	57.935	59.544	61.153	62.763
40	64.372	65.981	67.591	69.200	70.809	72.418	74.028	75.637	77.246	78.856
50	80.465	82.074	83.684	85.293	86.902	88.511	90.121	91.730	93.339	94.949
60	96.558	98.167	99.777	101.39	102.99	104.60	106.21	107.82	109.43	111.04
70	112.65	114.26	115.87	117.48	119.08	120.69	122.30	123.91	125.52	127.13
80	128.74	130.35	131.96	133.57	135.17	136.78	138.39	140.00	141.61	143.22
90	144.85	146.44	148.05	149.66	151.26	152.87	154.48	156.09	157.70	159.31
100	160.93	162.53	164 14	165 75	167.35	168.96	170.57	172.18	173.79	175.40

CONVERSION OF KILOMETRES INTO ENGLISH STATUTE-MILES.

Kilom.	0	1	2	3	4	5	6	7	8	9
	Miles.	Miles.	Miles.	Miles.	Miles.	Miles.	Miles.	Miles.	Miles.	Miles.
0	0.0000	0.6214	1.2427	1.8641	2.4855	3.1069	3.7282	4.3497	4.9711	5.5924
10	6.2138	6.8352	7.4565	8.0780	8.6994	9.3208	9.9421	10.562	11.185	11.805
20	12.427	13.049	13.670	14.292	14.913	15.534	16.156	16.776	17.399	18.019
30	18.641	19.263	19.884	20.506	21.127	21.748	23.370	22.990	23.613	24.233
40	24.855	25.477	26.098	26.720	27.341	27.962	28.584	29.204	29.827	30.447
50	31.069	31.690	32.311	32.933	33.554	34.175	34.797	35.417	36.040	36 660
60	37.282	37.904	38.525	39.147	39.768	40.389	41.011	41.631	42.254	42.874
70	43.497	44.118	44.739	45.361	45.982	46.603	47.225	47.845	48.468	49.088
80	49.711	50.332	50.953	51.575	52.196	52.817	53.439	54.059	54.682	55.302
90	55.924	56.545	57.166	57.788	58.409	59.030	59.652	60.272	60.895	61.515
100	62.138	62.759	63.380	64.002	64.623	65.244	65.866	66.486	67.109	67.729

TABLE XXVI.

LENGTH IN FEET OF 1′ ARCS OF LATITUDE AND LONGITUDE.

Lat.	1′ Lat.	1′ Long.	Lat.	1′ Lat.	1′ Long.
1°	6045	6085	31°	6061	5222
2°	6045	6083	32°	6062	5166
3°	6045	6078	33°	6063	5109
4°	6045	6071	34°	6064	5051
5°	6045	6063	35°	6065	4991
6°	6045	6053	36°	6066	4930
7°	6046	6041	37°	6067	4867
8°	6046	6027	38°	6068	4802
9°	6046	6012	39°	6070	4736
10°	6047	5994	40°	6071	4669
11°	6047	5975	41°	6072	4600
12°	6048	5954	42°	6073	4530
13°	6048	5931	43°	6074	4458
14°	6049	5907	44°	6075	4385
15°	6049	5880	45°	6076	4311
16°	6050	5852	46°	6077	4235
17°	6050	5822	47°	6078	4158
18°	6051	5790	48°	6079	4080
19°	6052	5757	49°	6080	4001
20°	6052	5721	50°	6081	3920
21°	6053	5684	51°	6082	3838
22°	6054	5646	52°	6084	3755
23°	6054	5605	53°	6085	3671
24°	6055	5563	54°	6086	3586
25°	6056	5519	55°	6087	3499
26°	6057	5474	56°	6088	3413
27°	6058	5427	57°	6089	3323
28°	6059	5378	58°	6090	3233
29°	6060	5327	59°	6091	3142
30°	6061	5275	60°	6092	3051

TABLE XXVII.—TRIGONOMETRIC AND MISCELLANEOUS FORMULAS.

TRIGONOMETRIC FORMULAS.

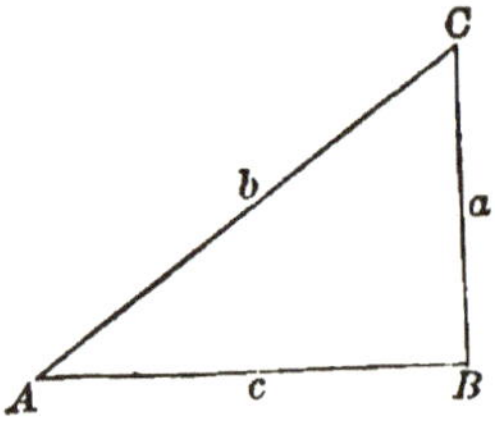

FIG. 98.

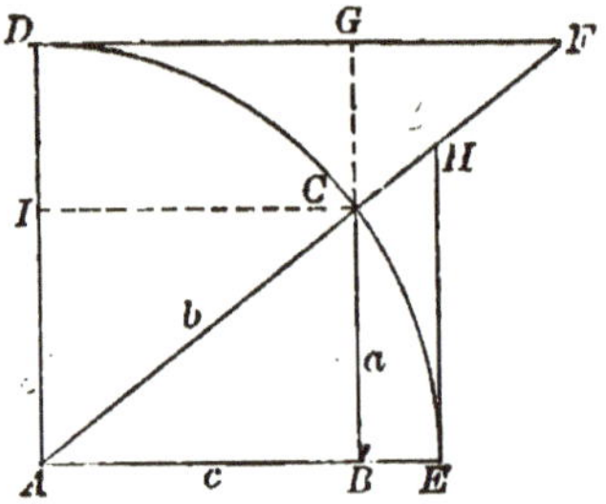

FIG. 99.

In Fig 99, let DCE be the arc of a quadrant, ABC a right triangle, the angle BAC subtended by the arc $CE = A$, and consider the radius $AC =$ unity. Then

$BC = \sin A.$	$AF = \operatorname{cosec} A.$
$AB = \cos A.$	$BE = \operatorname{versin} A.$
$HE = \tan A.$	$DI = \operatorname{coversin} A.$
$DF = \cot A.$	$CH = \operatorname{exsec} A.$
$AH = \sec A.$	$CF = \operatorname{coexsec} A.$

Using the small letters a, b, c, to represent the sides of a right triangle in Fig. 98 or 99, we may write

$$\sin A = \frac{a}{b}; \quad \operatorname{cosec} A = \frac{b}{a}; \quad \therefore \sin A = \frac{1}{\operatorname{cosec} A}.$$

$$\cos A = \frac{c}{b}; \quad \sec A = \frac{b}{c}; \quad \therefore \cos A = \frac{1}{\sec A}.$$

$$\tan A = \frac{a}{c}; \quad \cot A = \frac{c}{a}; \quad \therefore \tan A = \frac{1}{\cot A}.$$

TABLE XXVII.—TRIGONOMETRIC AND MISCELLANEOUS FORMULAS.

SOLUTION OF RIGHT TRIANGLES.

Required.	Given.	Formulas.
A, C, c	a, b	$\sin A = \cos C = \frac{a}{b}$; $c = \sqrt{(b+a)(b-a)}$.
A, C, b	a, c	$\tan A = \cot B = \frac{a}{c}$; $b = \sqrt{a^2 + c^2}$.
C, b, c	A, a	$C = 90° - A$; $c = a \cot A$; $b = a \operatorname{cosec} A$.
C, a, c	A, b	$C = 90° - A$; $a = b \sin A$; $c = b \operatorname{cosin} A$.
C, a, b	A, c	$C = 90° - A$; $a = c \tan A$; $b = c \sec A$.

SOLUTION OF OBLIQUE TRIANGLES.

Required.	Given.	Formulas.
b	A, B, a	$b = \frac{a \sin B}{\sin A}$
B	A, a, b	$\sin B = \frac{b \sin A}{a}$
$\frac{1}{2}(A+B)$	a, b, C	$\frac{1}{2}(A+B) = \frac{1}{2}(180 - C)$
$\frac{1}{2}(A-B)$		$\tan \frac{1}{2}(A-B) = \frac{a-b}{a+b} \tan \frac{1}{2}(A+B)$
A		$A = \frac{1}{2}(A+B) + \frac{1}{2}(A-B)$
B		$B = \frac{1}{2}(A+B) - \frac{1}{2}(A-B)$
A	a, b, c	If $s = \frac{1}{2}(a+b+c)$, $\sin \frac{1}{2}A = \sqrt{\frac{(s-b)(s-c)}{bc}}$
		$\cos \frac{1}{2}A = \sqrt{\frac{s(s-a)}{bc}}$,
		$\tan \frac{1}{2}A = \sqrt{\frac{(s-b)(s-c)}{s(s-a)}}$
		$\sin A = \frac{2\sqrt{s(s-a)(s-b)(s-c)}}{bc}$
Area		Area $= \sqrt{s(s-a)(s-b)(s-c)}$
Area	A, b, c	Area $= \frac{1}{2} bc \sin A$
Area	A, B, c	Area $= \frac{c^2 \sin A \sin B}{2 \sin(\quad + B)}$

TABLE XXVII.—TRIGONOMETRIC AND MISCELLANEOUS FORMULAS.

GENERAL FORMULAS.

$$\sin A = \sqrt{1-\cos^2 A} = \tan A \cos A.$$

$$\sin A = 2 \sin \tfrac{1}{2} A \cos \tfrac{1}{2} A.$$

$$\sin A = \frac{1}{\operatorname{cosec} A} = \sqrt{\tfrac{1}{2}(1-\cos A)}.$$

$$\cos A = \frac{1}{\sec A} = \sqrt{1-\sin^2 A} = \cot A \sin A.$$

$$\cos A = 1 - 2\sin^2 \tfrac{1}{2} A = 1 - \operatorname{vers} A.$$

$$\cos A = \sqrt{\tfrac{1}{2} + \tfrac{1}{2}\cos 2A} = \cos^2 \tfrac{1}{2} A - \sin^2 \tfrac{1}{2} A.$$

$$\tan A = \frac{\sin A}{\cos A} = \sqrt{\sec^2 A - 1}.$$

$$\tan A = \frac{\sqrt{1-\cos^2 A}}{\cos A} = \frac{\sin 2A}{1+\cos 2A}.$$

$$\tan A = \frac{1}{\cot A} = \frac{1-\cos 2A}{\sin 2A}.$$

$$\cot A = \frac{1}{\tan A} = \frac{\cos A}{\sin A} = \sqrt{\operatorname{cosec}^2 A - 1}.$$

$$\cot A = \frac{\sin 2A}{1-\cos 2A} = \frac{1+\cos 2A}{\sin 2A}.$$

$$\sec A = \frac{1}{\cos A} = \text{the reciprocal of any expression for } \cos A.$$

$$\operatorname{cosec} A = \frac{1}{\sin A} = \text{the reciprocal of any expression for } \sin A.$$

$$\operatorname{vers} A = 1 - \cos A = 2\sin^2 \tfrac{1}{2} A.$$

$$\operatorname{exsec} A = \sec A - 1 = \frac{\operatorname{vers} A}{\cos A}.$$

$$\sin \tfrac{1}{2} A = \sqrt{\frac{1-\cos A}{2}} = \sqrt{\frac{\operatorname{vers} A}{2}}.$$

TABLE XXVII.—TRIGONOMETRIC AND MISCELLANEOUS FORMULAS.

$$\cos \tfrac{1}{2} A = \sqrt{\frac{1 + \cos A}{2}}.$$

$$\tan \tfrac{1}{2} A = \frac{\tan A}{1 + \sec A} = \frac{1 - \cos A}{\sin A} = \frac{\sin A}{1 + \cos A}.$$

$$\cot \tfrac{1}{2} A = \frac{1 + \cos A}{\sin A} = \frac{\sin A}{1 - \cos A}.$$

$$\sin 2A = 2 \sin A \cos A.$$

$$\cos 2A = \cos^2 A - \sin^2 A = 2 \cos^2 A - 1.$$

$$\tan 2A = \frac{2 \tan A}{1 - \tan^2 A}.$$

$$\cot 2A = \frac{\cot^2 A - 1}{2 \cot A}.$$

$$\sin (A \pm B) = \sin A \cos B \pm \cos A \sin B.$$

$$\cos (A \pm B) = \cos A \cos B \mp \sin A \sin B.$$

$$\tan (A \pm B) = \frac{\tan A \pm \tan B}{1 \mp \tan A \tan B}.$$

$$\sin A + \sin B = 2 \sin \tfrac{1}{2}(A + B) \cos \tfrac{1}{2}(A - B).$$

$$\sin A - \sin B = 2 \cos \tfrac{1}{2}(A + B) \sin \tfrac{1}{2}(A - B).$$

$$\cos A + \cos B = 2 \cos \tfrac{1}{2}(A + B) \cos \tfrac{1}{2}(A - B).$$

$$\cos B - \cos A = 2 \sin \tfrac{1}{2}(A + B) \sin \tfrac{1}{2}(A - B).$$

$$\sin^2 A - \sin^2 B = \cos^2 B - \cos^2 A = \sin (A + B) \sin (A - B).$$

$$\cos^2 A - \sin^2 B = \cos (A + B) \cos (A - B).$$

$$\tan A \pm \tan B = \frac{\sin (A \pm B)}{\cos A \cos B}.$$

$$\cot A \pm \cot B = \frac{\pm \sin (A \pm B)}{\sin A \sin B}.$$

TABLE XXVII.—TRIGONOMETRIC AND MISCELLANEOUS FORMULAS.

MISCELLANEOUS FORMULAS.

Required.	Given.	Formulas.
Area of		
Trapezoid	Parallel sides = m and n Perp. dist. bet. them = p	$\frac{p}{2}(m+n)$
Regular Polygon	Length of side = l Number of sides = n	$\frac{nl^2}{4}\cot\frac{180^\circ}{n}$
Circle	Radius = r	πr^2 [$\pi = 3.1416$]
Ellipse	Semi-axes = a and b	πab
Parabola	Base = b, height = h	$\frac{2}{3}bh$
Surface of		
Cone	Radius of base = r Slant height = s	πrs
Cylinder	Radius = r, height = h	$2\pi rh$
Sphere	Radius = r	$4\pi r^2$
Zone	Height = h Radius of its sphere = r	$2\pi rh$
Volume of		
Prism or cylinder	Area of base = b Height = h	bh
Pyramid or cone	Area of base = b Height = h	$\frac{bh}{3}$
Frustum of		
Pyramid or cone	Area of bases = b and b' Height = h	$\frac{h}{3}(b + b' + \sqrt{bb'})$
Sphere	Radius = r	$\frac{4}{3}\pi r^3$

www.ingramcontent.com/pod-product-compliance
Lightning Source LLC
Chambersburg PA
CBHW021349210326
41599CB00011B/813
9783744678834